MW01492821

Lectures on Symmetry-Assisted Computation

Lectures on Symmetry-Assisted Computation

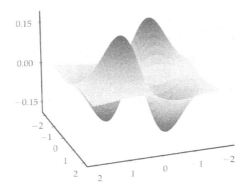

D. Pescia
ETH Zurich

World Scientific

NEW JERSEY · LONDON · SINGAPORE · BEIJING · SHANGHAI · HONG KONG · TAIPEI · CHENNAI · TOKYO

Published by

World Scientific Publishing Co. Pte. Ltd.

5 Toh Tuck Link, Singapore 596224

USA office: 27 Warren Street, Suite 401-402, Hackensack, NJ 07601

UK office: 57 Shelton Street, Covent Garden, London WC2H 9HE

British Library Cataloguing-in-Publication Data
A catalogue record for this book is available from the British Library.

About the cover image: Mathematical rendering of an eigenmode of a spin nanodot as observed in M. Buess *et al.* "Fourier transform imaging of spin vortex eigenmodes", *Phys. Rev. Lett.* **93**, 077207 (2004), https://journals.aps.org/prl/abstract/10.1103/PhysRevLett.93.077207. Courtesy of G. Pescia.

LECTURES ON SYMMETRY-ASSISTED COMPUTATION

ISBN 978-981-12-8011-5 (hardcover)
ISBN 978-981-12-8012-2 (ebook for institutions)
ISBN 978-981-12-8013-9 (ebook for individuals)

For any available supplementary material, please visit
https://www.worldscientific.com/worldscibooks/10.1142/13511#t=suppl

Desk Editor: Rhaimie Wahap

Typeset by Stallion Press
Email: enquiries@stallionpress.com

PREFACE

These "lecture notes" document the topics presented during an elective course at ETH Zurich. The course is offered at the master's and PhD candidate levels. The scope of the course was to provide fundamental knowledge about using symmetry to understand and solve scientific problems. The course was frequented mainly by students from the Departments of Physics, Chemistry, and Material Science. In the final format, it spanned about 12 weeks (corresponding to one semester at ETH Zurich), with an average of four hours per week dedicated to the lectures and two hours per week dedicated to exercises and what we call as "Student Projects".

The material covered in these notes is more comprehensive than the actual knowledge communicated during the oral lectures. Proofs of theorems and other technical, more mathematical details, for instance, were not discussed explicitly during the oral presentations. There, the emphasis was on the significance and practical use of the mathematical tools. Accordingly, in the lecture notes, mathematical definitions, propositions, and theorems are immediately followed by relevant "Comments" on their applications rather than by their proofs. It is left to the reader to decide how far to go with the more technical details.

The adjective "practical" summarizes precisely the spirit of these lectures: They were assembled with the explicit aim of teaching how symmetry can be used to understand, manipulate, and, ultimately, solve scientific problems. Those lectures that are dedicated to the more "artistic" aspects of symmetry (Lecture 5, on crystallography, for instance) are also written with the intent of training students to recognize the symmetry aspects of a problem before they can solve it. A further key ingredient of these lectures pertains to an aspect that, in my opinion, is often neglected. All problems, even those that are apparently "easy" or usually cast off into some referenced literature, are worked out here to a great extent and depth.

In these lectures, students not only learn how to solve scientific problems using symmetry arguments but also learn that they must do it themselves, from the beginning to the end. In this way, they are also trained to become productive in their future careers in industry, research, or business.

A final caveat: This manuscript refers to orally conducted lectures and, as a result, adopts a storytelling approach. References, for instance, are included in the text for immediate use and are, when possible, accompanied by their web addresses. Equations and figures are not numbered as their use is very local and intimately intertwined with the text.

The titles of the various lectures provide some information about the topics presented in these notes:

- Lecture 1: Geometrical, Algebraic, and Analytic aspects of symmetry
- Lecture 2: Representation Theory of Groups With Averages
- Lecture 3: SO_3 and the Method of Infinitesimal Transformations
- Lecture 4: The Symmetry Group of the Operator and Some Practical Applications
- Lecture 5: Point Groups and Space Groups of Crystallography
- Lecture 6: Applications to Solid-State Physics
- Lecture 7: The Relativistic Electronic Structure of Atoms and Solids
- Lecture 8: Young Diagrams and Particle Physics
- Lecture 9: The Permutation Group and its Applications to Many-Body Problems
- Lecture 10: Group Theory and Phase Transitions

For preparing these lectures, I have consulted a large number of books on the subject of symmetry. Among them, I quote those chapters in the set of books *Course of Theoretical Physics* by L. D. Landau and E. M. Lifshitz that explicitly refer to the use of symmetry in quantum mechanics. Here, one can learn how to dispense with the more formal aspects and "get to the point". I have learned some more mathematical aspects by reading the book (written in 1932 and revised in 1974) B. L. van der Waerden, *Group Theory and Quantum Mechanics*, Springer Verlag, https://link.springer.com/book/10.1007/978-3-642-65860-0 and the book by C. Isham, *Lectures on Group and Vector Spaces for Physicists*, World Scientific, https://www.worldscientific.com/worldscibooks/10.1142/0893#t=aboutBook. A classical book on practical group theory from the formidable ETH Zurich school is the one by A. Fässler and E. Stiefel, *Group Theoretical Methods and Their Applications*, Birkhäuser Verlag (the lectures by the late Prof. Stiefel are simply unforgettable). The geometrical aspects of symmetry and crystallography (a very broad and complex topic) are presented in a highly tutorial manner by Andrew Baker, Groups and Symmetry, Department of Mathematics, University of Glasgow (http://www.maths.gla.ac.uk/~ajb/dvi-ps/2q-notes.pdf) and by P. J. Morandi, The Classification of Wallpaper Patterns: From Group Cohomology to Escher's Tessellations, Department of Mathematical Sciences, New Mexico State University, Las Cruces, NM 88003, free on-line book, https://drive.google.com/file/d/12f ti6FKkwqlCyntRGAkXtjbrfUyw3p-g/view.

The current lectures are accompanied by a set of exercises (with solutions) meant to enhance the content of the various lectures. The exercises are conceived to train the technical skills of the students and their technical understanding of the lectures. In addition to the exercises, the students were guided to work out, within small groups, a well-defined scientific topic and perceive the role of symmetry in it. The projects have a stronger conceptual value than the exercises. In fact, some of them even introduce topics that go beyond the lectures themselves. Readers of this book (in particular, the teachers) are invited to expand on the topics contained in the projects and propose some new ones. The projects ended with manuscripts and presentations — a very important step for both students and teachers. Parts of these manuscripts have been edited into these notes and are incorporated as an essential component besides the lectures and exercises.

The exercises and projects were tutored by some collaborators of my former research group, whom I would like to explicitly mention and thank, together with those students who, by working on a project, fulfilled part of their oral examination:

- Dr. O. Portmann, Dr. A. Zanin, Th. Gumbsch, Dr. A. Bellissimo and Y. Forrer contributed to the solutions of the exercises and provided parts of their LaTeX versions (Y. Forrer also provided some figures).
- The project on graphene: coordinated by J. Wei; students involved: Joel Biederman, Enrico Della Valle, Michele Masseroni, Danish Nabi, and Paul Wegmann.
- The project on the mathematical and numerical aspects of the eigenvalue problem: coordinated by M. Holst; students involved: Lara Turgut, Daniel Tay, Ao Chen, and Elias X. Huber.
- The project on crystal splitting and Jahn–Teller effect: coordinated by J. Wei.; students involved: H. -Y. Kuo and X. H. Verbeek.

- The project on vibrational modes: coordinated by J. Wei; students involved: Tianqi Zhu, Jacob F. Fricke, Otto T. P. Schmidt, and Timo S. G. Niepel.
- The project on quantum chemistry: coordinated by Dr. A. Bellissimo; students involved: Robin Feldmann, Yorick Lassmann, Luzian Lebovitz, and Danylo T. Matselyukh.
- I also thank the student who worked out, in his oral exam, some of the topics included in the project on empty-lattice band structure. I have not been able to find his name.

I finally thank O. Schmidt and D. Scheiwiller for providing the preliminary draft of many figures.

This book is dedicated to Noah, Gabriel, Viola, Gereon and Hedi.

Prof. em. Dr. D. Pescia
Laboratory for Solid State Physics, ETH Zurich,
CH-8093 Zurich
Zürich, April 2023

PREAMBLE: THE SCIENTIFIC PROBLEM AND THE ROLE OF SYMMETRY ARGUMENTS

Scientific problems can often be mapped onto certain equations that must be solved by analytical and numerical methods. To provide a concrete example, consider a circular object made of some material. Like the membrane of a drum, it can be set into vibration by some external stimulus. In mathematics, the vibrational motion (let us consider only the motion of the membrane perpendicular to its plane, for simplicity) is described by a function of time and two spatial variables that must be introduced to describe the geometry of the membrane (regarded as a continuum of material). The variables we choose are the radial variable r, locating the distance of one point of the membrane from the center of the membrane, and the angular variable φ, locating the angle spanned by the position vector $\mathbf{r}$ and some in-plane fixed direction (let us call it "the x-axis"), see the following figure.

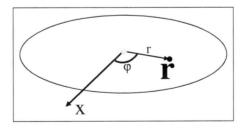

Circular membrane: The variable r is the distance of a point of the membrane (black dot) from its center (light gray). The variable φ is the angle between the position vector $\mathbf{r}$ (indicated by an arrow and designated by a lowercase Latin letter in bold) and some fixed direction in the plane of the membrane (the x-axis).

The so-called "amplitude of vibration" is, generally, a complicated function $u(r, \varphi, t)$ of the variables t (the time) and (r, φ). Classical mechanics tells us that the function $u(r, \varphi, t)$ can be expressed as the superposition of well-defined "modes" of vibration. These are special functions — let us designate them $u_\lambda(r, \varphi, t)$ — called "eigenmodes". An eigenmode, for instance, is a very intuitive one: It is one where the membrane moves perpendicular to its plane with constant velocity ($u(r, \varphi, t) = \text{const} \cdot t$). This mode is the so-called translational mode of "vibration" (if one insists on using the word "vibration" for this mode as well). Other eigenmodes are more subtle functions of time and the variables r, φ. Their (r, φ)-dependence, for instance, is provided by solving a special equation of classical physics, called the Helmholtz equation. For example, the one specifying the φ-dependence is

$$\frac{\partial^2 u(\varphi)}{\partial \varphi^2} = -\lambda \cdot u(\varphi)$$

where λ are numbers — called the eigenvalues of the equation — that must be determined together with the eigenmode $u_\lambda(\varphi)$ of vibration. The λs play an important role in the solution of the scientific problem: They are typically measurable quantities (the eigenfrequencies of vibrations in the specific example or the energy levels of an atomic system in quantum mechanical systems). λ and the corresponding $u_\lambda(\varphi)$ define one eigenmode ($\lambda = 0$ and $u_0(\varphi) = \text{const.}$ for the translational eigenmode). The equation itself is called an **eigenvalue equation**.

This equation, which is a differential equation because it involves the desired function and its derivatives, has a solution for any complex value of λ, and the resulting eigenfunction is, generally, a complex-valued function. This situation is somewhat unphysical: We expect to find real values for λ — only real-valued quantities are measurable quantities. We note that the eigenvalue equation itself does not entirely describe the physical problem of the vibrating membrane. It is compelling to demand that any solution be periodic in φ, with the period being 2π. This requirement is justified by the fact that, if we go around once along a circular object, we return to the same place; therefore, the vibrational amplitude must be recovered after the turn. This requirement represents a so-called **boundary condition**. In the specific example, requiring the 2π periodicity restricts λ to be n^2, $n = 0, \pm 1, \pm 2, \ldots$, i.e. real eigenfrequencies. The eigenvalue equation and the boundary conditions together defines what is known as the **eigenvalue problem**. The physically relevant eigenmodes are, therefore, the solutions to an eigenvalue problem and not simply those of an eigenvalue equation.

One can safely state that, as long as linear equations are involved, one always expects that solving a scientific problem boils down to solving some eigenvalue problem, similar to the one we have just discussed.

In the context of an eigenvalue problem, one might ask what role the symmetry arguments play. The first question that can arise is, what do we mean by "symmetry" of a problem? Referring to the eigenvalue problem for the vibrational amplitude $u(\varphi)$ of a circular membrane, there is a set of transformations (of "maps") that transform the object "membrane" into itself. Among these are certainly all rotations by any angle about an axis perpendicular to the membrane and crossing the membrane in its center. Additionally, all reflections at planes perpendicular to the plane of the membrane and passing through the rotation axis are transformations that map the membrane onto itself. Finally, the horizontal plane — the one containing the membrane — provides a symmetry element: A reflection at this plane transforms the membrane into itself. These transformations are what we call "the symmetry transformations of the membrane", and when we speak of the "symmetry of the problem" of the vibrating membrane, we refer to this set. Note that all the transformations we have just found have a common property: They leave the distance between two points in space unchanged, i.e. they are part of a family of transformations of the plane called isometries. Transformations that change the distance (called dilations) would change the size of the membrane or deform it. They also play an important role in a certain class of scientific problems but are not considered "symmetry operations" because they do not leave the object invariant.

The set of symmetry operations of the membrane has a very important mathematical structure: Symmetry operations can be "multiplied" to obtain another symmetry operation that is within the set. "Multiplication" means that we first execute one symmetry operation, followed by the second one. The operation of "multiplication" can be rendered mathematically by using the rule of composition of functions. The set of symmetry transformations is therefore endowed with a rule of multiplication that is closed, i.e. successive symmetry transformations lead to a transformation that is also within the set. The existence of a "closure" with respect to "multiplication" is given a special name in mathematics: The set is said to constitute a "group".[1] This leads to the result that the set of symmetry transformations of a physical object constitutes a "group",

---

[1] There are more properties that are required to be satisfied for a set to be a group: We will be more precise in Lecture 1.

called the "symmetry group" of the object. A significant portion of these lectures will be dedicated to the mathematics and terminology of symmetry elements and symmetry groups, as these are fundamental aspects required for understanding the role of symmetry and using it. Such mathematical and terminological aspects can be particularly challenging when dealing with, for instance, the physics of crystals and their phase transitions.[2]

The symmetry group of an object certainly represents an important aspect of knowledge about the object and also has significance in domains other than science, such as architecture, art, and design. Human history is strewn with examples of our interest in the elements of symmetry in paintings, graphics, and ornaments, and one can find a rich literature about it (see, for example, H. Weyl, *Symmetry*, Princeton University Press, Princeton New Jersey, 1952). These more general aspects are, however, not the focus of the current set of lectures. Its primary scope is the use of symmetry for solving scientific problems.

Returning again to our membrane, is there any relevance to our knowledge of the symmetry elements of a membrane to solving the eigenvalue equation governing the vibrational motion? One could be tempted to conclude that the vibrational modes — the functions that solve the eigenvalue equation — must realize the same symmetry as the object, i.e. they are invariant with respect to all symmetry elements. This conclusion is, however, incorrect. Suppose that we have a solution that breaks the symmetry of the system, i.e. the function is only invariant with respect to a reduced number of symmetry operations. This solution will appear on the right-hand side of the eigenvalue equation. For the eigenvalue equation to be satisfied, it is necessary that the left-hand side transform like the right-hand side, i.e. the left-hand side must transform as the solution itself. However, the left-hand side is a mathematical expression that typically involves the function and some mathematical device — in the current case, the second derivative — that acts upon the function. This mathematical device is called an **operator**. The second derivative is an example of an operator.

---

[2]Crystallography was independently developed by J. S. Fjodorow and A. M. Schönflies in around 1890 and culminated with the classification of solids in terms of Bravais lattices, crystal systems, crystal classes, and space groups. There are several books on the subject of crystallography, mostly reporting complete tables and a large summary of crystallographic data which can be referred to for specific applications. The one book which is not as comprehensive but is closer to our "educational" approach is the book by D. E. Sands, *Introduction to Crystallography*, Dover Publications Inc., Mineola, New York, USA, 1993.

How can an expression that not only involves the function itself but also some complicated operator acting upon it transform like the function itself? We will learn how to transform functions and operators in the course of the lectures, but we can already guess the answer to this question: An operator acting upon a function can only transform like the function itself if the operator itself is left invariant by the transformations! Here lies the key to understanding how the symmetry of an object plays a role in a physical problem: The operators appearing in the eigenvalue equations related to a given physical system must be invariant with respect to those transformations that map a system onto itself. It is not the solutions of the eigenvalue equation that have the symmetry of the object; it is the operators that act upon the functions that carry the symmetry, which one can infer by observing the spatial shape of the object.

It was E. Wigner (E. Wigner, "Einige Folgerungen aus der Schrödingerschen Theorie für die Termstrukturen", *Z. Phys.* **43**, 624–652 (1927), https://doi.org/10.1007/BF01397327), in his seminal paper on the role of symmetry in quantum mechanics (but similar arguments hold for any linear operator appearing in an eigenvalue problem), who discovered the consequences of this invariance for solving the eigenvalue problem. Here is a brief outline of Wigner's idea: First, let us explain the "numerical way" of solving the eigenvalue problem. The aim of solving the eigenvalue problem is primarily to find the eigenvalues. The eigenvalues are often those quantities (such as the eigenfrequencies of a vibrating membrane or the energy levels of a quantum mechanical system) that can be directly measured with appropriate instruments. Technically, eigenvalue problems are typically mapped onto some determinantal equation. This is then solved mostly numerically, but the solution of a determinantal equation is a formidable problem even for supercomputers, in view of the large size of the matrices involved. After finding the eigenvalues, one can compute the functions that solve the original eigenvalue equation to obtain an idea of the eigenmodes. This second step is often neglected, although modern experiments are capable of mapping the eigenmodes with the highest spatial and temporal resolutions.[3] Wigner taught us that symmetry can be used to solve the eigenvalue problem the other way around: First, the eigenmodes are found using symmetry arguments alone. Subsequently, the

---

[3]See, for example, M. Buess *et al.* "Fourier transform imaging of spin vortex eigenmodes", *Phys. Rev. Lett.* **93**, 077207 (2004), https://journals.aps.org/prl/abstract/10.1103/PhysRevLett.93.077207.

corresponding eigenvalues are computed by inserting the eigenmodes into the equation. "Inserting" into an eigenvalue equation is much easier than "solving" it, so symmetry can prove to be a valuable tool. For the specific case of the membrane, the differential equation is an elementary one and can be solved analytically by elementary means. However, using symmetry, we do not have to solve it. We use some mathematical tools (given the fact that the set of symmetry elements of the objects has the structure of a group) to find the solutions without solving the equation. These mathematical tools are collectively known as the "representation theory of groups" and will be a central component of these lectures. By virtue of these tools, and considering the entire set of symmetry elements of the membrane, we know that the eigenmodes are given by

$$u_n(\varphi) = e^{i \cdot n \cdot \varphi}$$

with $n = 0, \pm 1, \pm 2, \ldots$. Since we know all possible solutions, we can insert them into the eigenvalue equation and find the eigenvalues, without the need to solve the eigenvalue equation explicitly. In the specific case, one does not need to be familiar with the technique of differential equations, only that one needs to be able to differentiate functions to obtain the eigenvalues as

$$\lambda = n^2$$

One could ask the following question: If, in order to use symmetry, we need some non-trivial mathematical tools, why not simply learn how to solve differential equations directly? The answer to this question is very simple: because the solutions we have found on the basis of symmetry are the same for all systems and equations where the same symmetry elements appear, regardless of the type of system involved or the form of the operator. For example, we have just found the eigenfunctions of an operator with full rotational symmetry around the z-axis. In quantum mechanics, the z-component of the angular momentum operator is associated with the operator $-i \cdot \hbar \cdot \frac{\partial}{\partial \varphi}$, which also has rotational symmetry. So, let us use the same eigenfunctions for this operator as well, even if it is associated with a system (a single atom) which is completely different from the vibrating drum. Inserting these solutions into the eigenvalue equation for the operator

$$-i \cdot \hbar \cdot \frac{\partial}{\partial \varphi} u(\varphi) = \lambda \cdot u(\varphi)$$

we find that the eigenvalues are

$$\lambda = \hbar \cdot n$$

These are the values of the z-component of the angular momentum of an atom. We have found them using the results originating from our handling of a classical circular membrane! Using the symmetry of the system, we have proven the quantized nature of the orbital angular momentum. This is not bad for a start.

The symmetry of physical objects is often the starting point for the solution of a physical problem involving the objects themselves. However, there are situations where a physical object itself does not exist and the symmetry properties refer only to the operators, not to the underlying physical object. This is the situation in most problems related to modern particle physics: The operator itself is assigned symmetry without referring to an underlying object. However, Wigner's ideas can be immediately translated to this situation as well, as their formulation is based on the invariance of some operators with respect to some symmetry elements, and this invariance can be formulated independently of whether the symmetry elements transform a concrete physical body or not. The existence of a physical body is, of course, of great help in figuring out what is going on, but Wigner's principle allows for a greater degree of abstraction. In this abstract situation, it is even possible for the relevant symmetry group to be of a strange nature, making it inapplicable to real physical bodies. For example, the set of complex numbers of absolute value "1" is an important set of symmetry elements in physics (it generates the conservation of charge!), but it is impossible to find a physical body which is invariant with respect to complex numbers!

In addition, this abstract situation of symmetry might entail the introduction of "solutions" to the "eigenvalue" equation that are not necessarily ordinary functions, just as the eigenvalue equation is not necessarily an ordinary differential equation (in the future, we might discover the operator underlying the symmetry groups that we are using; for now though, we use symmetry without knowing the nature of its operator). For example, protons and neutrons are "solutions" to a hypothetical eigenvalue equation that has a set of complex numbers with an absolute value of 1 as symmetry elements, but protons and neutrons are particles and not ordinary functions.

Such abstract situations illustrate an important principle of modern physics: One can proceed toward the understanding of a physical problem using symmetry arguments alone, detached from any concrete and detailed knowledge of the system itself. Symmetry itself allows for concrete and often exact computations, while the exact knowledge of the detailed interaction is often redundant. For example, we do not know exactly the interaction that appears during a proton–neutron scattering experiment; however, we can predict its outcome by knowing that the interaction has a certain symmetry group. A further example is the interactions between atoms that finally lead to phase transitions in matter, which are very complex and whose details are still unknown. Yet, symmetry arguments allow us to predict critical and universal aspects of phase transitions without having to go through the painful and hopeless process of computing the trajectory of millions of interacting atoms. The accuracy of the predictions based on symmetry conveys an even more important principle: Most complex processes in nature are governed by their symmetry alone, not by the microscopic details. We will encounter a further example of the key role of symmetry in modern science when symmetry considerations lead us to analyze situations involving **topological protected** states and **quantum entangled states** — two key concepts of current research in quantum technologies.

A final word on the style of these lectures notes: The introduction of a set of purely mathematical tools that provide the basis rules for the use of symmetry is unavoidable. The mathematical tools boil down to a set of important, general theorems and other more "local" propositions. It is not the aim of these notes to provide rigorous proofs: We refer to more specialized literature that is much better qualified to offer them. We rather provide plausibility arguments (occasionally, these "plausibility arguments" will be rigorous). For example, when we deal with integrals, we always assume that they "behave properly", i.e. they are well defined in the sense of a Riemann integral (or, in some circumstances, distributions), excluding therefore "pathological" situations. The same holds true when we deal with operators; we assume that they behave properly within the context in which we are using them, excluding the rare "pathological" circumstances. Ultimately, we aim at defining a set of rules that allow for concrete computations based on symmetry, and we assume that all necessary restrictions for the validity of these rules are fulfilled. In practice, the fact that a certain computation produces a reasonable result serves as the best proof that the theorems and propositions we used are applicable.

CONTENTS

Preface   v

Preamble: The Scientific Problem and the Role of
Symmetry Arguments   ix

Part I: Lectures   1

Lecture 1.   Geometrical, Algebraic, and Analytic
Aspects of Symmetry   3

   1.1   The Geometrical Aspects of Symmetry   3
        1.1.1   The algebraic structure of the
               Euclidean space   4
        1.1.2   Symmetry transformation and
               symmetry groups   7
        1.1.3   Examples of symmetry elements and
               their notation   8
        1.1.4   Simple examples of symmetry groups and their
               notation 11
   1.2   Formal Aspects of Isometries of the Euclidean Space . . . 17
        1.2.1   Isometries of the line 20
        1.2.2   Isometries of the plane 20
        1.2.3   Isometries of the space 21
   1.3   Algebraic Properties of Groups 23
   1.4   Analytic Properties of Continuous Groups 28
        1.4.1   Group average 31

**Lecture 2.   Representation Theory of Groups
with Averages**   **39**

  2.1   Introduction   39
  2.2   Representation Theory of Groups (H. Weyl,
         1885–1955)   41
       2.2.1   Representations of groups   41
       2.2.2   Examples of finite- and infinite-dimensional
            representations   44
       2.2.3   Reducible and irreducible representations   49
       2.2.4   Theorems on irreducible representations   57
  2.3   Symmetry-Adapted Vectors   75

**Lecture 3. SO_3 and the General Method of
Infinitesimal Transformations**   **81**

  3.1   Irreducible Representations of the Circle Group C_∞ . . .   81
  3.2   The Irreducible Representations of SO_3   87
       3.2.1   Algebraic construction of irreducible
            representations: First attempt   87
       3.2.2   Analytic approach to finding the Cartan
            weights   89
  3.3   Representations of Continuous Groups Through
         Infinitesimal Transformations: Algebraic Method   97
       3.3.1   Lie groups and infinitesimal
            transformations   98
       3.3.2   Infinitesimal transformations and group
            representations 102

**Lecture 4.   The Symmetry Group of the Operator
and Some Practical Applications**   **111**

  4.1   The Relation Between Symmetry and Scientific
         Problems 111
       4.1.1   Introduction 111
       4.1.2   Theorem I: Quantum numbers theorem 117
       4.1.3   Theorem II: The matrix elements theorem . . . 118
       4.1.4   Theorem III: Symmetry-assisted
            computations 121

4.2 Application 1: Infinitesimal Transformations and
Quantum Mechanical Conservation Laws 125
4.3 Application 2: Symmetry-Breaking Interactions 126
4.4 Application 3: Helmholtz Equation on a Square
Membrane 135
4.5 Application 4: Molecular Vibrations 139
4.5.1 General aspects 139
4.5.2 Numerical example: The H_2O molecule 141

**Lecture 5. Point Groups and Space Groups of
Crystallography** **151**

5.1 Point Groups 151
5.1.1 Point groups in one-dimensional systems 153
5.1.2 Point groups in two-dimensional systems 153
5.1.3 Point groups in three-dimensional systems 154
5.1.4 Enumeration of the proper point groups
(Hessel) 156
5.1.5 Improper point groups 159
5.1.6 Improper point groups of type a 159
5.1.7 Improper point groups of type b 163
5.1.8 Non-crystallographic point groups 166
5.2 Constructive crystallography 167
5.2.1 An introductory example 168
5.2.2 A. Bravais lattices and crystal systems 172
5.2.3 B. Crystal classes 177
5.2.4 C. Space groups 179
5.2.5 One-dimensional space groups 181
5.2.6 Two-dimensional space groups 182
5.2.7 Three-dimensional space groups 192

Lecture 6. Applications to Solid-State Physics **197**

6.1 Irreducible Representations of the Translation
Group T 197
6.1.1 One-dimensional lattice 198
6.1.2 Two-dimensional lattice (square lattice for
the sake of illustration) 200
6.1.3 Three-dimensional lattice (face-centered cubic
lattice for the sake of illustration) 201

6.2   Irreducible Representations of Space Groups
       (C. Herring)   203
       6.2.1   Introduction   203
       6.2.2   Irreducible representations of the group
                of **k**   204
       6.2.3   The star operation   210
6.3   Symmetry Analysis of the Band Structure of Solids . . .   213
       6.3.1   Example 1. Empty-lattice band structure of the
                one-dimensional primitive lattice: Accidental
                versus essential degeneracy   213
       6.3.2   Example 2. The two-dimensional primitive square
                lattice: Essential degeneracy and avoided
                crossing   219
6.4   Non-symmorphic Space Groups   223
       6.4.1   Cosets   223
       6.4.2   Irreducible representations of the factor group
                of **k**   227

Lecture 7.   The Relativistic Electronic Structure of
Atoms and Solids   **233**

7.1   The Spin of the Electron and the Bethe Hypothesis . . .   233
       7.1.1   Introduction   233
       7.1.2   The $D^{\frac{1}{2}}$ representation of SO_3   236
7.2   Symmetry of Composite Systems   240
       7.2.1   The direct product of groups   241
       7.2.2   The tensor (or Kronecker) product of vectors and
                matrices   242
7.3   The Coupling of Degrees of Freedoms   249
       7.3.1   The symmetry group of the coupling operator
                $\mathbf{I}_p \otimes \mathbf{I}_q$ 249
       7.3.2   Clebsch–Gordan series   251
       7.3.3   An application: The fine structure of atomic
                levels   254
7.4   Clebsch–Gordan Coefficients   257
       7.4.1   Computation of Clebsch–Gordan coefficients
                for SO_3^D   260

7.5   Double Point Groups and the Relativistic Band
Structure 263
     7.5.1   Symmetry analysis along the Λ-direction 265
     7.5.2   The group C_{4v}^{D} 269
     7.5.3   The Clebsch–Gordan coefficients for double point
groups 271
7.6   The Wigner–Eckart–Koster Theorem 274

Lecture 8.   Young Diagrams and Particle Physics   **285**

8.1   Irreducible Representations of $GL(n,\mathscr{C})$ and Some
Important Subgroups 285
     8.1.1   Scalars, vectors, and tensors 286
     8.1.2   An application: Constructing representations
of $GL(2,\mathscr{C})$ 291
     8.1.3   Generalizations and general theorems 295
     8.1.4   The group SU_2 and its irreducible
representations 299
     8.1.5   Irreducible representations of SU_3 305
8.2   Applications to Particle Physics 306
     8.2.1   Introduction 307
     8.2.2   Experimental proof of SU_2-symmetry 309
     8.2.3   The hadron particle multiplets:
The early 1960s 312
     8.2.4   The eightfold way 315
     8.2.5   The quark hypothesis 317

**Lecture 9.   The Permutation Group and Its
Applications to Many-Body Problems**   **321**

9.1   Definitions and Notations 321
9.2   The Irreducible Representations of S_N 325
9.3   Applications to Physical Problems 328
     9.3.1   Example I: Two spin-$\frac{1}{2}$ particles 328
     9.3.2   Example II: Three spin-$\frac{1}{2}$ particles 330
9.4   The Schur–Weyl Duality 332
     9.4.1   An introduction to the problem 332
     9.4.2   The Schur–Weyl duality and the Young diagram
technique 334

9.5   Multiplets in Atomic Physics   337
         9.5.1   Schur–Weyl duality: Multiplet analysis   338
         9.5.2   Slater determinants   342

Lecture 10.   Group Theory and Phase Transitions   347

10.1   Introduction   347
10.2   The Landau–Lifshitz Symmetry Rules   350
10.3   Invariant Polynomials   356
10.4   Magnetic Phase Transitions   358
10.5   Landau's Model of the Liquid–Solid Phase
         Transition   360
10.6   The Method of Invariant Polynomials Applied to the
         C_{4v} Symmetry Group   364

Part II: Exercises   369

11.   Exercises to Lecture 1   371

12.   Exercises to Lecture 2   383

13.   Exercises to Lecture 3   395

14.   Exercises to Lecture 4   403

15.   Exercises to Lecture 5   415

Part III: Projects   427

**16.   Project for Lecture 2: Functional Analytic and
Numerical Aspects of Eigenvalue Problems   429**

16.1   Introduction   429
16.2   Functional Analytic Aspects of the Eigenvalue
         Problem   430
         16.2.1   Hilbert spaces   430
         16.2.2   Spectral theorem for finite-dimensional Hilbert
                  spaces   436
         16.2.3   Hilbert–von Neuman spectral theorem
                  (for laymen)   444

      16.2.4   Commuting self-adjoint matrices 445
  16.3   Numerical Approaches to Eigenvalue Problems 446
      16.3.1   Finite difference method 446
      16.3.2   Finite element method 450

**17.   Project for Lecture 4: Symmetry Arguments in
Classical Mechanics**   **455**

  17.1   The Symmetry Groups of Classical Mechanics 455
  17.2   The Relation Between Symmetry, Galilei Invariance, and
        Conservation Laws 457
  17.3   Scaling Transformations 462

**18.   Project for Lecture 4: Crystal Field Splitting and
the Jahn–Teller Effect**   **465**

  18.1   Crystal Field Splitting 465
      18.1.1   The role of symmetry in determining the
            potential 466
      18.1.2   Symmetry-assisted solution of the eigenvalue
            problem 471
  18.2   Jahn–Teller Effect 473
      18.2.1   Introduction 473
      18.2.2   A simple example 475

**19.   Project for Lecture 4: Vibrational Modes of the
NH_3 Molecule**   **479**

  19.1   Symmetry analysis of the vibrational problem of NH_3 . . 480

20.   Project for Lecture 5: Frieze Patterns   **489**

**21.   Project for Lecture 5: An Algebraic Proof of the
Hessel Theorem**   **497**

22.   Project for Lecture 6: Empty-Lattice Band Structure **503**

  22.1   The Γ- and X-points of the Square Planar Lattice 503
  22.2   Model Band Structure for a Tetragonal Lattice 509
      22.2.1   Empty-lattice eigenvalues 509
      22.2.2   Symmetry analysis 511
      22.2.3   Model calculation 515

23. Project for Lecture 6: Symmetry Analysis of Graphene Band Structure and Dirac Cones   **523**

23.1   Description of Structure and Symmetry Elements of the Graphene Lattice 523

23.2   The π-bands of Graphene 525

     23.2.1 π-energy levels at Γ point 526

     23.2.2 π-energy levels at K-point 528

     23.2.3   The π-bands in the vicinity of K 531

23.3   The sp^2-bands at the Γ-point within the Tight-Binding Model: Symmetry Analysis 537

     23.3.1   Part A: The labeling of energy levels with irreducible representations 537

     23.3.2   Part B: Symmetry-adapted tight-binding orbitals at the Γ-point 540

23.4   In-plane Band Structure with Symmetrized Plane Waves 542

     23.4.1   Part A: Symmetry analysis of the energy levels at the Γ-point 542

24. Project for Lecture 6: Topological Protection by Non-symmorphic Degeneracy   **547**

24.1   Introduction 547

24.2   The Space Group p2mg 548

24.3   The Space Group D_{6h}^4 $(P6_3/mmc)$ 553

25. Project for Lecture 7: Topological Aspects of Continuous Groups and the Universal Covering Groups   **555**

25.1   Connectivity of Groups 555

     25.1.1   Some concepts of algebraic topology 556

25.2   The Universal Covering Set and the Universal Covering Group 561

     25.2.1   Representations of universal covering groups 562

25.3   The Universal Covering Group of SO_3 564

26. Project for Lecture 7: Optical Spin Orientation in Atoms and Solids   **567**

26.1   Introduction 567

26.2   Optical Orientation at the Γ-point of GaAs 567

27. Project for Lecture 9: Quantum Chemistry   **577**

27.1   Introduction: The H_2-Molecule 577

27.2   The Eigenvalue Problem of Quantum Chemistry 581

      27.2.1   Two-electron RH algorithm 583

      27.2.2   The LCAO computation of the molecular

            orbitals 588

27.3   Symmetry Arguments in Quantum Chemistry

      Computations 590

27.4   The Borane Molecule BH_3 593

      27.4.1   Jahn–Teller-type distortion 601

Index   605

PART I

Lectures

Lecture 1

GEOMETRICAL, ALGEBRAIC, AND ANALYTIC ASPECTS OF SYMMETRY

In Lecture 1, we develop the geometrical, algebraic, and analytic properties of a set of symmetry transformations that underlies the solution to scientific problems. The technique developed in this lecture is called "group theory" in the literature. The lecture is accompanied by a set of exercises designed to train students in using group theoretical methods.

1.1 The Geometrical Aspects of Symmetry

The symmetry elements of a physical body capture the essential aspects of its geometry. This section is aimed at introducing the fundamental geometrical principles underlying symmetry elements in one, two, and three dimensions (and, occasionally, in more than three dimensions if conceptually important). We also learn how to perform elementary computations with symmetry operations and some useful related terminology. Finally, we provide examples of sets of symmetry elements of a physical body and recognize that these sets build a mathematical structure called "group". The scope of this section is therefore to get the reader acquainted with the elementary concepts and use of symmetry. A more comprehensive and detailed treatment of the geometrical aspects can be found in the lecture dedicated to the symmetry groups of finite systems (so-called "point groups") and crystals (so-called "space groups") (Lecture 5). The more mathematical aspects of groups and their "representation" (i.e. their rendering in a mathematical framework that can be used for computations) is discussed in Lecture 2.

1.1.1 *The algebraic structure of the Euclidean space*

Physical objects (including patterns) exist in a three-dimensional Euclidean space. The Euclidean space is a set of geometrical points, and any object occupies some subset of this space that extends, typically, along three different directions.

Comment. We will also introduce one- and two-dimensional objects and patterns. These occupy subsets of the Euclidean space that extend along a line (one-dimensional) or a plane (two-dimensional). In our three-dimensional world, the extra dimensions are still surrounding the objects with reduced dimensionality — think of a line perpendicular to a planar object or of two planes crossing at one line). The system is rendered low-dimensional when the transformations that displace the low-dimensional system from its dimensions are artificially eliminated from the symmetry considerations. For example, a twofold rotation around an axis perpendicular to a line transforms the line into itself. However, in doing this, it moves the line away from his location and is not considered to be a symmetry element of one-dimensional systems.[1] For instance, a square molecule, consisting of four atoms placed at the corner of a square, is made two-dimensional by excluding any symmetry element that, during the transformation, displace the square from the plane. The horizontal plane is also excluded as a mirror plane: In a two-dimensional world, it is identical with the identity transformation. These restrictions imply that some degrees of freedom, such as the orbitals perpendicular to the plane, are explicitly eliminated from the problem.

An alternative way of thinking about the condition of two-dimensional objects is by considering the planar shape to be at the base of some suitable artificial body extending along one direction in the third dimension. The symmetry elements of the artificial body are the relevant ones for the base as well. For example, one can think of the square molecule as being at the base of a pyramid with its apex along the perpendicular direction, or one can extend the entire square molecule along one perpendicular direction.

The set of points in the Euclidean space is endowed with an algebraic structure that is suitable for computations when one point is chosen as the origin O and each point P is connected by a vector $\overrightarrow{OP} \equiv \mathbf{p}$ starting at

---

[1]The extra dimensions surrounding a low-dimensional system, however, may carry important degrees of freedom — think of the atomic orbitals that are centered at one atom (a "zero"-dimensional object). By artificially cutting down on the symmetry elements, these degrees of freedom are eliminated from the problem.

O and ending at P. For such vectors, one can define the operation of the linear combination

$$\lambda_p \mathbf{p} + \lambda_q \mathbf{q}$$

(the sign of multiplication of a vector with a scalar is usually omitted) and the operation of the scalar product

$$\mathbf{p} \cdot \mathbf{q}$$

or, alternatively written,

$$(\mathbf{p}, \mathbf{q})$$

The distance between two points P and Q is then

$$\sqrt{(\mathbf{p} - \mathbf{q}) \cdot (\mathbf{p} - \mathbf{q})}$$

One usually introduces also a coordinate system (typically a set of mutually orthogonal vectors $\{\mathbf{e}_1, \mathbf{e}_2, \mathbf{e}_3\}$[2] with unit length, called a "basis set"), so that any other vector can be expressed as a linear combination of the basis vectors:

$$\mathbf{p} = \sum_i p_i \mathbf{e}_i$$

The set $\{p_i\}$ includes the coordinates of the point P (or of the vector $\mathbf{p}$) with respect to the origin. It is customary to order them along a vertical column:

$$\begin{pmatrix} p_1 \\ p_2 \\ p_3 \end{pmatrix}$$

However, for space consideration, we often will order them horizontally. By virtue of this construction, the Euclidean space is also designated as $\mathscr{R}^3$. The linear combination and the scalar product in terms of coordinates read

$$\lambda_p \begin{pmatrix} p_1 \\ p_2 \\ p_3 \end{pmatrix} + \lambda_q \begin{pmatrix} q_1 \\ q_2 \\ q_3 \end{pmatrix} = \begin{pmatrix} \lambda_p p_1 + \lambda_q q_1 \\ \lambda_p p_2 + \lambda_q q_2 \\ \lambda_p p_3 + \lambda_q q_3 \end{pmatrix}$$

---

[2] A Euclidean space is n-dimensional if a basis set with n-mutually orthogonal unit vectors can be established.

The scalar product is written as

$$\mathbf{p} \cdot \mathbf{q} = (p_1, p_2, p_3) \cdot \begin{pmatrix} q_1 \\ q_2 \\ q_3 \end{pmatrix} = \sum_i p_i \cdot q_i$$

A linear operator T in the Euclidean space is a map

$$T : \mathbf{p} \longrightarrow \mathbf{p}' = T\mathbf{p}$$

associating to every vector $\mathbf{p}$ a vector $\mathbf{p}'$ with the property that

$$T(\lambda_1 \mathbf{p}_1 + \lambda_2 \mathbf{p}_2) = \lambda_1 T\mathbf{p}_1 + \lambda_2 T\mathbf{p}_2$$

To compute the action of an operator on any vector, it is enough to know its action on the basis vectors. In fact, from

$$T\mathbf{e}_i = \mathbf{e}'_i = \sum_j T_{ji} \mathbf{e}_j$$

and

$$\mathbf{p} = \sum_i p_i \mathbf{e}_i \quad ; \quad \mathbf{p}' = \sum_i p'_i \mathbf{e}_i$$

one obtains

$$T\mathbf{p} = T \sum_i p_i \mathbf{e}_i = \sum_i p_i T\mathbf{e}_i = \sum_{i,j} T_{ji} p_i \mathbf{e}_j = \sum_i \left(\sum_j T_{ij} p_j \right) \mathbf{e}_i = \sum_i p'_i \mathbf{e}_i$$

Comparing the coefficients leads to

$$p'_i = \sum_j T_{ij} p_j$$

These results can be summarized using the square matrix $[T_{ij}] \equiv (\mathbf{e}_i, T\mathbf{e}_j)$:

$$\begin{pmatrix} \mathbf{e}'_1 & \mathbf{e}'_2 & \mathbf{e}'_3 \end{pmatrix} = \begin{pmatrix} \mathbf{e}_1 & \mathbf{e}_2 & \mathbf{e}_3 \end{pmatrix} \begin{pmatrix} T_{11} & T_{12} & T_{13} \\ T_{21} & T_{22} & T_{23} \\ T_{31} & T_{32} & T_{33} \end{pmatrix}$$

For the vector components, we have

$$\begin{pmatrix} p'_1 \\ p'_2 \\ p'_3 \end{pmatrix} = \begin{pmatrix} T_{11} & T_{12} & T_{13} \\ T_{21} & T_{22} & T_{23} \\ T_{31} & T_{32} & T_{33} \end{pmatrix} \begin{pmatrix} p_1 \\ p_2 \\ p_3 \end{pmatrix}$$

The square matrix $[T_{ij}]$ is said to be a representation of the operator T in the basis $\mathbf{e}_i$. The matrix element $T_{ij} \equiv (\mathbf{e}_i, T\mathbf{e}_j)$ is a component of the vector $T\mathbf{e}_j$ along the vector $\mathbf{e}_i$. The convention we have chosen is that the map of the basis vectors can be read out from the **columns** of the matrix.

1.1.2 *Symmetry transformation and symmetry groups*

Definition. A linear map that transforms a physical body (a "pattern") into itself is called a symmetry operation (symmetry transformation, or symmetry element).

Symmetry operations have four important properties. First, the composition of symmetry operations is also a symmetry operation. Composition means that one operation is performed first, followed by the next one. Our convention is that, when the symmetry operations are written in juxtaposed symbols, the operation symbolized on the right-hand side is performed first. Second, the identity is, trivially, a symmetry operation. Third, the inverse operation is also a symmetry element. Fourth, more than two operations can be composed. This is done by composing two of them first (the two on the left or the two on the right). The operation resulting from this composition is composed of the remaining operation. This fourth law is called "associativity". Technically, the set of symmetry operations constitutes a mathematical structure called "group".

Definition. A set G endowed with a law of composition $\circ$ with the properties that:

1. $g_1 \circ g_2 \in G$,
2. an element g_0 exists with $g \circ g_0 = g_0 \circ g = g$,
3. an element g^{-1} exists with $g^{-1} \circ g = g \circ g^{-1} = g_0$, and
4. $(g_1 \circ g_2) \circ g_3 = g_1 \circ (g_2 \circ g_3) \doteq g_1 \circ g_2 \circ g_3$

is called a **group**.

The fact that the set of symmetry transformations possesses the mathematical structure of a group is very useful: For any set with a group structure, mathematicians have developed a bouquet of general theorems that often find applications in the solutions to concrete scientific problems involving a system with a given symmetry group. These general theorems will be treated in Lecture 2.

For a subset of symmetry operations and symmetry groups, a special notation and symbology have been developed. We refer to Lecture 5 and the specialized literature for more detailed information. Here, we provide an introductory overview of this notation and symbology.

1.1.3   *Examples of symmetry elements and their notation*

Rotations: Consider a molecule H_2O (the following figure, left), and suppose that the molecule is rotated by 180° about the axis represented by the thick vertical arrow in the figure.[3]

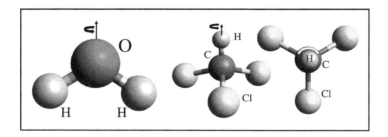

Left: A water molecule, where the thick vertical arrow is the rotation axis, and the side arrow indicates the sense of rotation. Right: (a) Side view and (b) top view of a chloroform molecule.

The position of the oxygen atom does not change upon this rotation, and the two hydrogen atoms will be interchanged, so that the rotation has left the molecule in a state indistinguishable from the original one. This rotation by 180° is a **symmetry operation** of the molecule, and the rotation axis is a **symmetry element**. This particular symmetry element is designated by the symbol C_2 in the Schönflies (S) notation (used mainly by spectroscopists) or by the symbol 2 in Hermann–Mauguin (HM) or international notation, which is often used by crystallographers. The symbol C_2 identifies both the symmetry element and the symmetry operation. A symmetry operation involving the rotation angle $\frac{360°}{n}$ is denoted by the S-symbol C_n (or the corresponding international symbol: n). As a further example, consider the chloroform molecule ($CHCl_3$, see the above figure). In this molecule, the Cl atoms form an equilateral triangle, the C atom is above this triangle, and the H atom resides exactly above the C atom. If the molecule is rotated about the axis joining the C and H atoms, it reaches

---

[3]Rotation axes are sometimes represented by arrows. The rotation is performed following a right screw that moves in the direction indicated by the arrow.

identical orientations after every 120° rotation. This defines a C_3 symmetry axis of rotation.

Identity: Any direction in any object is a C_1 axis: This symmetry is called the identity, and its symbol is C_1 or, correspondingly, 1.

Mirror planes: The H_2O and $CHCl_3$ molecules have symmetry elements other than rotation axes, called the **mirror planes**. The plane where the planar water molecule resides is a mirror plane: One half of the molecule is a mirror image of the other half. The chloroform molecule has three mirror planes, corresponding to each of the planes defined by the direction along the C–H bond and the direction along the C–Cl bond. The HM notation for a mirror plane is m. The Schönflies notation uses σ to indicate a mirror plane, with a subscript that indicates the position of the mirror plane in relation to the main rotation axis. Accordingly, σ_h indicates a mirror plane which is perpendicular to the main rotation axis (h stands for horizontal), and σ_v is a vertical mirror plane, which includes the main rotation axis. σ_d is a diagonal mirror plane, which includes the main rotation axis and bisects the angle between a pair of C_2 axes which are normal to the main rotation axis.

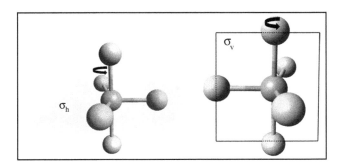

Left: The dashed plane is a σ_h symmetry element. Right: The plane including the rectangle is a σ_v symmetry element.

Center of inversion: A further symmetry element is the center of inversion. A straight line going through the center of inversion will meet its equivalent points at the same distance from the inversion center.

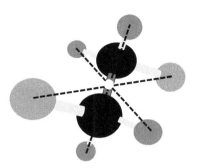

A centrosymmetric molecule: The inversion center is the white point where the dashed lines cross. The dashed lines join the atoms that transform into similar atoms upon inversion. In the specific case, the small spheres are hydrogen atoms, the two central spheres are carbon atoms, and the remaining spheres are bromine atoms ($C_2Br_2H_4$).

The symbol used in the Schönflies notation in i, and that in the HM notation is $\bar{1}$. Note the equation

$$C_2\sigma_h = i$$

which means that the inversion can always be considered a C_2 rotation followed by a reflection at a plane perpendicular to the rotation axis.

Improper rotation axes: In the S-notation, an improper rotation axis is designated by S_n and is an axis of rotatory reflection, i.e. the operation consists of a rotation by $\frac{360°}{n}$, followed by a reflection at a plane perpendicular to the axis.

In the HM notation, an improper rotation axis is an axis of rotatory inversion: The improper rotation is a combination of an n rotation with an inversion, and it is indicated by the symbol $\bar{n}$. In general, therefore, S_n and $\bar{n}$ are different symmetry elements. The following table summarizes some of the symmetry elements and symmetry operations that we will encounter in later chapters.

Table summarizing symmetry elements forthcoming in crystallography.

Type of element	Schönflies symbol	Hermann–Mauguin symbol
Rotation axis	C_1, C_2, C_3, C_4, C_6	$1, 2, 3, 4, 6$
Identity	C_1	1
Mirror plane	σ	m
Center of inversion	i	$\bar{1}$
Rotatory reflection axis	S_1, S_2, S_3, S_4, S_6	
Rotatory inversion axis		$\bar{1}, \bar{2}, \bar{3}, \bar{4}, \bar{6}$

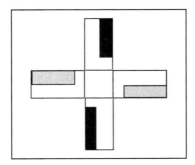

Illustration of S_4-symmetry. The rectangles consist of an identical "material", except that those in black reside above the plane of the paper while those in gray reside below it. The pattern is transformed into itself by the symmetry element S_4.

1.1.4 *Simple examples of symmetry groups and their notation*

1. Square molecule: Let a system consist of identical objects occupying the vertices $A, B, C,$ and D of a square.[4] Let the center of the square be O.

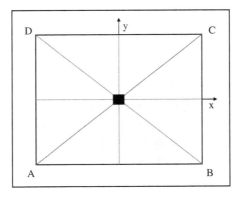

Graphic of the symmetry of a square molecule. The full square symbol at the center indicates the presence of a fourfold rotational axis perpendicular to the plane of the square object and passing through the origin O (the z-axis is not shown).

---

[4]The pattern **dual** to the square is obtained by joining the middle points of the square with lines: This produces a cross that has the same symmetry elements as the square.

Provided there is no symmetry operation relating the points residing above and below the plane, the symmetry elements of the square consist of those maps that interchange the vertices among themselves. One can describe the symmetry elements by various means. One of them is by introducing the permutations of the vertices. The following are the eight elements of the symmetry group of the square:

$$
\mathbb{1} = \begin{pmatrix} A & B & C & D \\ A & B & C & D \end{pmatrix} \qquad C_{4z} = \begin{pmatrix} A & B & C & D \\ B & C & D & A \end{pmatrix}
$$

$$
C_{2z} = \begin{pmatrix} A & B & C & D \\ C & D & A & B \end{pmatrix} \qquad C_{4z}^3 = \begin{pmatrix} A & B & C & D \\ D & A & B & C \end{pmatrix}
$$

$$
m_y = \begin{pmatrix} A & B & C & D \\ B & A & D & C \end{pmatrix} \qquad m_x = \begin{pmatrix} A & B & C & D \\ D & C & B & A \end{pmatrix}
$$

$$
m_{xy} = \begin{pmatrix} A & D & C & B \\ A & D & C & B \end{pmatrix} \qquad m_{\bar{x}y} = \begin{pmatrix} A & B & C & D \\ C & B & A & D \end{pmatrix}
$$

The new order of the letters $ABCD$ after performing the symmetry operation can be read out from the second line of the permutation symbol

$$
\begin{pmatrix} \cdots \\ \cdots \end{pmatrix}
$$

For the labeling of symmetry elements, it is useful to introduce further symbols. One possible convention is to label the rotation axis as an additional subscript ("z" stands for a vector pointing along z). Accordingly, C_{4z}, C_{4z}^3, C_{2z} denote a quarter turn about the rotation axis z, a quarter turn repeated three times, and a half turn, respectively. m denotes the reflections in lines. The line that remains fixed by the reflection is along a vector specified by the variable in the subscript. For instance, the line of reflection for the element m_{xy} is a line containing the unit vector $\frac{1}{\sqrt{2}}(\mathbf{e}_x + \mathbf{e}_y)$. One can also describe the symmetry elements by their actions on the xy-coordinates. The following table summarizes the symmetry elements and their possible descriptions.

Type	Operation	Coordinate	Permutation
e	e	xyz	ABCD
C_4	C_{4z}	$y\bar{x}z$	BCDA
C_4^2	$C_{4z}^2 = C_{2z}$	$\bar{x}\bar{y}z$	CDAB
C_4^3	C_{4z}^3	$\bar{y}xz$	DABC
$2\sigma_v$	$IC_{2y}(m_x)$	$x\bar{y}z$	DCBA
	$IC_{2x}(m_y)$	$\bar{x}yz$	BADC
$2\sigma_v'(2\sigma_d)$	$IC_{2xy}(m_{\bar{x}y})$	$\bar{y}\bar{x}z$	CBAD
	$IC_{2x\bar{y}}(m_{xy})$	yxz	ADCB

2. The two-dimensional dihedral group: Consider a regular n-gon (a regular polygon with n vertices $V_1, \ldots, V_n$ and n sides with equal length). Set the origin of the coordinate system at the center of the n-gon, and let the x-axis contain the vertex V_1.

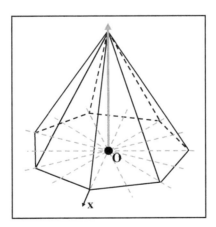

A regular planar eptagon at the base of a pyramid. The main rotation axis through the center is indicated by a thick gray arrow.

To describe the symmetry elements, we think of two mutually orthogonal in-plane unit vectors $(\mathbf{e}_x, \mathbf{e}_y)$ along the x-axis and, respectively, perpendicular to it. We then act on each of these unit vectors with the symmetry elements. The coordinates of the map of $\mathbf{e}_x$ are arranged along the first column of a 2×2 matrix. The coordinates of the map of $\mathbf{e}_y$ are arranged along the second column of the 2×2 matrix. The 2×2 matrix describes ("represents") the symmetry element in such a way that it can be used, for example, in computations. The rotational elements of the n-gon are described by the matrices

$$C_n^k \equiv \begin{pmatrix} \cos \frac{2\pi \cdot k}{n} & -\sin \frac{2\pi \cdot k}{n} \\ \sin \frac{2\pi \cdot k}{n} & \cos \frac{2\pi \cdot k}{n} \end{pmatrix}$$

($k = 0, 1, 2, \ldots, n-1$). These n-elements are anticlockwise rotations by $\frac{2\pi}{n}$ repeated k-times (or rotations by an angle $\frac{2\pi \cdot k}{n}$). The matrices have the property that

$$C_n^k (C_n^k)^t = \mathbb{1}$$

(the upper index "t" indicates the transposed matrix, and $\mathbb{1}$ is the unit matrix). Such matrices are called orthogonal matrices. In addition, their determinant is $= 1$: This is a property identifying "proper" rotations.

In addition, the group of the n-gon contains the n-reflections

$$\sigma_{v,k} \equiv \begin{pmatrix} \cos \frac{2\pi \cdot k}{n} & \sin \frac{2\pi \cdot k}{n} \\ \sin \frac{2\pi \cdot k}{n} & -\cos \frac{2\pi \cdot k}{n} \end{pmatrix}$$

$k = 0, 1, 2, 3, \ldots, n-1$. These matrices describe reflections in the lines

$$y = \tan \frac{\frac{2\pi \cdot k}{n}}{2} \cdot x$$

i.e. the lines joining the center of the polygon with the vertices and with the middle points between vertices (dashed gray lines in the figure). The reflection matrix can also be written as

$$\begin{pmatrix} \cos \frac{2\pi \cdot k}{n} & -\sin \frac{2\pi \cdot k}{n} \\ \sin \frac{2\pi \cdot k}{n} & \cos \frac{2\pi \cdot k}{n} \end{pmatrix} \begin{pmatrix} 1 & 0 \\ 0 & \bar{1} \end{pmatrix}$$

Accordingly, the polygon has the $2n$ symmetry elements

$$\mathbb{1}, C_n, \ldots, C_n^k, \ldots, C_n^{n-1}, \sigma_v, C_n\sigma_v, \ldots, C_n^{n-1}\sigma_v$$

where σ_v is a reflection in the horizontal x-line. The symmetry group is called the (two-dimensional) *dihedral group* D_n. Later, we will extend the action of the symmetry elements to points outside the plane. In that context, the group is called C_{nv} because it has a main n-fold rotation axis and n-mirror planes containing the main rotation axis. The symmetry group of a square is D_4 when points outside the plane are not considered; otherwise, it is called C_{4v} (4mm in the HM notation). By extending the construction of a dihedral group to $n = 1$, one can also define D_1, which contains the identity and a reflection line.

3. The group of the octahedron (O_h): This is the symmetry group of a cube (or of an octahedron[5]) called O_h (Schönflies) ($\frac{4}{m}\bar{3}\frac{2}{m}$, Int.). For example, the symmetry group of the molecule SF_6 is O_h.

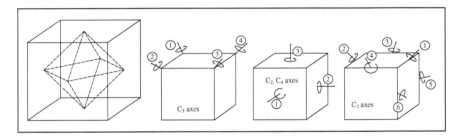

Left: Octahedron inscribed in a cube. Right: The various proper rotation axes in O_h.

With the aid of the figure, we can illustrate all 48 symmetry elements of O_h:

- the identity e;
- three rotations by π about the axes x, y, z ($3C_4^2$);
- Six rotations by $\pm\pi/2$ about the axes x, y, z ($6C_4$);
- Six rotations by π about the bisectrices in the planes; xy, yz, xz ($6C_2$)
- Eight rotations by $\pm2\pi/3$ about the diagonals of the cube ($8C_3$);
- the combination of the inversion I with the listed 24 proper rotations.

The corresponding table indicating the transformation of the coordinates x, y, z is given as follows:

---

[5]The cube and the octahedron (see the figure just above) are said to be dual to each other: The octahedron is e.g. obtained by connecting the face centers of a cube.

Type	Operation	Coordinate	Type	Operation	Coordinate
C_1	C_1	xyz	I	I	$\bar{x}\bar{y}\bar{z}$
C_4^2	C_{2z}	$\bar{x}\bar{y}z$	IC_4^2	IC_{2z}	$xy\bar{z}$
	C_{2x}	$x\bar{y}\bar{z}$		IC_{2x}	$\bar{x}yz$
	C_{2y}	$\bar{x}y\bar{z}$		IC_{2y}	$x\bar{y}z$
C_4	C_{4z}^{-1}	$\bar{y}xz$	IC_4	IC_{4z}^{-1}	$y\bar{x}\bar{z}$
	C_{4z}	$y\bar{x}z$		IC_{4z}	$\bar{y}x\bar{z}$
	C_{4x}^{-1}	$x\bar{z}y$		IC_{4x}^{-1}	$\bar{x}z\bar{y}$
	C_{4x}	$xz\bar{y}$		IC_{4x}	$\bar{x}\bar{z}y$
	C_{4y}^{-1}	$zy\bar{x}$		IC_{4y}^{-1}	$\bar{z}\bar{y}x$
	C_{4y}	$\bar{z}yx$		IC_{4y}	$z\bar{y}\bar{x}$
C_2	C_{2xy}	$yx\bar{z}$	IC_2	IC_{2xy}	$\bar{y}\bar{x}z$
	C_{2xz}	$z\bar{y}x$		IC_{2xz}	$\bar{z}y\bar{x}$
	C_{2yz}	$\bar{x}zy$		IC_{2yz}	$x\bar{z}\bar{y}$
	$C_{2x\bar{y}}$	$\bar{y}\bar{x}\bar{z}$		$IC_{2x\bar{y}}$	yxz
	$C_{2\bar{x}z}$	$\bar{z}\bar{y}\bar{x}$		$IC_{2\bar{x}z}$	zyx
	$C_{2y\bar{z}}$	$\bar{x}\bar{z}\bar{y}$		$IC_{2y\bar{z}}$	xzy
C_3	C_{3xyz}^{-1}	zxy	IC_3	IC_{3xyz}^{-1}	$\bar{z}\bar{x}\bar{y}$
	C_{3xyz}	yzx		IC_{3xyz}	$\bar{z}\bar{x}\bar{y}$
	$C_{3x\bar{y}z}^{-1}$	$z\bar{x}\bar{y}$		$IC_{3x\bar{y}z}^{-1}$	$\bar{z}xy$
	$C_{3x\bar{y}z}$	$\bar{y}\bar{z}x$		$IC_{3x\bar{y}z}$	$yz\bar{x}$
	$C_{3x\bar{y}\bar{z}}^{-1}$	$\bar{z}\bar{x}y$		$IC_{3x\bar{y}\bar{z}}^{-1}$	$zx\bar{y}$
	$C_{3x\bar{y}\bar{z}}$	$\bar{y}z\bar{x}$		$IC_{3x\bar{y}\bar{z}}$	$y\bar{z}x$
	$C_{3xy\bar{z}}^{-1}$	$\bar{z}x\bar{y}$		$IC_{3xy\bar{z}}^{-1}$	$z\bar{x}y$
	$C_{3xy\bar{z}}$	$y\bar{z}\bar{x}$		$IC_{3xy\bar{z}}$	$\bar{y}zx$

4. The symmetry group of a cone: This group is a so-called continuous group: It has an infinite number of elements which are not countable but are labeled by a set of parameters that assume continuous values in a certain interval.

The cone has rotational symmetry around its axis, with elements

$$C(\varphi) = \begin{pmatrix} \cos\varphi & -\sin\varphi \\ \sin\varphi & \cos\varphi \end{pmatrix}$$

$\varphi \in [0, 2\pi]$. Any vertical plane including the conical axis is a mirror plane. These elements are given by

$$C(\alpha)\sigma_v = \begin{pmatrix} \cos\alpha & -\sin\alpha \\ \sin\alpha & \cos\alpha \end{pmatrix} \cdot \begin{pmatrix} 1 & 0 \\ 0 & \bar{1} \end{pmatrix}$$

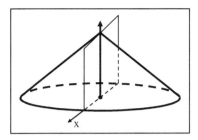

An object with conical shape: The rotation axis is indicated by the vertical arrow. A mirror plane including the x-axis is indicated.

$\alpha \in [0, 2\pi]$. The parameter space is therefore a square in the (φ, α)-plane with a side length of 2π. The group is called $C_{\infty,v}$.

1.2 Formal Aspects of Isometries of the Euclidean Space

In this section, we introduce some more formal statements on the geometrical aspects of symmetry. Later, we will use some of the technical aspects presented in this section for practical computation. Wherever necessary, we will reiterate them.

As symmetry operations, i.e. the transformations of a body that map the body into itself, must preserve the distance between two points, the following definition is appropriate:

Definition. Maps that preserve the distance are called isometries.

Proposition. *Any isometry of the Euclidean space mapping* $\mathbf{r}$ *to* $\mathbf{r}'$ *is a linear map*

$$\mathbf{r}' = R\mathbf{r} + \mathbf{t} \Leftrightarrow x_i' = \sum_j R_{ij} x_j + t_i$$

with R *being a* 3×3 *orthogonal matrix, i.e.* $R^t R = \mathbb{1}$. $\mathbf{t}$ *is called a translation.*

Comments.

1. The isometry $R\mathbf{r} + \mathbf{t}$ can be represented by the so-called *Seitz*[6] symbol $(R \mid \mathbf{t})$:

$$R\mathbf{r} + \mathbf{t} \doteq (R \mid \mathbf{t})\mathbf{r}$$

---

[6]Frederick Seitz, see e.g. https://en.wikipedia.org/wiki/Frederick_Seitz.

2. For Seitz symbols, we have the following computation rules that are a direct consequence of the law of compositions of mappings:

$$(R_1 \mid \mathbf{t}_1)(R_2 \mid \mathbf{t}_2) = (R_1 R_2 \mid \mathbf{t}_1 + R_1 \mathbf{t}_2)$$

and

$$(R \mid \mathbf{t})^{-1} = (R^{-1} \mid -R^{-1}\mathbf{t})$$

In fact,

$$(R_1 \mid \mathbf{t}_1)(R_2 \mid \mathbf{t}_2)\mathbf{r} = (R_1 \mid \mathbf{t}_1)(R_2\mathbf{r} + \mathbf{t}_2) = R_1(R_2\mathbf{r} + \mathbf{t}_2) + \mathbf{t}_1$$
$$= R_1 R_2 \mathbf{r} + R_1 \mathbf{t}_2 + \mathbf{t}_1 = (R_1 R_2 \mid \mathbf{t}_1 + R_1 \mathbf{t}_2)\mathbf{r}$$

This proves the first rule. Further, if $R_1 = R_2^{-1}$ and $\mathbf{t}_1 = -R_2\mathbf{t}_2$, then the composition gives the identity map; *ergo*, we have constructed the inverse map.

3. Successive Seitz symbols can also be computed according to

$$[(R_1 \mid \mathbf{t}_1)(R_2 \mid \mathbf{t}_2)] (R_3 \mid \mathbf{t}_3) = (R_1 \mid \mathbf{t}_1) [(R_2 \mid \mathbf{t}_2)(R_3 \mid \mathbf{t}_3)]$$

(the so-called law of associativity).

4. The following proof constructs all isometries of the space and, therefore, provides important practical knowledge.

Proof of the Proposition.

Part A: We show first that this map preserves the distance between points. In fact,

$$(\mathbf{p}' - \mathbf{q}') \cdot (\mathbf{p}' - \mathbf{q}') = R(\mathbf{p} - \mathbf{q}) \cdot R(\mathbf{p} - \mathbf{q}) = R^t R(\mathbf{p} - \mathbf{q}) \cdot (\mathbf{p} - \mathbf{q})$$
$$= (\mathbf{p} - \mathbf{q}) \cdot (\mathbf{p} - \mathbf{q})$$

Part B: We show now that if a map $F(\mathbf{r})$ preserves the distance between points, it must have the proposed expression. For the proof, we take $F(\mathbf{0}) = \mathbf{t}$ so that the composite map $(\mathbb{1} \mid -\mathbf{t}) \circ F(\mathbf{r}) \doteq F_O$ fixes the origin. Accordingly, F_O cannot contain translations. If we prove that $F_O = (R \mid \mathbb{1})$ with R an orthogonal matrix, then we have proved the claim, as the original isometry F is the composition $(\mathbb{1} \mid \mathbf{t}) \circ F_O$.

To construct the general map F_O with $F_O(\mathbf{0}) = \mathbf{0}$, imagine a point P mapped onto P' by the map F_O. By suitably choosing the coordinate system, P and P' can be considered to have simultaneously vanishing z-coordinates. The polar planar coordinates of the point P are $(r \cdot \cos\alpha, r \cdot \sin\alpha)$. Let us map the coordinate x to $r \cdot \cos(\alpha + \varphi)$ (an operation which fixes the origin). In order to preserve the distance from the origin, the coordinate y must change to $\pm r \cdot \sin(\alpha + \varphi)$. The mapping can be cast in terms of two 2×2 matrices acting onto the coordinates (x, y) arranged as a column vector:

$$F_O^+(\mathbf{r}) = \begin{pmatrix} \cos\varphi & -\sin\varphi \\ \sin\varphi & \cos\varphi \end{pmatrix} \begin{pmatrix} x \\ y \end{pmatrix}$$

and

$$F_O^-(\mathbf{r}) = \begin{pmatrix} \cos\varphi & \sin\varphi \\ \sin\varphi & -\cos\varphi \end{pmatrix} \begin{pmatrix} x \\ y \end{pmatrix}$$

Both matrices $[F_O^\pm]$ are orthogonal. The first mapping is a rotation by an angle φ. If the plane is spanned by the xy-coordinates and z is oriented to build a right-hand coordinate system, the angle φ is considered positive if the rotation follows the right-hand corkscrew rule. The second mapping is a reflection in the line through the origin defined by the equation

$$y = \tan\frac{\varphi}{2} \cdot x$$

The matrix describing a rotation by φ has a determinant of $+1$, whereas the matrix describing a reflection has a determinant of -1.

These isometries transfer points with a non-vanishing z-coordinate to a point with the same z-coordinate. However, a change to $-z$ is also a preserving isometry compatible with $F_O^\pm$, i.e. the most general form of an isometry that fixes the origin for a given axis z, is the 3×3 orthogonal matrix

$$F_O^{\pm,\pm} = \begin{pmatrix} F_O^\pm & 0 \\ 0 & \pm 1 \end{pmatrix}$$

The same general isometry can be constructed about any axis (which is then chosen to the z-axis), so that the most general isometry of the Euclidean space can be written as

$$(F_O^{\pm,\pm} \mid \mathbf{t})$$

with $F_O^{\pm,\pm}$ being orthogonal by construction. Our proof entails that, by a suitable choice of the z-axis, the matrix has the block form $F_O^{\pm,\pm}$. This concludes the proof of the proposition about isometries.

1.2.1 *Isometries of the line*

1. The operations $(\mathbb{1} \mid t)$ are the Seitz symbol for the mapping

$$(\mathbb{1} \mid t) : z \to (\mathbb{1} \mid t)(z) = z + t$$

It describes pure translations along a line, given by the z-axis. Translations are examples of isometries where all points of the line are moved, i.e. a translation has no fixed point.

2. The operation $(I \mid t)$ is the Seitz symbol for the mapping

$$(I \mid t) : z \to (I \mid t)(z) = -z + t$$

$(I \mid t)$ is an inversion in the origin, followed by a translation. It can also be visualized as an inversion in the point $\frac{t}{2}$.

The set of isometries $\{(\pm\mathbb{1} \mid t)\}$ builds the Euclidean group Euc(1).

1.2.2 *Isometries of the plane*

1. $(\mathbb{1} \mid \mathbf{t})$ are pure translations.

2. $(F_0^+ \mid \mathbf{t})$: Provided F_0^+ is not the identity, one can find a vector $\mathbf{c}$, indicating a point C, for which $(F_0^+ \mid \mathbf{t})\mathbf{c} = \mathbf{c}$. Then, $(F_0^+ \mid \mathbf{t})$ describes a pure rotation about C by an angle φ, i.e. the rotation axis crosses the plane not in O but in C. By shifting the rotation axis to C, the translation is eliminated, and a "proper" rotation appears.

3. The matrix $(F_0^- \mid \mathbf{t})$ represents a reflection in the line $y = \tan\frac{\varphi}{2} \cdot x$. Consider the unit vector $\mathbf{e}_\varphi = (\sin\frac{\varphi}{2}, \cos\frac{\varphi}{2})$ pointing along the reflection line.

Case 1: First, choose $\mathbf{t}$ perpendicular to $\mathbf{e}_\varphi$. Then, a line exists which is fixed by the map $(F_0^- \mid \mathbf{t})$, and the map is a pure reflection at this line. The line is parallel to $\mathbf{e}_\varphi$, and its distance from the origin is $\frac{|\mathbf{t}|}{2}$.
Case 2: Now, choose $\mathbf{t}$ parallel to $\mathbf{e}_\varphi$. Then, no point is left fixed by the map $(F_0^- \mid \mathbf{t})$, and the map is a *glide reflection*, i.e. a reflection at a line, followed by translation parallel to the line.
Case 3: For general $\mathbf{t}$, one can decompose $\mathbf{t}$ into a component parallel to $\mathbf{e}_\varphi$ and a component perpendicular to it to find that the map $(F_0^- \mid \mathbf{t})$ is a reflection at a suitable line, followed by a translation parallel to this line.

The set of isometries $\{(F_0^\pm \mid \mathbf{t})\}$ constitutes the Euclidean group Euc(2).

1.2.3 *Isometries of the space*

1. $(\mathbb{1} \mid \mathbf{t})$: pure translations, similar to # 1 of the planar case.
2. $(F_O^{+,+} \mid \mathbf{t})$: Provided $\mathbf{t}$ is in the xy-plane, these Seitz symbols indicate pure rotations about an axis parallel to the z-axis and passing through a point $C \neq O$. C is found by solving the equation $F_O^{+,+}\mathbf{c} + \mathbf{t} = \mathbf{c}$. These operations fixes a plane and are therefore similar to # 2 of the planar case.
3. $(F_O^{-,+} \mid \mathbf{t})$: Provided $\mathbf{t}$ is in the xy-plane, these Seitz symbols describe reflections or glide reflections that fix a plane and a similar to case 3 of the planar isometries.

   There are new types of isometries that have no analogon in two-dimensions.
4. The Seitz symbol $(F_O^{+,+} \mid \mathbf{t})$, with $\mathbf{t}$ parallel to the rotation axis, describes a rotation about z, followed by a translation parallel to z — the so-called screw rotations. Screw rotations do not fix a plane.
5. $(F_O^{+,-} \mid \mathbf{t})$: If $\mathbf{t}$ is parallel to the z-axis, this Seitz symbol represents the composition of a reflection in a plane and a rotation about a line perpendicular to the plane, followed by a translation parallel to this line. In contrast to glide reflections, screw reflections do not fix a plane.

The set of isometries $\{(F_0^{\pm\pm} \mid \mathbf{t})\}$ constitutes the Euclidean group Euc(3). The Euclidean group in n dimensions, Euc(n), consists of all elements of the type $(R \mid \mathbf{t})$, R being an orthogonal $n \times n$ matrix.

Similarity: In this paragraph, we show how to mathematically implement the intuitive fact that "similar objects" (roughly, objects that have a different size but the same geometry) have the same symmetry group. The key to this implementation is the operation of **similarity**. A similarity operation can be one of the isometries we are acquainted with but might also entail an operation called *dilation* or *scaling*.

Definition. A dilation is a map defined by

$$D(\mathbf{r}) = \delta \cdot (\mathbf{r} - \mathbf{c}) + \mathbf{c} = \delta \cdot \mathbf{r} + (1 - \delta)\mathbf{c}$$

$\delta > 0$ is a scalar and represents the scaling factor, and $\mathbf{c}$ is the center of dilation.

Comments.

1. The expression on the right-hand side allows us to write the dilation map using the Seitz symbol $(\delta \cdot \mathbb{1} \mid (1 - \delta)\mathbf{c})$.

2. The scaling map has the property that $D(\mathbf{c}) = \mathbf{c}$ and $\mid D(\mathbf{r}) - \mathbf{c} \mid= \delta \mid \mathbf{r} - \mathbf{c} \mid$, i.e. the scaling map is not an isometry. However, using the appropriate Seitz symbol, isometries can be composed of dilations for a more general set of operations.

Definition. The two isometries $F_1, F_2 : \mathscr{R}^3 \to \mathscr{R}^3$ are said to be similar if there exists a "similarity" transformation $H : \mathscr{R}^3 \to \mathscr{R}^3$ such that $F_2(\mathbf{r}) = H \circ F_1 \circ H^{-1}(\mathbf{r})$ ($\circ$ is the composition of mappings). We also use the term "conjugation" for the similarity operation and call F_1 and F_2 "conjugate".

Examples.

1. As an example of the similarity operation, we consider the two squares $\square_1$ and $\square_2$, centered at different points O_1 and O_2 in the plane and with different side lengths l_1 and l_2. Let their vertices be $V_1 - V_4$ and $U_1 - U_4$, taken in order in the anticlockwise direction. We aim to establish that their symmetry groups are similar, so that they can be considered "the same object" from the perspective of symmetry considerations. For this purpose, we set $\delta = \frac{l_2}{l_1}$ and define the following sequence of maps in $\mathscr{R}^2$:

   H_1 is the translation that moves O_2 to O_1.
   H_2 is the dilation by δ centered at O_1.
   H_3 is the rotation about O_1 that transfers U_i to V_i.

   Define now $H \doteq H_3 \circ H_2 \circ H_1$. Let $F(2)$ be an operation of D_4 that transforms $\square_2$ into itself. Then,

   $$H \circ F(2) \circ H^{-1} = F(1)$$

   is the "same" symmetry operation that transforms $\square_1$ into itself. Hence, the group $D_4(1)$ is similar to $D_4(2)$. Generally, if two groups are related by a similarity operation, they can be considered to be "the same". This is an expression of the fact that, to solve physical problems with symmetry arguments, the position and size of an object are often irrelevant.

2. Further examples of similarity are translations by two vectors $\mathbf{a}$ and $\mathbf{b}$ with different lengths and directions: $(\mathbb{1} \mid \mathbf{a})$ is similar to $(\mathbb{1} \mid \mathbf{b})$. The similarity transformation is provided by the map $H \doteq (\delta R_\theta \mid \mathbf{0})$, with $\delta =\mid \mathbf{b} \mid \cdot \mid \mathbf{a} \mid^{-1}$ and θ the angle between $\mathbf{b}$ and $\mathbf{a}$. The inverse similarity transformation reads $H^{-1} = (\delta^{-1} R_{-\theta} \mid \mathbf{0})$.

3. Two rotations by the same angle around two parallel rotation axes, one crossing the plane at the origin and the other crossing the plane at the

point P located by the vector $\mathbf{p}$ are also similar. In fact,

$$(\mathbb{1} \mid \mathbf{p})(R_\theta \mid \mathbf{0})(\mathbb{1} \mid -\mathbf{p}) = (R_\theta \mid \mathbf{p} - R\mathbf{p})$$

This last operation fixes P and is thus a rotation by an angle θ around an axis going through P.

4. Finally, $R_{\theta,\mathbf{n}}$ and $R_{\theta,\mathbf{m}}$ — two rotations by the same angle θ but about different rotation axes, $\mathbf{n}$ and $\mathbf{m}$, respectively — are also similar. The similarity operation is provided by the rotation that transfers $\mathbf{n}$ to $\mathbf{m}$.

1.3 Algebraic Properties of Groups

Groups have a number of algebraic properties which are very useful for performing practical computations based on symmetry arguments. In this section, we aim to understand these properties.

$\rightarrow$ **Definition.** A set $G = \{g_i\}$ endowed with a product $\circ$ obeying the following four rules is called a group:

1. If g_i and g_j are elements of G, so is the product $g_i \circ g_j$.
2. The rule of computation for the product of more than two elements (associativity) is $g_i \circ g_j \circ g_k = g_i \circ (g_j \circ g_k) = (g_i \circ g_j) \circ g_k$
3. There is an element e such that for any other element, $e \circ g = g \circ e = g$. e is called the identity element
4. Given an element g, there exists a unique element g^{-1} with $g \circ g^{-1} = g^{-1} \circ g = e$. g^{-1} is the inverse element.

Comment. The character "$\circ$" indicating the composition of elements is often neglected to simplify the notation.

$\rightarrow$ Order of a group: The number of elements of a group indicated, for example, by $\mid G \mid$ is called the order of the group.

$\rightarrow$ Subgroup: A subset of G which is closed with respect to the operation $\circ$ and contain the identity element and the inverse elements is a subgroup of G.

$\rightarrow$ Examples of Groups:

- The real numbers $\mathscr{R}$ constitute a group under the operation of addition. The identity element is the number 0, and the inverse of x is $-x$.
- The additive group of integers: The set of integers $\mathscr{Z}$ under the usual addition is a group. It is a subgroup of $\mathscr{R}$.
- The multiplicative group of non-zero real numbers: The set of the real numbers without the number zero $(\mathscr{R} - \{0\} \equiv \mathscr{R}_0$ is a group under

the usual multiplication. The identity element is the number 1, and the inverse element of x is $1/x$.

- The set consisting of $(1, -1)$ under the standard rule of multiplication.
- The set of complex numbers $(1, i, -1, -i)$ under the standard rule of multiplication.
- The set of all matrices of order $n \times n$ under addition.
- Complex numbers: The complex numbers $\mathscr{C}$ under addition and the non-zero complex numbers under multiplication form groups.
- The general linear group: The set of complex matrices of size $n \times n$ with non-zero determinants is a group under ordinary matrix multiplication. It is denoted by $GL(n, \mathscr{C})$. Similarly, we define $GL(n, \mathscr{R})$ as the set of real $n \times n$ matrices with non-zero determinants.
- The special linear groups: The sets of real or complex matrices of size $n \times n$ with determinants equal to 1 are groups. These groups are denoted by $SL(n, \mathscr{R})$ and $SL(n, \mathscr{C})$, respectively. They are subgroups of $GL(n, \mathscr{R})$ and $GL(n, \mathscr{C})$, respectively.
- The orthogonal groups: An orthogonal matrix R is an element of $GL(n, \mathscr{R})$ that satisfies the condition $R^T \cdot R = \mathbb{1}$, where R^T is the transpose of R. The set of all orthogonal matrices defines a group under matrix multiplication. It is denoted by $O(n) \subset SL(n, \mathscr{R})$.
- The special orthogonal groups: The set of all orthogonal matrices $R \in O(n)$ with determinants equal to 1 is a group, the special orthogonal group $SO(n) \subset O(n)$.
- The unitary groups: A complex matrix U is unitary if $U^\dagger U = \mathbb{1}$, where $U^\dagger$ is the transposed complex conjugate of U. The set of all $n \times n$ unitary matrices forms a group under matrix multiplication. This group is denoted by $U(n)$.
- The special unitary groups: The set of all unitary matrices of size $n \times n$ with determinants equal to 1 is also a group, the special unitary group. This group is denoted by $SU(n) \subset U(n)$.
- The symmetric (permutation) group of n-objects S_n (or Perm(n)). Assume that M is a set consisting of n elements. A bijective map $\sigma : M \to M$ is called a permutation of M. The permutations of M form a group under the usual composition of functions. This group is called the symmetric or permutation group and is denoted by S_n. We use the notation

$$\sigma = \begin{pmatrix} 1 & 2 & \ldots & n \\ \sigma(1) & \sigma(2) & \ldots & \sigma(n) \end{pmatrix}$$

to specify the list of n numbers resulting from the changing of the order of $1, 2, \ldots, n$ without repetition. With this notation, we can count the number of permutations: We have n choices for $\sigma(1)$, $n - 1$ choices for $\sigma(2)$, and so on, giving finally $\mid S_n \mid = n!$.

- The set $T(3) \doteq \{(\mathbb{1} \mid \mathbf{t})\}$ consisting of all translations is a proper subgroup of Euc(3) and is called the translation group.
- The set $O(3) \doteq \{(A \mid \mathbf{0}) : A^t A = \mathbb{1}\}$ consisting of all orthogonal matrices, is a subgroup of Euc(3) and is called the orthogonal group. It consists of all isometries that fix the origin.
- The group $SO(3)$ consisting of all orthogonal matrices with determinants equal to 1 is a subgroup of $O(3)$ and is called the special orthogonal group. It consists of all proper rotations that fix the origin.

$\rightarrow$ Abelian group: A group is called commutative (or Abelian) if $g_1 \circ g_2 = g_2 \circ g_1 \ \forall g \in G$.[7]

Comment. There are simple rules that allow to establish the commutativity of two symmetry operations:

- two consecutive rotations about the same axis, which implies that, for example, $C_n = \{C_1^0, \ldots, C_{n-1}\}$ is commutative;
- two reflections at orthogonal planes (which are equivalent to a rotation by π at the axis forming the intersection of the two planes);
- two rotations by π about two orthogonal axes (equivalent to a rotation by the same angle about a third axis, orthogonal to both);
- A rotation followed by a reflection at a plane orthogonal to the rotation axis.

$\rightarrow$ **Theorem (Rearrangement).** *If $\{e, g_1, \ldots, g_n\}$ are the elements of a group G and g_k is an arbitrary element, then*

$$g_k G \doteq \{g_k e, g_k g_1, \ldots, g_k g_n\}$$

and

$$G g_k \doteq \{e g_k, g_1 g_k, \ldots, g_n g_k\}$$

contain each group element once and only once.

---

[7] Niels Henrik Abel (1802–1829) was a norwegian mathematician. He contributed to different fields of mathematics. See e.g. https://en.wikipedia.org/wiki/Niels_Henrik_Abel.

Proof. $g_k g_j$ is certainly an element of the group by definition of the composition law. Suppose $g_i g_k = g_j g_k$ with $g_i \neq g_j$. Then, multiplying both sides of the equation by g_k^{-1} from the right, we obtain $g_i = g_j$, which is contrary to the assumption.   ◇

→ Multiplication table: The rearrangement theorem has an important application for finite groups: The result of two consecutive operations can be represented within a *multiplication table*. Suppose that the elements of a finite group are to be ordered along the top line of a square table and along its first left-hand column. The bulk of the table can be filled with the result of multiplying the top element with one element of the left-hand column, whereby the top operation (by convention) is performed first. Note that, in virtue of the rearrangement theorem, each column contains all elements only once. The ordering of the top and left-hand column elements is immaterial. In the following table, the ordering chosen is such that the operation e is located along the principal diagonal.

Multiplication table for C_{4v}.

C_{4v}	e	C_4	C_4^2	C_4^3	m_y	m_x	m_{xy}	$m_{\bar{x}y}$
e	e	C_4	C_4^2	C_4^3	y	x	xy	$\bar{x}y$
C_4^3	C_4^3	e	C_4	C_4^2	xy	$\bar{x}y$	x	y
C_4^2	C_4^2	C_4^3	e	C_4	x	y	$\bar{x}y$	xy
C_4	C_4	C_4^2	C_4^3	e	xy	$\bar{x}y$	y	x
m_y	y	xy	x	$\bar{x}y$	e	C_4^2	C_4	C_4^3
m_x	x	$\bar{x}y$	y	xy	C_4^2	e	C_4^3	C_4
m_{xy}	xy	x	$\bar{x}y$	y	C_4^3	C_4	e	C_4^2
$m_{\bar{x}y}$	$\bar{x}y$	y	xy	x	C_4	C_4^3	C_4^2	e

→ Homomorphism: A *homomorphism* is a map $\phi : G_1 \to G_2$ that associates to each element g_i of a group one element and only one element $\phi(g_i)$ of a group such that

$$\phi(g_l \circ g_k) = \phi(g_l) \bullet \phi(g_k)$$

$\circ$ is the group operation in G_1, and $\bullet$ is the group operation in G_2. The element $\phi(g_i)$ is called the image or map of the element g_i under the homomorphism. The set of images constitutes a subgroup of G_2. Although each element has only one image, several elements of G_1 may be mapped onto the same image. Thus, it might happen that $\phi(g_i) = \phi(g_j)$ although $g_i \neq g_j$. For example, one can always associate to each element of any group the number "1", which is a group under multiplication. "1" is then the image of the homomorphism. This homomorphism plays a prominent role in science. The *kernel* of ϕ is the set of elements in G_1 that are mapped onto the unit element in G_2.

$\rightarrow$ Isomorphism: Two groups G_1 and G_2 are said to be isomorphic if there is a map $i : G_1 \rightarrow G_2$ that implements a one-to-one correspondence between their elements (the map is bijective, i.e. it is surjective and injective) and the map preserves the group combination law, i.e.

$$i(g_1) \circ i(g_2) = i(g_1 \bullet g_2)$$

The map $i : G_1 \rightarrow G_2$ is an isomorphism. If two finite groups are isomorphic, then they have, for example, the same multiplication table, i.e. they are essentially the same, although the concrete significance of their elements may be different.

$\rightarrow$ **Proposition.** *If the kernel of the homomorphism ϕ consists only of e_1, then the map between G_1 and $Im(G_1)$ is a one-to-one map.*

Proof. Assume that $\phi(g_1) = \phi(g_2)$ for $g_1 \neq g_2$. Then,

$$\phi(g_1^{-1}g_1) = e_2 = \phi(g_1)^{-1}\phi(g_1) = \phi(g_1^{-1})\phi(g_2) = \phi(g_1^{-1}g_2)$$

$g_1^{-1}g_2$ belongs to the kernel of ϕ, which was assumed to consist only of e_1. Therefore, $g_1^{-1}g_2 = e_1$, i.e. $g_2 = g_1$. $\diamond$

$\rightarrow$ Cyclic group: A group G is said to be cyclic if all its elements can be written as the powers of one element g, which is then called as the generator of the group: $G_n = \{g^0 = e, g^1 = g, g^2, \ldots, g^{n-1}, g^n = e\}$. Cyclic groups are Abelian. An example of cyclic groups is the group C_n.

$\rightarrow$ Conjugacy classes: Two elements g_1 and g_2 of a group G are **conjugate** if there exists a group element x such that $g_1 = xg_2x^{-1}$ (or, equivalently, $g_1 = x^{-1}g_2x$). Applying the conjugation operation to one element g_1 using every possible element x of the group produces classes of mutually conjugate elements (the identity element e constitutes a class by itself).

Proposition. *All elements of one class are conjugated to each other but two elements belonging to different classes are not conjugate to each other.*

Proof. Consider, for example, g_1 as a conjugate to g_2 and g_3, then g_2 and g_3 are also conjugate to each other since, from $g_1 = x^{-1}g_2x$ and $g_1 = y^{-1}g_3y$, we have $g_2 = (yx^{-1})^{-1}g_3(yx^{-1})$. $\diamond$

Each class is therefore completely determined by considering one single element of the class, which might be taken as the representative of the class. Each element of the group can only belong to one class. Each element of a commutative group constitutes a class on its own. In fact, from $x \circ a \circ x^{-1} = a$ (because of commutativity), it follows that the conjugacy operation applied to an element a leads to the same element a.

$\rightarrow$ Rules for conjugacy: In the case of groups of symmetry operations consisting of rotations, reflections, and inversion of a physical system, there are some simple rules which allow us to determine the classes of a non-Abelian group without having to perform explicit calculations for all elements:

- Rotations through angles of different magnitudes must belong to different classes. Thus, for example, C_4 and C_4^2 belong to different classes.
- Rotations by $\pm$ the same angle about an axis belong to the same class only if there is an element in the group that changes the handiness of the coordinate system. Thus, C_4 and C_4^3 belong to the same class of C_{4v} because of the existence of a vertical reflection plane.
- Rotations through the same angle about different axes or reflections with respect to two different planes belong to the same class only if the two different axes or planes can be brought into each other by an element of the group. Thus, m_x and m_y belong to the same class of C_{4v} as they can be brought into each other by C_4.

$\rightarrow$ **Example:** The classes of C_{4v} are

$$\{e\}, \{C_4, C_4^3\}, \{C_4^2\}, \{m_x, m_y\}, \{m_{xy}, m_{\bar{x}y}\}$$

1.4 Analytic Properties of Continuous Groups

Definition. When the elements of a group can be mapped into one-to-one correspondence with the r real parameters $\mathbf{a} = (a_1, \ldots, a_r)$ that vary continuously within some given subset S_r of an r-dimensional space, then the group is called a continuous group. We write the elements of the continuous group as $R(\mathbf{a})$.

We refer to S_r as the parameter space, and r is called the order of the continuous group. There also exist mixed continuous groups, where $\mathbf{a}$ varies continuously within a certain domain but changes discontinuously when transitioning to a different domain.

Example 1. The group SO_2 consisting of the matrices

$$\begin{pmatrix} \cos\varphi & -\sin\varphi & 0 \\ \sin\varphi & \cos\varphi & 0 \\ 0 & 0 & 1 \end{pmatrix}$$

with $\varphi \in [0, 2\pi]$ is a continuous group of order 1. The matrices describe the transformation of the coordinates x, y, z upon rotations around the z-axis. It is also called C_∞.

Example 2. The group $C_{\infty h}$ is described by the matrices

$$\begin{pmatrix} \cos\varphi & -\sin\varphi & 0 \\ \sin\varphi & \cos\varphi & 0 \\ 0 & 0 & \alpha \end{pmatrix}$$

In the α, φ-parameter plane, the parameter space is described by the two disjoint segments with coordinates $(1, [0, 2\pi])$ and $(-1, [0, 2\pi])$. This is an example of a mixed continuous group.

Example 3. $T(1) = \{(0 \mid t)\}$ $(t \in \mathscr{R})$ is a one-dimensional translation group. It is a continuous group of order 1.

Example 4. The group of linear transformations in one dimension is defined by $x' = a_1 \cdot x + a_2$. The parameter a_1 is a dilation operation, and the parameter a_2 is a translation. a_1, a_2 are both real numbers. This is a group of order 2.

Topological and Lie groups: The multiplication rule

$$R(\mathbf{a})R(\mathbf{b}) = R(\mathbf{c})$$

defines a set of real functions in the parameter space

$$c_1 = f_1(a_1, \ldots, a_r, b_1, \ldots, b_r)$$

$$\vdots$$

$$c_r = f_r(a_1, \ldots, a_r, b_1, \ldots, b_r)$$

which gives the parameter $\mathbf{c}$ in terms of the parameters $\mathbf{a}$ and $\mathbf{b}$. These are the composition laws in the parameter space and are a kind of "multiplication table for continuous groups".

Definition. Let us denote $\mathbf{e}$ as the value of the identical element in the parameter space. If the functions $f_1, \ldots, f_r$ are continuous and the equation

$$e_1 = f_1(a_1, \ldots, a_r, \bar{a}_1, \ldots, \bar{a}_r)$$

$$\vdots$$

$$e_r = f_r(a_1, \ldots, a_r, \bar{a}_1, \ldots, \bar{a}_r)$$

can be solved to find the parameter $\bar{a}$ for the inverse element, then the group is said to be a **topological** group.

Definition. If the functions $f_1, \ldots, f_r$ also have derivatives of all orders, the group is said to be a **Lie group** (the name originates from Sophous Lie, a Norwegian mathematician).

Example 1. The translation group in one dimension $T(1)$: The law of composition in the parameter space is written as $z = x + y$, i.e. $f(x, y) = x + y$. The function is continuous. The equation $0 = x + \bar{x}$ has the solution $\bar{x} = -x$. The group is a Lie group, as the function $f(x, y)$ has derivatives of all orders.

Example 2. The group SO_2: It is a one-parameter group with the law of composition $\varphi_z = \varphi_x + \varphi_y$, i.e. $f(\varphi_x, \varphi_y) = \varphi_x + \varphi_y$. The inverse element is given in the parameter space by $-\varphi$. It is a Lie group of order 1.

Example 3. It can be shown (B. L. van der Waerden, *Group Theory and Quantum Mechanics*, Springer Verlag, 1980, p. 83) that in any Lie group of order 1 with elements $\{R(x)\}$, it is possible to write the laws of composition as $z = x + y$ and $f^{-1}(x) = -x$, so that we can assume that $R(x)R(y) = R(x + y)$, $R(x)^{-1} = R(-x)$.

Example 4. We find the law of composition for the group defined by $x' = a_1 \cdot x + a_2$. From $x' = a_1 \cdot x + a_2$ and $x'' = b_1 \cdot x' + b_2 = b_1 \cdot a_1 \cdot x + b_1 \cdot a_2 + b_2$, the law of composition in the parameter space is written as

$$c_1 = b_1 \cdot a_1$$

$$c_2 = b_1 \cdot a_2 + b_2$$

The unit element is identified by the parameters $(1, 0)$, and the parameters of the inverse element are found by solving the coupled equations

$$1 = \bar{a} \cdot a_1$$

$$0 = \bar{a} \cdot a_2 + \bar{b}$$

i.e.

$$\bar{a} = \frac{1}{a_1}$$

$$\bar{b} = -\frac{a_2}{a_1}$$

The inverse element exists only if the line $a_1 = 0$ is excluded from the a_1, a_2-planar parameter space. In all other regions of the a_1, a_2-plane, the group is a Lie group.

1.4.1 *Group average*

The operation of "averaging over the group" is used in the proofs of important theorems. This operation implies the existence of a **group average** (in german "Mittelwert", after H. Weyl or "Haar" measure).[8]

Definition. Let g be an element of a group G (discrete or continuous) and $F(g)$ be a continuous function defined based on the elements g. A functional that associates the scalar $\mathscr{M}_{g \in G} F(g)$ to any function $F(g)$ is called a group average if

1. it is linear:

$$\mathscr{M}_{g \in G} [\alpha F(x) + \beta G(x)] = \alpha \mathscr{M}_{g \in G} F(x) + \beta \mathscr{M}_{g \in G} G(x)$$

2. it is positive: $\mathscr{M}_{g \in G} F(g) \geq 0$ if $F(x) \geq 0$, being 0 only if $F(x) \equiv 0$
3. the group volume is finite, i.e. if $F(x) = 1$, then $\mathscr{M}_{x \in G} F(x)$ can be chosen to be 1.
4. it has left and right invariances with respect to group operations: $\forall y$

$$\mathscr{M}_{g \in G} F(y \circ g) = \mathscr{M}_{g \in G} F(g \circ y) = \mathscr{M}_{g \in G} F(g)$$

5. $\mathscr{M}_{g \in G} F(g^{-1}) = \mathscr{M}_{g \in G} F(g)$
6. $\mathscr{M}_{g \in G} \overline{F(g)} = \overline{\mathscr{M}_{g \in G} F(g)}$

Group average for finite groups: For a finite group,

$$\mathscr{M}_{g \in G} F(g) \doteq \frac{1}{|G|} \sum_g F(g)$$

---

[8]Hermann Weyl, (1885–1955), see e.g. https://de.wikipedia.org/wiki/Hermann_Weyl; Alfréd Haar, (1885–1933), hungarian mathematician.

is the group average. The rearrangement theorem ensures that when one sums over all elements of the group, the average is invariant with respect to left and right multiplication by any element of the group.

Group average for continuous groups: For continuous groups, we expect that the average takes the form of an integral over an appropriate differential element of the parameter space:

$$\int_{\{\mathbf{a}\}} dV_a \cdot g(\mathbf{a}) \cdot F(\mathbf{a})$$

where dV_a is short for $da_1 \cdot da_2 \cdots da_r$ (the infinitesimal volume in the parameter space) and $g(\mathbf{a})$ represents the density of symmetry elements in the parameter space. There is a general method to computing $g(\mathbf{a})$, starting from the function $f_1, \ldots, f_r$. The interested reader is invited to continue reading the following paragraphs, where the general formula for $g(\mathbf{a})$ and some examples are explicitly computed. Later in the lectures, when necessary, we will refer to some of the formulas obtained here. For the time being, we only need to know that there are important groups (such as finite groups and SO_3) for which formulas of the average exist.

As an option for the reader, we now show how to compute $g(\mathbf{a})$ for general groups and provide some examples.

For the integral to be an average, we require

$$\int_{\{\mathbf{a}\}} dV_a \cdot g(\mathbf{a}) \cdot F(\mathbf{a}) = \int_{\{\mathbf{a}\}} dV_a \cdot g(\mathbf{a}) \cdot F(\mathbf{f}(\mathbf{a}, \mathbf{b}))$$

for *any* parameter $\mathbf{b}$. $\mathbf{f}(\mathbf{a}, \mathbf{b})$ describes the parameter $\mathbf{c}$ arising from a right translation (the left translation is treated *mutatis mutandis*). We define, accordingly, a new variable over which the integral on the right-hand side can be performed as

$$\mathbf{c} = \mathbf{f}(\mathbf{a}, \mathbf{b})$$

The Jacobi matrix of this variable transformation reads

$$J = \begin{pmatrix} \frac{\partial f_1}{\partial a_1} & \frac{\partial f_1}{\partial a_2} & \cdots & \\ \frac{\partial f_2}{\partial a_1} & \frac{\partial f_2}{\partial a_1} & \cdots & \\ \vdots & \vdots & & \\ \frac{\partial f_1}{\partial a_1} & \frac{\partial f_1}{\partial a_2} & \cdots & \frac{\partial f_r}{\partial a_r} \end{pmatrix}$$

After evaluating the derivative, the variables $\mathbf{a}$ is obtained as a function of $\mathbf{c}, \mathbf{b}$ by solving the equation $\mathbf{c} = \mathbf{f}(\mathbf{a}, \mathbf{b})$ with respect to the desired variable $\mathbf{a}$. The matrix elements are finally a function of the variables $\mathbf{c}, \mathbf{b}$. With

$$dV_a = \frac{dV_c}{\det J\left(\mathbf{a}(\mathbf{c},\mathbf{b}),\mathbf{b}\right)}$$

the translation invariance reads

$$\int_{\{\mathbf{c}\}} dV_c \cdot g(\mathbf{c}) \cdot F(\mathbf{c}) = \int_{\{\mathbf{c}\}} \frac{dV_c}{\det J\left(\mathbf{a}(\mathbf{c},\mathbf{b}),\mathbf{b}\right)} \cdot g(\mathbf{a}(\mathbf{c},\mathbf{b})) \cdot F(\mathbf{c})$$

As this equality must hold for any function $F(\mathbf{c})$, the requirement of translation invariance puts a constraint in the density $g(\mathbf{c})$, which must be constructed to obey the relation

$$g(\mathbf{c}) = \frac{1}{\det J\left(\mathbf{a}(\mathbf{c},\mathbf{b}),\mathbf{b}\right)} \cdot g(\mathbf{a}(\mathbf{c},\mathbf{b}))$$

This relation can be simplified by setting $\mathbf{a}$ to be the parameter corresponding to the identity: $\mathbf{a} = \mathbf{e}$. Then, $\mathbf{c} = \mathbf{b}$ and

$$g_R(\mathbf{a}) = g(\mathbf{e}) \cdot \left[\det\left(\frac{\partial f_i(\mathbf{e}+\epsilon,\mathbf{a})}{\partial(\epsilon_j)}\bigg|_{\epsilon=0}\right)\right]^{-1}$$

We denote the density with a suffix "R" to indicate that it is obtained by right translations. The density for the left translation invariance reads

$$g_L(\mathbf{a}) = g(\mathbf{e}) \cdot \left[\det\left(\frac{\partial f_i(\mathbf{a},\mathbf{e}+\epsilon)}{\partial(\epsilon_j)}\bigg|_{\epsilon=0}\right)\right]^{-1}$$

Comment. A group is said to be **compact** if the parameter space is bounded and closed (a set is closed if every Cauchy sequence of elements of the the set has a limit element which also belongs to the set). For example, the axial group C_∞ is a compact group, as the parameter space is the unit circle in the plane. The translation group, however, is not compact, as the parameter space is unbounded. For compact groups, a theorem (A. Weil, *L'integration dans les groupes topologiques et ses applications*, Actualites Scientifiques et Industrielles, Vol. 869, Hermann, Paris, 1940) proves the existence of an average. The compactness of the parameter space is key to the existence of a finite volume integral.

We now work out some examples of the procedure for computing the function $g(\mathbf{a})$.

Example 1. Axial rotation group C_∞: For this group, the equation for determining $g(\varphi)$ in the parameter space reads

$$\varphi' = \epsilon + \varphi$$

so that

$$f(\epsilon, \varphi) = \epsilon + \varphi$$

The Jacobi determinant at $\epsilon = 0$ reads

$$\frac{\partial f(\epsilon, \varphi)}{\partial \epsilon} \Big|_{\epsilon=0} = 1$$

and the sought for density reads

$$g(\varphi) = 1 \cdot g(0)$$

Accordingly, the average of the group C_∞ writes

$$\int_0^{2\pi} f(\varphi) \cdot g(0) \cdot d\varphi$$

Requiring that the volume integral to be just 1 determines $g(0)$ to be $\frac{1}{2\pi}$ and the average over the group $C_\infty \equiv SO_2$ reads

$$\mathscr{M}_{SO_2} f(\varphi) = \int_0^{2\pi} f(\varphi) \cdot \frac{1}{2\pi} \cdot d\varphi$$

Example 2. linear transformations in one dimension:

$$x' = a_1 \cdot x + a_2$$

This is a two parameter continuous group. The transformation in parameter space linking

$$\begin{pmatrix} 1 + \epsilon_1 \\ \epsilon_2 \end{pmatrix}$$

to

$$\begin{pmatrix} c_1 \\ c_2 \end{pmatrix}$$

writes

$$\begin{pmatrix} c_1 = b_1 \cdot (1 + \epsilon_1) \\ c_2 = b_1 \cdot \epsilon_2 + b_2 \end{pmatrix}$$

and the Jacobi matrix at $\epsilon_{1,2} = 0$ reads

$$\begin{pmatrix} b_1 & 0 \\ 0 & b_1 \end{pmatrix}$$

The sought for density is, accordingly,

$$g_R(a_1, a_2) = \frac{g(1,0)}{a_1^2}$$

For left translations we find

$$\begin{pmatrix} c_1 = (1 + \epsilon_1) \cdot a_1 \\ c_2 = (1 + \epsilon_1) \cdot a_2 + \epsilon_2 \end{pmatrix}$$

and the Jacobi matrix at $\epsilon_{1,2} = 0$ reads

$$\begin{pmatrix} a_1 & 0 \\ a_2 & 1 \end{pmatrix}$$

The density we are interested in is, accordingly,

$$g_L(a_1, a_2) = \frac{g(1,0)}{a_1}$$

Group average of SO_3

The axis-angle representation of three-dimensional rotations, given by the Rodrigues formula[9]

$$\begin{pmatrix} n_1^2 + (1 - n_1^2)\cos(\varphi) & n_1 n_2 (1 - \cos(\varphi)) - n_3 \sin(\varphi) & n_1 n_3 (1 - \cos(\varphi)) + n_2 \sin(\varphi) \\ n_1 n_2 (1 - \cos(\varphi)) + n_3 \sin(\varphi) & n_2^2 + (1 - n_2^2)\cos(\varphi) & n_2 n_3 (1 - \cos(\varphi)) - n_1 \sin(\varphi) \\ n_1 n_3 (1 - \cos(\varphi)) - n_2 \sin(\varphi) & n_2 n_3 (1 - \cos(\varphi)) + n_1 \sin(\varphi) & n_3^2 + (1 - n_3^2)\cos(\varphi) \end{pmatrix}$$

provides a convenient parametrization of all elements of $SO(3)$: Every element of $SO(3)$ can be represented by a unit vector with components (n_1, n_2, n_3) and $n_1^2 + n_2^2 + n_3^2 = 1$ (corresponding to the rotation axis) and a scalar φ corresponding to the rotation angle. Accordingly, the elements of $SO(3)$ are given by the three parameters

$$n_1 \cdot \varphi, n_2 \cdot \varphi, n_3 \cdot \varphi$$

residing within a sphere with a radius of π. The classes of $SO(3)$ (rotations about different axes by the same angle belong to the same class if there

---

[9]We will obtain this formula later in the lectures. After Benjamin Olinde Rodrigues, (1795–1851), French mathematician, see e.g. https://en.wikipedia.org/wiki/Olinde_Rodrigues.

is an element of the group that joins the rotation axes) occupy concentric spherical surfaces with a radius of φ.

We now proceed to find the density $g(n_1 \cdot \varphi, n_2 \cdot \varphi, n_3 \cdot \varphi)$ by the general method illustrated above.

We use the Rodriguez formula to find the matrix describing a rotation by a small angle around an arbitrary axis n_1, n_2, n_3:

$$R(\epsilon_1 = n_1 \cdot \epsilon, \epsilon_2 = n_2 \cdot \epsilon, \epsilon_3 = n_3 \cdot \epsilon) = \begin{pmatrix} 1 & -\epsilon_3 & +\epsilon_2 \\ \epsilon_3 & 1 & -\epsilon_1 \\ -\epsilon_2 & +\epsilon_1 & 1 \end{pmatrix}$$

The number of elements remains unchanged over a spherical surface in parameter space, so that we can find, without loss of generality, the density of elements for rotations around the x-axis. There,

$$R(1 \cdot \varphi, 0, 0) = \begin{pmatrix} 1 & 0 & 0 \\ 0 & \cos\varphi & -\sin\varphi \\ 0 & \sin\varphi & \cos\varphi \end{pmatrix}$$

The product matrix

$$\begin{pmatrix} 1 & 0 & 0 \\ 0 & \cos\varphi & -\sin\varphi \\ 0 & \sin\varphi & \cos\varphi \end{pmatrix} \cdot \begin{pmatrix} 1 & -\epsilon_3 & +\epsilon_2 \\ \epsilon_3 & 1 & -\epsilon_1 \\ -\epsilon_2 & +\epsilon_1 & 1 \end{pmatrix} = \begin{pmatrix} 1 & -\epsilon_3 & +\epsilon_2 \\ \epsilon_3\cos\varphi+\epsilon_2\sin\varphi & \cos\varphi-\epsilon_1\sin\varphi & 1 -\epsilon_1\cos\varphi-\sin\varphi \\ \epsilon_3\sin\varphi-\epsilon_2\cos\varphi & \sin\varphi+\epsilon_1\cos\varphi & 1 -\epsilon_1\sin\varphi+\cos\varphi \end{pmatrix}$$

describes a rotation with transformed parameters $(\varphi' \cdot n_1', \varphi' \cdot n_2', \varphi' \cdot n_3')$. In order to read these parameters from the rotation matrix, we consider that:

- The trace of the rotation matrix is

$$1 + 2\cos\varphi' = 1 + 2\cos\varphi - 2\epsilon_1\sin\varphi$$

To first order in ϵ_1, the solution of this equation is

$$\varphi' = \varphi + \epsilon_1$$

- The ratios $\frac{n_2'}{n_1'}$ and $\frac{n_3'}{n_1'}$ are related to the coefficients of a rotation matrix by the relations (to the lowest order in ϵ_i)

$$\frac{n_2'}{n_1'} = \frac{a_{31} - a_{13}}{a_{23} - a_{32}} \approx -\frac{1}{2}\epsilon_3 + \frac{1}{2}\epsilon_2\frac{1 + \cos\varphi}{\sin\varphi}$$

$$\frac{n_3'}{n_1'} = \frac{a_{12} - a_{21}}{a_{23} - a_{32}} \approx \frac{1}{2}\epsilon_2 + \frac{1}{2}\epsilon_3\frac{1 + \cos\varphi}{\sin\varphi}$$

Accordingly, the components of the rotation axis (normalized to 1 as the lowest order) are

$$n_1' = 1 \quad n_2' \approx -\frac{1}{2}\epsilon_3 + \frac{1}{2}\epsilon_2 \frac{1+\cos\varphi}{\sin\varphi} \quad n_3' \approx \frac{1}{2}\epsilon_2 + \frac{1}{2}\epsilon_3 \frac{1+\cos\varphi}{\sin\varphi}$$

The law of transformation we are after now reads

$$\varphi' n_1' = \epsilon_1 + \varphi$$

$$\varphi' n_2' = \frac{1}{2}\varphi\left(-\epsilon_3 + \epsilon_2 \frac{1+\cos\varphi}{\sin\varphi}\right)$$

$$\varphi' n_3' = \frac{1}{2}\varphi\left(\epsilon_2 + \epsilon_3 \frac{1+\cos\varphi}{\sin\varphi}\right)$$

From the inverse of the Jacobi determinant, we obtain

$$g(\varphi) = \frac{2}{\varphi^2}(1-\cos\varphi) \cdot g(\varphi=0) = \frac{4}{\varphi^2} \cdot \sin^2\frac{\varphi}{2} \cdot g(\varphi=0)$$

Using $g(\varphi = 0)$ to normalize the group integral and expressing the components of the rotation axis using spherical coordinates (ϑ, ϕ), one finally obtains

$$\mathscr{M}_{SO_3} f(\varphi, \vartheta, \phi) = \frac{1}{2\pi^2} \int_0^\pi \int_0^{2\pi} \sin\vartheta \, d\vartheta \, d\phi \int_0^\pi \sin^2\frac{\varphi}{2} d\varphi \cdot f(\varphi, \vartheta, \phi)$$

Lecture 2

REPRESENTATION THEORY
OF GROUPS WITH AVERAGES

In Lecture 2, we develop the mathematical instruments needed to solve concrete scientific problems using symmetry arguments. The technique developed in this lecture is called "representation theory of groups" in the literature. This lecture is accompanied by a student project, "Functional analytic and numerical aspects of eigenvalue problems", which provides some more insights into the more mathematical aspects of the eigenvalue problem. A set of exercises allows the student to become acquainted with the tools of representation theory.

2.1 Introduction

Many problems in the natural and social sciences[1] produce an **eigenvalue equation** of the type

$$L(q)u(q) = \lambda \cdot u(q)$$

Comments.

1. q is a set of variables that describe the configuration space of the system, such as the position vector of an electron moving in the field of one nucleus or a distribution of nuclei (molecule, crystal). They can also represent a momentum or a spin variable.
2. $L(q)$ is a linear differential or integral operator relating to an observable of the system. We have in mind, for example, the Hamilton operator of an electron in the field of one nucleus or a distribution of nuclei (molecule, crystal), comprising, say, a kinetic energy $-\frac{i\hbar}{2m}\triangle$ and a potential energy

---

[1]The original Google search algorithm also used the eigenvector corresponding to the largest eigenvalue of a matrix.

$V(\mathbf{r})$. The set of equations describing a system undergoing harmonic vibrations is a further example.

3. $u(q)$ are the functions that solve the eigenvalue equation. They represent the so-called "amplitude" (or wave function) in quantum mechanical problems or the amplitude of vibration in classical, vibrating systems. In Lecture 8, $u(q)$ will be actual "particles", such as electrons and protons (but only the "symmetry properties" of the operator $L(q)$ are known).

4. λ are the desired "eigenvalues". These are the real numbers that can be measured in an experiment. Accordingly, they represent the main goal of solving the scientific problem. In classical mechanics, λ are, for example, the eigenfrequencies of the system.

5. Some **boundary conditions**, which impose, for example, that the solutions $u(q)$ of the problems be bounded or square integrable or have some well-defined values in some region of the configuration space, transform the eigenvalue equation into an **eigenvalue problem**. Eigenvalue problems (not eigenvalue equations) often have a discrete set of solutions λ and $u(q)$. The solutions $u_\lambda(q)$ of the eigenvalue equations are then called "eigenstates".

6. A formal way of solving the eigenvalue problem is to cast the problem into an eigenvalue problem of a matrix L with matrix elements (u_n, Lu_m). u_n are often standard functions (or "the basis states" of the system) and are obtained from textbooks or suitable libraries. In this way, the problem is cast into the language of linear algebra. Typically, however, the matrix representing L is infinite-dimensional, and the determinantal equation is impossible to solve. A practical way of solving the eigenvalue problem of the matrix L is then to limit the basis functions used to build the matrix to a finite set — as many as possible to improve precision and as little as possible to make the problem numerically manageable. Then, one proceeds to solving by "brute force" the determinantal equation $\det(L - \lambda \mathbb{1}) = 0$ within the finite subspace in order to compute the eigenvalues.

7. The Role of Symmetry: The role of symmetry is to replace the brute force method with a more manageable analytic thinking. Typically, the system under study — producing, say, some potential energy landscape — has some symmetry group G, such as a set of symmetry elements g that transforms the system into itself. The presence of a symmetry group for the system implies that, for instance, a particle with coordinate q will find itself in the same physical situation if the

coordinates are changed by any of the symmetry elements g, so that a very important property of $L(q)$ emerges: The "invariance of L" with respect to all the symmetry elements g. One underlines this invariance by stating that "G is the symmetry group of the operator L". We will specify in Lecture 4 how this invariance is properly formulated from a mathematical perspective. This invariance allows symmetry arguments to be used for solving the eigenvalue problem. It will turn out in Lecture 4 that symmetry arguments help in classifying the various eigenvectors (in terms of "quantum numbers" in quantum mechanics or, respectively, "eigenmode" numbers in classical physics) and, what is most relevant, computing the eigenvalues themselves, at least in an approximate way.

2.2 Representation Theory of Groups (H. Weyl, 1885–1955)

2.2.1 *Representations of groups*

Definition. Consider a group G and a set of linear operators acting onto a vector space V endowed with a Hermitic metric ("scalar product") and a complete set of (ideally) orthonormal basis functions.[2] A map that associates to every element g a linear operator $T(g)$ acting in V is called a (linear) **representation** of the group G, provided the rule

$$T(g_i)T(g_j) = T(g_i \circ g_j)$$

is fulfilled (the so-called "homomorphism rule") for any elements g_i, g_j in the group G. We assume that the mapping is continuous, but we drop the word "continuous" when we speak of "representations".

Comments.

1. V is called "the representation space". If the vector space is finite-dimensional, the representation is finite or finite-dimensional. Otherwise, it is called infinite or infinite-dimensional. The dimension of V is called the dimension of the representation.
2. If the kernel of the homomorphism only consists of the identity element, the representation is said to be faithful.

---

[2]In the student project, "Functional analytic and numerical properties of eigenvalue problems", we will discuss the mathematical details of the eigenvalue problems, their vector space, and the relevant type of operators involved.

3. Symmetry transformations, such as isometries, have the property of preserving the scalar product in the vector space they act upon. It is therefore meaningful to use, for representing symmetry elements, operators T with the special property of preserving the scalar product in the vector space they act upon.

Definition. An operator T such that

$$(Tu, Tv) = (u, v)$$

for any vectors $u, v \in V$ is called a "unitary" operator.[3] If the operators $T(g)$ are unitary, the representation T is said to be a unitary representation.

Proposition. *A finite-dimensional representation $T(g)$ can be described by a matrix. The law of composition is the* **the matrix multiplication.** *In this case, we refer to it as the matrix representation of a group.*

Proof.

- Choose from V a basis set $e_1, \ldots, e_n$.
- Find

$$T(g)e_i = \sum_j T_{ji}(g)e_j$$

- Place along the i-th column of the matrix

$$[T_{ji}(g)] = [(e_j, T(g)e_i)] \doteq T(g)$$

the coefficients used for representing the transformed basis vector $\mathbf{e}_i'$ as the linear combination of the basis vectors $\mathbf{e}_j$.

$$\begin{pmatrix} \mathbf{e}_1' & \cdots & \mathbf{e}_n' \end{pmatrix} = \begin{pmatrix} \mathbf{e}_1 & \cdots & \mathbf{e}_n \end{pmatrix} \begin{pmatrix} T_{11}(g) & T_{12}(g) & \cdots & T_{1n}(g) \\ T_{21}(g) & T_{22}(g) & \cdots & \\ \vdots & \vdots & & \vdots \\ T_{n1}(g) & & \cdots & T_{nn}(g) \end{pmatrix}$$

---

[3]This definition is exact for finite-dimensional vector spaces V. For infinite-dimensional vector spaces, the domain of definition of T must be specified further. We refer to the project "Functional analytic and numerical properties of eigenvalue problems" for this subtle issue.

- Let us find out now which composition rule between the matrices realizes the homomorphism. On one side,

$$T(g_1)T(g_2)e_i = T(g_1)\sum_k T_{ki}(g_2)e_k = \sum_{kl} T_{lk}(g_1)T_{ki}(g_2)e_l$$

$$= \sum_k \left(\sum_l T_{kl}(g_1)T_{li}(g_2)\right) e_k$$

On the other side,

$$T(g_1 \circ g_2)e_i = \sum_k T_{ki}(g_1 \circ g_2)e_k$$

As we require that the left-hand side be equal to the right-hand side, we obtain

$$T_{ki}(g_1 \circ g_2) = \sum_l T_{kl}(g_1) \cdot T_{li}(g_2)$$

i.e. the matrix representation of the element $g_1 \circ g_2$ is the matrix representing g_1 **multiplied** by the matrix representing g_2. ◇

Comments.

1. **Practical problems are often formulated within finite-dimensional vector spaces. Therefore, in these lectures, we continue with finite-dimensional representations (also called "matrix representations"). We explicitly mention when and how definitions and theorems apply to infinite-dimensional representation, provided this is relevant for practical use.**

2. Unitary operators in a finite vector space are represented by unitary matrices, i.e. matrices for which

$$U^\dagger = U^{-1}$$

$U^\dagger$ is the adjoint matrix, i.e. the complex conjugate transpose.

3. We do not use a special character to distinguish between finite matrices and general operators. When necessary, we point out explicitly which entity we refer to.

4. We recognize that matrix representations are useful because the algebra of matrices is generally simpler to work with compared to abstract symmetry operations. Furthermore, physical operators (in particular, within some approximation scheme) are often written as matrices, so we mostly deal with matrix representation of groups when solving practical problems.

2.2.2   Examples of finite- and infinite-dimensional representations

1. The simplest representation of a group is obtained when we associate the number "1" with every element of the group. This representation is one-dimensional and unfaithful and is called the identical or trivial representation. Every group has a trivial representation.

2. A one-dimensional representation of the group $C_2 = \{e, C_2\}$ is

$$\Gamma(e) = 1$$

$$\Gamma(C_2) = -1$$

   The representation space is $\mathscr{C}$. The representation is unitary and faithful.

3. A representation of C_2 in $\mathscr{C}^2$ is

$$\Gamma(e) = \begin{pmatrix} 1 & 0 \\ 0 & 1 \end{pmatrix}$$

$$\Gamma(C_2) = \begin{pmatrix} \bar{1} & 0 \\ 0 & \bar{1} \end{pmatrix}$$

   The representation is unitary and faithful.

4. The natural representation of the circle group C_∞ is provided by the matrices of SO_2:

$$\begin{pmatrix} \cos\varphi & -\sin\varphi \\ \sin\varphi & \cos\varphi \end{pmatrix}$$

5. A two-dimensional matrix representation of C_{4v} is provided by the eight matrices

$$e = \begin{pmatrix} 1 & 0 \\ 0 & 1 \end{pmatrix}; \quad C_{4z} = \begin{pmatrix} 0 & \bar{1} \\ 1 & 0 \end{pmatrix}; \quad C_{2z} = \begin{pmatrix} \bar{1} & 0 \\ 0 & \bar{1} \end{pmatrix}; \quad C_{4z}^3 = \begin{pmatrix} 0 & 1 \\ \bar{1} & 0 \end{pmatrix}$$

$$m_y = \begin{pmatrix} \bar{1} & 0 \\ 0 & 1 \end{pmatrix}; \quad m_x = \begin{pmatrix} 1 & 0 \\ 0 & \bar{1} \end{pmatrix}; \quad m_{\bar{x}y} = \begin{pmatrix} 0 & \bar{1} \\ \bar{1} & 0 \end{pmatrix}; \quad m_{xy} = \begin{pmatrix} 0 & 1 \\ 1 & 0 \end{pmatrix}$$

6. The one-dimensional translation group is isomorphic to $(\mathscr{R}, +)$ and is defined by the transformation

$$x' = x + t$$

   with $t \in \mathscr{R}$. Any representation $T(t)$ obeys the multiplication rule

$$T(t+s) = T(t) \cdot T(s) \quad T(-t) = T(t)^{-1}$$

A unitary representation of the translation group in the representation space $\mathscr{C}$ is therefore given by the isomorphisms

$$t \to T(t) = e^{-ikt}$$

with $k \in \mathscr{R}$. There are, however, an infinite number of non-unitary one-dimensional representations, given by $e^{-i\alpha \cdot t}$ with α being any complex number. Unless α is a real number, the representation is not unitary.

7. Representation of the translation group in a function space: Most physical problems are defined within a function space, so it is important to be able to define a representation of any group within a function space. Let us define the operator $\Gamma(t)$ acting on a function such that (t a real number)

$$[\Gamma(t)u](x) \doteq u(x - t)$$

A. The operator $\Gamma(t)$ is a homomorphism, as

$$\Gamma(t_2)\Gamma(t_1)u(x) = \Gamma(t_2)u(x - t_1) = u((x - t_2) - t_1) = u(x - (t_2 + t_1))$$
$$= \Gamma(t_2 + t_1)u(x)$$

Comment. In writing the third term, we have used the Wigner convention that the operator acts on the variable x and not on the argument of the function. In this second case, $\Gamma(t_2)u(x-t_1) = u((x-t_1)-t_2)$. For the specific case of the translation group, which is commutative, both the definitions of the operator (acting on x or on the variable of the function) would produce the same result. For non-commutative groups, instead, we observe an essential difference.

B. The operator $\Gamma(t)$ is unitary, i.e.

$$(\Gamma(t)u(x), \Gamma(t)v(x)) = (u(x - t), v(x - t))$$

In fact, with the variable substitution $x - t = y$, we obtain

$$\int_{-\infty}^{\infty} \bar{u}(x - t) \cdot v(x - t) \cdot dx = \int_{-\infty}^{\infty} \bar{u}(y) \cdot v(y) \cdot dy = (u(x), v(x))$$

Comment. For this unitary representation of the translation group in a function space, one might find a pictorial explanation. Suppose we have a certain body and we have found some physical characteristics related to the body which can be assigned a scalar field $u(\mathbf{x})$. $u(\mathbf{x})$ could be a spatially dependent temperature distribution or a spatially dependent density. Suppose now we instantly move the body to a translated position in space by the vector **t**. By the instant translation, nothing happens to the temperature or density of the body, except that, if one would like to describe the temperature or density distribution, one must take into account that the body is somewhere else. The new function describing the properties of the body is $\Gamma(\mathbf{t})u$, but this new function is of course related to the original one by a simple relation, which is specified by computing the function at some position **x**:

$$\Gamma(\mathbf{t})u(\mathbf{x}) = u(\mathbf{x} - \mathbf{t})$$

and this is true because, as we have specified, the instant translation did not affect the actual values of the function — the object did not change or have any possibility of being affected by the environment by the instant, purely "mathematical" movement.

8. A further representation of the one-dimensional translation group is defined by

$$[\Gamma'(t)u](x) \doteq e^{-i\cdot t\cdot x} \cdot u(x)$$

$\Gamma'(t)$ is a homomorphism, as

$$\Gamma'(t_2)\Gamma'(t_1)u](x) = \Gamma'(t_2)e^{-i\cdot t_1\cdot x} \cdot u(x) = e^{-i\cdot t_2\cdot x}e^{-i\cdot t_1\cdot x}u(x)$$
$$= e^{-i\cdot(t_2+t_1)\cdot x} \cdot u(x)$$
$$= [\Gamma'(t_2 + t_1)u](x)$$

The representation is a unitary representation, and the representation space is infinite-dimensional.

9. For the translation group, we have found two infinite-dimensional representations in a function space:

$$[\Gamma(t)u](x) \doteq u(x - t)$$

(known as the translation or Schrödinger representation) and

$$[\Gamma'(t)u](x) \doteq e^{-i\cdot t\cdot x} \cdot u(x)$$

known as the modulation representation. The interested reader might continue reading to discover that the two representations are related by a Fourier-transform ($\mathscr{F}$) similarity transformation.

Proposition. $\mathscr{F}^{-1}\Gamma'\mathscr{F} = \Gamma$.

Proof. To prove the claim, we define the Fourier transform by

$$(\mathscr{F}u)(x) \doteq \frac{1}{\sqrt{2\pi}} \int dy e^{-ixy} u(y)$$

and

$$(\mathscr{F}^{-1}u)(x) \doteq \frac{1}{\sqrt{2\pi}} \int dy e^{ixy} u(y)$$

Accordingly,

$$(\Gamma'(t)\mathscr{F})u(x) = \frac{1}{\sqrt{2\pi}} \int dy e^{-ix(y+t)} u(y)$$

and

$$(\mathscr{F}^{-1}\Gamma'\mathscr{F}u)(x) = \frac{1}{\sqrt{2\pi}} \int dz e^{izx} \frac{1}{\sqrt{2\pi}} \int dy e^{-iz(y+t)} u(y)$$

$$= \int dy \underbrace{\underbrace{\frac{1}{2\pi} \int dz e^{-iz(y-(x-t))}}_{\delta(y-(x-t))} u(y)}_{u(x-t)}$$

$$= [\Gamma(t)u](x)$$

We will later see that the representations Γ and Γ' are related to the quantum mechanical momentum and position operators, respectively, so that we have just shown that the momentum operator is the Fourier transform of the position operator.

10. The von Neumann–Stone theorem[4]: Note that a famous theorem by von Neumann and Stone states that in the space of square-integrable functions there is, up to unitary equivalence, only one representation of the translation group, i.e. all representations are equivalent to the Schrödinger representation Γ (Jonathan Rosenberg, *A Selective History of the Stone-von Neumann Theorem*, in Contemporary Mathematics, Editors: Robert S. Doran and Richard V. Kadison, 2004, https://www.math.umd.edu/~jmr/StoneVNart.pdf.

---

[4] After J. von Neumann, see e.g. https://en.wikipedia.org/wiki/John_von_Neumann and Marshall Stone, see e.g. https://en.wikipedia.org/wiki/Marshall_H._Stone.

11. The representations Γ and Γ' also appear as unitary infinite representations of the Heisenberg group

$$H(a, b, c) = \left\{ \begin{pmatrix} 1 & a & c \\ 0 & 1 & b \\ 0 & 0 & 1 \end{pmatrix} \mid a, b, c \in \mathscr{R} \right\}$$

A map that associates to any such matrices the unitary operator
$\Pi(a, b, c)$ with

$$[\Pi(a, b, c)u]u(x) = e^{ic} \cdot e^{-ibx} \cdot u(x - a)$$

is a representation of the Heisenberg group into the set of linear
operators acting on the space of square-integrable functions. Again (by
the von Neumann–Stone theorem), this is, up to equivalent ones, the
only unitary representation of the Heisenberg group with operators
acting on a function space.

12. The rotation group: A representation of O_3 in a function space is defined
by the equation

$$O_R u(\mathbf{r}) \doteq u(R^{-1}\mathbf{r})$$

with R being a 3×3 (proper or improper) rotation matrix. The operator
is a homomorphism of O_3 into the set of linear operators acting on a
scalar function of the variable $\mathbf{r}$:

$$[O_{R_2} O_{R_1} u](\mathbf{r}) = [O_{R_2} u](R_1^{-1}\mathbf{r}) = u(R_1^{-1} R_2^{-1}\mathbf{x}) = [O_{R_2 R_1} u](\mathbf{r})$$

The third term arises from the Wigner rule of transforming the
variable $\mathbf{r}$ and not the argument of the function. By this definition
of the acting of an operator onto a function, the use of the inverse
matrix is necessary in order to obtain a homomorphism. Other
authors, such as J. C. Slater (J. C. Slater, (1900–1976) was an influential
American physicist, see e.g. https://en.wikipedia.org/wiki/John_C._
Slater), transform the variable in the function by the matrix R,
thereby achieving an "antihomomorphism" (see e.g. J.C. Slater, Rev.
Mod. Phys. **37**, 68 (1965); S.L. Altmann, C.J. Bradley, Rev. Mod.
Phys. 37, 33 (1965)). Physical applications are not affected by Slater's
unconventional definition, but the Wigner convention is more generally
adopted. The plausibility argument that explains the significance of
this representation follows the lines developed for the translation group.
The instant transformation one deals with in this case is a (proper or
improper) rotation of the physical body.

2.2.3 *Reducible and irreducible representations*

Definition. Let T be a finite- or infinite-dimensional representation of the group G within a space V. A linear subspace V_1 of V is said to be G-invariant if, for any element $g \in G$ and any vector u in V_1, $T(g)u$ also belongs to V_1, i.e. the subspace V_1 is transformed into itself by the symmetry operations. The representation of G on V_1 is called the restriction of T on V_1.

Definition. A finite- or infinite-dimensional representation T of the group G on the space V is *irreducible* if the only invariant subspace is $\{0\}$ or V itself. Otherwise, the representation is said to be *reducible*.

Direct sum of vectors and matrices: A question arises as to whether any representation can be decomposed into irreducible representations: Irreducible representations would become particularly important as the building blocks of representation theory of groups. In order to perform the "decomposition", we need to define a suitable operation. This requires introducing the direct sum of vector spaces and operators.

For this purpose, we consider two vector spaces V_a and V_b with the basis states $\{e_1, e_2, \ldots, e_a\}$ and $\{f_1, f_2, \ldots, f_b\}$. The number of basis states can be infinite, so that the operation of direct sum can be performed in infinite-dimensional spaces as well. As a simple example, we take V_a to be a plane with basis vectors e_x, e_y and V_b a line with the basis vector e_z which does not lie in the x, y-plane (it is not necessary that the e-set and f-set are orthogonal, but they must be linearly independent). We also assume that we can decompose any vector in V_a and V_b into a linear combination of the basis vectors, so that we know their coordinates

$$\mathbf{x} = (x_1, \ldots, x_n, \ldots, x_a); \quad \mathbf{y} = (y_1, \ldots, y_n, \ldots, y_b)$$

Definition. The set of coordinates

$$(x_1, \ldots, x_n, \ldots, x_a, y_1, \ldots, y_n, \ldots, y_b)$$

obtained by juxtaposing ("pairing") the initial two sets of coordinates define a new vector with respect to the basis set

$$\{e_1, e_2, \ldots, e_a, f_1, f_2, \ldots, f_b\}$$

This new vector is called the direct sum of the vectors $\mathbf{x}$ and $\mathbf{y}$ and is designated by

$$\mathbf{x} \oplus \mathbf{y}$$

Comments.

1. The definition of direct sum allows for the following computation rules:

   a.

   $$\lambda(\mathbf{x} \oplus \mathbf{y}) = \lambda\mathbf{x} \oplus \lambda\mathbf{y}$$

   b.

   $$\mathbf{x}_1 \oplus \mathbf{y}_1 + \mathbf{x}_2 \oplus \mathbf{y}_2 = (\mathbf{x}_1 + \mathbf{x}_2) \oplus (\mathbf{y}_1 + \mathbf{y}_2)$$

   These properties show that the direct sum space is a vector space $V_a \oplus V_b$, with

   $$\dim(V_a \oplus V_b) = \dim(V_a) + \dim(V_b)$$

   c. It is endowed with a Hermitic metric:

   $$(\mathbf{x}_1 \oplus \mathbf{y}_1, \mathbf{x}_2 \oplus \mathbf{y}_2) = (\mathbf{x}_1, \mathbf{x}_2) + (\mathbf{y}_1, \mathbf{y}_2)$$

   (in this equation, the brackets indicate the scalar product).

2. In conjunction with the direct sum of spaces, one can introduce the *direct sum of operators and matrices*. We start with two matrices

$$A = \begin{pmatrix} \alpha_{11} & & \cdots & \alpha_{1a} \\ \alpha_{21} & \alpha_{22} & \cdots & \alpha_{2a} \\ \vdots & \vdots & & \vdots \\ \alpha_{a1} & \alpha_{a2} & \cdots & \alpha_{aa} \end{pmatrix}$$

and

$$B = \begin{pmatrix} \beta_{11} & & \cdots & \beta_{1b} \\ \beta_{21} & \beta_{22} & \cdots & \beta_{2b} \\ \vdots & \vdots & & \vdots \\ \beta_{b1} & \beta_{a2} & \cdots & \beta_{bb} \end{pmatrix}$$

and transform with them the vectors in V_a and V_b, respectively:

$$x'_i = \sum_j \alpha_{ij} x_j$$

$$y'_n = \sum_m \beta_{nm} y_m$$

The pair (x_i', y_n') in the direct sum transform as

$$(x_i', y_n') = \left(\sum_j \alpha_{ij} x_j, \sum_m \beta_{nm} y_m \right)$$

This transformation property can be written as

$$
\begin{pmatrix} x_1' \\ \vdots \\ x_i' \\ \vdots \\ x_a' \\ y_1' \\ \vdots \\ y_n' \\ \vdots \\ y_b' \end{pmatrix}
=
\begin{pmatrix}
\alpha_{11} & \cdots & \alpha_{1a} & 0 & \cdots & 0 \\
\vdots & & \vdots & 0 & \cdots & 0 \\
\alpha_{i1} & \cdots & \alpha_{ia} & 0 & \cdots & 0 \\
\vdots & & \vdots & 0 & \cdots & 0 \\
\alpha_{a1} & \cdots & \alpha_{aa} & 0 & \cdots & 0 \\
0 & \cdots & 0 & \beta_{11} & \cdots & \beta_{1b} \\
0 & \cdots & \vdots & \vdots & & \vdots \\
0 & \cdots & 0 & \beta_{n1} & \cdots & \beta_{nb} \\
0 & \cdots & 0 & \vdots & & \vdots \\
0 & \cdots & 0 & \beta_{b1} & \cdots & \beta_{bb}
\end{pmatrix}
\begin{pmatrix} x_1 \\ \vdots \\ x_i \\ \vdots \\ x_a, \\ y_1 \\ \vdots \\ y_n \\ \vdots \\ y_b \end{pmatrix}
$$

(These matrices might have an infinite number of lines and columns when the spaces are infinite-dimensional.)

Definition. The matrix

$$
\begin{pmatrix}
\alpha_{11} & \cdots & \alpha_{1a} & 0 & \cdots & 0 \\
\vdots & & \vdots & 0 & \cdots & 0 \\
\alpha_{i1} & \cdots & \alpha_{ia} & 0 & \cdots & 0 \\
\vdots & & \vdots & 0 & \cdots & 0 \\
\alpha_{a1} & \cdots & \alpha_{aa} & 0 & \cdots & 0 \\
0 & \cdots & 0 & \beta_{11} & \cdots & \beta_{1b} \\
0 & \cdots & \vdots & \vdots & & \vdots \\
0 & \cdots & 0 & \beta_{n1} & \cdots & \beta_{nb} \\
0 & \cdots & 0 & \vdots & & \vdots \\
0 & \cdots & 0 & \beta_{b1} & \cdots & \beta_{bb}
\end{pmatrix}
$$

in short

$$\begin{pmatrix} A & 0 \\ 0 & B \end{pmatrix}$$

acting on the direct sum space $V_a \oplus V_b$ is the direct sum of the matrices A and B and is designated as $A \oplus B$.

Comments.

1. Rules of computation:

   a.

   $$(A \oplus B)(\mathbf{x} \oplus \mathbf{y}) = A\mathbf{x} \oplus B\mathbf{y}$$

   b.

   $$(A_1 \oplus B_1)(A_2 \oplus B_2) = A_1 A_2 \oplus B_1 B_2$$

   i.e. the matrix multiplication reduces to multiplying the blocks as a whole.

2.

$$\det (A \oplus B) = \det A \cdot \det B$$

$$\operatorname{tr} (A \oplus B) = \operatorname{tr} A + \operatorname{tr} B$$

   (where det = determinant and tr = trace).

3. Such a matrix which has non-vanishing elements in rectangular blocks along the main diagonal and zeros elsewhere is said to be in a *block-diagonalized* form.

Proposition. *Suppose that T_1 and T_2 are two matrix representations of a group G onto the representation spaces V_1 and V_2, respectively. Then, $T_1 \oplus T_2$ is a matrix representation of the group G in $V_1 \oplus V_2$.*

Proof.

$$(T_1 \oplus T_2)(g_1)(T_1 \oplus T_2)(g_2) = T_1(g_1)T_1(g_2) \oplus T_2(g_1)T_2(g_2)$$

$$= T_1(g_1 g_2) \oplus T_2(g_1 g_2)$$

$$= (T_1 \oplus T_2)(g_1 g_2)$$

Comments.

1. In terms of block matrices

$$T_1 \oplus T_2(g) = \begin{pmatrix} T_1(g) & 0 \\ 0 & T_2(g) \end{pmatrix} \doteq T(g)$$

the homomorphism is written as

$$T(g_1)T(g_2) = \begin{pmatrix} T_1(g_1)T_1(g_2) & 0 \\ 0 & T_2(g_1)T_2(g_2) \end{pmatrix} = \begin{pmatrix} T_1(g_1g_2) & 0 \\ 0 & T_2(g_1g_2) \end{pmatrix}$$
$$= T(g_1g_2)$$

2. The operation of the direct sum of representations produces by construction a set of invariant subspaces, each of which sustains a representation of the group. The direct sum of representations is therefore an example of how a **reducible** representation could look like, with the vector space V being the direct sum of invariant subspaces, each carrying some representation of the group.

Proposition. *Suppose we have a matrix representation $T_1(g)$ of a group G with elements g. Then, the set of matrices $T_2(g) \equiv ST_1(g)S^{-1}$ also constitutes a representation, with S being a square matrix with an inverse.*

Proof.

$$T_2(g_1)T_2(g_2) = ST_1(g_1)S^{-1}ST_1(g_2)S^{-1} = ST_1(g_1g_2)S^{-1} = T_2(g_1g_2) \quad \diamond$$

The two representations T_2 and T_1 are said to be equivalent, and the transformation between the two is a similarity transformation. Using a similarity transformation, one can produce an infinite number of equivalent representations.

Definition. Consider a matrix representation T of a group and assume that the vector space can be decomposed — in the sense of the direct sum of spaces — into irreducible invariant subspaces $V = \oplus V_i$, each carrying an irreducible representation of the group. Then, the representation T is said to be *completely reducible*. For a completely reducible representation, there exists a non-singular operator S such that $STS^{-1}u$ is in V_i if $u \in V_i$, $\forall i$.

Comments.

1. For matrix representations, these definitions can be cast in terms of block-diagonalized matrices. In fact, given a matrix representation T, if the same similarity transformation brings all the matrices of the representation T into the same block-diagonalized form, then T is said to be completely reducible. Note that a similarity transformation could obscure the block-diagonalized form by making the zero matrix elements disappear, but the representation would still be considered completely reducible, i.e. a representation is said to be completely reducible if it can be put into a block-diagonalized form by an appropriate similarity transformation.

2. It might be possible that the representations T_i appearing in the reduction of T are further reducible, i.e. a similarity transformation exists which transforms all matrices $T_i(g)$ into block forms themselves. Clearly, this process of reduction can be carried out until we find no similarity transformation which reduces all matrices of the representation further. Thus, the final form of the matrices of a representation T may look like

$$
T(g) = \begin{pmatrix}
T_1(g) & 0 & \cdots & 0 \\
0 & T_2(g) & \cdots & 0 \\
\vdots & \vdots & & \vdots \\
0 & 0 & \cdots & T_s(g)
\end{pmatrix}
$$

with all the matrices of T having the same reduced structure and all the blocks of T being no further reducible. We have achieved a complete reduction of a representation into the so-called irreducible representations. Note that an irreducible representation may occur more than once in the reduction of a reducible representation: In general,

$$
T = \oplus_i n_i T_i
$$

Clearly, the irreducible representations are the "basic" representations from which all other completely reducible representations can be constructed.

3. There are representations which are reducible but not completely reducible. For example, consider the group of integer numbers $n \in \mathscr{Z}$ under the composition law of the sum. The matrix

$$
T(n) \doteq \begin{pmatrix} 1 & n \\ 0 & 1 \end{pmatrix}
$$

is a representation of the group. The representation is reducible, as the vectors of the form

$$\begin{pmatrix} a \\ 0 \end{pmatrix}$$

build an invariant subspace. The representation is not completely reducible: The direct complement to the vector space given by

$$\begin{pmatrix} 0 \\ b \end{pmatrix}$$

is not invariant. The non-complete reducibility of this representation can also be verified by showing that there is no similarity transformation that block-diagonalizes all matrices simultaneously:

$$\begin{pmatrix} a & b \\ c & d \end{pmatrix} \begin{pmatrix} 1 & n \\ 0 & 1 \end{pmatrix} \begin{pmatrix} a & b \\ c & d \end{pmatrix}^{-1} = \begin{pmatrix} \alpha & 0 \\ 0 & \beta \end{pmatrix}$$

has no solution for any choice of the complex numbers α, β.

It is of great importance to have guidelines about which representations are completely reducible and, in that case, to have a complete characterization of the irreducible representations. The following theorems provide the required general guidelines.

Theorem (Maschke[5]). *Every unitary finite- or infinite-dimensional representation $g \to U(g)$ of any group on a vector space with the Hermitic metric is completely reducible.*

Proof. Assume that the subspace $V_{\parallel}$ is invariant with respect to $U(g)$, $V = V_{\parallel} \oplus V_{\perp}$. We need to prove that $V_{\perp}$ is also invariant with respect to $U(g)$. Let u belong to $V_{\parallel}$ and v to $V_{\perp}$. Then,

$$(U(g)v, u) = \big(v, U^{\dagger}(g)u\big) = \big(v, U^{-1}(g)u\big) = \big(v, U(g^{-1})u\big)$$

However, $U(g^{-1})u$ is certainly in $V_{\parallel}$, so that its scalar product with v is vanishing. This proves that $U(g)v$ also belongs to $V_{\perp}$.   ◇

Examples.

1. The non-unitary representation $\left(\begin{smallmatrix} 1 & n \\ 0 & 1 \end{smallmatrix} \right)$ of the additive integer group is *not* completely reducible.

---

[5] Heinrich Maschke (1853–1908) was a German mathematician.

2. A unitary representation of the one-dimensional translation group in the Hilbert space of square-integrable functions is given by

$$U(t)u(x) = u(x - t)$$

According to the theorem, it is completely reducible. In fact, assume that u belongs to a subspace M which is invariant with respect to $U(t)$, and consider a vector v perpendicular to the subspace M, so that

$$(v, U(t)u) = 0$$

Accordingly,

$$0 = \frac{1}{2\pi} \int dt e^{-ikt} (v, U(t)u) = \frac{1}{2\pi} \int dt e^{-ikt} \int dx \bar{v}(x) \cdot f(x - t)$$
$$= \bar{\tilde{v}}(k) \cdot \tilde{u}(k)$$

in virtue of the fact that the operation $(v, U(t)u)$ is equivalent to the convolution of $\bar{v}$ and u and of the fact that the Fourier transform of the convolution of $\bar{v}$ and u is the product of the Fourier transforms of $\bar{v}$ and u. The equation

$$\bar{\tilde{v}}(k) \cdot \tilde{u}(k) = 0 \quad \forall k$$

has the solution

$$\tilde{v}(k) \equiv 0$$

and, accordingly,

$$v(x) \equiv 0$$

so that M coincides with the entire space and $M_\perp$ is trivially invariant. It is therefore correct to conclude that $U(t)$ is completely reducible, but it is also correct to state that there is no invariant space other than the whole Hilbert space. This means that $U(t)$ is an infinite-dimensional irreducible representation of the one-dimensional translation group.

Theorem (Maschke). *Any finite-dimensional representation of a group with an average is equivalent to a unitary representation.*

Corollary. *Any finite-dimensional representation of a group with an average is completely reducible.*

Proof of the Theorem. Let us consider a vector space with the Hermitic metric (x, y) and a representation $h \to T(h)$. We need to find a non-singular matrix S such that

$$\big(ST(h)S^{-1}x, ST(h)S^{-1}y\big) = (x, y)$$

We consider the matrix

$$A = M_g T^\dagger(g)T(g)$$

and study its properties. First, the matrix is Hermitic, as $A^\dagger = A$. Second, it is a positive matrix, as, for any vector x, $(Ax, x) \geq 0$. As such, it has a square root. We therefore write $A \doteq S^2$. Third, A fulfills the following identity:

$$T^\dagger(h)S^2 T(h) = M_g T^\dagger(h)T^\dagger(g)T(g)T(h) = M_g T^\dagger(hg)T(hg)$$
$$= M_c T^\dagger(c)T(c) = S^2$$

S is also Hermitic, i.e. $S^\dagger = S$. We now compute

$$\big(ST(h)S^{-1}x, ST(h)S^{-1}y\big) = \big(T(h)S^{-1}x, S^2 T(h)S^{-1}y\big)$$

$$= \left(S^{-1}x, \underbrace{T^\dagger(h)S^2 T(h)}_{S^2} S^{-1}y \right)$$

$$= \big(S^{-1}x, Sy\big) = \big(SS^{-1}x, y\big)$$

$$= (x, y) \qquad\qquad \diamond$$

2.2.4   *Theorems on irreducible representations*

Owing to the fact that representations can be decomposed into the direct sum of irreducible representations, we need to learn as much as possible about these "building blocks". In this section, we provide theorems for

- constructing all irreducible representations of a group,
- testing any representation for its irreducibility,

for groups with an average. Groups that do not have an average, such as the translation group, will be considered, if necessary, separately as they lack the requirements on which the proof of the theorems is based. The construction of the irreducible representations of a group as being the building blocks of every unitary representation follows not only a strictly mathematical rationale but also has physical implications. In Lecture 4, we prove a precise relation between the irreducible representations of a group and the "quantum numbers" (e.g. in quantum physics) or numbers labeling eigenmodes (in classical wave propagation problems). Establishing the properties of irreducible representations is therefore a key element toward fulfilling one of the key scopes of symmetry, namely its use for **classifying** the various eigenstates of a system in terms of characteristic numbers.

Almost all results of irreducible representations are based on the lemma of Schur.

Lemma 1 (Schur[6]). *Let T be an irreducible matrix representation of a group G in a complex finite-dimensional Hilbert space V and A a matrix that commutes with all matrices $T(g)$. Then,*

$$A = \lambda \cdot \mathbb{1}$$

in V.

Proof. A has certainly an eigenvalue λ in the complex vector space V, with an eigenspace that we call E_λ. This is because of the fundamental theorem of algebra, which states that the determinantal equation of a matrix has at least one solution different from zero. Within this eigenspace, we can imagine the set of eigenfunctions $\{u_\lambda\}$, so that A restricted to E_λ can be written as $\lambda \cdot \mathbb{1}$. As A commutes with all matrices $T(g)$, $T(g)u_\lambda$ is also in the eigenspace E_λ. This is because of the theorem of commuting matrices, stating that commuting matrices have common eigenspaces. Accordingly, E_λ is G-invariant. However, as $T(g)$ was assumed to be irreducible, E_λ must coincide with V, and the lemma is proved.   ◇

Comment. The assumption that A has an eigenvalue is certainly true if V is finite-dimensional, as the determinantal equation always has at least one solution. If V is infinite-dimensional, then the proof is still valid if A is a "reasonable" operator with at least one value λ and a set $\{u_\lambda\}$ with $Au_\lambda = \lambda u_\lambda$.

Lemma 2 (Schur). *Consider now two irreducible matrix representations $T_1(g) : V_1 \to V_1$ and $T_2(g) : V_2 \to V_2$ and a rectangular matrix $A : V_1 \to V_2$ with the property*

$$T_2(g)A = AT_1(g)$$

---

[6]Issai Schur (1875–1941), Russian mathematician, see e.g. https://en.wikipedia.org/wiki/Issai_Schur.

The claim is that either $A = 0$ or A is an isomorphism; therefore, T_1 is equivalent to T_2.

Proof. Take a vector $u \in V_1$ which is mapped by A onto the vector 0 in V_2. u belongs to the kernel of A. Because of the commutativity condition, $AT_1(g)u = 0$, i.e. $T_1(g)u$ also belongs to the kernel of A, so that the kernel of A is a subspace of V_1 which is invariant with respect to T_1 and thus must coincide with V_1, as T_1 was supposed to be irreducible. Therefore, either the operator A, applied to all elements of V_1, gives 0 and is to be identified with the 0-operator or else u is zero and A is injective. Consider now an element v as the image of the matrix A, i.e. there exists a vector u in V_1 such that $Au = v$. As $T_2 v = T_2 A u = A T_1 u$, $T_2 v$ also belongs to the image of A. As the space V_2 was assumed to be G-invariant, the image of A coincides with V_2, so that, if it is not the zero operator, A is both injective and surjective and thus an isomorphism. $\diamond$

The two lemmas by Schur lead, for instance, to the great orthogonality theorem. This theorem concerns unitary irreducible representations of a group with an average. The representation space can be finite or infinite but is supposed to be a vector space with the Hermitic metric and a complete orthonormal basis set.

Great Orthogonality Theorem (GOT).

Claim a: Any unitary irreducible representation U of a group with an average is finite-dimensional (even if it is established in an infinite-dimensional vector space).

Claim b: The matrix elements of U fulfill the relation

$$\mathcal{M}_g \bar{U}_{ik}(g) U_{lm}(g) = \frac{1}{\dim U} \cdot \delta_{il} \delta_{km}$$

Claim c: If $U^{(1)}$ and $U^{(2)}$ are two inequivalent irreducible representations, then

$$\mathcal{M}_g \bar{U}_{ik}^{(1)}(g) U_{lm}^{(2)}(g) = 0$$

Comment. Because of the GOT, knowing the matrix elements of the representations allows us, in principle, to find out whether the representation is irreducible without having to search for the similarity transformation that puts the representation in a block form. However, this theorem has a little practical application, as the matrix elements of representations are generally not available in the standard literature. Like Schur's lemma, the GOT is mainly used in the proof of further, more useful, theorems.

Proof of Claim a. The strategy for the proof involves assuming that the irreducible representation has the desired dimension N. By means of an inequality, one then shows that N is a finite number.

Let us start by introducing a complete set of basis vectors $\{e_i\}$. The index i extends from 1 to N. Let X be a non-singular matrix. We then define the matrix

$$A \doteq \mathscr{M}_g U(g) X U(g^{-1})$$

and show that it commutes with any matrix $U(a)$

$$\begin{aligned} U(a)A &= \mathscr{M}_b U(a)U(b)XU(b^{-1}) \\ &= \mathscr{M}_b U(ab)XU(b^{-1})U(a^{-1})(U(a)) \\ &= \mathscr{M}_b U(ab)XU((ab)^{-1}U(a) \\ &= AU(a) \end{aligned}$$

by virtue of the translational invariance of $\mathscr{M}$. Accordingly, $A = \lambda \mathbb{1}$ (Schur's lemma 1), with $\lambda \neq 0$ being dependent on the choice of X. Explicitly,

$$\lambda \cdot \delta_{ij} = \mathscr{M}_g \sum_{k,m} (e_i, U(g)e_k) \cdot (e_k, Xe_m) \cdot \left(e_m, U(g^{-1})e_j\right)$$

We then choose $i = j$ and $(e_k, Xe_m) = \frac{\delta_{km}}{N}$ for $k = m = i$ or else 0. This gives

$$\lambda = \mathscr{M}_g \frac{1}{N} \mid (e_i, U(g)e_i) \mid^2$$

We now sum both sides over the set $e_1, \ldots, e_N$ (note that we can interchange the "sum" operation and the average operation as both have finite results) and obtain

$$N \cdot \lambda = \sum_i \mathscr{M}_g \frac{1}{N} \mid (e_i, U(g)e_i) \mid^2 = \mathscr{M}_g \frac{1}{N} \sum_i \mid (e_i, U(g)e_i) \mid^2$$

We use the Bessel inequality

$$\mid (e_i, U(g)e_i) \mid^2 \leq \parallel U(g)e_i \parallel^2$$

and the fact that $U(g)$ is unitary: $\| U(g)e_i \|^2 = \| e_i \|^2 = 1$ to write

$$\lambda \cdot N \leq \mathscr{M}_g \frac{1}{N} \cdot N = 1$$

in virtue of the normalization of the average to "1". As $\lambda > 0$, the inequality $\lambda N \leq 1$ implies that the largest possible value for N (which is the dimension of the representation space) must be finite. Hereinafter, we can use matrices for representing the symmetry elements of the irreducible representation of groups with an average. ⋄

Proof of Claim b. In terms of matrix elements, we have the equation

$$\lambda \delta_{ij} = \mathscr{M}_g \sum_{k,m} U(g)_{ik} X_{km} U(g^{-1})_{mj}$$

in which the matrix is quite arbitrary, but λ will be determined by the choice of X. In particular, taking the trace on both sides of

$$\lambda \cdot \mathbb{1} = \mathscr{M}_g U(g) X U(g^{-1})$$

we obtain the general relation

$$\lambda \cdot \dim(U) = \mathrm{Tr}(X)$$

The GOT follows by making use of the freedom of choosing the appropriate matrix X. Let us take, for example, a matrix X with zero elements everywhere except for the element in the row k and column m, which is taken to be 1. In this special case, we have

$$\lambda = \frac{\delta_{km}}{\dim(U)}$$

Considering that

$$U(g^{-1})_{mj} = \bar{U}(g)_{jm}$$

we obtain

$$\mathscr{M}_g \bar{U}_{ik}(g) U_{jm}(g) = \frac{1}{\dim U} \cdot \delta_{ij} \delta_{km}$$

Proof of Claim c: If the irreducible representations are inequivalent, set $\lambda = 0$ in the relation above (Schur's lemma 2). ⋄

Comment. When writing these orthogonality relations, it is convenient to introduce the vector space $\mathrm{Map}(G, \mathscr{C})$ of functions defined onto G with an image in $\mathscr{C}$ and endowed with a scalar product

$$(f_1(g), f_2(g)) \doteq \mathscr{M}_g \bar{f}_1(g) \cdot f_2(g)$$

The orthogonality relation is written as a scalar product of the matrix elements:

$$\left(U_{i,j}^\alpha(g), U_{kl}^\beta(g)\right) = \frac{1}{\dim U^\alpha} \cdot \delta_{\alpha\beta} \delta_{ik} \delta_{jl}$$

The character of a representation

Because representations related by a similarity transformation are equivalent, there is a considerable degree of arbitrariness in the actual form of the matrices. However, there are quantities which do not change under similarity transformations and which therefore provide a unique way of characterizing a representation. These quantities relate to the trace of the matrices representing the elements of the group.

Definition. The trace of a matrix A is defined as

$$\mathrm{Tr}\ A = \sum_i (e_i, Ae_i)$$

i.e. it is the sum of the diagonal matrix elements.

The trace is independent of the choice of the basis functions. In fact, we have

$$\mathrm{Tr}\ (AB) = \sum_i \sum_k A_{ik} B_{ki} = \sum_i \sum_k B_{ik} A_{ki} = \mathrm{Tr}\ (BA)$$

and

$$\mathrm{Tr}(SAS^{-1}) = \mathrm{Tr}(ASS^{-1}) = \mathrm{Tr}(A)$$

Rule of computation:

$$\mathrm{Tr}\ (\alpha A + \beta B) = \alpha \mathrm{Tr}\ (A) + \beta \mathrm{Tr}\ (B)$$

Definition. Let U be a matrix representation of a group G. The *character* of U is the function

$$\chi_U(g) := TrU(g)$$

Comments.

1. As the character is invariant under the similarity transformation, equivalent representations have the same characters.

2. The dimension of a representation is the character of the matrix representing the identity element of the group.
3. Conjugate elements have the same character, i.e. the character is identical within the same class of conjugate elements (in other words, the character is a map associating to every **class** a complex number, i.e. it is a class function).
4. Each class is therefore completely determined, as far as its character is concerned, by considering one single element of the class, which might be taken as the representative of the class.

There is a number of important and very "practical" theorems involving characters, which can be used to find and classify irreducible representations and to find out whether a representation is irreducible or not without requiring the explicit knowledge of the matrices.

Proposition A. *Orthogonality theorem for characters. The orthogonality relation*

$$\left(U_{i,j}^{\alpha}, U_{kl}^{\beta}\right) = \frac{1}{\dim U^{\alpha}} \cdot \delta_{\alpha\beta}\delta_{ik}\delta_{jl}$$

can be transformed into an orthogonality relation for the character of U.

Proof. Setting $i = k$ and $j = l$ and summing both sides over the indices i and j, one obtains

$$\mathscr{M}_g \bar{\chi}^{\alpha}(g)\chi^{\beta}(g) = \delta_{\alpha\beta} \Leftrightarrow (\chi^{\alpha}, \chi^{\alpha}) = \delta_{\alpha\beta} \qquad \diamond$$

This theorem allows for an immediate test of irreducibility because the average of the square of the absolute value of the characters of an irreducible representation of a group with an average is just 1 if U^{α} is irreducible:

$$\mathscr{M}_g \mid \chi^{\alpha}(g) \mid^2 = 1$$

As each group admits the 1-representation, a further consequence of this orthogonality theorem is the equation

$$\mathscr{M}_g \chi^{\alpha}(g) = 0$$

provided U^{α} is not the 1-representation.

Proposition B. *Decomposition theorem.* It very often happens that we have a representation of the group which is a reducible one. Such a representation U may be written as the direct sum of various irreducible representations, according to

$$U = \oplus_\alpha n_\alpha U^\alpha$$

where U^α are irreducible representations of the group. n_α is the number of times the irreducible representation U^α is contained in the direct sum. To find n_α, we take the traces of both sides in the expression for the direct sum. Assume that $\chi(g)$, the trace of the representation to be decomposed, exists and $\chi^{(\alpha)}(g)$ is the character of the element g in the desired irreducible representation U^α, then

$$\chi(g) = \sum_\gamma n_\gamma \chi^\gamma(g)$$

for all $g \in G$. Multiplying both sides by $\bar{\chi}^\alpha(g)$ and averaging over g, we obtain $\mathscr{M}_g \bar{\chi}^\alpha(g)\chi(g) = n_\alpha$. This means that

$$n_\alpha = \mathscr{M}_g \bar{\chi}^\alpha(g)\chi(g)$$

or $n_\alpha = (\chi^\alpha, \chi)$. This last equation gives a method for obtaining the coefficients n_α from solely knowing the characters of the representation to be reduced and those of the irreducible component without explicitly requiring the matrix elements. A condition for the application of this formula is that the character of the to-be-decomposed representation is a finite number for all elements g.

Proposition C. *The number of irreducible representations for finite groups (Burnside[7] theorem):*

(I) *The decomposition of the regular representation of a finite group contains all irreducible representations with a multiplicity that corresponds to their dimension.*

(II) *The dimensions $dim(U^i)$ of the irreducible representations satisfy the relation*

$$\sum_i (dim(U^i))^2 = |G|$$

This formula states that the sum of the squares of the dimensions of the inequivalent irreducible representations is equal to the order of the group and is useful for constructing all irreducible representations of a finite group.

---

[7]William Burnside (1852–1927), an English mathematician known for his contributions to group theory.

(III) *The number of irreducible representations U^α of a finite group G is equal to the number of conjugacy classes of G. This is again useful information for constructing all irreducible representations of a finite group.*

Proof of (I). The (right) regular representation of a finite group is a linear map that associates to every element $h \in G$ an operator $T_r(h)$ acting onto the elements of the group itself and defined as

$$T_r(h)(g) = g \circ h$$

(right translation). Alternatively, one can define the left regular representation of G as

$$T_l(h)(g) = h^{-1} \circ g$$

For a finite group, the matrix representation of T_r or T_l is produced in two steps: As a first step, one associates by a one-to-one map to every element of the group a basis vector in a space $R^{|G|}$. Let us, for instance, associate the vector

$$\begin{pmatrix} 1 & 0 & 0 & \ldots & 0 \end{pmatrix}$$

to the element e, the vector

$$\begin{pmatrix} 0 & 1 & 0 & \ldots & 0 \end{pmatrix}$$

to the element g_1, and so on (with the rule being the element g_n produces a "1" only in the nth component of the vector). The operation of multiplying the elements with h shifts the elements according to the entries of the column of the multiplication table residing under the element h. The vectors associated to the elements in this column are then ordered as columns of a matrix, which becomes the regular matrix representation of the element h.

This representation has some important properties. Each element appears only once in a column of the representation table, and all diagonal elements are 0, except for the identity operation, which has exactly "1" along the diagonal. In other words, the regular representation of e is the unit matrix in $R^{|G|}$ and has character $\chi_r(e) = |G|$. The character for the representation of all other elements is zero, as their matrices all have zero on their diagonals. We now find the coefficients n_i in

$$T_r = \oplus n_i U_i$$

Using the decomposition theorem, we obtain

$$n_i = \frac{1}{|G|} \sum_a \chi^{i*}(a)\chi_r(a) = \frac{1}{|G|}\chi^{i*}(e) \cdot |G| = d_i$$

with d_i being the dimension of the ith irreducible representation. We thus have the equation

$$T_r = \oplus_i d_i U_i.$$

This equation means that the decomposition of the regular representation contains all irreducible representations with a multiplicity corresponding to their dimension.

Proof of (II). Taking the traces on both side of this last equation for the matrix representing e leads to

$$\sum_i d_i^2 = |G|$$

which is the important formula we anticipated.

Proof of (III). We first prove a useful orthogonality theorem for characters. Take the relation

$$\mathcal{M}U^\alpha(a)X(U^\alpha(a))^{-1} = \frac{\mathrm{Tr}(X)}{d_\alpha}E$$

and setting $X = U^\alpha(b)$. Then, we have

$$\mathcal{M}U^\alpha(a)U^\alpha(b)(U^\alpha(a))^{-1} = \frac{\chi^\alpha(b)}{l_\alpha}E$$

We multiply both sides with a matrix $U^\alpha(c)$:

$$\mathcal{M}U^\alpha(c)U^\alpha(a)U^\alpha(b)(U^\alpha(a))^{-1} = \frac{\chi^\alpha(b)}{d_\alpha}U^\alpha(c)$$

Taking the trace of both sides, we obtain the multiplication theorem for characters:

$$\chi^\alpha(b) \cdot \chi^\alpha(c) = d^\alpha \cdot \mathcal{M}_a \chi^\alpha(caba^{-1})$$

We now proceed with the proof of (III). Rewrite the orthogonality relation between characters as

$$\sum_{k=1}^{C} n_k \chi^\alpha(k)^* \chi^\beta(k) = |G|\,\delta_{\alpha\beta}$$

with C being the number of classes and n_k the number of elements within each class. This relation can be further written as

$$\sum_k \left[\sqrt{\frac{n_k}{|G|}} \chi^\alpha(k)^* \cdot \sqrt{\frac{n_k}{|G|}} \chi^\beta(k) \right] = \delta_{\alpha\beta}$$

We then introduce the vectors

$$\chi_\alpha \doteq \frac{1}{\sqrt{|G|}} \left(\sqrt{n_1}\chi_\alpha(1), \sqrt{n_2}\chi_\alpha(2), \ldots, \sqrt{n_C}\chi_\alpha(C) \right)$$

to write the orthogonality relation as

$$\chi_\alpha \cdot \chi_\beta = \delta_{\alpha\beta}$$

The χ_α are mutually orthogonal vectors in a space whose dimension is the number of classes C in the group. There can be a maximum number of C of such mutually orthogonal vectors. But these vectors are labeled by an index α corresponding to the irreducible representations of the group. Hence, the number of irreducible representations must be less than or equal to the number of classes. This means that the rows of the character matrix

$$\begin{bmatrix} G & 1 & 2 & \cdots & C \\ U_{\alpha_1} & \sqrt{\frac{n_1}{|G|}}\chi_{\alpha_1}(1) & \sqrt{\frac{n_2}{|G|}}\chi_{\alpha_1}(2) & \cdots & \sqrt{\frac{n_C}{|G|}}\chi_{\alpha_1}(C) \\ \vdots & \vdots & \vdots & & \vdots \\ U_{\alpha_n} & \sqrt{\frac{n_1}{|G|}}\chi_{\alpha_n}(1) & \sqrt{\frac{n_2}{|G|}}\chi_{\alpha_n}(2) & \cdots & \sqrt{\frac{n_C}{|G|}}\chi_{\alpha_n}(C) \end{bmatrix}$$

are mutually orthogonal and that $n \leq C$.

On the other side, using the multiplication theorem for characters

$$\chi_\alpha(b) \cdot \chi_\alpha(c)^* = d^\alpha \cdot \mathscr{M}_a\chi_\alpha(c^{-1}aba^{-1})$$

one can show that the columns of the character matrix are also mutually unitary orthogonal. In fact,

$$\sum_\alpha \chi^\alpha(b) \cdot \chi^\alpha(c)^* = \sum_\alpha d^\alpha \cdot \mathscr{M}_a\chi^\alpha(c^{-1}aba^{-1})$$

$$= \mathscr{M}_a\chi_{\text{reg}}(c^{-1}aba^{-1})$$

If we assume that c^{-1} is not in the conjugacy class of b, then $aba^{-1} \neq c^{-1}$ and $c^{-1}aba^{-1} \neq e$. As all characters of the regular representation are zero (except for the character of the element e), we conclude that

$$\sum_\alpha \chi^\alpha(b) \cdot \chi^\alpha(c)^* = 0$$

for b and c in different classes. This means that the columns of the character matrix are indeed mutually orthogonal. They build a set of C orthogonal vectors in an n-dimensional space, i.e. $C \leq n$.

We obtain, accordingly, the inequality

$$C \leq n \leq C$$

which has the solution

$$n = C \qquad\qquad\qquad \diamond$$

Character table for finite groups: Based on part (III) of the Burnside theorem, the characters of the irreducible representations of a finite group are conveniently displayed in the form of **character tables**. The classes of the group (at least one representative) are usually listed along the top line of the table and the irreducible representations on the left-hand side column. The bulk contains the characters themselves. As a consequence of this theorem, this table is always a square and the corresponding rows are normalized to the order of the finite group and orthogonal to each other.

We summarize now some simple rules for constructing the character table of a finite group:

1. It is convenient to display in table form the characters of the irreducible representations. Such a table gives less information than a complete set of matrices, but we will see that it is sufficient to define quantum numbers and eigenmodes.
2. The number of irreducible representations is equal to the number of classes in the group.
3. The sum of the squares of the dimensions d_α of the irreducible representations is equal to the number of elements of the group

$$\sum_i d_i^2 = |G|$$

4. The characters of the irreducible representation must be mutually orthogonal and normalized to the order of the group:

$$\sum_a \chi^\alpha(a)^* \chi^\beta(a) = |\, G\,| \cdot \delta_{\alpha\beta}$$

5. Every group admits the one-dimensional identical representation in which each element of the group is represented by the number 1. The orthogonality relation between characters then shows that for any

irreducible representation other than the identity representation,

$$\sum_a \chi(a) = 0$$

Example. As an example, we construct the character table of C_{4v}.

There will be five irreducible representations, with one of them being the trivial one. Specializing SO_2 to the rotations of C_{4v} and computing the trace of the matrices for the m-operations, we arrive at the conclusion that there is an irreducible representation of dimension 2 (called the representation Δ_5). The equation

$$d_{\Delta_1}^2 + d_{\Delta_{1'}}^2 + d_{\Delta_2}^2 + d_{\Delta_{2'}}^2 + d_{\Delta_5}^2 = 8$$

then has a unique solution given in the column on the table corresponding to the element e. A further one-dimensional representation is given by the determinant of the rotation matrices — the determinant of matrices obeys the "homomorphism rule"

$$\det (AB) = \det (A) \cdot \det (B)$$

The remaining entries of the character table can be constructed by trial and error, taking into account the orthogonality relations of both rows and columns (and also taking care that the multiplication table is fulfilled).

Character table of C_{4v}, the symmetry group of a pyramid with a square base. The label of the symmetry elements is taken from the table compiled for the group O_h in Lecture 1 and describes the map of the coordinates (x, y, z). The irreducible representations are labeled according to Bouckaert, L. P., Smoluchowski, R. and Wigner, E. "Theory of Brillouin zones and symmetry properties of wave functions in crystals", *Phys. Rev.* **50**, 58–67 (1936), https://journals.aps.org/pr/abstract/10.1103/PhysRev.50.58.

C_{4v}	E	C_{2x}	$C_{4x}C_{4x}^{-1}$	$IC_{2z}IC_{2y}$	$IC_{2yz}IC_{2y\bar{z}}$
Δ_1	1	1	1	1	1
Δ_2	1	1	$\bar{1}$	1	$\bar{1}$
$\Delta_{2'}$	1	1	$\bar{1}$	$\bar{1}$	1
$\Delta_{1'}$	1	1	1	$\bar{1}$	$\bar{1}$
Δ_5	2	$\bar{2}$	0	0	0

Further character tables for important groups can be found in the review paper by G. F. Koster, "Space groups and their representations", in *Solid*

State Physics Vol. 5, Editors: F. Seitz and D. Turnbull, Academic Press, New York, 1957, pp. 173–256, https://doi.org/10.1016/S0081-1947(08)60 103-4. This paper contains all relevant information about the 32 point groups of crystallography and their character tables, using a notation which is common in solid-state physics. Some tables, most commonly used in these lectures, are reported in the following. The irreducible representations are labeled using the convention in the paper by Bouckaert, L. P. and Smoluchowski, R. and Wigner, E. "Theory of Brillouin zones and symmetry properties of wave functions in crystals", *Phys. Rev.* **50**, 58–67 (1936).

Character table of the group called C_{3v}. This is the symmetry group of a pyramid with an equilateral triangle as base. For the labeling, see the caption of C_{4v}.

C_{3v}	E	$C_{3xyz}C_{3xyz}^{-1}$	$IC_{2x\bar{y}}IC_{2\bar{x}z}IC_{2y\bar{z}}$
Λ_1	1	1	1
Λ_2	1	1	$\bar{1}$
Λ_3	2	$\bar{1}$	0

Character table of the group called C_{2v}. This is the symmetry group of a pyramid with a rectangle as base. For the labeling, see the caption of C_{4v}.

C_{2v}	E	C_{2xy}	IC_{2z}	$IC_{2x\bar{y}}$
Σ_1	1	1	1	1
Σ_2	1	1	$\bar{1}$	$\bar{1}$
Σ_3	1	$\bar{1}$	$\bar{1}$	1
Σ_4	1	$\bar{1}$	1	$\bar{1}$

Character table of the group called T_d. This is the symmetry group of a regular tetrahedron. For the labeling, see the caption of C_{4v}.

T_d	E	$3C_4^2$	$8C_3$	$6IC_4$	$6IC_2$
Γ_1	1	1	1	1	1
Γ_2	1	1	1	$\bar{1}$	$\bar{1}$
Γ_{12}	2	2	$\bar{1}$	0	0
Γ_{15}	3	$\bar{1}$	0	$\bar{1}$	1
Γ_{25}	3	$\bar{1}$	0	1	$\bar{1}$

Character table of the group O_h. This is the symmetry group of a regular octahedron (or its dual body, the cube). For the labeling, see the caption of C_{4v}.

O_h	E	$3C_4^2$	$6C_4$	$6C_2$	$8C_3$	I	$3IC_4^2$	$6IC_4$	$6IC_2$	$8IC_3$
Γ_1	1	1	1	1	1	1	1	1	1	1
Γ_2	1	1	$\bar{1}$	$\bar{1}$	1	1	1	$\bar{1}$	$\bar{1}$	1
Γ_{12}	2	2	0	0	$\bar{1}$	2	2	0	0	1
Γ_{15}	3	$\bar{1}$	1	$\bar{1}$	0	3	1	$\bar{1}$	1	0
Γ_{25}	3	$\bar{1}$	$\bar{1}$	1	0	$\bar{3}$	1	1	$\bar{1}$	0
$\Gamma_{1'}$	1	1	1	1	1	$\bar{1}$	$\bar{1}$	$\bar{1}$	$\bar{1}$	$\bar{1}$
$\Gamma_{2'}$	1	1	$\bar{1}$	$\bar{1}$	1	$\bar{1}$	$\bar{1}$	1	1	$\bar{1}$
$\Gamma_{12'}$	2	2	0	0	$\bar{1}$	$\bar{2}$	$\bar{2}$	0	0	1
$\Gamma_{15'}$	3	$\bar{1}$	1	$\bar{1}$	0	3	$\bar{1}$	1	$\bar{1}$	0
$\Gamma_{25'}$	3	$\bar{1}$	$\bar{1}$	1	0	3	$\bar{1}$	$\bar{1}$	1	0

Proposition D. *The number of irreducible representations for continuous groups (Peter–Weyl theorem).*

When dealing with continuous groups, we expect the number of classes to be a **non-countable** infinite set. We also expect, in general, an infinite number of irreducible representations. The question is one about whether they are countable or not. The countability of the irreducible representations has a profound significance for quantum mechanics: As we will see, irreducible representations are used to label the states of a quantum mechanical system, so that their countability is equivlent to the quantized character of the eigenvalues of many relevant operators of quantum mechanics.

The theorem by Peter and Weyl proves the countability of the irreducible representations of a continuous group **with an average**. This is typically done in two steps. In the first step, an infinite-dimensional unitary representation of the group is established (called the regular representation), which is then certainly completely reducible because it is unitary and, accordingly, can be split into a direct (discrete) sum of some irreducible representations. In the second step, it is demonstrated that this sum contains *all* irreducible representations, which is equivalent to showing that all irreducible representations of a group with an average are indeed countable. For a rigorous proof of the theorem, see for example https://te rrytao.wordpress.com/2011/01/23/.

The theorem was originally published as F. Peter and H. Weyl, "Über die Vollständigkeit der primitiven Darstellungen einer geschlossenen kontinuierlichen Gruppe", *Math. Ann.* **97**, 737–755 (1927), https://link.sp ringer.com/article/10.1007/BF01447892.

A simple version of the Peter–Weyl theorem is as follows:

Theorem (Peter and Weyl). *The decomposition of the regular representation of a continuous group with an average contains all irreducible representations with a multiplicity that corresponds to their dimension. In other words, the Burnside formula for finite groups*

$$T_r = \oplus_{\lambda=1}^{C} d_\lambda \cdot T_\lambda$$

with C being the finite number of classes is generalized by the Peter–Weyl theorem to

$$T_r = \oplus_{\lambda=1}^{\infty} d_\lambda \cdot T_\lambda$$

with d_λ being the dimension of the irreducible representation T_λ.

Plausibility. We provide now an intuitive proof of the theorem that serves more the purpose of enhancing its importance than being rigorous.

Step 1. We introduce the regular representation of a continuous group. Let G be a group with an average. One defines $L^2(G)$ as the set of complex-valued functions f that satisfy the following conditions:

1. f is measurable under the Haar measure.
2. $\mathscr{M} \mid f(g) \mid^2 < \infty$.
3. The Haar measure induces on $L^2(G)$ a scalar product

$$(f_1, f_2) \doteq \mathscr{M}_g \bar{f}_1(g) f_2(g)$$

With these definitions, $L^2(G)$ becomes a Hilbert space with a complete orthonormal system of basis vectors (CONS).

Within this Hilbert space, one defines the regular representations of G by associating to every element h of G an operator $T_r(h)$ such that

$$T_r(h) f(g) = f(g \circ h)$$

This operator is called the right regular representation of G. Alternatively, on can define the left regular representation $T_l(h)$ by

$$T_l(h) f(g) = f(h^{-1} \circ g)$$

Both representations are

i. unitary:
$$(T_r(h)f_1, T_r(h)f_2) = \mathscr{M}_g \bar{f}_1(gh)f_2(gh) = \mathscr{M}_g \bar{f}_1(g)f_2(g) = (f_l, f_2)$$

and

ii. unitary equivalent, i.e. there is a one-to-one mapping such that $AT_l(h) = T_r(h)A$ for all $h \in G$. In fact, define $Af(g) = f(g^{-1})$. Then,
$$T_r(h)Af(g) = T_r(h)f(g^{-1}) = f(g^{-1}h) = f((h^{-1}g)^{-l}) = Af(h^{-1}g)$$
$$= AT_l(h)f(g)$$

so that considering either one of the two regular representations is enough to draw conclusions. As the regular representation is a unitary one, it is completely reducible into the orthogonal sum of finite-dimensional irreducible representations. The resulting invariant subspaces carrying the irreducible representations can be counted by construction, so the irreducible representations occurring in the decomposition of the regular representations are certainly countable. However, are they all possible irreducible representations of the group, or are there more irreducible representations which are not in the decomposition of the regular representation and might even be non-countable? Here is where step 2 comes into play.

Step 2. The problem with the formula
$$T_r = \oplus_{\lambda=1}^{\infty} d_\lambda \cdot T_\lambda$$

for continuous infinite groups is that the decomposition theorem cannot be applied straightforwardly. The decomposition theorem contains the characters of the irreducible representations. These is a well- defined quantity for groups with averages, as their irreducible representations are finite. But it also contains the character of the to-be-decomposed regular representation, and this quantity is at first glance ill-defined. In fact, the character of the element g in the regular representation is written as
$$\sum_n (e_n, T_r(g)e_n)$$

and is typically divergent for an infinite-dimensional representation as the regular one, if considered point-wise, even if the group has an average. Accordingly, our theorems involving characters cannot be straightforwardly extended, for example, to the reduction of the regular representation in continuous groups. Harish-Chandra[8] (see M. F. Atiyah, "The Harish-Chandra character", in *Representation Theory of Lie Groups*, Editors: M. F. Atiya *et al.* DOI: http://dx.doi.org/10.1017/CBO9780511662683) was able to define the notion of characters for infinite-dimensional representations by showing that

$$\sum_n (e_n, T_r(g)e_n)$$

converges **in the sense of distributions**.[9] This distribution is the Harisch-Chandra character of $T_r(g)$.

We sketch the ideas of Harisch-Chandra by computing the character of the regular representation for groups with averages. We first consider $T_r(g)$ and find, within $L^2(G)$, the eigenvalues of $T_r(g)$, i.e. those matrix elements which satisfy the equation

$$T_r(g)f(h) = f(hg) = \tau_n(g)f(h)$$

with τ being the desired eigenvalue. Accordingly, in the basis set provided by the eigenfunctions, the eigenvalues appear on the diagonal of the infinite matrix required to represent the regular representation. As T_r is unitary, we have $\mid \tau(g) \mid = 1$, i.e. $\tau_n(g)$ lies on the unit circle in the $\mathscr{C}$ plane. It is plausible that τ can be written as

$$\tau_n(g) = e^{i \cdot \Theta_n(g)}$$

with $\Theta_n(g)$ being a real function of the parameters used to describe the element g ($\Theta_n(e) = 0$) and the index $n \in \mathscr{N}$ describing the eigenfunction used to build the diagonal matrix element $T_r(g)_{nn}$. We can now use these matrix elements, à la Fourier series, to write any function $f(g)$ as

$$f(g) = \sum_n a_n e^{i \cdot \Theta_n(g)} \quad a_n = \mathscr{M}_g e^{-i \cdot \Theta_n(g)} \cdot f(g)$$

We are now ready to compute the character of the regular representation of a group with an average, taken as a distribution over a suitable test

---

[8]Harish-Chandra Mehrotra (1923–1983), mathematical physicist, see e.g. https://en.wikipedia.org/wiki/Harish-Chandra.

[9]For a practical introduction to the theory of distributions see e.g. M. J. Lighthill, "An Introduction to Fourier Analysis and Generalised Functions", (1958), Cambridge University Press; https://doi.org/10.1017/CBO9781139171427.

function $f(g)$:

$$\chi_{T_r}(f) = \mathscr{M}_g \sum_n \overline{(e_n, T_r(g)e_n)} \cdot f(g)$$

$$= \mathscr{M}_g \sum_n e^{-i \cdot \Theta_n(g)} \cdot f(g) = \sum_n \mathscr{M}_g e^{-i \cdot \Theta_n(g)} f(g) = \sum_n a_n$$

$$= f(0)$$

From the relation

$$\chi_{T_{\text{reg}}}(f) = \mathscr{M}_g \chi_{T_{\text{reg}}}(g) \cdot f(g) = f(0)$$

we deduce the formal identity

$$\chi_{T_{\text{reg}}}(g) = \delta(g)$$

where $\delta(g)$ is the Dirac distribution at the identity e. This result is a generalization of the analogous one for finite groups, namely

$$\chi_r(e) = |G| \quad \chi_r(g) = 0, g \neq e$$

We use the properties of the delta function to compute, formally, the multiplicity of the irreducible representation T_α in the reduction of the regular representation and obtain

$$n_\alpha = \mathscr{M}_g \bar{\chi}_\alpha(g) \chi(g) = \mathscr{M}_g \bar{\chi}_\alpha(g) \delta(g) = \chi_\alpha(e) = d_\alpha$$

In words, every finite irreducible representation of G appears in the decomposition of the regular representation d_α times.

2.3 Symmetry-Adapted Vectors

The next task we aim to solve by using symmetry arguments is to compute the eigenvalues themselves, at least in an approximate way. For this purpose, we need to compute the so-called symmetry-adapted vectors. In this section, we provide the fundamental mathematical theorems about the subject of symmetry-adapted vectors. One can also refer to this section as developing the "technology" of Clebsch–Gordan coefficients for general groups with averages.

Consider a vector space V carrying a representation U of a group G. We foresee a set of basis vectors $\{e_k\}$, $(k = 1, \ldots)$ in V that we have taken, for example, from a library or that we know from textbooks on special functions. These basis vectors do not take into account, in general, the special symmetry of a physical system but are generally available for any problem. Regarding the representation U, we often have the situation that U is not necessarily irreducible but can be brought into a block-diagonalized form:

$$U = \oplus_l n_l D_l$$

D_l are irreducible representations, labeled by the index l. This index is related to some characteristic ("quantum") number of the physical system under investigation. In this specific section, we use a labeling which originates from the solution to the Schrödinger equation with spherical symmetry, so that the reader might find himself more comfortable with the labeling. n_l is computed using the decomposition theorem, if possible.

Once n_l is computed, we know that U can be, in principle, reduced to a block-diagonalized form: There is a set of subspaces which are invariant with respect to all operations of G, i.e. the operations transform a vector lying within this subspace into another vector in the same subspace. We also know that these putative invariant subspaces are pairwise orthogonal, i.e. the original vector space V is the orthogonal sum of G-invariant subspaces, each carrying an irreducible representation D_l:

$$V = \oplus_l n_l V_l$$

Let us formulate now precisely the G-invariance of the d_l-dimensional space D_l. In the original basis set $\{e_k\}$, the space V_l is not necessarily G-invariant: The special linear combinations of the basis vectors $\{e_k\}$ must be constructed that block-diagonalize U and render V_l G-invariant.

Definition. The linear combination of basis vectors that block-diagonalize the representation U are called "the symmetry-adapted vectors".

Comments.

1. These vectors are characterized by the index l that signifies that they are related to the irreducible representation D_l. In addition, they need a further index (let us call it m) that varies from 1 to d_l, as their number must be equal to the dimensionality of D_l. Let us therefore call these symmetry-adapted functions Y_l^m (the notation used refer to the practical and familiar example of atoms, where Y_l^m are the spherical harmonics).
2. The same irreducible representation appears n_l times in the reduction of U so that we have further sets of symmetry-adapted vectors which are orthogonal to Y_l^m but have the same transformation properties.

3. One can take this multiplicity into account by, for example, introducing a further index m' varying from 1 to n_l,[10] so that the symmetry-adapted vectors are sets of the type

$$\{Y_l^{m,m'}\} \quad m = 1, \ldots, d_l \quad m' = 1, \ldots, n_l$$

4. Mathematically, the symmetry-adapted vectors fulfill the relation

$$U(g)Y_l^{m,m'} = \sum_k D_{km}^l Y_l^{k,m'}$$

i.e. the symmetry-adapted vector $Y_l^{m,m'}$ transforms according to the mth-column of the lth irreducible matrix representation when $U(g)$ is applied to it.

For performing practical computations of the eigenvalues, we must be able to find precisely the symmetry-adapted vectors (we will discuss the reason later). The following theorem is fundamental to constructing symmetry-adapted vectors.

Theorem. *Consider a vector* **e** *in the space* V. *The vector*

$$P_m^{(l)}\mathbf{e} \doteq d_l \cdot \mathcal{M}_g \overline{D_{mm}^l}(g)U(g)\mathbf{e}$$

transforms according to the mth column of the irreducible representation D_l.

Proof. We think of the expansion of **e** as a linear combination of the symmetry-adapted vectors $\{Y_l^{m,m'}\}$ i.e.

$$\mathbf{e} = \sum_{p,q,m'} a_{p,q,m'} Y_p^{q,m'}$$

and compute

$$P_m^{(l)}\mathbf{e} = d_l \cdot \mathcal{M}_g \overline{D_{mm}^l}(g)U(g) \sum_{p,q,m'} a_{p,q,m'} Y_p^{q,m'}$$

$$= d_l \cdot \sum_{p,q,m'} a_{p,q,m'} \mathcal{M}_g \overline{D_{mm}^l}(g) \sum_k D_{kq}^p Y_p^{k,m'}$$

$$= d_l \cdot \sum_{p,q,m',k} a_{p,q,m'} \underbrace{[\mathcal{M}_g \overline{D_{mm}^l} D_{kq}^p]}_{\frac{1}{d_l}\delta_{lp}\delta_{mq}\delta_{mk}} Y_p^{k,m'}$$

$$= \sum_{m'} Y_l^{m,m'} a_{l,m,m'} \qquad \diamond$$

---

[10] This situation does not arise in atomic physics because $n_l = 1$ and a further index m' is not necessary. We will discuss this later in the lectures.

This last result shows that the projector $P_m^{(l)}$ projects out of any vector the sum of those components that transform according to the column m of the lth irreducible representation contained in the reduction of U. Thus, this projector can be used to create a set of symmetry-adapted vectors. Of course, it is possible that the desired symmetry-adapted wave functions are not contained in the trial vector $\mathbf{e}$, so that the result of applying the projector is zero. In that case, in order to generate symmetry-adapted wave functions, one must repeat the procedure with another "trial" vector.

Corollary. *Given a set $\{e_i\}$ of basis functions in the representation space V, the desired symmetry-adapted vectors are the eigenvectors of the projector matrix*

$$\left(e_i, d_l \cdot \mathscr{M}_g \overline{D_{mm}^l}(g) U(g) e_j\right)$$

to non-zero eigenvalues.

This projector requires an explicit knowledge of the (diagonal) matrix elements of the various representations and not merely a knowledge of their characters (which is the usual information given in the published literature). For one-dimensional representations, the characters are the matrix elements themselves, so that this projector provides the complete answer for the symmetry-adapted basis functions of one-dimensional representations. For representations with larger dimensions, there is a procedure to obtain the d_l-partners of a basis set for an irreducible representation from the characters themselves. This procedure involves the projector

$$P^{(l)} \doteq \sum_m P_m^{(l)} = d_l \mathscr{M}_g \overline{\chi_l}(g) U(g)$$

which projects out of any normalizable vector the sum of all basis functions transforming according to the columns of the lth irreducible representation. This last projector requires only the knowledge of the characters of the irreducible representation of which one would like to construct symmetry-adapted wave functions. Note that the projector does not produce all partners within the irreducible subspace. However, having determined one Y_l, we can then apply to Y_l the operations of the symmetry group to obtain exactly d_l linearly independent vectors which can be made orthonormal and thus can be used as a basis set for the irreducible representation D^l (remember that the application of the operations of the group to a vector belonging to an invariant subspace carrying the irreducible representation D^l does not carry us outside this subspace). Note that, after having generated the basis functions $\{Y_l^m\}$ transforming according to the irreducible representation D^l, it is possible to construct the matrix representation of $D^l(g)$ (in practice, this is the procedure most commonly adopted to construct matrix representations of the irreducible representations), using

$$D(g)Y_l^m = \sum_k D_{km}^l(g)Y_l^k$$

This procedure requires only the initial knowledge of the characters. In summary, the algorithm for obtaining symmetry-adapted vectors reads:

- One starts with a trial vector e.
- P^l applied to e possibly produces one symmetry-adapted vector Y_l.
- $D(g)Y_l$ produces all partner vectors transforming according to D^l (take, for this purpose, a sufficiently large number of $D(g)$).
- If $P^l e = 0$, use a different trial function.

Example. As a way of illustrating the technical aspects of the use of the projector technique, we study the tower (or permutation) representation of C_{4v}. One can obtain this representation by associating to each corner of a square a "weight" or scalar, thus producing unit vectors in a four-dimensional space. The operations of C_{4v} are then represented as 4×4 matrices along the column of which one can find the map of the four basis vectors under the action of the symmetry operation. As an example, take C_{4z}. Apply it to the four unit vectors. Place their maps along the columns of a matrix, which, accordingly, is written as

$$\begin{pmatrix} 0 & 0 & 0 & 1 \\ 1 & 0 & 0 & 0 \\ 0 & 1 & 0 & 0 \\ 0 & 0 & 1 & 0 \end{pmatrix}$$

All other matrices can be constructed in a similar way. The following character table for C_{4v} reports, in the bottom line, the characters of the tower representation just obtained.

In order to find the irreducible components of the tower representation Δ_T, we can use the decomposition theorem:

$$\Delta_T = \Delta_5 \oplus \Delta_{2'} \oplus \Delta_1$$

Character table of the group C_{4v}. Along the bottom line, the characters of the tower representations are given. The column on the right reports the symmetry-adapted vectors (computed in the following) that block-diagonalize the tower representation on the square.

C_{4v}	e	$2C_4$	C_2	$2m$	$2m'$	Basis
Δ_1	1	1	1	1	1	$(1,1,1,1)$
$\Delta_{1'}$	1	1	1	$\bar{1}$	$\bar{1}$	
Δ_2	1	$\bar{1}$	1	1	$\bar{1}$	
$\Delta_{2'}$	1	$\bar{1}$	1	$\bar{1}$	1	$(1,\bar{1},1,\bar{1})$
Δ_5	2	0	$\bar{2}$	0	0	$\left\{ \begin{array}{l} (1,0,\bar{1},0) \\ (0,1,0,\bar{1}) \end{array} \right\}$
Δ_T	4	0	0	0	2	

We find the symmetry-adapted vectors using the projector technique:

$$P^{\Delta_1}(1,1,1,1) \propto (1,1,1,1) + \cdots + (1,1,1,1) \propto (1,1,1,1)$$

$$P^{\Delta'_2}(1,0,0,0) \propto (1,0,0,0) - (0,1,0,0) - (0,0,0,1) + (0,0,1,0) -$$
$$(0,1,0,0) - (0,0,0,1) + (1,0,0,0) + (0,0,1,0)$$
$$\propto (1,\bar{1},1,\bar{1})$$

$$P^{\Delta_5}(1,0,0,0) \propto 2 \cdot (1,0,0,0) - 2 \cdot (0,0,1,0) \propto (1,0,\bar{1},0)$$

$$C_4(1,0,\bar{1},0) \propto (0,1,0,\bar{1})$$

The symmetry-adapted vectors found by applying the projector technology are reported in the right-hand column of the character table above. This column, often headed in the literature with the words "basis functions", hosts typical symmetry-adapted vectors and functions characteristic of each irreducible representation.

Lecture 3

SO_3 AND THE GENERAL METHOD OF INFINITESIMAL TRANSFORMATIONS

In this lecture, we use the mathematical tools established in Lecture 2 to construct the irreducible representations of important groups such as SO_2 and SO_3 — important not only because they appear in scientific problems but because they provide general background knowledge. SO_2 is an Abelian group, so we will learn how to proceed with the representation theory of this important class of groups. SO_3 is a group with a non-trivial Abelian subgroup, so we will learn how to handle the representation theory of this class of groups. In particular, the study of the representation theory of these two groups will serve as a tutorial way for exploring the general method of infinitesimal transformations for continuous groups. This last subject will enable us, later in the course, to deal with further continuous groups that are relevant in atomic, chemical, and solid-state physics, as well as in particle physics. The lecture is accompanied by a set of exercises.

3.1 Irreducible Representations of the Circle Group C_∞

The axial rotation group (or circle group) C_∞ is the group of rotations around a fixed axis. The group is Abelian, so each element builds a class on its own. Each class is characterized by a rotation angle $\alpha \in [0, 2\pi]$. The group is continuous and has an average; accordingly, by virtue of the Peter–Weyl theorem, its irreducible representations, which might be an infinite number (as the number of conjugacy classes is infinite) are countable and finite-dimensional, i.e. it can be labeled by some discrete index (contrary to the elements themselves, which are labeled by a continuous variable). The natural representations of C_∞ is provided by the matrix group SO_2. It is a faithful representation, indicating that there is an isomorphism between C_∞ and SO_2. We now obtain the irreducible representations using two different methods.

Algebraic construction of the irreducible representations

a. **Theorem.** The finite unitary irreducible representations of an Abelian group are one-dimensional.

Proof. Consider a matrix representation $T(g)$. The matrices have the property that

$$[T(g_1), T(g_2)] = 0$$

for any element g_1, g_2. According to a theorem of linear algebra, commuting unitary matrices can be diagonalized simultaneously, i.e. they can be block-diagonalized to exactly diagonal matrices. The irreducible representations can be read out from the diagonal matrix elements and are, by construction, one-dimensional.   ◇

b. The diagonal matrix elements must be complex numbers of absolute value 1, as the representation is unitary. For the specific case of C_∞, we can set

$$T_m(\alpha) = e^{-i \cdot m \cdot \alpha}$$

(α the rotation angle) as the matrix representation of the irreducible representation T_m. The matrix defined in this way is a homomorphism since

$$e^{-im(\alpha_1 + \alpha_2)} = e^{-im\alpha_1} \cdot e^{-im\alpha_2}$$

Because of the requirement of periodicity, $m \in \mathscr{Z}$. The integer numbers appearing in the exponent of the representation T_m are the so-called **Cartan**[1] **weights** of the representation.

c. The set $\{T_m\}$ are all irreducible unitary representations of C_∞.

To prove this, assume there exists an irreducible $\Gamma \neq T_m$. In virtue of the orthogonality of the characters, we must have

$$\frac{1}{2\pi} \int_0^{2\pi} e^{im\alpha} \cdot \Gamma(\alpha) = 0$$

which holds $\forall m$. However, the set $\{e^{-im\alpha}\}$ is complete in the interval $[0, 2\pi]$, so that $\Gamma(\alpha)$ must be identical to zero, i.e. the representation has a zero average and cannot be irreducible, contrary to the assumption.

---

[1] Élie Joseph Cartan (1869–1951), influential French mathematician, see e.g. https://de.wikipedia.org/wiki/Élie_Cartan.

On the base of these results, we can envisage a kind of "character table" for C_∞.

C_∞	α
$T_{m \in \mathscr{Z}}$	$e^{-i \cdot m \cdot \alpha}$

Comment. The ("natural") SO_2 representation of C_∞

$$\begin{pmatrix} \cos\varphi & -\sin\varphi \\ \sin\varphi & \cos\varphi \end{pmatrix}$$

constitutes a 2×2 representation of the circle group. The representation is reducible:

$$SO_2 = T_1 \oplus T_{-1}$$

Analytic construction of the irreducible representations

There is an alternative method of constructing irreducible representations of a continuous group, which is often used when the group is more complex than C_∞. We illustrate this method with the example of C_∞. The method is based on starting from a representation of the group in an infinite-dimensional function space and finding the irreducible components.

a. The representation space chosen for C_∞ is the space of 2π periodic functions, and the representation of C_∞ is defined by the unitary operator $T_r(\alpha)$ with

$$T_r(\alpha)f(\varphi) \doteq f(\varphi - \alpha)$$

It so happens that this is the regular representation of C_∞, implying that we know that it contains all finite irreducible representations of C_∞ once and only once (Peter–Weyl theorem).

b. We search for the irreducible representations of C_∞ by searching for invariant subspaces, i.e. by solving the functional equation

$$T_r(\alpha)f(\varphi) = \lambda f(\varphi)$$

Once we have the invariant subspaces, we construct the representation carried by these by computing the matrix elements of the operator $T_r(\alpha)$.

c. There is a method for solving this functional equation which consists of linearizing the operator for small angles α:

$$T_r(\alpha)f(\varphi) = f(\varphi - \alpha) \cong f(\varphi) - i \cdot \alpha \cdot \left(-i \cdot \frac{\partial}{\partial \varphi}\right) f(\varphi)$$

The finding of the invariant subspaces reduces to finding the solution of the differential equation

$$-i \cdot \frac{\partial}{\partial \varphi} f(\varphi) = \lambda' f(\varphi)$$

with the boundary conditions that the solution must be 2π periodic. The linearization procedure replaces a (difficult to solve) functional equation with a more manageable differential equation (where introducing the imaginary number i is a matter of convention).

d. The differential equation has a complete set of solutions

$$\left\{ \frac{e^{i \cdot \lambda' \cdot \varphi}}{\sqrt{2\pi}} \right\}$$

with $\lambda' = m \in \mathscr{Z}$. These are the desired invariant subspaces. They are one-dimensional, so they provide one-dimensional representations, which are, by definition, irreducible.

e. The matrix elements of the operator $T_r(\alpha)$ with these functions provide the desired irreducible matrix representations of the circle group:

$$\left(\frac{e^{i \cdot m \cdot \varphi}}{\sqrt{2\pi}}, T_r(\alpha) \frac{e^{i \cdot m \cdot \varphi}}{\sqrt{2\pi}} \right) = e^{-i \cdot m \cdot \alpha}$$

On the base of these results, we can envisage a character table for C_∞, including some basis functions.

C_∞	α	Basis
$T_{m \in \mathscr{Z}}$	$e^{-i \cdot m \cdot \alpha}$	$e^{i \cdot m \cdot \varphi}$

The irreducible representations constructed in this way are all the finite irreducible representations of the circle group, as the set of basis functions is a complete set. Accordingly, any periodic function can be expressed as a linear combination of these invariant basis states, in agreement with the generalized Fourier theorem provided by the Peter–Weyl theorem.

$$f(\varphi) = \sum_m c_m e^{i \cdot m \varphi}$$

Comments.

1. The rotation operator about the z-axis for small angles is given by

$$T_r(\alpha) \approx \mathbb{1} - i \cdot \left(-i \cdot \frac{\partial}{\partial \varphi} \right) \cdot \alpha$$

We recognize on the right-hand side the z-component of the orbital angular momentum, quantized according to the correspondence principle (except for the factor $\hbar^{-1}$):

$$-i \cdot \frac{\partial}{\partial \varphi} = \frac{L_z}{\hbar}$$

and

$$T_r(\alpha) \approx \mathbb{1} - i \cdot \alpha \cdot \frac{L_z}{\hbar}$$

This relation identifies the invariant subspaces carrying the irreducible representations of the rotation group with the eigenspaces of L_z and the Cartan weights of the irreducible representations T_m as the eigenvalues of $\frac{L_z}{\hbar}$. Using this analytic method, one recognizes that Cartan weights, which are representative of group theoretical arguments, are immediately related to quantum numbers, which are representative of the physical properties of a system. This relation was a remarkable achievement of E. Wigner in his seminal paper.

2. The relation

$$T_r(\alpha) \approx \mathbb{1} - i \cdot \frac{L_z}{\hbar} \cdot \alpha$$

can be interpreted as L_z generating the operator of infinitesimal rotations about the z-axis. L_z is called, accordingly, the "infinitesimal transformation (generator, operator)", and the Cartan weights of the representation T_m are the matrix elements of the infinitesimal transformation.

3. T_r is a unitary operator so that the infinitesimal generator is a Hermitic operator:

$$\left(\mathbb{1} + i \cdot \frac{L_z^\dagger}{\hbar} \cdot \alpha \right) \cdot \left(\mathbb{1} - i \cdot \frac{L_z}{\hbar} \cdot \alpha \right) \approx \mathbb{1} + \frac{L_z^\dagger - L_z}{\hbar} \cdot \alpha \overset{!}{=} \mathbb{1} \Leftrightarrow L_z^\dagger = L_z$$

Note that the purpose of using the imaginary constant i while writing the linearized rotation operator is exactly that of rendering the infinitesimal generator Hermitic. This strengthens the link between rotation group and quantum mechanical angular momentum.

4. The operator of exponential mapping: Starting from the infinitesimal operator, which is used to express the representation of small angle rotations about the z-axis in a function space, one can construct a representation of the rotations by any finite angle in the function space:

$$\frac{dT(\alpha)}{d\alpha} = \frac{T(\alpha + \epsilon) - T(\alpha)}{\epsilon} = \frac{T(\alpha)T(\epsilon) - T(\alpha)}{\epsilon}$$

and

$$T(\epsilon) = \mathbb{1} - i \cdot \frac{L_z}{\hbar} \cdot \epsilon$$

We obtain a differential equation for $T(\alpha)$:

$$\frac{dT(\alpha)}{d\alpha} = -i \cdot \frac{L_z}{\hbar} \cdot T(\alpha)$$

with a formal solution of

$$T(\alpha) = e^{-i \cdot \frac{L_z}{\hbar} \cdot \alpha}$$

This operator is the so-called exponential mapping of a circle group into the function space of periodic functions. It is a unitary operator, as L_z is Hermitic. The exponential function of the operator is defined by the Taylor expansion of the exponential

$$e^{-i \cdot \frac{L_z}{\hbar} \cdot \alpha} = \sum_n \frac{(-i\alpha)^n}{n!} \cdot \frac{L_z}{\hbar}^n$$

with

$$L_z^n = (-i \cdot \hbar)^n \cdot \frac{d^n}{d\alpha^n}$$

Comment. The application of the Taylor expansion of the operator to $f(\varphi)$ gives $f(\varphi - \alpha)$. Accordingly, the operators $e^{-i \cdot \frac{L_z}{\hbar} \cdot \alpha}$ and $T_r(\alpha)$ are formally identical, but they are not the same. Their domains of definition are different: $e^{-i \cdot \frac{L_z}{\hbar} \cdot \alpha}$ is applicable to analytic functions, whereas T_r has the entire Hilbert space as the domain of definition. The exponential mapping of an infinitesimal operator remains, however, a practical way of extending the representation also to elements not infinitesimally close to the identity element.

3.2 The Irreducible Representations of SO_3

The group of proper rotations in the three-dimensional space contains all proper rotations about any axis. Every rotation has a rotation axis, characterized by a unit vector, $\mathbf{n}$, and a rotation angle, φ. By the corkscrew rule one direction can be defined as positive. The number of parameters required to describe a rotation is 3 — two for the unit vector along the rotation axis and the rotation angle, so that this group is a three-dimensional continuous group. The "natural" representation of the group of proper rotations in E_3 is the group of orthogonal real 3×3 matrices with a determinant of 1, called SO_3. In order to solve problems involving a system with full proper rotational symmetry (non-relativistic electrons), one needs to construct irreducible representations and invariant subspaces of SO_3. The group has an average and is a compact group, indicating that we know that its irreducible representations are denumerable.

3.2.1 *Algebraic construction of irreducible representations: First attempt*

The strategy for constructing irreducible representations of SO_3 using algebraic methods is one which is common to constructing irreducible representations of continuous groups: It is based on starting from an appropriate Abelian subgroup, if one exists. In the specific case of SO_3, the Abelian subgroup at hand is the one consisting of all rotations about a fixed axis (which we set as the z-axis) by an angle φ. This subgroup is isomorphic to SO_2. This subgroup has, in the specific case of SO_3, a further specific property. As all rotations by φ about any axis belong to the same conjugacy class of SO_3, the elements of the Abelian subgroup are also representatives of the conjugacy classes of SO_3, so that the character of their irreducible representations are the desired characters for the irreducible representations of SO_3. The rationale behind the Abelian subgroup strategy is the following. The elements of the subgroup commute. Accordingly, the sought-after matrix representations of this subgroup can be simultaneously diagonalized. We can think of using the basis set that diagonalizes the representations of the Abelian subgroup for constructing the matrix representations of the remaining elements of the group. These are, for SO_3, rotations about an axis different from z. They might produce a mix of the basis states and, accordingly, require larger matrices for their representation. In this situation, the matrices of the Abelian subgroup will remain diagonal, but not necessarily be 1×1, so that we anticipate representations of the elements of SO_3 with larger matrices. At this point of reasoning, therefore, we can envisage that the unitary matrix representing the rotations about z, in the chosen basis set, will have the following general form:

$$X_n(z, \varphi) = \begin{pmatrix} e^{-i \cdot x_1 \cdot \varphi} & 0 & 0 & \cdots & 0 \\ 0 & e^{-i \cdot x_2 \cdot \varphi} & 0 & 0 & \cdots \\ \vdots & \vdots & \vdots & \vdots & \vdots \\ 0 & 0 & & \cdots & e^{-i \cdot x_n \cdot \varphi} \end{pmatrix}$$

Definition. The numbers $(x_1, \ldots, x_n)$ appearing in the exponents are the so-called Cartan weights of the desired representation X_n.

We can make some educated guesses about their value:

- They must be some integer numbers, which we refer to as m_i hereinafter. This is required by the 2π periodicity of rotations. Later, we introduce groups that do not exhibit this periodicity, so we will encounter non-integer numbers.
- If m_i is a weight, $-m_i$ is also one.

Proof. Consider that a rotation about the axis $-z$ by an angle φ is identical with a rotation about z by an angle $-\varphi$, i.e.

$$X_n(-z, \varphi) = \begin{pmatrix} e^{+i \cdot m_1 \cdot \varphi} & 0 & 0 & \cdots & 0 \\ 0 & e^{-i \cdot m_2 \cdot \varphi} & 0 & 0 & \cdots \\ \vdots & \vdots & \vdots & \vdots & \vdots \\ 0 & 0 & & \cdots & e^{+i \cdot m_n \cdot \varphi} \end{pmatrix}$$

However, the rotation about $-z$ by φ must have the same character as the rotation about z by the same angle, i.e.

$$\sum_i e^{+i \cdot m_i \cdot \varphi} = \sum_i e^{-i \cdot m_i \cdot \varphi} = \overline{\sum_i e^{+i \cdot m_i \cdot \varphi}}$$

This is an identity, i.e. it must hold for any φ. Accordingly, it has either the solution $m_i \equiv 0$ (which we discard, as it rules out irreducible representations with dimensionality larger than one) or the weights m_i and $-m_i$ appear in pairs.

- We are seeking all finite-dimensional representations, so that we can set $l \geq 0$ as the largest weight. The smallest one will then be $-l$.

Their multiplicity and the multiplicity of the Cartan weights between $-l$ and l cannot be determined with simple arguments. However, an important, advanced theorem of Lie groups (originally discussed by E. Cartan and H. Weyl, "Theorem of the highest weight", see Theorem 12.6 in Brian C. Hall, *Lie Groups, Lie Algebras, and Representations: An Elementary Introduction* (2nd edn.), Graduate Texts in Mathematics, Vol. 222, Springer, 2015, https://link.springer.com/book/10.1007/978-3 -319-13467-3) tells us that the largest Cartan weight identifies uniquely the finite-dimensional irreducible representations of a Lie group. We call from here onward the representation that we are interested in with the symbol D_l (which is also often found in the literature).

The algebraic method, as we have used so far, cannot completely determine the weights. We need to either expand the algebraic technique (which we will do later) or else find a different method. The most appropriate one is the analytic one.

3.2.2 *Analytic approach to finding the Cartan weights*

We define an infinite-dimensional representation of SO_3 in a function space by means of the relation

$$O_R f(\mathbf{e}) \doteq f(R^{-1}\mathbf{e})$$

with R being a 3×3 rotation matrix and $\mathbf{e}$ a vector on the unit sphere. The operator O_R is a homomorphism, as

$$O_{R_2} O_{R_1} f(\mathbf{e}) = O_{R_2 R_1} f(\mathbf{e})$$

Therefore, the operator O_R defines a unitary representation in the function space.

To construct the irreducible representations and the invariant subspaces, we are interested in rotations with a small angle ε around an arbitrary rotation axis $\mathbf{n}$. Such rotations transform $\mathbf{e}$ into $\mathbf{e} + \mathbf{n} \times \mathbf{e} \cdot \varepsilon$ and the operator $O_{\mathbf{n},\varepsilon}$ transforms $f(\mathbf{e})$ into

$$f(\mathbf{e} - \mathbf{n} \times \mathbf{e} \cdot \varepsilon) = f(\mathbf{e}) - i \cdot n_x \cdot \varepsilon \cdot i \left[z\frac{\partial f}{\partial y} - y\frac{\partial f}{\partial z} \right]$$

$$- i \cdot n_y \cdot \varepsilon \cdot i \left[x\frac{\partial f}{\partial z} - z\frac{\partial f}{\partial y} \right] - i \cdot n_z \cdot \varepsilon \cdot i \left[y\frac{\partial f}{\partial x} - x\frac{\partial f}{\partial y} \right]$$

This establishes the linearized rotation operator as

$$O_{\mathbf{n},\varepsilon} = \mathbb{1} - i \cdot \mathbf{n} \cdot \frac{\mathbf{L}}{\hbar} \cdot \varepsilon$$

with $\mathbf{L}$ being the quantum mechanical operator for the orbital angular momentum. Note that a rotation about a fixed rotation axis does not change the length of a vector, so that we can consider the function $f(\mathbf{e})$ as being defined on the unit sphere.

The problem of finding invariant (hopefully irreducible) subspaces has now been reduced to finding the set of functions that are transformed into themselves by the L_x, L_y, and L_z operators. This task is still a difficult one as it involves solving coupled partial differential equations. However, a trick simplifies the task: Instead of working with the infinitesimal operators, one searches for one operator that commutes with all of them — the so-called **Casimir operator**. The Casimir operator has eigenspaces that are also common to all infinitesimal transformations and should provide the desired invariant subspaces. According to a theorem by Racah (G. Racah, "Group theory and spectroscopy", in *Springer Tracts in Modern Physics*, Editor: G. Höhler, Vol. 37, Springer, Berlin, Heidelberg, 1965, https://doi.org/10.1007/BFb0045770), the number of independent Casimir operators of a Lie group is given by the number of infinitesimal operators that commute. In the specific case of SO_3, this number is just one. The one Casimir operator of SO_3 is

$$L_x^2 + L_y^2 + L_z^2 \doteq \mathbf{L}^2$$

Within a function space,

$$\frac{\mathbf{L}^2}{\hbar^2} = -\Lambda \doteq -\frac{1}{\sin\vartheta}\frac{\partial}{\partial\vartheta}\left(\sin\vartheta\frac{\partial}{\partial\vartheta}\right) - \frac{1}{\sin^2\vartheta}\frac{\partial^2}{\partial\varphi^2}$$

and the problem of finding invariant subspaces carrying the representations of the group SO_3 reduces to solving the eigenvalue problem

$$-\Lambda f(\vartheta,\varphi) = \lambda f(\vartheta,\varphi) \quad f(\vartheta + 2\pi, \varphi + 2\pi) = f(\vartheta,\varphi)$$

This is a well-known differential equation of mathematical physics, solved by

$$\lambda = l(l+1)$$

$(l = 0, 1, 2, 3, \ldots)$ and the eigenspaces

$$E_l = \{Y_l^l(\vartheta,\varphi), Y_l^{l-1}(\vartheta,\varphi), \ldots, Y_l^{-l}(\vartheta,\varphi)\}$$

of the so-called spherical harmonics.[2] On these invariant subspaces, we can construct the matrix representations of $I_x \doteq \frac{L_x}{\hbar}, I_y \doteq \frac{L_y}{\hbar}, I_z \doteq \frac{L_z}{\hbar}$. For instance,

$$
I_l^z = \begin{pmatrix} l & 0 & & \cdots & \\ 0 & l-1 & 0 & \cdots & \\ \vdots & \vdots & \vdots & & \\ 0 & 0 & & \cdots & -l \end{pmatrix}
$$

i.e.

$$
\langle Y_l^m, I_z Y_l^{m'} \rangle = m \cdot \delta_{mm'}
$$

We summarize the matrix representation of I_z, I_x, and I_y according to the convention of Condon and Shortley (only those matrix elements which are different from 0 are given):

$$
(Y_{l,m}, I_z Y_{l,m}) = m
$$

$$
(Y_{l,m\pm1}, I_x Y_{l,m}) = \frac{1}{2}\sqrt{(l \mp m)(l \pm m + 1)}
$$

$$
(Y_{l,m\pm1}, I_y Y_{l,m}) = \mp\frac{i}{2}\sqrt{(l \mp m)(l \pm m + 1)}
$$

The exponential mapping: Once these matrix elements are known, we can construct the matrices that represent rotations by a small rotation angle:

$$
D_l(\mathbf{n}, \varphi) = \mathbb{1} - i \cdot \left(\mathbf{n}, \frac{\mathbf{L}}{\hbar}\right) \cdot \varphi
$$

In this relation, $\mathbf{L}$ has to be considered as a vector of matrices, although we use the same notation as the differential vector operator. This relation can then be used to find the matrices representing a rotation by a finite angle φ about any fixed axis $\mathbf{n}$ by means of the exponential map

$$
D_l(\mathbf{n}, \varphi) = e^{-i \cdot \left(\mathbf{n}, \frac{\mathbf{L}}{\hbar}\right) \cdot \varphi}
$$

---

[2]We use the spherical harmonics following the convention of E. U. Condon and G. Shortley, *The Theory of Atomic Spectra*, Cambridge University Press, Cambridge, England, 1951. See e.g. D. A. Varshalovich, A. N. Moskalev, and V. K. Khersonskii, *Quantum Theory of Angular Momentum*, Chapter 5, World Scientific, 1988, https://doi.org/10.1142/0270.

The exponential function of a matrix is defined by the Taylor expansion:

$$e^A = \sum_n \frac{A^n}{n!}$$

The computation rules with the exponential function of matrices are:

1. $e^{A^\dagger} = (e^A)^\dagger$ ($\dagger$ is the transpose complex conjugate).
2. Baxter–Campbell–Haussdorf (BCH) formula:

$$e^A \cdot e^B = \exp(C)$$

$$C = A + B + \frac{1}{2}[A, B] + \frac{1}{12}\left(\big[[A, B], B\big] + \big[[B, A], A\big]\right) + \cdots$$

where $[A, B] \doteq AB - BA$ is the "commutator" of the two matrices A, B. This is an important technical formula with broad applications in mathematics. A historical perspective of the development of this formula can be found in, for example, R. Achilles and A. Bonfiglioli, "The early proofs of the theorem of Campbell, Baker, Hausdorff, and Dynkin", *Arch. Hist. Exact Sci.* **66**, 295–358 (May 2012), https://link.springer.com/art icle/10.1007/s00407-012-0095-8. The proof of the BCH formula shows, as the main result, that **all** summands appearing in C can be expressed as iterated commutators of A and B.

3. $\det e^A = e^{\mathrm{tr}(A)}$.

For instance,

$$D_l(z, \varphi) = \begin{pmatrix} e^{-il\varphi} & 0 & \cdots & \\ 0 & e^{-i(l-1)\varphi} & 0 & \cdots \\ \vdots & \vdots & \vdots & \\ 0 & 0 & \cdots & e^{il\varphi} \end{pmatrix}$$

The representation D_l, carried by the space E_l, is a $2l + 1$-dimensional representation of SO_3. Along the diagonal of the matrix for I_z, we can find the Cartan weights that we are interested in.

Proposition. *The invariant subspaces E_l are irreducible. This means that the representations D_l of SO_3, $l = 0, 1, 2, \ldots$ constructed in the previous paragraph are irreducible representations of SO_3.*

Proof. Assume that there is an invariant subspace of E_l. Within this subspace, select the highest value of m, which by assumption is smaller than l. An operator $I_x + iI_y$ can be defined, which transforms $Y_l^{m<l}$ to Y_l^{m+1}, so that $I_x + iI_y$ brings us outside the subspace we assumed to be invariant. Accordingly, $m < l$ is not possible. Choose now the lowest $m > -l$. The operator $I_x - iI_y$ brings us to $m - 1$, and also, $m > -l$ is not possible. Accordingly, E_l is irreducible.

Proposition. *The irreducible representations D_l represent all finite-dimensional irreducible representations of SO_3.*

Proof. $\{E_l\}$, $l = 0, 1, 2, \ldots$ constitutes a complete set of basis functions on the sphere.

Comment. To provide these central theorems of irreducible representations of SO_3, we have only used the matrices of the infinitesimal transformations, not the actual rotation matrices. It is often the case that most problems and computations only require the algebra of the infinitesimal generators.

Proposition. *The irreducible representations D_l appear exactly once in the decomposition of the infinite-dimensional representation O_R. In mathematical terms,*

$$O_R = \oplus_l D_l(R)$$

Proof. The set $\{E_l\}$, $l = 0, 1, 2, \ldots$ constitutes a complete set of basis functions on the unit sphere. Accordingly, the Hilbert space $L_2(S^2)$ is the direct sum of E_l. This proves the claim.

Using a matrix representation, we can express this result as the matrix of the operator O_R, constructed using spherical harmonics as basis functions, which is block diagonalized:

$$O_R = \begin{pmatrix} D_0(R) & 0 & & \cdots \\ 0 & D_1(R) & 0 & \cdots \\ 0 & 0 & D_2(R) & 0 & \cdots \\ \vdots & \vdots & \vdots & \vdots \end{pmatrix}$$

Character table of SO_3: The character table of the irreducible representation D_l can be constructed directly from the trace of the matrix of a representative element, for instance the trace of the matrix $D_l(z, \varphi)$:

$$\sum_{m=-l}^{l} e^{-im\varphi} = \frac{\sin\left[(2l+1)\frac{\varphi}{2}\right]}{\sin\frac{\varphi}{2}}$$

The character table is as follows:

SO_3	φ	Basis functions
D_0	1	$f(r)Y_0^0$
D_1	$\frac{\sin(3\frac{\varphi}{2})}{\sin\frac{\varphi}{2}}$	$g(r) \cdot \{Y_1^1, Y_1^0, Y_1^{-1}\}$
D_2	$\frac{\sin(5\frac{\varphi}{2})}{\sin\frac{\varphi}{2}}$	$h(r) \cdot \{Y_2^2, Y_2^1, Y_2^0, Y_2^{-1}, Y_2^{-2}\}$
$\vdots$	$\vdots$	$\vdots$

Under the heading "Basis functions", one finds a collection of symmetry-adapted spherical harmonics. The spherical harmonics can be multiplied by any radial-dependent function, so that the number of symmetry-adapted basis functions for each irreducible representation is infinite. It can happen that the radial functions also constitute a countable set with the index corresponding to, say, the main quantum number in atomic physics.

Remark. One can prove the irreducibility of the D_l representation and their completeness using the theorems on characters. In fact,

$$\mathscr{M}_{SO_3} f(\varphi, \vartheta, \phi) = \frac{1}{2\pi^2} \int_0^\pi \int_0^{2\pi} \sin\vartheta d\vartheta d\phi \int_0^\pi \sin^2\frac{\varphi}{2} d\varphi \cdot \frac{\sin^2\left[(2l+1)\frac{\varphi}{2}\right]}{\sin^2\frac{\varphi}{2}} = 1$$

We can also prove that D_l are all irreducible representations. For example, let $\Delta \neq D_l$ be an irreducible representation of SO_3. In virtue of the orthogonality of characters, we must have

$$\mathscr{M}_{SO_3} \Delta(\varphi) \frac{\sin\left[(2l+1)\frac{\varphi}{2}\right]}{\sin\frac{\varphi}{2}} = 0$$

The integral on the left-hand side amounts to

$$\frac{4}{\pi} \int_0^{\pi/2} d\tau \Delta(\tau) \cdot \sin\tau \cdot \sin\left((2l+1)\tau\right)$$

But the set of functions $\sin\tau, \sin 3\tau, \sin 5\tau, \ldots$ forms a complete set in the interval $[0, \frac{\pi}{2}]$,[3] so that the integral can only vanish if $\Delta(\tau) \equiv 0$, which contradicts the assumption of irreducibility of the hypothetical representation Δ.

Application of the exponential mapping: The Rodrigues formula (Rodrigues, 1816)

A famous three-dimensional representation of SO_3 is the Rodrigues rotation formula, named after the 19th century French banker, mathematician, and social reformer, Benjamin Olinde Rodrigues. We aim to obtain this useful formula by means of the exponential mapping technique. In the Rodrigues representation, the matrix elements of the SO_3-rotation matrices are real, so that we cannot use the spherical harmonics $\{Y_1^1, Y_1^0, Y_1^{-1}\}$ to compute the exponential mapping, as they would lead to complex matrix elements. However, we are free of taking suitable linear combinations of them as basis states. We chose, as new basis states, the functions

$$\sqrt{\frac{1}{2}} \cdot (-Y_1^1 + Y_1^{-1}); \quad \frac{i}{\sqrt{2}} \cdot (Y_1^1 + Y_1^{-1}); \quad Y_1^0$$

These new functions are the functions x, y, and z, respectively, orthonormalized on the unit sphere. We now use the matrix elements in the convention of Condon and Shortley to recompute the matrices for I_x, I_y, I_z in these new basis states. For example, we find

$$\left(\sqrt{\frac{1}{2}} \cdot (-Y_1^1 + Y_1^{-1}), I_z \frac{i}{\sqrt{2}} \cdot (Y_1^1 + Y_1^{-1}) \right) = -i$$

All other matrix elements can be computed in a similar way to produce

$$I_z = \begin{pmatrix} 0 & -i & 0 \\ i & 0 & 0 \\ 0 & 0 & 0 \end{pmatrix} \quad I_x = \begin{pmatrix} 0 & 0 & 0 \\ 0 & 0 & -i \\ 0 & i & 0 \end{pmatrix} \quad I_y = \begin{pmatrix} 0 & 0 & i \\ 0 & 0 & 0 \\ -i & 0 & 0 \end{pmatrix}$$

---

[3]To appreciate this, consider any function $f_1(x)$ defined in the interval $1 \equiv [0, \frac{\pi}{2}]$. Think of continuing this function to the interval $-1 \equiv [0, -\frac{\pi}{2}]$ so that for the new branch, we have $f_{-1}(x) = -f_1(x)$. Return to the original function and continue it to the interval $2 \equiv [\frac{\pi}{2}, \pi]$, so that for the new branch, we have $f_2(x) = f_1(\pi - x)$. Finally, create a new branch in the interval $-2 \equiv [-\frac{\pi}{2}, \pi]$, so that $f_{-2}(x) = f_2(-x - \pi)$. The Fourier theorem tells us that the function consisting of the four branches $\{f_1, f_{-1}, f_2, f_{-2}\}$ can be approximated by the sum of $\sin\tau, \sin 3\tau, \sin 5\tau, \ldots$, so that this set of basis functions is certainly complete in the original interval $[0, \frac{\pi}{2}]$.

The orthogonal Rodrigues matrix is computed by the exponential map:

$$-i(\mathbf{I},\mathbf{n})\varphi = n_x \cdot \varphi \cdot \begin{pmatrix} 0 & 0 & 0 \\ 0 & 0 & -1 \\ 0 & 1 & 0 \end{pmatrix} + n_y \cdot \varphi \cdot \begin{pmatrix} 0 & 0 & 1 \\ 0 & 0 & 0 \\ -1 & 0 & 0 \end{pmatrix}$$

$$+ n_z \cdot \varphi \cdot \begin{pmatrix} 0 & -1 & 0 \\ 1 & 0 & 0 \\ 0 & 0 & 0 \end{pmatrix}$$

and

$$D_1(\mathbf{n},\varphi) = e^{\begin{pmatrix} 0 & \bar{n}_z & n_y \\ n_z & 0 & \bar{n}_x \\ \bar{n}_y & n_x & 0 \end{pmatrix}\cdot\varphi}$$

We compute the exponential of the matrix by means of the Taylor expansion. This requires computing the powers of the matrix

$$U \doteq \begin{pmatrix} 0 & \bar{n}_z & n_y \\ n_z & 0 & \bar{n}_x \\ \bar{n}_y & n_x & 0 \end{pmatrix}$$

The task of computing the powers is accomplished using the *Cayley–Hamilton theorem*. This theorem states that a square matrix satisfies its own characteristic equation. The characteristic equation of the matrix U reads

$$\det(U - \lambda\mathbb{1}) = \det\begin{pmatrix} -\lambda & \bar{n}_z & n_y \\ n_z & -\lambda & \bar{n}_x \\ \bar{n}_y & n_x & -\lambda \end{pmatrix} = \lambda^3 + \underbrace{(n_1^2 + n_2^2 + n_3^2)}_{=1}\lambda = 0,$$

Applying the Cayley–Hamilton theorem with $\lambda = U$ produces

$$U^3 + U = 0 \quad \Rightarrow \quad U^3 = -U.$$

We find, accordingly,

$$U^0 = E; \quad U^1 = U; \quad U^2; \quad U^3 = -U;$$
$$U^4 = -U^2; \quad U^5 = -U^3 = U; \quad U^6 = \dots.$$

We can now sum the various terms of the Taylor series:

$$D_1(\mathbf{n}, \varphi) = \sum_{n=0}^{\infty} \frac{\varphi^n}{n!} U^n$$

$$= \mathbb{1} + U \left(\varphi - \frac{\varphi^3}{3!} + \frac{\varphi^5}{5!} \pm \cdots \right)$$

$$+ U^2 \left(\frac{\varphi^2}{2!} - \frac{\varphi^4}{4!} \pm \cdots \right)$$

$$= \mathbb{1} + U \sin(\varphi) + U^2 (1 - \cos(\varphi))$$

This expression for the rotation matrix is called the Euler–Rodrigues rotation formula (Leonhard Euler (1707–1783)). Explicitly inserting the matrices U and U^2 produces the Rodrigues rotation matrix:

$$D_1(\mathbf{n}, \varphi) = \begin{pmatrix} n_x^2 + (1 - n_x^2) \cos(\varphi) & n_x n_y (1 - \cos(\varphi)) - n_z \sin(\varphi) \\ n_x n_y (1 - \cos(\varphi)) + n_z \sin(\varphi) & n_y^2 + (1 - n_y^2) \cos(\varphi) \\ n_x n_z (1 - \cos(\varphi)) - n_y \sin(\varphi) & n_y n_z (1 - \cos(\varphi)) + n_x \sin(\varphi) \end{pmatrix}$$

$$\begin{pmatrix} n_x n_z (1 - \cos(\varphi)) + n_y \sin(\varphi) \\ n_y n_z (1 - \cos(\varphi)) - n_x \sin(\varphi) \\ n_z^2 + (1 - n_z^2) \cos(\varphi) \end{pmatrix}.$$

Comments.

1. This matrix has the property of being an orthogonal one with a determinant of 1 and is specified by three parameters. It is therefore a general element of SO_3. This result shows explicitly that the exponential map is able to generate the entire group, starting from the appropriate infinitesimal generators.

2. There is a familiar way of using the matrix U. Consider that

$$U\mathbf{r} = \mathbf{n} \times \mathbf{r}$$

so that the Euler–Rodrigues rotation formula is also written as

$$D_1(\mathbf{n}, \alpha)\mathbf{r} = \mathbf{r} + \mathbf{n} \times \mathbf{r} \cdot \sin(\alpha) + \mathbf{n} \times (\mathbf{n} \times \mathbf{r}) (1 - \cos(\alpha))$$

3.3 Representations of Continuous Groups Through Infinitesimal Transformations: Algebraic Method

The analytic method we have used to find the irreducible representations of the continuous groups SO_2 and SO_3 is based on the following. First, we define an infinite-dimensional representation of these groups into a function space. Then, we find the operator that describes rotations by a small angle about any direction (z in the case of SO_2) (infinitesimal operator). Finally, we find the eigenstates of these infinitesimal operators (which is equivalent to solving a differential equation). These eigenspaces are then used as invariant subspaces over which matrix representations of the infinitesimal transformations could be constructed. The eigenspaces turned out to be irreducible, so that the matrices provided by this method were completely reduced and no longer block-diagonalizable. Using the exponential map, the matrix representations of the infinitesimal transformations are translated to rotations by a finite angle.

There is an alternative technique for constructing the matrix representations of SO_3: This method aims at constructing the matrix representations of the infinitesimal transformations by **algebraic methods** (the exponential map is thereafter used to translate the matrix representation of the infinitesimal transformations to any element of the group). By introducing this method, we focus on SO_3, but we have in mind more general continuous matrix groups.

3.3.1   *Lie groups and infinitesimal transformations*

Lie groups are defined by the analyticity of the functions $f_1, \ldots, f_r$ appearing in the composition law

$$c_1 = f_1(a_1, \ldots, a_r, b_1, \ldots, b_r)$$

$$\vdots$$

$$c_r = f_r(a_1, \ldots, a_r, b_1, \ldots, b_r)$$

The composition laws are not immediately apparent even for simple groups such as SO_3 and are not often used in practical application. We have used them, for example, to compute the Haar measure. One practical use of the composition laws in the parameter space is developing the infinitesimal generator technique for general Lie groups.

Composition laws to first order and the existence of infinitesimal generators: We assume that the unit element of the group is specified by the parameters $(e_1, e_2, \ldots, e_r)$, which we set as the origin of the coordinate system in parameter space and write the composition law for small a_m and b_m as

$$c_1 = f_1(0, 0, \ldots, 0) + \sum \left[\frac{\partial f_1(\mathbf{a}, \mathbf{e})}{\partial a_m} \Big|_{\mathbf{a}=\mathbf{0}} a_m + \frac{\partial f_1(\mathbf{0}, \mathbf{b})}{\partial b_m} \Big|_{\mathbf{b}=\mathbf{0}} b_m \right]$$

$$\vdots$$

$$c_r = f_r(0, 0, \ldots, 0) + \sum \left[\frac{\partial f_r(\mathbf{a}, \mathbf{0})}{\partial a_m} \Big|_{\mathbf{a}=\mathbf{0}} a_m + \frac{\partial f_r(\mathbf{0}, \mathbf{b})}{\partial b_m} \Big|_{\mathbf{b}=\mathbf{0}} b_m \right]$$

Setting $a_m = b_m = 0$ must necessarily lead to $c_m = 0$, which constrains $f_m(0, \ldots, 0)$ to be exactly 0. Setting $a_m = 0$ must necessarily lead to $c_m = b_m$, which specifies $\frac{\partial f_m(\mathbf{0}, \mathbf{b})}{\partial b_i} |_{\mathbf{b}=\mathbf{0}}$ to be exactly δ_{im}. Setting $b_m = 0$ must lead necessarily to $c_m = a_m$, and this specifies $\frac{\partial f_m(\mathbf{a}, \mathbf{0})}{\partial a_i} |_{\mathbf{a}=\mathbf{0}}$ to be exactly δ_{im}. The composition law of the Lie groups in the vicinity of the parameter $\mathbf{0}$ reads, for small as and bs,

$$c_m = a_m + b_m$$

$m = 1, \ldots, r$. These linearized composition laws are valid for any Lie group. We show that they imply the existence of infinitesimal generators for a matrix Lie group $G = \{F(\mathbf{a})\}$. Setting $b_m \equiv 0$ in the linearized composition laws constrains any element $\mathbf{a}$ in the vicinity of the identity to be identified by small parameters, so its matrix writes as

$$F(\mathbf{a}) = \mathbb{1} + \sum_k \frac{\partial F(\mathbf{a})}{\partial a_k} \Big|_{\mathbf{a}=\mathbf{e}} \cdot a_k \doteq \mathbb{1} + i \sum_k A_k \cdot a_k$$

The

$$A_k \doteq -i \cdot \frac{\partial F(\mathbf{a})}{\partial a_k} \Big|_{\mathbf{a}=\mathbf{e}}$$

are **infinitesimal transformation matrices** (or infinitesimal generators) that are used to represent the unitary matrices $F(\mathbf{a})$ to first order.

Example. SO_2: The matrix group SO_2 is written as

$$F(\varphi) = \begin{pmatrix} \cos\varphi & -\sin\varphi \\ \sin\varphi & \cos\varphi \end{pmatrix}$$

There is only one parameter, $a_1 \equiv \varphi$, and one infinitesimal transformation, obtained by

$$-i \cdot \frac{d}{d\varphi} \begin{pmatrix} \cos\varphi & -\sin\varphi \\ \sin\varphi & \cos\varphi \end{pmatrix} @_{\varphi=0} = \begin{pmatrix} 0 & i \\ -i & 0 \end{pmatrix}$$

Example. SO_3: The parameter space of SO_3 is chosen to be a sphere with a radius of π. Each rotation is a point of this sphere with coordinates of

$$(a_1 = n_x\varphi, a_2 = n_y\varphi, a_3 = n_z\varphi)$$

$(n_x^2 + n_y^2 + n_z^2 = 1)$. All rotation matrices (take, for example, the Rodrigues formula) can be written as a function of the parameters (a_1, a_2, a_3). In order to compute A_1, we set $a_2 = a_3 = 0$. This results in a rotation by an angle a_1 about the x-axis, leading to the rotation matrix

$$\begin{pmatrix} 1 & 0 & 0 \\ 0 & \cos a_1 & -\sin a_1 \\ 0 & \sin a_1 & \cos a_1 \end{pmatrix}$$

The derivative of this matrix with respect to a_1 for $a_1 = 0$ produces the required infinitesimal transformation:

$$A_1 = -i \cdot \begin{pmatrix} 0 & 0 & 0 \\ 0 & 0 & \bar{1} \\ 0 & 1 & 0 \end{pmatrix}$$

Similarly, we obtain

$$A_2 = -i \cdot \begin{pmatrix} 0 & 0 & 1 \\ 0 & 0 & 0 \\ \bar{1} & 0 & 0 \end{pmatrix}$$

and

$$A_3 = -i \cdot \begin{pmatrix} 0 & \bar{1} & 0 \\ 1 & 0 & 0 \\ 0 & 0 & 0 \end{pmatrix}$$

For small values of $\mathbf{a}$, we obtain

$$F(\mathbf{a}) = \mathbb{1} + \begin{pmatrix} 0 & \bar{a}_3 & a_2 \\ a_3 & 0 & \bar{a}_1 \\ \bar{a}_2 & a_1 & 0 \end{pmatrix}$$

Using the parametrization $\mathbf{a} = \mathbf{n} \cdot \varphi$, containing explicitly the small rotation angle φ, we obtain, for small φ,

$$F(\mathbf{n}, \varphi) = \mathbb{1} + \begin{pmatrix} 0 & \bar{n}_z & n_y \\ n_z & 0 & \bar{n}_x \\ \bar{n}_y & n_x & 0 \end{pmatrix} \cdot \varphi$$

Composition laws to second order and the structure constants:
We continue the Taylor expansion of the functions $f_m(a_1, \ldots, a_r, b_1, \ldots, b_r)$
to include the second-order terms:

$$c_m = a_m + b_m + \sum_k \frac{\partial^2 f_m(\mathbf{a}, \mathbf{b} = 0)}{\partial a_k \partial a_k}\bigg|_{a_k=0} a_k^2 + \sum_k \frac{\partial^2 f_m(0, \mathbf{b})}{\partial b_k \partial b_k}\bigg|_{b_k=0} b_k^2$$

$$+ \sum_{i,k} \frac{\partial^2 f_m(\mathbf{a}, \mathbf{b})}{\partial a_i \partial b_k}\bigg|_{a_i=b_k=0} a_i \cdot b_k + \cdots$$

Setting $\mathbf{b} = 0$, one must have $c_m = a_m$, so that

$$\frac{\partial^2 f_m(\mathbf{a}, \mathbf{b} = 0)}{\partial a_k \partial a_k}\bigg|_{a_k=0} = 0$$

Setting $\mathbf{a} = 0$, one must have $c_m = b_m$, so that

$$\frac{\partial^2 f_m(0, \mathbf{b})}{\partial b_k \partial b_k}\bigg|_{b_k=0} = 0$$

and the Taylor expansion, including the second order, is written as

$$c_m = a_m + b_m + \sum_{i,k} \underbrace{\frac{\partial^2 f_m(\mathbf{a}, \mathbf{b})}{\partial a_i \partial b_k}\bigg|_{a_i=b_k=0}}_{\doteq \alpha_{ik}^m} a_i \cdot b_k + \cdots$$

where α_{ik}^m are numerical coefficients that are characteristic of the group but
depend on the choice of the parameters. The sum over i and k extends from
1 to r. For a commutative group, they are identical to zero and therefore
represent a measure of the deviation from commutativity.

Definition. The numbers

$$C_{ik}^m \doteq a_{ik}^m - a_{ki}^m$$

$(C_{ki}^m = -C_{ik}^m)$ are called the **structure constants** of the group for the
given parameters $a_1, \ldots, a_r$.

Theorem (structure constants theorem). *The infinitesimal transformations of a matrix group G fulfill the relations*

$$A_p A_q - A_q A_p \doteq [A_p, A_q] = i \cdot \sum_{m=1}^r C_{pq}^m A_m$$

The C_{pq} in this relation are exactly the structure constants of the group!

Proof. See V. Smirnov, A Course of Higher Mathematics, Vol. 3, Parte I, Pergamon Press, London, 1964, pp. 291–298, Library of Congress Catalog Card Number 63-10134.

Comments.

1. For Hermitic infinitesimal generators, this relation means that the structure constants are real numbers. This follows from

$$[A_p, A_q]^\dagger = -i \cdot \sum_{m=1}^{r} \overline{(C_{pq}^m)} A_m^\dagger = -i \cdot \sum_{m=1}^{r} \overline{(C_{pq}^m)} A_m$$

$$= -[A_p, A_q] = -i \sum_{m=1}^{r} C_{pq}^m A_m$$

from which $\overline{C_{pq}^m} = C_{pq}^m$.

2. This relation shows that the operation of $[.,.]$ (the so-called commutator) leads to an infinitesimal transformation, i.e. the set of infinitesimal transformations of a group is not only closed with respect to linear superposition but also with respect to a second operation — the commutator. Technically, the set is said to constitute an algebra.

Definition. The set of infinitesimal operators constitutes a vector space and is closed with respect to the commutator operation. It therefore is said to constitute an algebra — the Lie algebra $\mathscr{L}$ of the group G.

3.3.2 *Infinitesimal transformations and group representations*

Let us suppose that we have found the infinitesimal transformations of a continuous matrix Lie group. We know, through the exponential map, that we use them to construct the matrices of the group itself. However, can we use them to find the representations of the group other than the "natural one"?

The following theorems (due to S. Lie) are at the center of the method of using infinitesimal transformations for finding representations of Lie groups.

Theorem (infinitesimal transformations for representations). *Let $F'(\mathbf{a}) \in G'$ be a group of continuous unitary analytic matrices representing a continuous matrix group G with elements $F(\mathbf{a})$, depending on the r parameters $a_1, \ldots, a_r$. Then, there exists a set of Hermitic matrices A'_k*

with the property that, for small a_k,

$$F'(\mathbf{a}) = \mathbb{1} + i \cdot A_k' \cdot a_k$$

or, equivalently,

$$A_k' = -i \cdot \frac{F'(\mathbf{a})}{da_k}\bigg|_{a_k=0}$$

Proof. The existence of infinitesimal transformations for the matrix group G was itself obtained from the law of composition in the vicinity of the unit element. For the group G' of the matrices $F'(\mathbf{a})$, one has the same group operations as for the group itself and therefore the same underlying law of composition. Accordingly, the matrices $F'(\mathbf{a})$ have also infinitesimal transformations.

Comments.

1. This theorem determines the existence of infinitesimal transformations for a Lie group G and for the group G' of matrices that constitute a representation of the elements of G.
2. For example, the existence of infinitesimal transformations for the irreducible representations D_l of the group SO_3 was shown by the explicit construction of the corresponding matrices over spherical harmonics.
3. The sum of infinitesimal transformations and their multiplication with a scalar are also infinitesimal transformations. They are used to express the linearized matrices $F(\mathbf{a})$ and $F'(\mathbf{a})$ when a different set of parameters is used. Accordingly, the set of infinitesimal transformations constructs a vector space of matrices.
4. Hereinafter, the matrices A_k' are the quantities we are seeking to compute, as their knowledge can be used to compute the matrix representations $F'(\mathbf{a})$.

Theorem (structure constant theorem for representations). *Let $F'(\mathbf{a}) \in G'$ be a group of continuous unitary analytic matrices representing a continuous matrix group G with elements $F(\mathbf{a})$, depending on the r parameters $a_1, \ldots, a_r$. Then, the infinitesimal transformations A_k', obtained by differentiating the matrices $F'(\mathbf{a})$, obey the same commutation relations as the matrices A_k.*

Proof. For the matrices $F'(\mathbf{a})$, one has the same group multiplication as the matrices $F(\mathbf{a})$ and, accordingly, the same law of composition, e.g. the same structure constants.

Comments.

1. Technically, the set of matrices resulting from the linear combinations of the A'_ks also constitute an algebra $\mathscr{L}'$.
2. This theorem is a central one, as it relates specific group parameters (the structure constants appearing in the composition law) with the commutation relations for the desired infinitesimal transformations A'_k. The commutation relations involving the structure constants represent a strong constraint that the infinitesimal transformations of the desired matrix representations $F'(\mathbf{a})$ must obey. This theorem provides therefore the basis for the following definition.

We recall that the matrices $F'(\mathbf{a})$ are a representation of the matrices $F(\mathbf{a})$, i.e

$$F'_{\mathbf{a}}F'_{\mathbf{b}} = (F_{\mathbf{a}}F_{\mathbf{b}})'$$

Definition. Consider a matrix group G with r-parameters and a set of infinitesimal generators $\{A_k\}$. The set of r-matrices $\{A'_k\}$ for which

$$(\alpha \cdot A_p + \beta \cdot A_q)' = \alpha \cdot A'_p + \beta \cdot A'_q; \quad [A_p, A_q]' = \left[A'_p, A'_q\right]$$

is called a representation of the Lie algebra of the group G (or, often, a representation of the "commutation relations").

Comment. The previous equation describe the "translation" of the homomorphism rule from the group elements into the set of infinitesimal generators.

Application of this definition: SO_3: We use the matrices

$$I_z = \begin{pmatrix} 0 & -i & 0 \\ i & 0 & 0 \\ 0 & 0 & 0 \end{pmatrix} I_x = \begin{pmatrix} 0 & 0 & 0 \\ 0 & 0 & -i \\ 0 & i & 0 \end{pmatrix} I_y = \begin{pmatrix} 0 & 0 & i \\ 0 & 0 & 0 \\ -i & 0 & 0 \end{pmatrix}$$

to compute the commutation relations and the structure constants:

$$I_x I_y - I_y I_x = i \cdot 1 \cdot I_z \quad I_y I_z - I_z I_y = i \cdot I_x = \quad I_z I_x - I_x I_z = i \cdot I_y$$

We now use these commutation relations for the second attempt of obtaining, purely on the base of algebraic methods, the Cartan weights of the irreducible representations D_l. For a suitable set of basis vectors, the desired infinitesimal transformation I_z^l is diagonal:

$$\begin{pmatrix} l & 0 & \cdots & \\ 0 & x_{l-1} & \cdots & \\ \vdots & \vdots & & \\ 0 & & \cdots & -l \end{pmatrix}$$

Along the diagonal, one expects the integer numbers x_m, with the index m running from 1 to n and counting the (yet unknown) dimensionality of the representation that we seeking. The integers x_m are between the highest and lowest Cartan weights, l and $-l$, respectively. Both integers x_m and $-x_m$ are expected to appear along the diagonal. In order to determine the actual composition of the diagonal of the $n \times n$ matrix, potentially representing I_z, we need to use further properties specific to the group SO_3: The commutation relations (note that, for the technical purpose of using these relations, one defines auxiliary operators $I_\pm = I_x \pm I_y$, which fulfill the following relations:

$$I_+I_- - I_-I_+ = 2 \cdot I_z; \quad I_zI_+ - I_+I_z = I_+; \quad I_zI_- - I_-I_z = -I_-$$

The commutation relations are used to prove the following proposition.

Proposition. *Consider a weight x_m and the eigenvector u_{x_m}. Then, $I_+u_{x_m}$ is an eigenstate of I_z to the eigenvalue $x_m + 1$ and $I_-u_{x_m}$ is an eigenstate of I_z to the eigenvalue $x_m - 1$.*

Proof.

$$I_z(I_\pm u_{x_m}) = I_\pm I_z u_{x_m} + [I_z, I_\pm]u_{x_m} = x_m I_\pm u_{x_m} \pm 1 \cdot I_\pm u_{x_m} = (x_m \pm 1)u_{x_m}$$

◇

Comments.

1. The commutation relations (and with them the structure constants) appear as "corrections" in the eigenvalue equation that provides the eigenvalues closest to any x_m.
2. This result organizes the diagonal matrix elements of I_z such that they change by "+ one" when moving "up" and "- one" when moving "down" along the diagonal. Taking into account that the highest Cartan weight is l, there will be $2l + 1$ Cartan weights, $l, l - 1, \ldots, -l$, providing a $2l + 1$-dimensional matrix and a $2l + 1$-dimensional representation. The quantum number l will assume the values $0, 1, 2, \ldots$, and the eigenvectors

are the set $\{u_{l,m=l}, u_{l,m=l-1}, \ldots, u_{l,m=-l}\}$. Two indices are required: l for labeling the representations, which we called D_l, and m for labeling the invariant vectors within each representation. Based on the concept pf invariant function spaces, we know that a possible representation of the basis vectors, $\{u_{l,m}\}$, is the spherical harmonics.

Theorem (exponential map). Once the matrix representations of the infinitesimal generators that fulfill the commutation relations have been determined, through analytical or algebraic methods,[4] one needs a tool to construct the matrices $F'(\mathbf{a})$ representing the group elements, starting from the set $\{A'_k\}$.

An important tool that we have already used is the exponential map. We have obtained the matrices of the subgroup of SO_3 consisting of rotations by any finite angle about a fixed rotation axis by means of the exponential map, leading to the Rodrigues formula. By adapting the exponential map to further rotation axes, the entire group SO_3 can be generated, subgroup by subgroup. The group of matrices D_l representing SO_3 can also be generated by using the exponential map of the suitable infinitesimal transformations.[5] The question now is how general is the possibility of using the exponential map to generate continuous groups themselves and their representations.

The answer of this question begins with a key theorem in mathematics, the proof of which can be found in C. Chevalley, *Theory of Lie Groups. I*, Princeton Mathematical Series, 8, Princeton University Press, 1946, ISBN 978-0-691-04990-8.

This theorem proceeds as follows. Let us suppose that we have a Lie group G of matrices $F(\mathbf{a})$ with

$$F(\mathbf{a}) = \mathbb{1} + i \cdot A_k \cdot a_k$$

We first chose a unit vector $\mathbf{n}$ in the parameter space and create the one-parameter infinitesimal group

$$F(a) = \mathbb{1} + i \cdot (\mathbf{n} \cdot \mathbf{A}) \cdot a$$

---

[4]This might be a really difficult task.

[5]We did only perform this operation explicitly for rotations about the z-axis. In principle, all representation matrices D_l can be generated using the exponential map, subgroup by subgroup, just as the group SO_3 itself is generated. In practice, however, the solution to concrete problems will only require the infinitesimal generators and not necessarily the entire set of rotation matrices.

The theorem by Chevalley states that any element of the group G in the neighborhood of $\mathbb{1}$ is a member of the one-parameter subgroup

$$D(a) = e^{i \cdot (\mathbf{n} \cdot \mathbf{A}) \cdot a}$$

i.e. it can be expressed using the exponential mapping. The entire group G can then be generated subgroup by subgroup.

Plausibility: For the set of subgroups to become a group, one needs to show that the multiplication of two elements belonging to different subgroups

$$e^{i \cdot (\mathbf{n_1} \cdot \mathbf{A}) \cdot a} \cdot e^{i \cdot (\mathbf{n_2} \cdot \mathbf{A}) \cdot b}$$

leads to an element that is also the exponential of an element of the Lie algebra. This is rendered fundamentally possible by the fact that the BCH formula, used to compute the product of matrix exponentials, contains only commutators which, in turn, can be written as linear combinations of elements of the Lie algebra.

On the base of this result, one can formulate an exponential mapping theorem for representations.

Theorem. *Let $\{A'_k\}$ be a set of $n \times n$ infinitesimal generators with*

$$F'(\mathbf{a}) = \mathbb{1} + i \sum_k A'_k \cdot a_k$$

and consider the subgroup defined by

$$F'(a) \doteq e^{i \cdot \mathbf{A}' \cdot \mathbf{n} \cdot a}$$

with $\mathbf{a} = a \cdot \mathbf{n}$. The claim is that the family of subgroups $\{F'(a)\}$ provides an n-dimensional representation of G.

Plausibility:
A. We first show that each subgroup $F'(a)$ is a representation of $F(a)$, i.e.

$$F'(a)F'(b)) = (F(a)F(b))'$$

We compute the left-hand side:

$$F'(a)F'(b) = e^{i \cdot \mathbf{A}' \cdot \mathbf{n} \cdot a} \cdot e^{i \cdot \mathbf{A}' \cdot \mathbf{n} \cdot b} = e^{i \cdot \mathbf{A}' \cdot \mathbf{n} \cdot (a+b)}$$

We compute the right-hand side:

$$(F(a)F(b))' = F'(a+b) = e^{i \cdot \mathbf{A}' \cdot \mathbf{n} \cdot (a+b)}$$

i.e. the homomorphism rule within each single subgroup is fulfilled.

This is the first step of the proof: As we are multiplying elements within the same Abelian one-parameter subgroup, we can use the simple version of the BCH formula, setting all commutators to zero.

B. The second step involves multiplying elements belonging to different subgroups. To simplify the expressions, let us introduce the notation

$$\mathbf{A} \cdot \mathbf{n}_1 \equiv A_1 \quad \mathbf{A} \cdot \mathbf{n}_2 \equiv A_2 \quad \mathbf{A}' \cdot \mathbf{n}_1 \equiv A_1' \quad \mathbf{A}' \cdot \mathbf{n}_2 \equiv A_2'$$

One needs to prove that

$$(F(A_1) \cdot F(A_2))' = F'(A_1)F'(A_2)$$

We limit ourselves to verifying this relation in the limit that

$$e^P e^Q \approx \mathbb{1} + P + Q + \frac{1}{2}[P, Q]$$

The left-hand side becomes (approximately)

$$\left(e^{iA_1 \cdot a} \cdot e^{iA_2 \cdot a}\right)' \approx \left(\mathbb{1} + (A_1 + A_2)i \cdot a - \frac{1}{2}a^2[A_1, A_2]\right)'$$

$$\approx \mathbb{1} + (A_1' + A_2')i \cdot a - \frac{1}{2}a^2[A_1', A_2']$$

The right-hand side becomes

$$e^{iA_1' \cdot a} \cdot e^{iA_2' \cdot a} \approx \mathbb{1} + (A_1' + A_2')i \cdot a - \frac{1}{2}a^2[A_1', A_2']$$

We conclude that the exponential mapping preserves the homomorphism between different subgroups as well.

The exponential mapping theorem provides a practical tool for using the representations of the infinitesimal operators to generate the representations of the original group. A general statement is certainly possible: **Finding the representations of a Lie group reduces to finding suitable matrices representing its infinitesimal generators.**

The final theorem concerns the invariant subspaces.

Proposition. *The matrix $e^{i \cdot A' \cdot \mathbf{n} \cdot a}$ and $A' \cdot \mathbf{n}$ have the same invariant subspaces.*

Proof. The exponential function, defined by its Taylor expansion, contains the powers of the matrix $A' \cdot \mathbf{n}$. The powers of $A' \cdot \mathbf{n}$ and $A' \cdot \mathbf{n}$ have the same invariant subspaces.

Comments.

1. By virtue of this theorem, the subspaces which are irreducible with respect to the infinitesimal transformations will provide irreducible representations of the group G.

2. This theorem introduces a possible strategy, as follows, for finding the representation matrices of the group elements: (a) First, find the subspaces that block-diagonalize A'_k; (b) construct the matrices A'_k themselves; and (c) translate the matrices A'_k to all elements of the group using the exponential map. We have followed this strategy to find the irreducible representations D_l of SO_3. A second strategy is provided by the method based on the structure constants theorem. This method dispenses of step (a) and provides a purely algebraic algorithm for finding the matrices A'_k directly.

Lecture 4

THE SYMMETRY GROUP OF THE OPERATOR AND SOME PRACTICAL APPLICATIONS

In this lecture, we introduce three theorems that establish the exact link between symmetry and scientific problems and use these theorems to solve examples of concrete scientific problems. This lecture is accompanied by three students projects that use the tools and algorithms discussed in the lecture: "Symmetry Arguments in Classical Mechanics", "Crystal Field Splitting and Jahn–Teller Effect" and "Vibrational Modes of the NH$_3$ Molecule". The more technical aspects are covered by a set of exercises.

4.1 The Relation Between Symmetry and Scientific Problems

4.1.1 *Introduction*

In Lecture 2, we learned the mathematical aspects of the representation theory of groups. We are currently faced with the following context:

- There is an eigenvalue problem for a Hermitic operator H of which we would like to compute eigenvalues and eigenvectors. We have in mind, for instance, the Hamilton operator of an electron in a potential energy landscape:

$$H(\mathbf{x}, \mathbf{p}) = -\frac{\hbar^2}{2m} \nabla^2 + V(\mathbf{x})$$

- The system is transformed into itself by a set of symmetry transformations $G = \{g\}$. Referring to the Hamilton operator above, this statement means that the distribution of centers producing the potential energy landscape is supposed to be transformed into itself by the elements of G.

- A set of operators $\{U(g)\}$ can be used to represent these symmetry transformations in the domain of definition of the operator H. For our specific example, the set of operators act onto the wave functions $\psi(\mathbf{x})$ residing in the domain of definition of $H(\mathbf{x}, \mathbf{p})$.
- A set of theorems allows us to establish irreducible representations and symmetry-adapted vectors of $\{U(g)\}$ in the domain of definition of H.

The link between these elements was established by E. Wigner in a seminal paper that was recognized by the Nobel Prize in 1963 (E. Z. Wigner, "Einige Folgerungen aus der Schrödingerschen Theorie für die Termstrukturen", *Physik*, **43**, 624 (1927), https://doi.org/10.1007/BF01 397327). The scope of this lecture is to give a tutorial (hopefully accurate) account of Wigner's ideas. For a precise understanding of these ideas, we need a "transformation theory" that links the transformation of the $\mathbf{x}$ coordinates to the transformation of scalar functions and operators.

Transformation theory: Suppose that we would like to use a different set of coordinates to describe the position of the particle. This is implemented by a coordinate transformation of the type

$$x_1' = f_1(x_1, x_2, \ldots, x_n)$$
$$x_2' = f_2(x_1, x_2, \ldots, x_n)$$
$$\vdots$$
$$x_n' = f_n(x_1, x_2, \ldots, x_n)$$

The functions f_i relating the primed coordinates to the original, unprimed coordinates can be nonlinear, e.g. in the case of the transformation from Cartesian to spherical coordinates. The coordinate transformation is invertible if the determinant of the Jacobi matrix

$$J = \begin{pmatrix} \frac{\partial f_1(\mathbf{x})}{\partial x_1} & \frac{\partial f_1(\mathbf{x})}{\partial x_2} & \cdots & \frac{\partial f_1(\mathbf{x})}{\partial x_n} \\ \frac{\partial f_2(\mathbf{x})}{\partial x_1} & \frac{\partial f_2(\mathbf{x})}{\partial x_2} & \cdots & \frac{\partial f_2(\mathbf{x})}{\partial x_n} \\ \vdots & \vdots & & \vdots \\ \frac{\partial f_n(\mathbf{x})}{\partial x_1} & \frac{\partial f_n(\mathbf{x})}{\partial x_2} & \cdots & \frac{\partial f_n(\mathbf{x})}{\partial x_n} \end{pmatrix}$$

is not vanishing. The inverse transformation can be written in a compact way as

$$x_i = \mathbf{f}_i^{-1}(\mathbf{x}')$$

For linear transformations, we have, for instance,

$$
\begin{pmatrix} x_1' \\ x_2' \\ \vdots \\ x_n' \end{pmatrix} = \begin{pmatrix} R_{11} & \cdots & R_{1n} \\ \vdots & & \vdots \\ R_{n1} & \cdots & R_{nn} \end{pmatrix} \begin{pmatrix} x_1 \\ x_2 \\ \vdots \\ x_n \end{pmatrix} + \begin{pmatrix} t_1 \\ t_2 \\ \vdots \\ t_n \end{pmatrix}
$$

where the matrix elements R_{ij} are some constants. The Jacobi matrix is the R matrix itself.

A. Transformation of scalar functions of the variables x: The central question of transformation theory is: Given a coordinate transformation, how does any scalar function $\psi(x_1, \ldots, x_n)$ transform? The principle ruling the transformation of the function is that the primed function — let us call it ψ' — as a function of the primed variables must acquire the same value of the unprimed function as a function of the unprimed variables, i.e.

$$
\psi'(\mathbf{x}') = \psi(\mathbf{x})
$$

This relation can be used to compute $\psi'(\mathbf{x}')$:

$$
\psi'(\mathbf{x}') \doteq \psi\Big(\mathbf{f}_1^{-1}(x_1', \ldots, x_n'), \ldots, \mathbf{f}_n^{-1}(x_1', \ldots, x_n') \Big)
$$

In the case of linear transformations, we obtain

$$
\psi'(\mathbf{x}') = \psi\Big(R^{-1}\mathbf{x}' - \mathbf{t} \Big)
$$

B. Transformation of the gradient: We now proceed to transform the gradient. We compute

$$
\frac{\partial}{\partial x_i}\psi(\mathbf{x}) = \frac{\partial}{\partial x_i}\psi'(\mathbf{x}'(\mathbf{x})) = \sum_j \frac{\partial \psi'}{\partial x_j'}(\mathbf{x}') \cdot \underbrace{\frac{\partial f_j}{\partial x_i}}_{J_{ji}} = \sum_j J_{ji} \frac{\partial}{\partial x_j'}\psi'(\mathbf{x}')
$$

By comparing the right- and left-hand sides, we can determine that

$$
\left(\frac{\partial}{\partial x_i} \right) = \sum_j J_{ji} \frac{\partial}{\partial x_j'}
$$

The matrix elements of the Jacobi matrix read

$$
J_{ji}(\mathbf{x}') = \frac{\partial f_j}{\partial x_i}\Big|_{\mathbf{x} = \mathbf{f}^{-1}(\mathbf{x}')}
$$

The transformation of the derivatives are written in a compact way as

$$\nabla = J^T \nabla'$$

For linear transformations, we have

$$\nabla = R^T \nabla' \quad \Leftrightarrow \quad \nabla' = (R^t)^{-1} \nabla$$

(translations leave the gradient operator invariant). The gradient and position vectors undergo a "contragredient" transformation. Their transformation property with respect to proper and improper rotations is the same only if the matrix is orthogonal.

C. Transformation of the Laplace operator upon isometries:

$$\triangle' = \left(\nabla', \nabla' \right) = \left((R^t)^{-1} \nabla, (R^t)^{-1} \nabla \right)$$

If R is a proper or improper orthogonal matrix, we have

$$\left((R^t)^{-1} \nabla, (R^t)^{-1} \nabla \right) = \left(\underbrace{(R^t R)^{-1}}_{\mathbb{1}} \nabla, \nabla \right) = \triangle$$

i.e. the Laplace operator is left invariant by isometries with orthogonal matrices.

D. Transformation of the Hamilton operator upon isometries: We have in mind a particle subject to a potential energy landscape $V(\mathbf{x})$:

$$H(\mathbf{x}, \mathbf{p}) = -\frac{\hbar^2}{2m} \triangle + V(\mathbf{x})$$

We compute

$$H'(\mathbf{x}', \mathbf{p}') = -\frac{\hbar^2}{2m} \triangle'(\mathbf{x}') + V'(\mathbf{x}') = -\frac{\hbar^2}{2m} \triangle(\mathbf{x}') + V\left(R^{-1}\mathbf{x}' - \mathbf{t} \right)$$

Suppose that there is a set of operations that transform the centers producing the potential energy landscape into itself. For the same operations, we also have

$$V\left(R^{-1}\mathbf{x}' - \mathbf{t} \right) = V(\mathbf{x}')$$

There is therefore a group of symmetry operations with the property that

$$H(\mathbf{x}, \mathbf{p}) = H(\mathbf{x}', \mathbf{p}')$$

i.e. for these specific operations, the Hamilton operator in the unprimed and primed variables has the same functional dependence! We refer to this set as "symmetry group of the Schrödinger equation" or "invariance group of the Hamilton operator".

The transformation operator: One can use the equation

$$\psi'(\mathbf{x}') = \psi\Big(R^{-1}\mathbf{x}' - \mathbf{t}\Big)$$

to define, formally, a transformation operator that acts on scalar functions,

$$\psi' \doteq U((R \mid \mathbf{t}))\psi$$

by means of the relation

$$U((R \mid \mathbf{t}))\psi(\mathbf{x}) \doteq \psi\Big(R^{-1}\mathbf{x} - \mathbf{t}\Big)$$

To simplify the expression, we call $((R \mid \mathbf{t})) \doteq g$. The operator $U(g)$ is a (faithful) representation of the transformation g in a function space. In fact,

$$(U(g_2)U(g_1))\psi(\mathbf{x}) = U(g_2)\psi(g_1^{-1}\mathbf{x}) = \psi(g_1^{-1}g_2^{-1}\mathbf{x})$$

$$= \psi((g_2 g_1)^{-1}\mathbf{x}) = U(g_2 g_1)\psi(\mathbf{x}) \qquad\qquad \diamond$$

$U(g)$ is a unitary operator in a function space endowed with Hermitic metric. In fact,

$$\Big(U(g)\psi, U(g)\phi\Big) = \int_{\mathscr{R}^3} dxdydz \cdot \overline{\psi(g^{-1}(\mathbf{x}))} \cdot \phi(g^{-1}(\mathbf{x}))$$

We introduce the variables

$$\mathbf{x}' \doteq g^{-1}\mathbf{x}$$

For $dx \cdot dy \cdot dz$, we have the relation

$$dx \cdot dy \cdot dz = \det(J) \cdot dx' \cdot dy' \cdot dz'$$

with J being the Jacobi matrix. For isometries, the Jacobi matrix is an orthogonal matrix with a determinant of ± 1. For odd rotations, there appears also an odd number of interchanges of the upper and lower integration limits. For even rotations, there appears a even number of interchanges. Accordingly, one can always write

$$\int_{\mathscr{R}^3} dxdydz \cdot \overline{\psi(g^{-1}(\mathbf{x}))} \cdot \phi(g^{-1}(\mathbf{x}) = \int_{\mathscr{R}^3} dxdydz \cdot \overline{\psi(\mathbf{x})} \cdot \phi(\mathbf{x}) = \Big(\psi, \phi\Big)$$

Upon introduction of this operator, the transformation properties of other operators acting on the function space are computed following the standard rules of operator transformation. For instance,

$$\underbrace{U(g)x_iU(g)^{-1}}\,\psi(\mathbf{x}) = U(g)x_i\psi(g\mathbf{x}) = \underbrace{\left(\sum_j R_{ij}^{-1}x_j - t_i\right)}\psi(\mathbf{x})$$

This relation can be summarized as the transformation property of the position operator under isometries:

$$U(g)\mathbf{x}U(g^{-1}) = (R^{-1} \mid -\mathbf{t})\mathbf{x}$$

In a similar way, one obtains

$$U(g)\nabla U(g^{-1}) = R^{-1}\nabla$$

In the formal framework of operators, the invariance of the Hamilton operator with respect to isometries is written as

$$U(g)HU(g^{-1}) = H \quad \forall g \;\Leftrightarrow\; \left[H, U(g)\right] = 0 \quad \forall g$$

The operation $\left[A, B\right]$ is defined as

$$\left[A, B\right] = AB - BA$$

and is called the "commutator", and A, B are said to "commute".

The advantage of using the transformation operator is that one can generalize the concept of the invariance group of an operator to any kind of situation, as well as those involving groups not related to coordinate transformations and isometries.

Definition. Given an operator H and some kind of group $G = \{g\}$, the set of unitary operators $\{U(g)\}$ representing the elements g in the domain of definition of H and for which $\left[H, U(g)\right] = 0$ is called the symmetry group of the operator H.

Comment. H can be any operator, not necessarily the Hamilton operator, of a quantum mechanical system. On the other hand, the elements g do not need to be isometries in the sense of transformations in the Euclidean space. Later, we provide examples where the "wave" functions are "particle states" and the symmetry group is some abstract group that transforms particle states into other particle states. However, the following theorems, which are based on this more general definition, apply generally.

4.1.2 *Theorem I: Quantum numbers theorem*

Given the following:

- an n-fold degenerate energy eigenvalue λ of H
- a set of eigenfunctions $\{f_1, f_2, \ldots, f_n\}$ such that $H f_i = \lambda f_i \; \forall i$
- a set of operators $U(g)$ (g indicating the elements of a group) such that $[U(g), H] = 0$ for any g.

Proposition A. *The subspace $\{f_i\}$ is invariant with respect to the operations $U(g)$.*

Proposition B. *The subspace $\{f_i\}$ carries an n-dimensional matrix representation of the symmetry group of H.*

Proof of A:

$$HU(g)f_i = U(g)Hf_i = U(g)\lambda f_i = \lambda U(g)f_i$$

This means that $U(g)f_i$ is also an eigenfunction of H with the same eigenvalue λ. As λ was assumed to have only n linear independent eigenfunctions, $U(g)f_i$ must be some linear combination of the eigenfunctions $\{f_i\}$: This means that any eigenspace of H is closed with respect to the operations of the symmetry group of H.   ◇

Proof of B: Define within $\{f_i\}$ an $n \times n$ matrix by means of the following equation:

$$U(g)f_i = \sum_j U(g)_{ji} f_j$$

As the space is closed, the matrix with matrix elements $U(g)_{ji}$ is a matrix representation of the symmetry element g within the eigenspace $\{f_i\}$.   ◇

Comments.

1. By construction, the function f_i transforms according to the ith column of the representation. They are called "symmetry adapted" functions. The coefficients of the linear combination determining the transformed functions are to be read out from the ith column of the matrix U_{ji}.
2. If the representation in the eigenspace to the eigenvalue λ is irreducible, the level is said to sustain an **essential degeneracy**: Small perturbations of H which do not break the symmetry cannot lift the degeneracy of the eigenspace.

3. If the representation is reducible, instead, a small, symmetry-conserving perturbation of the original operator is typically enough to split the original reducible representation into its irreducible components. In the case of a reducible representation, we refer to it as **accidental degeneracy** because the original degeneracy d_λ of the eigenvalue λ is removed by a small, symmetry-preserving perturbation of the original operator.

4. This exact theorem associates to any eigenspace of H one or more irreducible representations of the symmetry group. The labels of these representations provide "quantum numbers" (or "eigenfrequencies" in classical problems) characterizing the various eigenspaces.

4.1.3   *Theorem II: The matrix elements theorem*

A second theorem provides an exact result on the use of symmetry to solve an eigenvalue problem.

Preliminary assumptions: We assume that the H-operator has a symmetry group, $\left[H, U(g)\right] = 0$, with $g \in G$ and G being a group with an average. We also assume to have established, using the projector technique, a CONS of symmetry-adapted basis functions $\{f_{n,l} \cdot Y_l^m\}$ which block-diagonalize $U(g)$. In a function space, we expect the CONS to have an infinite number of elements. The basis functions consist of the product of two functions:

- a function Y_l^m transforming according to the mth column of the lth irreducible representation U_l.[1] The functions Y_l^m are defined on an appropriate set of variables. In the language of atomic physics (used here for the sake of remaining concrete, although the lettering is intended to be used for general purposes), the set of variables would correspond to the spherical angular coordinates.
- The second component of the CONS basis states is f_{nl} and is a symmetry-invariant function that depends on a coordinate that is left invariant by any transformation g (in the atomic physics language, we think of the radial coordinate).

---

[1] The basis functions Y_l^m have a further index p that takes into account the fact that, for a given U_l, there might be more orthogonal functions transforming according to the same column m of the representation U_l. This situation does not happen in atomic physics, but one encounters it, for example in solid-state physics or particle physics. For simplicity of writing, we include this index tacitly in the index n and make it explicit only when needed.

Comments.

The invariant functions f_{nl} typically depend on l *and* a further index n (or further indices $n, n', \ldots$) which are used to label those basis functions covering the sector of the Hilbert space containing the invariant variables. Notably, $f_{n,l}$ are independent of m, i.e. the symmetry-invariant function $f_{n,l}$ is common to all partners m. This is because a symmetry-invariant function for each individual m would corrupt the transformation behavior with respect to U_l.

Proposition (Wigner–Eckart–Koster (WEK)[2])

$$\left(f_{nl} Y_l^m, H f_{n',l'} Y_{l'}^{m'} \right) = c(n, n', l) \cdot \delta_{ll'} \cdot \delta_{mm'}$$

Comments.

1. The symmetry-adapted functions block-diagonalize the H operator. The Hilbert space is divided into subspaces that are invariant with respect to a given irreducible representation U_l. Matrix elements with functions that transform according to different representations $l' \neq l$ are exactly vanishing. Within the block $l = l'$, only those matrix elements remain finite for which $m = m'$. Their value is given by a constant $c(n, n', l)$, which must be computed either analytically or numerically, i.e. it is outside the range of symmetry arguments. Accordingly, the WEK theorem causes a reduction of the size of the matrices that must be diagonalized numerically.

2. Let us explain this important content in terms of block matrices. If a representation U_l occurs only once, then the symmetry-adapted functions diagonalize both U_l and H within the corresponding block. Referring to the schematic block-diagonalized matrices in the following, both U_l and H_l are diagonal matrices, provided one uses symmetry - adapted basis functions. Often, a representation occurs more than once (in atomic physics, for instance, the representation D_l carrying the orbital momentum quantum number l occurs an infinite number of times in the representation of rotations in the Hilbert space of the atom, with each l block having a different radial wave function). If we call this representation $U_{l'}$, then the symmetry-adapted functions still diagonalize

---

[2]G. F. Koster, "Matrix elements of symmetric operators", *Phys. Rev.* **109**, 227 (1958), https://doi.org/10.1103/PhysRev.109.227.

all $U_{l'}$ blocks. For example, in the following block matrix, the two $U_{l'}$ blocks must be regarded as both diagonal. In contrast, the $H_{l'}$ blocks have non-diagonal matrix elements. In the following block matrix, the $H_{l'}$ block is written as composed of four non-diagonal blocks. Each has the dimensionality of $U_{l'}$. Within each $H_{l'}$ block, some matrix elements are necessarily zero, namely those involving different m!

3. In summary, the WEK for scalar operators tells us that symmetry arguments do **block-diagonalize** a scalar operator but **do not diagonalize it**.

$$
\begin{pmatrix}
[U_l] & [0] & [0] & [0] & \cdots \\
[0] & [U_{l'}] & [0] & [0] & \cdots \\
[0] & [0] & [U_{l'}] & [0] & \cdots \\
\vdots & \vdots & \vdots & \vdots &
\end{pmatrix}
\underbrace{}_{U(g)}
\rightarrow
\begin{pmatrix}
[H_l] & [0] & [0] & [0] & \cdots \\
[0] & [H_{l',1}] & [H_{l',2}] & [0] & \cdots \\
[0] & [H_{l',3}] & [H_{l',4}] & [0] & \cdots \\
\vdots & \vdots & \vdots & \vdots &
\end{pmatrix}
\underbrace{}_{H}
$$

Proof of the proposition: As the scalar product is invariant under unitary transformations, we have

$$
\left(f_{nl} Y_l^m, H f_{n',l'} Y_{l'}^{m'} \right) = \left(U(g) f_{nl} Y_l^m, U(g) H f_{n',l'} Y_{l'}^{m'} \right)
$$
$$
\underbrace{=}_{[H,U]} \left(U(g) f_{nl} Y_l^m, H U(g) f_{n',l'} Y_{l'}^{m'} \right)
$$

By definition,

$$
U(g) f_{nl} Y_l^m = \sum_r U_{l_{rm}} f_{nl} Y_l^r; \quad U(g) f_{n'l'} Y_{l'}^{m'} = \sum_s U_{l'_{sm'}} f_{n'l'} Y_{l'}^s
$$

Inserting and building the average on both sides of the equation and using the great orthogonality theorem for the matrix elements of the irreducible representations produces the claim:

$$
\left(f_{nl} Y_l^m, H f_{n',l'} Y_{l'}^{m'} \right) = \sum_{r,s} \left(f_{nl} Y_l^r, H f_{n',l'} Y_{l'}^s \right) \underbrace{\mathscr{M}_g \overline{U_{l_{rm}}(g)} \cdot U_{l'_{sm'}}(g)}_{\text{GOT}: \frac{1}{d_l} \cdot \delta_{ll'} \delta_{rs} \delta_{mm'}}
$$
$$
= \underbrace{\sum_r \left(f_{nl} Y_l^r, H f_{n',l'} Y_{l'}^r \right)}_{c(n,n',l)} \delta_{ll'} \delta_{mm'} \qquad \diamond
$$

4.1.4   *Theorem III: Symmetry-assisted computations*

The two theorems just discussed define the search for symmetry-adapted functions to be the key step toward the symmetry-assisted solution of eigenvalue problems. For practical computations, however, the two theorems are of little use because the number of basis functions involved, albeit strongly reduced, might still be infinite. Practical computations are usually performed on a finite set of pre-selected basis functions; therefore, the two theorems introduced above must be suitably adapted. The Ritz theorem and the associated variational algorithm provide the proper adaptation of the two previous theorems to practical (albeit approximate) computations. We first provide a summary of the Ritz–Rayleigh theorem and the associated variational principle.

Theorem (Ritz and Rayleigh[3]).
For the ground-state energy E_0 of a Hamiltonian H, the following inequality applies:

$$E_0 \leq \mathrm{min}_{\forall \psi} \frac{\left(\psi, H\psi\right)}{\left(\psi, \psi\right)}$$

For the proof of this theorem, we refer to standard textbooks on the spectrum of self-adjoint operators.

Variational Principle (Ritz and Rayleigh): Relevant for applications is the variational principle that originates from this fundamental theorem. In practice, one selects within the domain of definition of H, an n-dimensional subspace V with basis functions $\{f_j\}$. This is known as trial subspace, and $\{f_j\}$ are the trial functions. They are not necessarily the exact eigenfunctions of H within the space V but are, usually, some functions that are made available by standard computational software. We want to keep the general aspect that appears in practical computations: The basis functions within the trial space V do not need to be orthonormalized, i.e.

$$\left(f_i, f_j\right) = S_{ij} \neq \delta_{ij}$$

The (Hermitic) matrix S_{ij} summarizes the so-called "overlap integrals". The diagonal matrix elements are usually "1", as the trial basis functions

---

[3]Walther Ritz, "Über eine neue Methode zur Lösung gewisser Variationsprobleme der mathematischen Physik", *J. Reine Angew. Math.* **135** 1–61 (1909), https://www.degruyter.com/document/doi/10.1515/crll.1909.135.1/html.

are usually taken to be normalized to "1'. The variational algorithm consists of the following steps:

Step 1: Build the trial wave function

$$\psi = \sum_k c_k f_k$$

The parameters $\{c_k\}$ are the quantities that must be determined by the variational principle.

Step 2: Insert the trial function into

$$\frac{\left(\psi, H\psi\right)}{\left(\psi, \psi\right)}$$

and transform this expression into an expression for the desired parameters $\{c_k\}$:

$$\frac{\left(\psi, H\psi\right)}{\left(\psi, \psi\right)} \doteq E(\{\overline{c_k}, c_{k'}\}) = \frac{\sum_{k,k'} \overline{c_k} \cdot c_{k'} \overbrace{\left(f_k, H f_{k'}\right)}^{\doteq M_{kk'}}}{\sum_{k,k'} \overline{c_k} \cdot c_{k'} S_{kk'}}$$

The matrix $M_{kk'}$ summarizes the matrix elements of H in the trial space V and is called the Ritz matrix.

Step 3: Minimize the functional $E\{\overline{c_k}, c_{k'}\}$ with respect to the parameters $\overline{c_k}$ and c'_k by setting its derivative to zero. For instance,

$$\frac{\partial E(\{\overline{c_k}, c_{k'}\})}{\partial \overline{c_i}} = \frac{\sum_{k'} c_{k'} M_{ik'} \cdot \sum_{k,k'} \overline{c_k} \cdot c_{k'} S_{kk'} - \sum_{k,k'} \overline{c_k} \cdot c_{k'} M_{kk'} \cdot \sum_{k'} c_{k'} \cdot S_{ik'}}{(\sum_{k,k'} \overline{c_k} \cdot c_{k'} S_{kk'})^2}$$

$$= \frac{\sum_{k'} c_{k'} M_{ik'}}{\sum_{k,k'} \overline{c_k} \cdot c_{k'} S_{kk'}} - E(\{\overline{c_k}, c_{k'}\}) \frac{\sum_{k'} c_{k'} \cdot S_{ik'}}{\sum_{k,k'} \overline{c_k} \cdot c_{k'} S_{kk'}}$$

$$! = 0$$

i.e.

$$\sum_{k'} c_{k'} (M_{ik'} - E(\{\overline{c_k}, c_{k'}\} \cdot S_{ik'}) = 0$$

or, using a more compact matrix, we can write

$$\left([\mathbf{M}] - E[\mathbf{S}]\right)\mathbf{c} = 0$$

The result of the minimization is an algebraic system of equations for the desired coefficients $\mathbf{c} = (c_1, \ldots, c_n)$ that, according to the standards of linear algebra, can only be fulfilled non-trivially for those values of $E(\mathbf{c})$ for which

$$\det\ ([\mathbf{M}] - E \cdot [\mathbf{S}]) = 0$$

In other words, the result of the minimization procedure is the eigenvalue problem of the matrix $S^{-1}M$.[4]

Step 4: From step 3, we obtain, typically, a set of n vectors $\mathbf{c}$ and a set of n values $\{E_k\}$. The coefficients originating from the solution of this system of equations minimize the functional $E\{\overline{c_k}, c_{k'}\}$, and the expression for the lowest energy eigenvalue among the set $\{E_k\}$ is an approximation of the ground-state eigenvalue. The remaining eigenvalues are an approximation of the excited states. Provided the trial functions are chosen as close as possible to the putative true eigenfunctions, the eigenstates obtained as local minima of the $E\{\overline{c_k}, c_{k'}\}$ landscape are a good approximation of the desired exact eigenvalues of the operator H. The central role within the Ritz algorithm is played by the Ritz matrix M and, if one works with $S \neq \mathbb{1}$, by the matrix $S^{-1}M$.

Theorem (The symmetry of the Ritz matrix).
With the aim of exploring the role of symmetry within the Ritz algorithm, we introduce now a symmetry group G for the Hamiltonian H, with U being some representation of G within the domain of definition of H, i.e. $[U(g), H] = 0$. H is the original Hamiltonian and its domain of definition contains the exact sought for eigenfunctions. In the domain of definition of H, we select a finite-dimensional vector space V of "trial" functions and select in it a set of basis functions $\{f_j\}$. Without loss of generality, we assume that the basis functions are orthonormalized (if not, we orthonormalize them with a Gram–Schmidt procedure). Within the trial space V, we build the Ritz matrix M with matrix elements

$$M_{ij} \doteq \left(f_i, H f_j \right)$$

To find the role of symmetry, we impose the condition that the trial space V carries a finite-dimensional matrix representation D of G. Then, the following important property about the matrices $D(g)$ and M can be demonstrated.

---

[4]Provided M is Hermitic, the derivative with respect to $c_{k'}$ produces the same eigenvalue problem.

Proposition.

$$MD(g) = D(g)M \quad \forall g \in G \quad \forall f \in V$$

Comments.

1. In virtue of this commutativity, Theorem II applies, but this time, its application has a practical consequence, as one deals with finite matrices. According to Theorem II, the matrix M is block-diagonalized by the symmetry-adapted functions constructed by applying the projector technology to the trial functions. As in the exact case of Theorem II, if an irreducible representation appears only once in the decomposition of D, then the matrix M has only diagonal matrix elements corresponding to one of its eigenvalues. Should the representation D_l appear $n_l > 1$ times in the decomposition of D, then the eigenvalue problem within the entire subspace carried by the D_l components must be solved numerically. However, as the space carrying the multiple D_l components is, in general, smaller than the original V space, the decomposition of D has effectively reduced the size of the eigenvalue problem.

2. In the perturbative approach, one might even go further, within a multiple symmetry block, and speculate that the matrix elements between the eigenfunctions belonging to a different l block are small and can be neglected. The eigenvalue problem of a multiple block reduces to the eigenvalue problem within each l block separately: In the perturbative approach, one really deals with the diagonalization of comparatively much smaller matrices.

Proof of the Proposition: We establish a complete orthonormal basis set $\{h_i\}$ in the domain of definition of H, with the basis set $\{f_i\}$ in V being a subset of it. We then write the commutation of the operators H and $U(g)$ as a matrix operation for the matrix elements, in particular

$$\sum_n \left(f_i, H h_n \right) \left(h_n, U(g) f_j \right) = \sum_n \left(f_i, U(g) h_n \right) \left(h_n, H f_j \right)$$

On the left-hand side, $U(g) f_j \in V$, as the space V is supposed to be closed with respect to $U(g)$ so that the sum on the left-hand side is only non-vanishing for $h_n = f_n$. On the right-hand side, $U(g) h_n$ is orthogonal to V for $h_n \neq f_n$, so that only those matrix elements with $h_n = f_n$ are non-vanishing and the sum is truncated. Accordingly, the initial identity of the matrix elements $(HU)_{ij}$ and $(U(g)H)_{ij}$ becomes

$$\sum_n \Big(f_i, Hf_n\Big)\Big(f_n, U(g)f_j\Big) = \sum_n \Big(f_i, U(g)f_n\Big)\Big(f_n, H)f_j\Big)$$

This equation express the commutativity of the **finite matrices** M and D.

◇

4.2 Application 1: Infinitesimal Transformations and Quantum Mechanical Conservation Laws

In classical physics, continuous symmetry groups determine the constants of the classical motion. For example, translational invariance of the potential along a certain direction produces conservation of the component of the momentum vector along that direction (among the student projects attached to these lectures, a project "Symmetry in Newtonian Mechanics" discusses the role of continuous symmetries in Newtonian mechanics). A similar situation exists in quantum mechanics.

We consider a Hermitic operator H with some set of symmetries $\{g\}$ which are represented in its domain of definition by the set of unitary operators $U(g)$. We have a situation in mind where the set $U(g)$ is a continuous one and that its elements are produced by the exponential map $e^{i\cdot(\mathbf{I},\mathbf{n})\cdot\varphi}$. $(\mathbf{I},\mathbf{n})$ is an infinitesimal transformation, and it is a Hermitic operator. In quantum mechanics, $(\mathbf{I},\mathbf{n})$ is, accordingly, an observable. The commutativity of $U(g)$ and H implies the commutativity between H and $(\mathbf{I},\mathbf{n})$.

Suppose that at a certain time t_0, the observable $(\mathbf{I},\mathbf{n})$ has been detected. The result of the measurement is one of its eigenvalues, say λ, and after the measurement, the system has been projected into one of the eigenstates u_λ in the eigenspace E_λ:

$$(\mathbf{I},\mathbf{n})u_\lambda = \lambda u_\lambda$$

One could ask the question: What will be the outcome of measuring again the observable $(\mathbf{I},\mathbf{n})$ at a later time?

We know that, at a later time, the system will be in a state that is propagated in time by the Hamilton operator according to

$$u(t) = e^{-i\frac{H}{\hbar}t}u_\lambda$$

The exponential of the operator is computed, as usually, by means of a power law expansion containing all the powers of H, so that $(\mathbf{I},\mathbf{n})$ commutes with $e^{-i\frac{H}{\hbar}t}$ as well. When we apply the operator $(\mathbf{I}, \mathbf{n})$ to $u(t)$, a remarkable result is observed:

$$(\mathbf{I}, \mathbf{n})u(t) = (\mathbf{I}, \mathbf{n})e^{-i\frac{H}{\hbar}t}u_\lambda = e^{-i\frac{H}{\hbar}t}(\mathbf{I}, \mathbf{n})u_\lambda = e^{-i\frac{H}{\hbar}t}\lambda u_\lambda = \lambda u(t)$$

The outcome of a later measurement of the observable $(\mathbf{I}, \mathbf{n})$ will be again the eigenvalue λ. One could state this result in the following form: "the observable $(\mathbf{I}, \mathbf{n})$ is a constant of the quantum mechanical motion".

4.3   Application 2: Symmetry-Breaking Interactions

(H. Bethe,[5] "Termaufspaltung in Kristallen", *Ann. Phys.* **395**, 133–208 (1929), https://onlinelibrary.wiley.com/doi/abs/10.1002/andp.19293950 202)

A typical situation arises in quantum mechanics by the appearance of symmetry-breaking interactions. Think, for example, of the eigenvalue problem of an atomic electron. The potential energy is spherically symmetric, and its solutions are atomic orbitals. They provide a complete set of basis states. When the atom becomes a constituent of a molecule or a crystal, the potential energy typically has a lower symmetry (often a point group symmetry), and the eigenvalues and eigenfunctions change. How do we use the complete set of atomic orbitals to solve the eigenvalue problem of the embedded atom? The answer, on a perturbative level, can be obtained by using the tools provided by group theory.

Formally, one considers a system with an unperturbed operator H_0. The problem of the unperturbed operator is considered solved, either analytically or numerically. This means that we have a set of eigenvalues and corresponding eigenfunctions at our disposal. We add now a static (i.e. non-time-dependent) perturbation V, producing a perturbed operator $H = H_0 + V$. The term "perturbed" refers to the expectation (which must be confirmed, *a posteriori*, by the results of the calculations) that the perturbation V changes the eigenvalues of H_0 only *a little*, i.e. their change is much smaller than the distance between the eigenvalues of H_0. This situation can be dealt with using symmetry arguments.

Let G_0 be the symmetry group of the operator H_0. If not all transformations of G_0 leave V invariant, then V is a symmetry-breaking perturbation, and the symmetry group of the total operator H is given by the symmetry group of V. Often, the symmetry group of V is a subgroup of

---

[5]Hans Bethe (1906–2005), influential theoretical physicist, see e.g. https://en.wikipedia.org/wiki/Hans_Bethe.

G_0. Let us call it G_V. G_V contains fewer number of elements. Accordingly, when limited to the operations of G_V, the irreducible representations of G_0 might become reducible. What is the significance of this fact for, say, the eigenvalue E_0 of H_0, carrying the representation Γ_0 of G_0? The equation

$$\Gamma_0 \, |_{G_V} = \oplus_i n_i \Gamma_i(G_V)$$

tells us that, because of the perturbation, new invariant subspaces carrying the irreducible representations of G_V appear: The degeneracy of the original eigenvalue E_0 is lifted accordingly. The eigenvalue E_0 splits into new eigenstates carrying the irreducible representations of G_V contained in the decomposition of Γ_0. These irreducible representations are said to be compatible with the irreducible representation of G_0 and the decomposition process can be summarized in the form of compatibility tables. The irreducible representations $\Gamma_i(G_V)$ can be used to label the split eigenvalues. In quantum mechanical terms, they provide a new set of "quantum numbers".

The decomposition of Γ_0 in terms of $\Gamma_i(G_V)$ has a further important significance. Starting from the set of basis functions transforming according to Γ_0 and building the eigenspace of E_0, one can construct, for example using the projector technology,[6] a set of symmetry-adapted vectors that transform according to the representations $\Gamma_i(G_V)$. These new set of basis functions block-diagonalize the matrix of the Hamiltonian $H_0 + V$ computed within the eigenspace of E_0. The resulting eigenvalues, computed over the new symmetry-adapted vectors, approximate the exact eigenvalues of the operator $H_0 + V$. In summary, the mere knowledge of the irreducible representations of G_V and which of these are contained in the decomposition of the representation Γ_0 allows for an exact labeling of the energy states arising from the splitting of the original level. The computation of the symmetry-adapted vectors allow for an approximate computation of the eigenvalue of the perturbed operator $H_0 + V$.

A short summary of atomic physics[7]: Symmetry-breaking interactions are a common situation in atoms embedded in some molecular or crystalline environment. Therefore, we need to introduce a short summary of the most relevant results of atomic physics, as the solution of the eigenvalue problem

---

[6]For continuous groups, it might be useful to use alternative methods, such as those based on the technique of infinitesimal operators.

[7]For a more extended summary of atomic physics and quantum mechanics see e.g. D. Pescia, "Lectures on Quantum Mechanics for Material Scientists", online available at https://doi.org/10.3929/ethz-b-000621567.

of atoms represents the "unperturbed" starting point to the symmetry-breaking situation. For a system with atomic number Z, there are Z electrons that interact via Coulomb interaction with the nucleus and with themselves. The problem of determining the energy levels of the electrons is a many-body one: The Coulomb repulsion among the electrons causes a complicated multi-particle problem whose solution can only be found approximately. There are many approximate methods for calculating the wave functions and energy states of atoms. These methods are described in advanced textbooks. The main result of these computational methods is the proof that, in atoms, the energy levels can be computed considering one single electron moving in the field of the nucleus *and* a mean field of the remaining electrons (single-electron approximation, or independent particle model), corresponding to the Hamiltonian

$$-\frac{\hbar^2}{2m}\triangle + \Phi_{\text{mean field}}(r)$$

The mean field is found to have, to a very good approximation, spherical symmetry. Accordingly, one can use the same terminology adopted for the hydrogen atom and label each energy level (commonly termed as "shell") by the quantum numbers n, l.

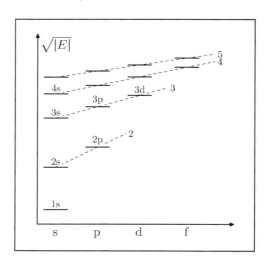

The energy levels of an electron in a spherically symmetric potential (schematic). Energy levels are characterized by two quantum numbers: l refers to the rotational degree of freedom, and n is the main quantum number. Dashed lines join levels belonging to the same main quantum number. The energy levels have an essential degeneracy with respect to the magnetic quantum number $m = l, \ldots, -l$.

Comments.

1. One important quantitative result of solving the eigenvalue problem of the mean field operator is that, in more complex atoms, the accidental degeneracy encountered in the hydrogen atom with pure Coulomb potential is lifted, producing an energy difference between levels with the same n but different l (see the figure above).

2. The mean field Hamiltonian provides, in formal terms, the "unperturbed operator" (H_0 in our terminology), and its eigenvalue problem is considered, in problems involving symmetry-breaking interactions, to be the one which has been solved. Its eigenspaces are the starting point for the perturbational approach inherent to the treatment of the symmetry-breaking perturbation problems in group theory.

3. We illustrate in the following some technical aspects of using symmetry in a situation of symmetry breaking by analyzing a few simple examples. The interested reader might find the subject of symmetry-breaking interactions more comprehensively explained in a student project called "Crystal Field Splitting and the Jahn–Teller Effect".

Example 1: Compatibility table for s, p, d atomic levels embedded in a C_{3v} symmetry-breaking environment: The compatibility table between SO_3 and C_{3v} is found by considering the character of the matrix representations of the operations of C_{3v} in the representations D_0, D_1, D_2 of the parent group SO_3.

- D_0: The character of D_0 when restricted to C_{3v} is 1 for all classes (the identical representation). The D_0 representation of SO_3 when restricted to C_{3v} becomes the Λ_1 irreducible representation of C_{3v}.

- D_1: The character of a rotation in this representation is $1 + 2\cos\varphi$. This means that the character of C_3 elements is 0 and the characters of C_2 elements is -1. The C_2 elements are accompanied by inversion (character -1 as well), and the product is 1. Thus, we have for the characters of the representation D_1 when restricted to C_{3v},

$$\chi(E) = 3, \chi(C_3) = 0, \chi(\sigma) = 1$$

Accordingly, the representation D_1 when restricted to C_{3v} is no longer irreducible but can be reduced into irreducible representations of C_{3v}. Specifically, $D_1 = \Lambda_1 \oplus \Lambda_3$.

  – D_2: The character of rotations in the D_2 representation is $\frac{\sin \frac{5}{2}\varphi}{\sin \frac{\varphi}{2}}$. The inversion is represented by $\mathbb{1}$. Summarizing,

C_{3v}	E	$C_{3xyz} C_{3xyz}^{-1}$	$IC_{2x\bar{y}} IC_{2\bar{x}z} IC_{2y\bar{z}}$	
Λ_1	1	1	1	
Λ_2	1	1	-1	
Λ_3	2	-1	0	
D_0	1	1	1	Λ_1
D_1	3	0	1	$\Lambda_1 \oplus \Lambda_3$
D_2	5	-1	1	$\Lambda_1 \oplus 2\Lambda_3$

The compatibility table reads as follows.

SO_3	D_0	D_1	D_2
C_{3v}	Λ_1	$\Lambda_1 + \Lambda_3$	$\Lambda_1 + 2\Lambda_3$

Example 2: Compatibility table for s, p, d, orbitals in a crystal field with O_h symmetry: The decomposition process is summarized in the following table.

O_h	E	$3C_4^2$	$6C_4$	$6C_2$	$8C_3$	I	$3IC_4^2$	$6IC_4$	$6IC_2$	$8IC_3$	
Γ_1	1	1	1	1	1	1	1	1	1	1	
Γ_2	1	1	-1	-1	1	1	1	-1	-1	1	
Γ_{12}	2	2	0	0	-1	2	2	0	0	1	
Γ_{15}	3	-1	1	-1	0	3	1	-1	1	0	
Γ_{25}	3	-1	-1	1	0	-3	1	1	-1	0	
$\Gamma_{1'}$	1	1	1	1	1	-1	-1	-1	-1	-1	
$\Gamma_{2'}$	1	1	-1	-1	1	-1	-1	1	1	-1	
$\Gamma_{12'}$	2	2	0	0	-1	-2	-2	0	0	1	
$\Gamma_{15'}$	3	-1	1	-1	0	3	-1	1	-1	0	
$\Gamma_{25'}$	3	-1	-1	1	0	3	-1	-1	1	0	
$D_0 \vert_{O_h}$	1	1	1	1	1	1	1	1	1	1	$= \Gamma_1$
$D_1 \vert_{O_h}$	3	-1	1	-1	0	3	1	-1	1	0	$= \Gamma_{15}$
$D_2 \vert_{O_h}$	5	1	-1	1	-1	5	1	-1	1	-1	$= \Gamma_{12} \oplus \Gamma_{25'}$

Example 3: Perturbation of O_h-symmetry by crystal potential with C_{4v} symmetry: The decomposition process is summarized in the following table.

C_{4v}	E	C_{2x}	$C_{4x}C_{4x}^{-1}$	$IC_{2z}IC_{2y}$	$IC_{2yz}IC_{2y\bar{z}}$	
Δ_1	1	1	1	1	1	
Δ_1'	1	1	1	-1	-1	
Δ_2	1	1	-1	1	-1	
Δ_2'	1	1	-1	-1	1	
Δ_5	2	-2	0	0	0	
$\Gamma_1\,\|_{C_{4v}}$	1	1	1	1	1	$=\Delta_1$
$\Gamma_{15}\,\|_{C_{4v}}$	3	-1	1	1	1	$=\Delta_1\oplus\Delta_5$
$\Gamma_{25'}\,\|_{C_{4v}}$	3	-1	-1	-1	1	$=\Delta_{2'}\oplus\Delta_5$
$\Gamma_{12}\,\|_{C_{4v}}$	2	2	0	2	0	$=\Delta_1\oplus\Delta_2$

A summary of compatibilities in Examples 1–3 is given in the following table.

SO_3	D_0	D_1	D_2	D_3	
O_h	Γ_1	Γ_{15}	$\Gamma_{12}\oplus\Gamma_{25'}$	$\Gamma_{2'}\oplus\Gamma_{25}\oplus\Gamma_{15}$	
O_h	Γ_1	Γ_2	Γ_{12}	Γ_{25}'	Γ_{15}'
C_{4v}	Δ_1	Δ_2	$\Delta_1+\Delta_2$	$\Delta_2'+\Delta_5$	$\Delta_1'+\Delta_5$
O_h	Γ_1'	Γ_2'	Γ_{12}'	Γ_{25}	Γ_{15}
C_{4v}	Δ_1'	Δ_2'	$\Delta_1'+\Delta_2'$	$\Delta_2+\Delta_5$	$\Delta_1+\Delta_5$

The following figure summarizes some compatibilities between SO_3, O_h, and C_{3v} in the form of energy level splitting.

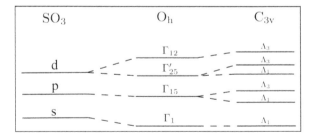

Summary of compatibilities.

In the following figure, we indicate how such crystal field splitting compatibilities manifest themselves in the band structure of solids as well. This is due to the fact that, as we will learn, the energy states appearing in the band structure of solids are often labeled by the point group irreducible

representations. In the figure, we observe that the band structure at the
Γ-point (a special point of the Brillouin zone of solids) develops along the
Λ-direction (a special direction in the Brillouin zone of solids) of some
materials, following the crystal field compatibility rules.

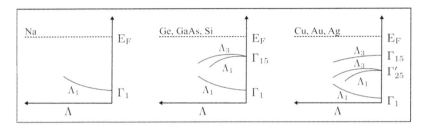

Summary of compatibilities at some characteristic points of the Brillouin zone in solids.
For more details see Chapter 6.

Example 4. Symmetry-adapted vectors: As an example of the
computation of symmetry-adapted vectors in a situation of symmetry
breaking, we compute explicitly the symmetry-adapted polynomials for
the C_{3v} symmetry-breaking crystal potential. The starting point are the
polynomials transforming according to D_0, D_1, and D_2. They can be found
in suitable tables of atomic physics:

- D_0: the polynomial is a constant.
- D_1: carries the monomials x, y, z (p-level)
- D_2: carries the five polynomials $\{(x^2 - y^2), 2z^2 - x^2 - y^2, xy, xz, yz\}$
  (d-levels).

Breaking the symmetry with a C_{3v} crystal potential was found, from the
compatibility table, to partially lift the degeneracy. New polynomials are
produced, which are symmetry adapted to the irreducible representations
of C_{3v}. To apply the projector technology, we need the transformation
properties of x, y, z under the operations of C_{3v}. The following figure embeds
a triangle in a cubic coordinate system and allows us to visualize the
operations of C_{3v} in a xyz-coordinate system.

- $D_1 = \Lambda_1 \oplus \Lambda_3$. We compute the symmetry-adapted polynomials:

$$P_{\Lambda_1} x \propto x + z + y + y + z + x \propto x + y + z$$

$$P_{\Lambda_3}(x) \propto 2 \cdot x - 1 \cdot z - 1 \cdot y$$

$$C_3(2x - z - y) = 2z - y - x$$

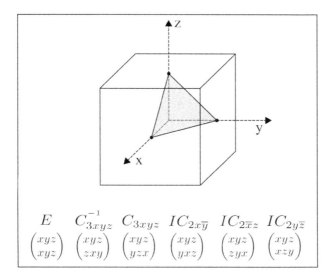

$$
\begin{array}{cccccc}
E & C_{3xyz}^{-1} & C_{3xyz} & IC_{2x\overline{y}} & IC_{2\overline{x}z} & IC_{2y\overline{z}} \\
\begin{pmatrix} xyz \\ xyz \end{pmatrix} & \begin{pmatrix} xyz \\ zxy \end{pmatrix} & \begin{pmatrix} xyz \\ yzx \end{pmatrix} & \begin{pmatrix} xyz \\ yxz \end{pmatrix} & \begin{pmatrix} xyz \\ zyx \end{pmatrix} & \begin{pmatrix} xyz \\ xzy \end{pmatrix}
\end{array}
$$

Transformation of the coordinates x, y, z under the operations of the group C_{3v}.

- $D_2 = \Lambda_1 \oplus 2\Lambda_3$. We compute the symmetry-adapted polynomials:

$$P_{\Lambda_1} xy \propto xy + zx + yz + yx + zy + xz \propto xy + xz + yz$$

$$P_{\Lambda_3}(x^2 - y^2) \propto 2x^2 - 2y^2 - z^2 + x^2 + y^2 + z^2$$

$$\propto x^2 - y^2$$

and so forth. Summarizing,

C_{3v}	E	$2C_3$	3σ	basis functions
Λ_1	1	1	1	$1, x + y + z, xy + xz + yz, \ldots$
Λ_2	1	1	-1	
Λ_3	2	-1	0	$2x - z - y, 2z - y - x, x^2 - y^2, 3z^2 - r^2, 2xy - zx - yz,$
				$2zx - yz - xy, \ldots$

Example 5: Tables of basis functions: The following tables get the reader acquainted with the way symmetry-adapted polynomials are presented in the literature. Many websites present the character tables of point groups and their polynomial basis functions: Among them is the website by Achim Gelessus at Jacobs University Bremen, http://symme try.jacobs-university.de/.

C_{3v}	E	$C_{3xyz}C_{3xyz}^{-1}$	$IC_{2x\bar{y}}IC_{2\bar{x}z}IC_{2y\bar{z}}$	Basis
Λ_1	1	1	1	1 $x+y+z$
Λ_2	1	1	-1	$x(y^2-z^2)+$ $y(z^2-x^2)+$ $z(x^2-y^2)$
Λ_3	2	-1	0	$2x-y-z$ $y-z$

C_{4v}	E	C_{2x}	$C_{4x}C_{4x}^{-1}$	$IC_{2z}IC_{2y}$	$IC_{2yz}IC_{2y\bar{z}}$	Basis
Δ_1	1	1	1	1	1	$1, x, 2x^2-y^2-z^2$
Δ_1'	1	1	1	-1	-1	$yz(y^2-z^2)$
Δ_2	1	1	-1	1	-1	y^2-z^2
Δ_2'	1	1	-1	-1	1	yz
Δ_5	2	-2	0	0	0	y, z, xy, xz

O_h	E	$3C_4^2$	$6C_4$	$6C_2$	$8C_3$	I	$3IC_4^2$	$6IC_4$	$6IC_2$	$8IC_3$	Basis
Γ_1	1	1	1	1	1	1	1	1	1	1	1
Γ_2	1	1	-1	-1	1	1	1	-1	-1	1	$x^4(y^2-z^2)+$ $y^4(z^2-x^2)+$ $z^4(x^2-y^2)$
Γ_{12}	2	2	0	0	-1	2	2	0	0	1	(x^2-y^2) $2z^2-x^2-y^2$
Γ_{15}	3	-1	1	-1	0	-3	1	-1	1	0	x y z
Γ_{25}	3	-1	-1	1	0	-3	1	1	-1	0	$z(x^2-y^2)$ $x(y^2-z^2)$ $y(z^2-x^2)$
$\Gamma_{1'}$	1	1	1	1	1	-1	-1	-1	-1	-1	$xyz[x^4(y^2-z^2)+$ $y^4(z^2-x^2)+$ $z^4(x^2-y^2)]$
$\Gamma_{2'}$	1	1	-1	-1	1	-1	-1	1	1	-1	xyz
$\Gamma_{12'}$	2	2	0	0	-1	-2	-2	0	0	1	$xyz(x^2-y^2)$ $xyz(2z^2-x^2-y^2)$
$\Gamma_{15'}$	3	-1	1	-1	0	3	-1	1	-1	0	$xy(x^2-y^2)$ $yz(y^2-z^2)$ $zx(z^2-x^2)$
$\Gamma_{25'}$	3	-1	-1	1	0	3	-1	-1	1	0	xy yz zx

4.4   Application 3: Helmholtz Equation on a Square Membrane

As a way of illustrating the use of symmetry for solving eigenvalue problems, we consider the eigenvalue problem of a vibrating square membrane clamped along the boundary. The displacement $u(x, y)$ perpendicular to the plane of the membrane satisfies the scalar Helmholtz equation[8]

$$-\triangle u(x,y) = \frac{\omega^2}{c^2} u(x,y)$$

where $\triangle$ represents the Laplace operator and ω are the frequencies of oscillation. The particular boundary condition that we use in the main section of this application is that the amplitude along the boundary of the membrane vanishes. The above equation, together with some boundary conditions, e.g. vanishing of the amplitude at the boundary (known as the Dirichlet boundary condition) or periodic boundary conditions (known as the Born–von Karman boundary condition), builds a well-defined eigenvalue problem for the negative of the Laplace operator (let us call this the M operator).

In order to numerically solve a partial differential equation, one strategy is to establish on the domain occupied by the object, a fine lattice mesh. The technical name of this strategy goes is the "Finite Difference Method",[9] and this will be illustrated more generally in the student project, "Functional Analytic and Numerical Aspects of Eigenvalue Problems", where numerical methods are discussed in greater depth. Here we illustrate the essential tools of this method. The precision and the number of eigenvalues obtained increases with the number of lattice points, but so does the size of the numerical problem, so that we limit ourselves — for the sake of keeping the matrices to a manageable size–to just approximating the square membrane with 12 points placed on a square with a lattice constant of a.

On a lattice, the amplitude $u(x, y)$ becomes a function at the lattice points, associating to every lattice point of the active lattice a "weight":

$$u(0,0) \quad u(1,0) \quad u(1,1) \quad u(0,1)$$

(the coordinates are in units of a).

---

[8]https://encyclopediaofmath.org/index.php?title=Helmholtz_equation.
[9]For more details on the Finite Difference Method see e.g. https://en.wikipedia.org/wiki/Finite_difference_method.

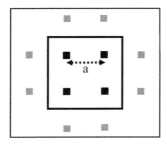

Square membrane on a lattice: The black squares represent active points, the gray squares are their nearest neighbors. The boundary of the square membrane is the continuous line surrounding the active point.

To translate the problem to a lattice, we need to define the Laplace operator on a lattice. This is done by noting that, with $(x_{\pm 1} = x_0 \pm a)$,

$$u(x_{\pm 1}, y) = u(x_0, y) + \frac{\partial u}{\partial x}\Big|_{x_0} \cdot (x_{\pm 1} - x_0) + \frac{1}{2}\frac{\partial^2 u}{\partial x^2}\Big|_{x_0} (x_{\pm 1} - x_0)^2$$

and accordingly,

$$\frac{\partial^2 u}{\partial x^2}\Big|_{x_0} = \frac{u(x_1, y) - 2 \cdot u(x_0, y) + u(x_{-1}, y)}{a^2}$$

A similar equation holds true along the y-coordinate, so that the lattice version of the operator

$$-\frac{\partial^2 u}{\partial x^2}\Big|_{x_0, y_0} - \frac{\partial^2 u}{\partial y^2}\Big|_{x_0, y_0}$$

is written as

$$\frac{+4 \cdot u(x_0, y_0) - u(x_1, y_0) - u(x_{-1}, y_0) - u(x_0, y_1) - u(x_0, y_{-1})}{a^2}$$

The initial Helmholtz equation, formulated as a partial differential equation, becomes a set of coupled linear equations for the desired amplitudes $u_1, \ldots, u_4$:

$$\frac{+4 \cdot u_{0,0} - u_{1,0} - u_{\bar{1},0} - u_{0,1} - u_{0,\bar{1}}}{a^2} = \frac{\omega^2}{c^2} u_{0,0}$$

$$\frac{+4 \cdot u_{1,0} - u_{0,0} - u_{2,0} - u_{1,1} - u_{1,\bar{1}}}{a^2} = \frac{\omega^2}{c^2} u_{1,0}$$

$$\frac{+4 \cdot u_{1,1} - u_{1,0} - u_{1,2} - u_{0,1} - u_{2,1}}{a^2} = \frac{\omega^2}{c^2} u_{1,1}$$

$$\frac{+4 \cdot u_{0,1} - u_{\bar{1},1} - u_{1,1} - u_{0,0} - u_{0,2}}{a^2} = \frac{\omega^2}{c^2} u_{0,1}$$

This set of equations include eight nearest-neighbor lattice points (see the figure above). They are necessary to implement the boundary conditions. At this point, we implement the Dirichlet boundary conditions, so that the amplitudes $u_{\bar{1},0}, u_{0,\bar{1}}, u_{2,0}, u_{1,\bar{1}}, u_{1,2}, u_{2,1}, u_{\bar{1},1}, u_{0,2}$ are set to 0. Writing the system of coupled homogeneous linear equations in matrix form, we obtain

$$\begin{pmatrix} 4 & \bar{1} & 0 & \bar{1} \\ \bar{1} & 4 & \bar{1} & 0 \\ 0 & \bar{1} & 4 & \bar{1} \\ \bar{1} & 0 & \bar{1} & 4 \end{pmatrix} \begin{pmatrix} u_{0,0} \\ u_{1,0} \\ u_{1,1} \\ u_{0,1} \end{pmatrix} = \frac{\omega^2 \cdot a^2}{c^2} \begin{pmatrix} u_{0,0} \\ u_{1,0} \\ u_{1,1} \\ u_{0,1} \end{pmatrix}$$

Finding the eigenvalues of the 4×4 M (Ritz) matrix requires, in principle, solving a quartic determinantal equation, which can be performed only numerically. The use of symmetry simplifies this numerics. The lattice established to represent the square membrane consist of four active sites and eight neighboring boundary sites, see the figure above. Accordingly, the lattice has the C_{4v} symmetry group of the continuous membrane. Our guess is that the Ritz matrix has this symmetry group as well. To check if this guess is correct and make use of this symmetry, we establish in the just defined lattice a tower (or "permutation") representation of this group, which we called Δ_T at the end of Chapter 2 (if we choose a larger number of points, the permutation representation involves larger matrices). We indeed observe that the matrices of Δ_T representing the elements of C_{4v} commute with M. We therefore work out the invariant subspaces of Δ_T, which are also invariant subspaces of the M matrix. We know from the example at the end of Lecture 2 that

$$\Delta_T = \Delta_1 \oplus \Delta_{2'} \oplus \Delta_5$$

and the symmetry-adapted vectors are

$$\Delta_1 = \begin{pmatrix} 1, & 1, & 1, & 1 \end{pmatrix} \quad \Delta_{2'} = \begin{pmatrix} 1, & \bar{1}, & 1, & \bar{1} \end{pmatrix}$$

$$\Delta_5 = \begin{pmatrix} 1, & 0, & \bar{1}, & 0 \\ 0, & 1, & 0, & \bar{1} \end{pmatrix}$$

We can use them to compute the desired eigenvalues of M, without solving any determinantal equation, by inserting these eigenvectors into one single line of the eigenvalue problem. The sought-after eigenvalues are 2 (the lowest one, corresponding to a uniform excitation of the membrane), 4 (twice degenerate, with nodes), and 6. Translating these values to proper units, we obtain

$$\omega_{\Delta_1} = \sqrt{2} \cdot 2\frac{c}{L}; \quad \omega_{\Delta_5} = 2 \cdot 2\frac{c}{L}; \quad \omega_{\Delta_2'} = \sqrt{6} \cdot 2\frac{c}{L}$$

Comments.

1. The lowest eigenfrequency corresponds to a mode that has the full C_{4v}-symmetry. The two highest eigenfrequencies correspond to higher-lying excitation modes that break the symmetry of the original operator (spontaneous symmetry breaking).
2. It is interesting to compare the solutions obtained for a lattice using symmetry arguments with the exact solutions of the Helmholtz equation for a square membrane, shown in the following figure.[10]

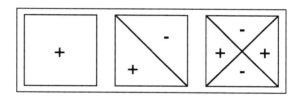

Exact eigenmodes of a square membrane: + and − indicate the signs of the amplitude of vertical (perpendicular to the plane of the paper) vibration in that region of the membrane. The vibration evolves in time changing from + to − and − to +, respectively. (a) Spatially uniform vibration with eigenfrequency is $\sqrt{2} \cdot \pi \frac{c}{L}$. (b) A node along the diagonal separates two regions with opposite amplitudes of vibration. The eigenfrequency is $\sqrt{5} \cdot \pi \frac{c}{L}$. (c) Two nodes separate regions with opposite amplitude of vibration. The eigenfrequency is $\sqrt{8} \cdot \pi \frac{c}{L}$.

Finally, we treat the case of a membrane with periodic boundary conditions. We go back to our original set of coupled linear equations and implement the periodic boundary conditions by setting $u_{\bar{1},0} = u_{1,0}$, $u_{\bar{1},1} = u_{1,1}$, $u_{2,0} = u_{0,0}$, $u_{2,1} = u_{0,1}$, $u_{0,\bar{1}} = u_{0,1}$, $u_{0,2} = u_{0,0}$, $u_{1,\bar{1}} = u_{1,1}$,

---

[10]M. D. Waller, *Proc. Phys. Soc. B* **67**, 895 (1954), https://iopscience.iop.org/articl e/10.1088/0370-1301/67/12/406/pdf. For a 3D demonstration, see Nasser M. Abbasi, "Vibration of a rectangular membrane", http://demonstrations.wolfram.com/Vibration OfARectangularMembrane/.

$u_{1,2} = u_{1,0}$. Accordingly, the M matrix is written as

$$
\begin{pmatrix}
4 & \bar{2} & 0 & \bar{2} \\
\bar{2} & 4 & \bar{2} & 0 \\
0 & \bar{2} & 4 & \bar{2} \\
\bar{2} & 0 & \bar{2} & 4
\end{pmatrix}
$$

Its eigenvalues are 0, corresponding to a gapless excitation mode of uniform vibration, 4 and 8, labeling excited states. The gapless excitation mode is a systematic result of periodic boundary conditions and occurs in the vibrational spectrum of solids as well.

4.5   Application 4: Molecular Vibrations

Group theoretical methods can be used for the solution of problems involving atomic vibrations in molecules and solids. These problems are treated within the framework of classical mechanics and provide a clear example on how numerics is simplified by using symmetry arguments.

4.5.1   *General aspects*

Think of a set of N atoms, each identified by an index k varying from 1 to N, and attach to each of them a Cartesian coordinate system $\{x_l\}$, with l varying from 1 to 3, describing the displacement of the atoms away from their equilibrium position.[11] Within the so-called harmonic approximation, the increase in the total energy Φ of the system when a displacement takes place is given by

$$
\sum_{h,k,l,m} \frac{\partial^2 \Phi}{\partial x_{k,l} \partial x_{h,m}} \Big|_{\mathbf{x}_0} x_{k,l} \cdot x_{h,m}
$$

($\mathbf{x}_0$ indicates the equilibrium coordinates). Within Newtonian mechanics, the motion of the k, l degree of freedom is given by the Newton equations:

$$
m_k \cdot \ddot{x}_{kl} = -\sum_{h,m} \frac{\partial^2 \Phi}{\partial x_{k,l} \partial x_{h,m}} \Big|_{\mathbf{x}_0} \cdot x_{hm}
$$

---

[11]One can also work with more symmetry-adapted coordinates, although numerical approaches to computing molecular vibrations typically use a Cartesian coordinate system that is common to all components.

m_k is the mass of the kth atom, and the double dots stand for the second derivative with respect to time.

$$\frac{\partial^2 \Phi}{\partial x_{k,l} \partial x_{h,m}}\bigg|_{\mathbf{x}_0} \equiv f^0_{kl,hm}$$

are the set of force constants whose arbitrariness is restricted only by the symmetry of the system and the fact that *action–reaction* implies

$$f^0_{hm,kl} = f^0_{kl,hm}$$

Such a system of coupled differential equations has a set of $3N$ linearly independent solutions, which are called eigenmodes. We seek them through the eigenmode *ansatz*:

$$x_{kl}(t) = x^0_{kl} \cdot e^{i\omega t}$$

Inserting this *ansatz*, which contains the desired eigenfrequencies ω, produces a system of $3N$ coupled algebraic equations for the eigenmode coefficients x^0_{kl}:

$$\omega^2 \cdot m_k \cdot x^0_{kl} = \sum_{h,m} f^0_{kl,hm} \cdot x^0_{hm}$$

which can be written in matrix form,

$$\omega^2 \begin{bmatrix} x^0_{1x} \\ x^0_{1y} \\ \vdots \\ x^0_{Nz} \end{bmatrix} = \underbrace{\begin{bmatrix} \frac{1}{m_1} & 0 & \cdots & \\ 0 & \frac{1}{m_1} & \cdots & \\ 0 & 0 & \cdots & \\ 0 & 0 & \cdots & \\ 0 & 0 & \cdots & \\ 0 & 0 & \cdots & \frac{1}{m_N} \end{bmatrix}}_{S^{-1}}$$

$$\cdot \underbrace{\begin{bmatrix} f^0_{1x,1x} & f^0_{1x,1y} & \cdots & f^0_{1x.Nz} \\ & \vdots & \vdots & & \vdots \\ f^0_{Nz,1x} & & \cdots & f^0_{Nz,Nz} \end{bmatrix}}_{F^0} \begin{bmatrix} x^0_{1x} \\ x^0_{1y} \\ \vdots \\ x^0_{Nz} \end{bmatrix}$$

or, in short,

$$\omega^2 \mathbf{x}_0 = S^{-1} F^0 \mathbf{x}_0$$

The desired eigenfrequencies are the solutions of the determinantal equation

$$\det\left[S^{-1}F^0 - \omega^2 \cdot \mathbb{1}\right] = 0$$

The solution of this determinantal equation provides the frequencies of the $3N$ eigenmodes. Note that included in these solutions are the trivial cases of pure displacement and pure rotations of the molecule as a whole. These particular kinds of motion are characterized by $\omega = 0$, as they are not influenced by restoring forces. In general, there will be $3N - 6$ non-zero solutions for ω. Each non-zero solution gives a specific normal mode of vibration.

4.5.2   *Numerical example: The H_2O molecule*

How can group theory help solve the problem of finding normal modes and eigenfrequencies? We illustrate this with the example of the vibrational modes of a H_2O molecule. This is a planar molecule, and for the sake of keeping the computations as simple as possible, we exclude the motion of the atoms along the direction perpendicular to the molecule. To capture the displacement of the atoms within the plane of the molecule, we need to attach to each atom a pair of coordinates. These can be a set of commonly oriented Cartesian coordinates. A different set of coordinates are those aligned along the direction of neighboring atoms, see the following figure.

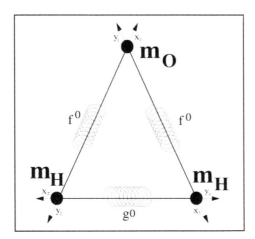

The water molecule with vectors describing the planar displacement of the atoms. "Springs" simulating the harmonic coupling along the atomic bonds are also drawn, with their respective spring constants.

In either set of coordinates, the force matrix is related to the second derivative of the potential landscape, but finding it is not a trivial problem (typically a numerical one). However, symmetry provides insights into the solution of the problem of diagonalizing the force matrix which do not require the exact knowledge of the force matrix itself. We therefore take the path of finding out how far one can go using symmetry arguments alone.

Symmetry Part I: The symmetry of the vibrationally active modes: The rationale for using symmetry is that the force matrix and the matrix S commute with all operations that transform the molecule into itself. This is because such transformations leave the potential energy landscape and its derivative invariant. For the specific case of the H_2O molecule, the symmetry group is C_{2v}. The vector space used to develop the symmetry arguments is the one hosting the molecule displacements: On this vector space, the force matrix acts. There are six basis displacement vectors that summarize the individual displacement vectors of the atoms (each atom is associated with a pair of displacement vectors, describing oscillations in the plane of the molecule):

$$x_1^0 = \begin{pmatrix} 1, & 0, & 0, & 0, & 0, & 0 \end{pmatrix}$$
$$x_2^0 = \begin{pmatrix} 0, & 1, & 0, & 0, & 0, & 0 \end{pmatrix}$$
$$x_3^0 = \begin{pmatrix} 0, & 0, & 1, & 0, & 0, & 0 \end{pmatrix}$$
$$x_4^0 = \begin{pmatrix} 0, & 0, & 0, & 1, & 0, & 0 \end{pmatrix}$$
$$x_5^0 = \begin{pmatrix} 0, & 0, & 0, & 0, & 1, & 0 \end{pmatrix}$$
$$x_6^0 = \begin{pmatrix} 0, & 0, & 0, & 0, & 0, & 1 \end{pmatrix}$$

Within the six-dimensional space hosting the atomic displacements, we build the "**vectorial tower (permutation) representation**" ($\equiv \Sigma_T^V$) of C_{2v}. The name "vectorial" is used to distinguish it from the "scalar" tower representation, where each component of the system is allocated a "scalar" amplitude. Such a scalar amplitude would find its use in solving the wave equation for scalar waves onto a lattice. In a vectorial representation, each atom is associated with a coordinate system, which hosts displacements with a "vectorial" character.

The symmetry group C_{2v} has symmetry elements: E, C_2, σ_u (reflection at the molecule plane), and σ_v. The vectorial matrix representation in the set of coordinates we have chosen for this problem reads

$$E = \begin{bmatrix} 1 & 0 & 0 & 0 & 0 & 0 \\ 0 & 1 & 0 & 0 & 0 & 0 \\ 0 & 0 & 1 & 0 & 0 & 0 \\ 0 & 0 & 0 & 1 & 0 & 0 \\ 0 & 0 & 0 & 0 & 1 & 0 \\ 0 & 0 & 0 & 0 & 0 & 1 \end{bmatrix} = \sigma_u \Rightarrow \chi_E = \chi_{\sigma_u} = 6$$

$$C_2 = \begin{bmatrix} 0 & 1 & 0 & 0 & 0 & 0 \\ 1 & 0 & 0 & 0 & 0 & 0 \\ 0 & 0 & 0 & 0 & 0 & 1 \\ 0 & 0 & 0 & 0 & 1 & 0 \\ 0 & 0 & 0 & 1 & 0 & 0 \\ 0 & 0 & 1 & 0 & 0 & 0 \end{bmatrix} = \sigma_v \Rightarrow \chi_{C_2} = \chi_{\sigma_v} = 0$$

By looking at the character table of C_{2v} and knowing the characters of the Σ_T^V representation, we can decompose it into a sum of irreducible representation of C_{2v}:

C_{2v}	E	C_2	σ_u	σ_v	
Σ_1	1	1	1	1	
Σ_2	1	−1	−1	1	
Σ_3	1	1	−1	−1	
Σ_4	1	−1	1	−1	
Σ_T^V	6	0	6	0	$3\Sigma_1 \oplus 3\Sigma_4$

Among the six vibrational modes expected, there will be such modes with a vanishing frequency: The two translational modes in the plane and the rotational mode. We can picture them on an intuitive basis using a diagram, see the following figure.

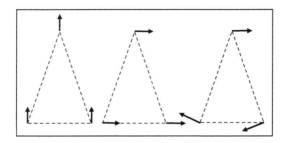

Sketch of the two in-plane translational modes (left and middle) and of the in-plane rotational mode (right) for the H_2O molecule.

The translational mode on the left transforms as Σ_1, and the remaining ones (middle and right) according to Σ_4. Accordingly, among the $3\Sigma_1 \oplus 3\Sigma_4$ modes, there will be only two active vibrational modes with Σ_1-symmetry and one vibrational mode with Σ_4-symmetry:

$$\Sigma_V^{\text{active}} = 2\Sigma_1 \oplus \Sigma_4$$

We expect therefore that the solution of the eigenvalue equation for the $S^{-1}F^0$ matrix will produce three distinct eigenfrequencies for the vibrational modes. This result concludes the first part of the symmetry analysis.

Symmetry Part II: The vibrational eigenmodes. The symmetry analysis can be developed further to find the two basis vectors transforming according to Σ_1 and the basis vector transforming to Σ_4 that sustain the active vibrational modes. This information can then be used to block-diagonalize the force matrix and simplify the eigenvalue problem. First, we find the general Σ_1 vector transforming according to Σ_1 using the projector method. From the general trial vector

$$(\lambda_1, \lambda_2, \lambda_3, \lambda_4, \lambda_5, \lambda_6)$$

we obtain its Σ_1 component:

$$P_{\Sigma_1}(\lambda_1, \lambda_2, \lambda_3, \lambda_4, \lambda_5, \lambda_6) = 2(\lambda_1, \lambda_2, \lambda_3, \lambda_4, \lambda_5, \lambda_6) + 2(\lambda_2, \lambda_1, \lambda_6, \lambda_5, \lambda_4, \lambda_3)$$

$$= (\lambda_1 + \lambda_2, \lambda_2 + \lambda_1, \lambda_3 + \lambda_6, \lambda_4 + \lambda_5, \lambda_5 + \lambda_4, \lambda_6 + \lambda_3)$$

i.e. the symmetry-adapted vectors transforming according to Σ_1 are of the type

$$(\lambda_1, \lambda_1, \lambda_3, \lambda_2, \lambda_2, \lambda_3)$$

Accordingly, possible basis vectors within the subspace hosting the Σ_1 modes are the mutually orthogonal vectors

$$(1,1,0,0,0,0); \quad (0,0,1,0,0,1); \quad (0,0,0,1,1,0)$$

Within this space, we construct the linear combination that represents the translational mode by subtracting the third basis vector from the first and obtain

$$(1,1,0,\bar{1},\bar{1},0)$$

(see left mode in the previous figure). The desired active vibrational modes reside within a two-dimensional space in the Σ_1 sector and are orthogonal to the translational mode. One vector orthogonal to the translational mode is, by inspection, the basis vector

$$(0,0,1,0,0,1)$$

We find the second one using the Gram–Schmidt algorithm:

$$v_2 = (0,0,0,1,1,0) + \frac{1}{2}(1,1,0,\bar{1},\bar{1},0)$$

$$\propto (1,1,0,1,1,0)$$

In summary, the subspace of Σ_1 carrying the active vibrational modes consists of the two vectors (to use them to block-diagonalize the force matrix, we must normalize them to unity)

$$\text{Mode I:} \quad \frac{1}{\sqrt{2}}(0,0,1,0,0,1) \quad \text{Mode II:} \quad \frac{1}{2}(1,1,0,1,1,0)$$

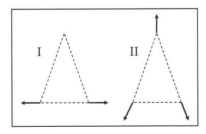

Sketch of the two in-plane vibrational modes with Σ_1 symmetry.

What is left to do for finding the two non-vanishing eigenfrequencies of the modes with Σ_1 symmetry is to compute the four matrix elements of the operator $S^{-1}F^0$ over these two weighted orthogonal basis vectors. This leads to a 2×2 matrix that must be diagonalized to find the eigenfrequencies we are interested in. In this specific case, symmetry arguments has reduced the problem in the Σ_1 sector to the (trivial) solution of a second-order algebraic equation.

We proceed in a similar way to analyze the Σ_4 sector. First, the application of the projector algorithm finds that the Σ_4 symmetry-adapted vectors must be of the type:

$$(\lambda_1, -\lambda_1, \lambda_3, \lambda_2, -\lambda_2, -\lambda_3)$$

For example, mutually orthogonal basis vectors of this type are

$$\frac{1}{\sqrt{2}}(1, \bar{1}, 0, 0, 0, 0); \quad \frac{1}{\sqrt{2}}(0, 0, 1, 0, 0, \bar{1}); \quad \frac{1}{\sqrt{2}}(0, 0, 0, 1, \bar{1}, 0)$$

We obtain a curling-like rotational mode through a linear combination of these three vectors:

$$(1, \bar{1}, 1, \bar{1}, 1, \bar{1})$$

(the rotational mode we pictured before). A further linear combination gives us a horizontal translational mode as

$$(1, \bar{1}, \bar{1}, \bar{1}, 1, 1)$$

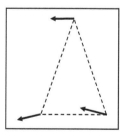

Horizontal translational mode with Σ_4 symmetry, computed with the projector method.

The remaining active vibrational mode is a linear combination of the basis vectors that is perpendicular to both the previous two modes:

$$\left(\alpha \begin{pmatrix} 1 \\ \bar{1} \\ 0 \\ 0 \\ 0 \\ 0 \end{pmatrix} + \beta \begin{pmatrix} 0 \\ 0 \\ 1 \\ 0 \\ 0 \\ \bar{1} \end{pmatrix} + \gamma \begin{pmatrix} 0 \\ 0 \\ 0 \\ 1 \\ \bar{1} \\ 0 \end{pmatrix} \right) \cdot \begin{pmatrix} 1 \\ \bar{1} \\ 1 \\ \bar{1} \\ 1 \\ \bar{1} \end{pmatrix} = 0$$

and

$$\left(\alpha \begin{pmatrix} 1 \\ \bar{1} \\ 0 \\ 0 \\ 0 \\ 0 \end{pmatrix} + \beta \begin{pmatrix} 0 \\ 0 \\ 1 \\ 0 \\ 0 \\ \bar{1} \end{pmatrix} + \gamma \begin{pmatrix} 0 \\ 0 \\ 0 \\ 1 \\ \bar{1} \\ 0 \end{pmatrix} \right) \cdot \begin{pmatrix} 1 \\ \bar{1} \\ \bar{1} \\ \bar{1} \\ 1 \\ 1 \end{pmatrix} = 0$$

The solution is

$$\beta = 0 \quad \text{and} \quad \alpha = \gamma$$

Accordingly, the mode carrying the active Σ_4 vibrational mode is

$$\text{Mode III:} \quad \frac{1}{2}(1, \bar{1}, 0, 1, \bar{1}, 0)$$

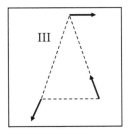

Vibrational mode with Σ_4 symmetry.

The mode builds a one-dimensional eigenspace and is therefore an eigenmode. The expectation value of the matrix $S^{-1}F$ in this eigenmodes gives the desired eigenfrequency level with Σ_4 symmetry.

A simple model of the force matrix: As an accurate force matrix can only be obtained numerically, we test our symmetry arguments with a simplified model, in which only harmonic coupling between the coordinates that are along the "spring" axis is allowed. In this simplified model, the Newton equations are written as

$$m_O \ddot{x}_1 = -f^0(x_1 + y_2)$$

$$m_O \ddot{y}_1 = -f^0(y_1 + x_3)$$

$$m_H \ddot{x}_2 = -g^0(x_2 + y_3)$$

$$m_H \ddot{y}_2 = -f^0(y_2 + x_1)$$

$$m_H \ddot{x}_3 = -f^0(x_3 + y_1)$$

$$m_H \ddot{y}_3 = -g^0(y_3 + x_2)$$

Using now the *ansatz*

$$\begin{pmatrix} x_i \\ y_i \end{pmatrix} = \begin{pmatrix} x_i^0 \\ y_i^0 \end{pmatrix} e^{i\omega t}$$

we obtain

$$\omega^2 \begin{pmatrix} x_1^0 \\ y_1^0 \\ x_2^0 \\ y_2^0 \\ x_3^0 \\ y_3^0 \end{pmatrix} = \begin{pmatrix} \frac{f^0}{m_O} & 0 & 0 & \frac{f_0}{m_O} & 0 & 0 \\ 0 & \frac{f^0}{m_O} & 0 & 0 & \frac{f^0}{m_O} & 0 \\ 0 & 0 & \frac{g^0}{m_H} & 0 & 0 & \frac{g^0}{m_H} \\ \frac{f^0}{m_H} & 0 & 0 & \frac{f^0}{m_H} & 0 & 0 \\ 0 & \frac{f^0}{m_H} & 0 & 0 & \frac{f^0}{m_H} & 0 \\ 0 & 0 & \frac{g^0}{m_H} & 0 & 0 & \frac{g^0}{m_H} \end{pmatrix} \begin{pmatrix} x_1^0 \\ y_1^0 \\ x_2^0 \\ y_2^0 \\ x_3^0 \\ y_3^0 \end{pmatrix}$$

Our symmetry rules predict that the eigenfrequencies of the Σ_1 modes can be found by solving the determinantal equation of the 2 matrix with the matrix elements

$$a_{11} = \frac{1}{\sqrt{2}} \cdot (0,0,1,0,0,1) \begin{pmatrix} \frac{f^0}{m_O} & 0 & 0 & \frac{f_0}{m_O} & 0 & 0 \\ 0 & \frac{f^0}{m_O} & 0 & 0 & \frac{f^0}{m_O} & 0 \\ 0 & 0 & \frac{g^0}{m_H} & 0 & 0 & \frac{g^0}{m_H} \\ \frac{f^0}{m_H} & 0 & 0 & \frac{f^0}{m_H} & 0 & 0 \\ 0 & \frac{f^0}{m_H} & 0 & 0 & \frac{f^0}{m_H} & 0 \\ 0 & 0 & \frac{g^0}{m_H} & 0 & 0 & \frac{g^0}{m_H} \end{pmatrix} \frac{1}{\sqrt{2}} \cdot \begin{pmatrix} 0 \\ 0 \\ 1 \\ 0 \\ 0 \\ 1 \end{pmatrix}$$

$$a_{22} = \frac{1}{2} \cdot (1,1,0,1,1,0) \cdot \begin{pmatrix} \frac{f^0}{m_O} & 0 & 0 & \frac{f_0}{m_O} & 0 & 0 \\ 0 & \frac{f^0}{m_O} & 0 & 0 & \frac{f^0}{m_O} & 0 \\ 0 & 0 & \frac{g^0}{m_H} & 0 & 0 & \frac{g^0}{m_H} \\ \frac{f^0}{m_H} & 0 & 0 & \frac{f^0}{m_H} & 0 & 0 \\ 0 & \frac{f^0}{m_H} & 0 & 0 & \frac{f^0}{m_H} & 0 \\ 0 & 0 & \frac{g^0}{m_H} & 0 & 0 & \frac{g^0}{m_H} \end{pmatrix} \frac{1}{2} \cdot \begin{pmatrix} 1 \\ 1 \\ 0 \\ 1 \\ 1 \\ 0 \end{pmatrix}$$

and

$$a_{12} = a_{21} = 0$$

After performing some algebraic matrix multiplications, we find the matrix

$$\begin{pmatrix} 2\frac{g_0}{m_h} & 0 \\ 0 & \left(\frac{f_0}{m_O} + \frac{f_0}{m_H} \right) \end{pmatrix}$$

This matrix is diagonal, so the diagonal elements are the eigenvalues themselves. We have two eigenmodes with Σ_1 symmetry and a frequency of

$$\omega_I^2 = 2\frac{g_0}{m_h} \qquad \omega_{II}^2 = \frac{f_0}{m_O} + \frac{f_0}{m_H}$$

The frequency of the eigenmode with Σ_4 symmetry is computed from

$$\frac{1}{2}\cdot(1,\bar{1},0,1,\bar{1},0)\cdot
\begin{pmatrix}
\frac{f^0}{m_O} & 0 & 0 & \frac{f_0}{m_O} & 0 & 0 \\
0 & \frac{f^0}{m_O} & 0 & 0 & \frac{f^0}{m_O} & 0 \\
0 & 0 & \frac{g^0}{m_H} & 0 & 0 & \frac{g^0}{m_H} \\
\frac{f^0}{m_H} & 0 & 0 & \frac{f^0}{m_H} & 0 & 0 \\
0 & \frac{f^0}{m_H} & 0 & 0 & \frac{f^0}{m_H} & 0 \\
0 & 0 & \frac{g^0}{m_H} & 0 & 0 & \frac{g^0}{m_H}
\end{pmatrix}
\frac{1}{2}\cdot
\begin{pmatrix}
1 \\ \bar{1} \\ 0 \\ 1 \\ \bar{1} \\ 0
\end{pmatrix}$$

The eigenfrequency of the Σ_4 mode is

$$\omega_{III} = \frac{f_0}{m_O} + \frac{f_0}{m_H}$$

Note that, in this model, modes II and III are accidentally degenerate. A more precise yet harmonic model is bound to lift this degeneracy.

One can find dynamical representations of the vibrational modes at https://www.chem.purdue.edu/jmol/vibs/h2o.html. In this website, the two eigenmodes with Σ_1 symmetry are labeled a_1. The eigenmode with Σ_4 symmetry is labeled b_2, according to a different labeling scheme of the irreducible representations of C_{2v}.

The interested reader can find a further example of the use of symmetry for studying molecular vibrations in the student project "Vibrational Modes of the NH_3-Molecule".

Lecture 5

POINT GROUPS AND SPACE GROUPS
OF CRYSTALLOGRAPHY

In this lecture, we provide the guidelines to understanding the symmetry group of finite objects (called "point groups") and the symmetry group of infinite objects with translational symmetry (called "patterns" or "crystals"). The lecture is accompanied by two projects. One is called "Frieze Patterns" and is designed to educate students in recognizing elements of symmetry of a pattern. The second one is an "Algebraic Proof of Hessel Theorem", an important theorem about the finite subgroups of SO_3. Finally, the lecture is complemented by a set of exercises, which are also designed for training students in recognizing symmetry elements of two-dimensional space groups.

5.1 Point Groups

It is not the scope of this section to provide a detailed review of the properties of point groups or to provide extended examples of molecules with their specific point group symmetry. There is a large number of publications dealing with point groups in physics and chemistry, both on paper and online. The following Wikipedia page provides precise and concise information, together with a number of examples: https://en.wikipedia.org/wiki/Molecular_symmetry. This page also contains a link to character tables of important point groups and their polynomial basis functions (e.g. http://symmetry.jacobs-university.de/). Our aim is to develop the **tools** for constructing point groups from fundamental principles. In doing this, we also learn the symbols used in the point-group method, as they are important for understanding the scientific literature.

The symmetry elements of a finite body must fix a common point. Alternatively, some other operation displaces the body as a whole and therefore cannot be a symmetry element because it does not transform the body into itself. Without loss of generality (more precisely, up to a similarity transformation), one can choose this common fixed point as the origin of the coordinate system, so that the symmetry group of a finite body consists of elements of the form $(F_0 \mid \mathbf{0})$, with F_O being an orthogonal matrix that fixes the origin. The largest subgroup of $Euc(3)$ that fixes the origin is called the group O_3 (the orthogonal group) and consists of all proper and improper rotations. It is the symmetry group of a spherical object and is a continuous group. It plays an important role in atomic physics. An important subgroup of O_3 is the group SO_3 (the special orthogonal group) of all proper rotations, which is naturally represented by all 3×3 orthogonal matrices with a determinant of $+1$. In two dimensions, the corresponding groups are O_2 and SO_2, which transform a circular planar object into itself.[1] In one dimension, the group that fixes the origin is the finite group $\{\mathbb{1}, I\}$. When one deals with a physical body having a discrete mass distribution or a finite number of constituents (for instance, a molecule), one has to search for the finite subgroups of SO_3 and O_3 (respectively, O_2 and SO_2). These are the so-called point groups. In this section, we construct and discuss all relevant point groups. Mathematically, one has two important introductory propositions supporting the existence of point groups.

Proposition. Let G be a finite subgroup of $\text{Euc}(n)$. Then, there is a point in $\mathcal{R}^n$ which is fixed by every element F_i of G. Accordingly, finite subgroups of $\text{Euc}(n)$ are called *point groups*.

Proof. Take any point defined by the position vector $\mathbf{p}$. Define

$$\mathbf{p}_0 = \frac{1}{\mid G \mid} \sum_i F_i(\mathbf{p})$$

as the sum taken over all elements of G. Then,

$$F_i(\mathbf{p}_0) = \mathbf{p}_0$$

This is because multiplying the sum over all elements by any element produces all elements again and only once.   ◇

---

[1]The circular planar object can be embedded in three dimensions while preserving the symmetry groups when considering it as the basis of a cone.

5.1.1 *Point groups in one-dimensional systems*

The only isometries of the line that leave a point fixed are $x' = \pm x$, corresponding to the symmetry elements $\mathbb{1}$ and I. Accordingly, the only point groups are $C_1 = \{\mathbb{1}\}$ and $C_i \doteq \{\mathbb{1}, I\}$.

5.1.2 *Point groups in two-dimensional systems*

Theorem (Leonardo da Vinci: According to Hermann Weyl, Leonardo discovered this theorem when he pursued, in his architectural studies, the question of how to attach chapels and niches to a building without destroying the symmetry of the core building): *The finite subgroups of $O(2)$ (i.e. any two-dimensional point group) are one of the proper rotation groups C_n ($n = 1, 2, \ldots$) or one of the (two-dimensional) dihedral groups D_n, with $n = 1, 2, \ldots$*

Proof.

- C_n: Consider first only proper rotations. Assume that there exists a symmetry element $R(\varphi)$, which must have the form of the matrix

$$\begin{pmatrix} \cos\varphi & -\sin\varphi \\ \sin\varphi & \cos\varphi \end{pmatrix}$$

with φ being some rotation angle. Using the group rule, $R^2(\varphi)$ must be also a symmetry element, but because of the particular shape of the matrix, $R^2(\varphi) = R(2\varphi)$. We can repeat this operation n times, but in order to have a finite group, we must have $n\varphi = 2\pi$ for some $n \geq 1$. This establishes

$$C_n = \{C_n^0 = \mathbb{1}, C_n^1, \ldots, C_n^{n-1}\}$$

as possible symmetry groups, $n = 1, 2, \ldots$
- D_n: If we add one reflection σ to a line, i.e.

$$\begin{pmatrix} \cos\phi & \sin\phi \\ \sin\phi & -\cos\phi \end{pmatrix}$$

then $C_n\sigma, C_n^2\sigma, \ldots, C_n^k\sigma$ establishes further reflection lines. However, successive reflections at two different reflection lines produce a rotation by twice the angle between the lines, which must be identical to one of the rotations C_n^k in order for the number of symmetry elements to be finite. This gives $\phi = \varphi$ and the symmetry groups

$$D_n = C_n^0 = \mathbb{1}, C_n^2, \ldots, C_n^{n-1}, C_n^0 \sigma, C_n^1 \sigma, \ldots, C_n^{n-1} \sigma$$

$n = 1, 2, \ldots$ There are no further elements available to expand the group while remaining in the plane and keeping the origin fixed. $\quad\diamond$

Comment.

The two-dimensional dihedral group D_n is the symmetry group of a planar polygon with n sides. When this polygon is considered the basis of a n-sided pyramid, the symmetry group D_n is denoted by the three-dimensional (also called pyramidal) point group C_{nv}.

5.1.3 *Point groups in three-dimensional systems*

Theorem. (Hessel: J. F. C. Hessel (1796–1872) proved its theorem in 1830, but his work did not receive any recognition among his contemporaries until it was published posthumously in 1897): *Any finite subgroup of SO_3 is one of the proper symmetry groups of an n-pyramid or a double n-pyramid or the proper symmetry group of one of the regular Platonic solids.*[2]

The proof of this theorem is part of "Project Hessel", which is one of the student projects accompanying these lecture notes. Here, we provide information about point groups that are of routine use in physical problems.

Stereographic projection: A useful way of symbolically representing symmetry elements is their stereographic projection. The stereographic projection projects the axes of rotation and mirror planes (and the position of a point and its maps under the elements of the symmetry group) from their positions on a sphere to the equatorial plane. To perform a stereographic projection of symmetry elements, we adopt the following conventions, referring to the accompanying figure:

---

[2]There are five regular Platonic solids: the tetrahedron, the cube, the octahedron, the icosahedron, and the dodecahedron. The cube and the octahedron (respectively, the dodecahedron and the icosahedron) are dual: if one connects the straight lines passing though the middle of the faces of one solid, one obtain the vertices of the other. Accordingly, the dual solids have the same symmetry groups. The tetrahedron is self-dual.

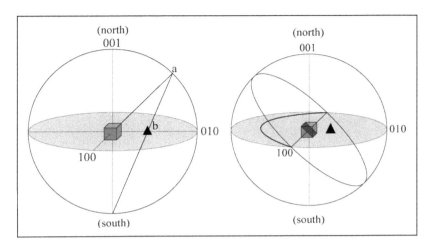

Left: A gray cube is placed at the origin of the sphere. A segment along its diagonal is intended to represent a symmetry axis (e.g. a threefold rotation axis). The segment is continued until it crosses the unit sphere at the point labeled "a". The point "a" is connected with the south pole and crosses the equatorial plane at "b". "b" is the stereographic projection of the rotation axis: It is marked with a full triangle that signals the threefold rotation axis. Right: A mirror plane (dark gray), crossing the sphere along a circle, is projected into an arc on the equatorial plane. The arc is the stereographic projection of the mirror plane.

1. Place an object at the center of a sphere, and draw all the symmetry axes and planes. If there is a principal rotation axis, it is usually oriented to cross the north pole.

2. From the sphere center, continue the symmetry axes or the mirror plane to intercept the sphere — the intercept of a rotation axis gives a point on the sphere, whereas the intercept of a mirror plane gives a continuous curved line.

3. Connect these points and lines with the south pole of the sphere, and mark the location where the connecting lines cross the equatorial plane.

4. A solid line is used to represent the projection of mirror planes on the equatorial plane. The outline of the stereogram is shown as a dashed line if it does not coincide with a mirror plane.

5. Use special symbols to represent the projection of two-, three-, four-, and sixfold axes on the equatorial plane. They are represented by elliptical, triangular, square, and hexagonal markers, respectively.

6. A general point and all its maps under the symmetry elements are represented by crosses and circles, with the number of symmetric points being the order of the group. Some of the elements will drive the general point to the northern hemisphere. Connect these positions with the south pole, and mark the intersection with the equatorial plane with a circle. Those points residing on the southern hemisphere are projected onto the equatorial plane with crosses.

5.1.4   *Enumeration of the proper point groups (Hessel)*

The cyclic groups C_n: We start by expanding on the results for the finite subgroups of SO_2. Take $\mathbf{e}$ as a unit vector in $\mathscr{R}^3$, and assume that it is along a rotation axis C_n, with $n = 1, 2, 3, \ldots$ (in 3D crystallography, we will show later that the so-called crystallographic restriction limits n to the values 1, 2, 3, 4, or 6). Let $\mathbf{u}$ be a unit vector perpendicular to it. Then, the points defined by the vectors $\mathbf{u}, C_n\mathbf{u}, \ldots, C_n^{n-1}\mathbf{u}$ are the vertices of a regular n-gon lying on a plane perpendicular to $\mathbf{e}$. By joining each vertex with the point at $2\mathbf{e}$, we obtain an n-pyramid, which is mapped into itself by the two-dimensional cyclic groups $C_n = \{C_n, C_n^2, \ldots, C_n^{n-1}\}$ with n elements.

A molecule with the symmetry group C_3 is shown in the following figure. Five of the C_n group are compatible with the crystallographic restriction and are listed here by their stereographic projection.

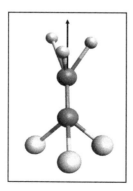

A molecule with C_3 point group symmetry: The main rotation axis going through the two central carbon atoms is indicated by an arrow. The top three atoms (H) are rotated with respect to the bottom three atoms (Cl).

The following figure shows the stereograms of the cyclic point groups of crystallography.

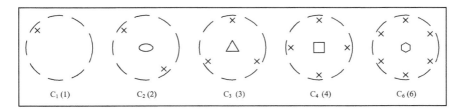

| C_1 (1) | C_2 (2) | C_3 (3) | C_4 (4) | C_6 (6) |

Stereograms of the cyclic point groups of crystallography. Left: the S-notation. In brackets: the HM-notation.

The three-dimensional dihedral groups D_n (order $2n$):

When the n-pyramid is expanded to a double n-pyramid, further symmetry groups appear, which are obtained by adding a twofold axis perpendicular to the main C_n-axis along the vector $\mathbf{u}$. The C_n main rotational axis introduces n of such twofold axes (which we call B_2). The symmetry elements are

$$C_n^0, C_n^1, \ldots, C_n^{n-1}, C_n^0 B_2, C_n^1 B_2, \ldots, C_n^{n-1} B_2$$

There are four new groups that are compatible with the crystallographic constraint: D_2, D_3, D_4, D_6. D_1 and C_2 are the same group. D_2 is also known as the "Klein's four group" and denoted by "V".

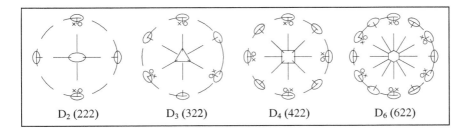

| D_2 (222) | D_3 (322) | D_4 (422) | D_6 (622) |

Stereograms of the D_n groups.

The proper rotation group of the tetrahedron (T):

It has all the proper rotational symmetry elements of a regular tetrahedron. The figure shows a regular tetrahedron inscribed in a cube for convenience.

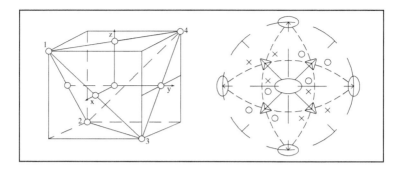

Left: A tetrahedron inscribed in a cube. The C_2 axes are along (x, y, z). Right: The stereogram of the group T (23).

Each of the four body diagonals of the cube provides C_3 and C_3^2 axes. In addition, T has three C_2 axes joining the center of the cube with the two opposite sides of the tetrahedron. In total, the group contains 12 elements:

$$\mathbb{1}, 4C_3, 4C_3^2, 3C_2$$

Referring to the table for O_h containing the transformation of the (x, y, z) coordinates (see Lecture 1), the elements of T are given by the sets

$$C_1 \quad C_4^2 \quad C_3$$

The proper rotation group of the octahedron (O): The group O has 24 symmetry elements containing all proper rotations that send an octahedron (in the figure inscribed into its dual object, the cube) into itself. This includes $\mathbb{1}, 8C_3, 6C_2, 6C_4, 3C_2$. With the aid of the following figure, we can illustrate all 24 symmetry elements of O.

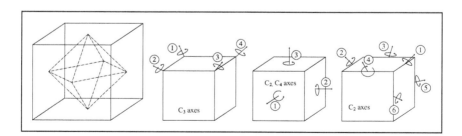

Left: Octahedron inscribed in a cube. Right: The various proper rotation axes in O.

Referring to the table of O_h containing the transformation of the coordinates (x, y, z) (see Lecture 1), the operations of O are the following sets:

$$C_1 \quad C_4^2 \quad C_4 \quad C_2 \quad C_3$$

Summary: We summarize the 11 proper point groups of 3D crystallography (in brackets, the Hermann–Mauguin notation):

- Cyclic: $C_1(1)$ $C_2(2)$ $C_3(3)$ $C_4(4)$ $C_6(6)$
- 3D dihedral: $D_2(222)$ $D_3(32)$ $D_4(422)$ $D_6(622)$
- Platonic: $T(23)$ $O(432)$

5.1.5 *Improper point groups*

The new element appearing in O_3 is the inversion I, which commutes with all other elements of the group O_3. There are two ways one can use the inversion to expand the proper rotation point groups.

5.1.6 *Improper point groups of type a*

Let the new group contain I as an element. Then, the new group has twice the number of elements as the original proper rotation group. In 3D crystallography, starting from the 11 proper rotation point groups, one obtains 11 new improper crystallographic point groups (on the left, the Schönflies nomenclature, and on the right, the Hermann–Mauguin labeling).

Table of the type-a improper point groups and their parent groups.

$C_1(1) \rightarrow C_i(\bar{1})$	$C_2(2) \rightarrow C_{2h}(\frac{2}{m})$	$C_3(3) \rightarrow S_6(\bar{3})$
$C_4(4) \rightarrow C_{4h}(\frac{4}{m})$	$6(= C_6) \rightarrow C_{6h}(\frac{6}{m})$	
$D_2(222) \rightarrow D_{2h}(\frac{2}{m}\frac{2}{m}\frac{2}{m})$	$D_3(32) \rightarrow D_{3d}(\bar{3}\frac{2}{m})$	$D_4(422) \rightarrow D_{4h}(\frac{4}{m}\frac{2}{m}\frac{2}{m})$
$D_6(622) \rightarrow D_{6h}(\frac{6}{m}\frac{2}{m}\frac{2}{m})$		
$T(23) \rightarrow T_h(\frac{2}{m}\bar{3})$	$O(432) \rightarrow O_h(\frac{4}{m}\bar{3}\frac{2}{m})$	

Comments.

1. Reflection groups:

- C_{2h} is built by multiplying both $\mathbb{1}$ and C_2 with I and contains the elements

$$\mathbb{1} \quad C_2 \quad I \quad IC_2$$

As $IC_2 = \sigma_h$, one can think of the group being made of the elements

$$\mathbb{1} \quad C_2 \quad \sigma_h = IC_2 \quad C_2\sigma_h = I$$

where σ_h is the reflection at the plane perpendicular to the main rotation axis from which the name C_{2h} originates.

- C_{4h} and C_{6h} (generated by C_4 and σ_h and C_6 and σ_h, respectively) also belong to this type-a improper point group. C_{2h}, C_{4h}, and C_{6h} are members of a class of improper point groups known as reflection point groups (C_{1h} and C_{3h} will be generated later as type-b improper point groups). The order of the group is $2n$. For instance, the $CHCl_2CHCl_2$ molecule has the symmetry group C_{2h}.

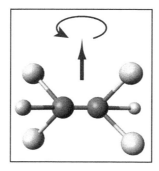

A molecule with C_{2h} point group. The horizontal reflection plane is perpendicular to the twofold rotation axis indicated by an arrow.

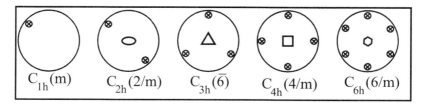

Stereograms of the five reflection groups.

2. Improper rotation point groups:

C_i consists of $\mathbb{1}$ and I. As $I = C_2\sigma_h \equiv S_2$, one can also think of

$$C_i = \{S_2^0, S_2^1\}$$

and the group is also called S_2. It belongs to a category called improper rotation point groups.

- S_6 contains the elements

$$C_3^0 \quad C_3^1 \quad C_3^2 \quad IC_3^0 \quad IC_3^1 \quad IC_3^2$$

One can think of the group as being made of

$$S_6^0 = \mathbb{1} \quad S_6^1 \quad S_6^2 \quad S_6^3 \quad S_6^4 \quad S_6^5$$

i.e. the group is obtained by the powers of the improper rotation S_6. S_6 is also an improper rotation point group (S_4 will be generated later as a type-b improper point group).

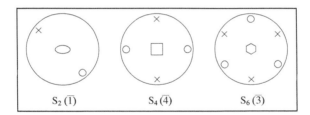

Stereograms of S_n groups of crystallography.

3. Prismatic point groups:

- The groups D_{2h}, D_{4h}, and D_{6h} belong to the class of improper point groups called prismatic groups. The elements with twofold rotations axis multiplied with the inversion can be rewritten as reflection at planes perpendicular to the rotation axis. This means, for instance, that the elements of D_{2h} can be written as

$$\{C_2^0 \quad C_2^1 \quad B_2 \quad B_2' \quad (IC_2 =)\sigma_h \quad (IB_2 =)\sigma_v \quad (IB_2' =)\sigma_v' \quad (I =)C_2\sigma_h = S_2^1\}$$

For example, the point group of benzene is D_{6h}.

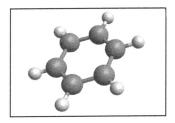

A molecule (benzene, C_6H_6) with D_{6h}-symmetry.

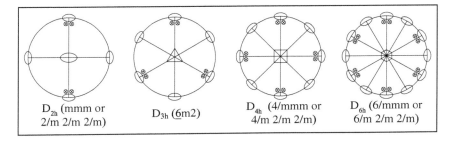

| D_{2h} (mmm or 2/m 2/m 2/m) | D_{3h} ($\bar{6}$m2) | D_{4h} (4/mmm or 4/m 2/m 2/m) | D_{6h} (6/mmm or 6/m 2/m 2/m) |

Stereograms of the D_{nh}-groups. The D_{3h}-point group will be discussed later as a type-b improper point group.

4. Antiprismatic point groups:

- The group D_{3d} contains the elements

$$C_3^0, C_3^1, C_3^2, B_2, B_2', B_2'', IB_2 = \sigma_d, IB_2' = \sigma_d', IB_2'' = \sigma_d'', B_2\sigma_d, B_2\sigma_d', B_2''\sigma_d''$$

It belongs to the class of the so-called antiprismatic groups. The mirror planes labeled with the index "d" are perpendicular to the horizontal plane but are "diagonal", as they bisect the twofold B_2 axes. The last operations can be written as improper rotations involving a rotation and a reflection.

5. The group T_h:

The elements of the group T_h are the proper rotations of T and the improper rotations obtained by multiplying all elements of the group T with the inversion. Referring to the table for O_h containing the transformation of the (x, y, z) coordinates (see Lecture 1), the elements of T_h are given by the sets

$$C_1 \quad C_4^2 \quad C_3 \quad I \quad IC_4^2 \quad IC_3$$

6. The group O_h:

The group O_h was discussed in Lecture 1.

5.1.7   *Improper point groups of type b*

There exist finite subgroups with improper rotations where I is not a symmetry element. To figure out how to construct these groups starting from the groups of proper rotations, consider that the proper rotations within these new groups must build a subgroup $\{R_1, R_2, \ldots, R_n\}$ and that their number is exactly half the total number of elements of the desired group, $\{R_1, R_2, \ldots, R_n, X_1, X_2, \ldots, X_n\}$. As $IX_i \cdot IX_j$ is a proper rotation, the set

$$\{R_1, R_2, \ldots, R_n, IX_1, IX_2, \ldots, IX_n\}$$

is one of the proper point groups. This result provides an algorithm for the construction of further rotation groups with improper symmetry elements:

- Take one of the proper rotation groups, and find all its subgroups with half the number of elements.
- Invert the remaining elements.
- The new set consisting of the elements of the proper subgroup and the inverted elements form the desired type-b improper point group.

Using this algorithm, we obtain 10 distinct further new improper crystallographic point groups of type b. In the following table, the left entry is the original proper point group, in the middle is the subgroup of proper rotations, and on the right is the new subgroup originated by inverting.

$432(\mathrm{O})$	$(23)(\mathrm{T})$	$\bar{4}3m\ (T_d)$
$622(D_6)$	$(6)(C_6)$	$6mm\ (C_{6v})$
$622(D_6)$	$(32)(D_3)$	$\bar{6}m2\ (D_{3h})$
$422(D_4)$	$(4)(C_4)$	$4mm\ (C_{4v})$
$422(D_4)$	$(222)(D_2)$	$\bar{4}2m\ (D_{2d})$
$6(C_6)$	$(3)(C_3)$	$\bar{6}\ (C_{3h})$
$32(D_3)$	$(3)(C_3)$	$3m\ (C_{3v})$
$4(C_4)$	$(2)\ (C_2)$	$\bar{4}\ (S_4)$
$222(D_2)$	$(2)(C_2)$	$mm2\ (C_{2v})$
$2(C_2)$	$(1)(C_1)$	$m\ (C_s = C_{1h})$

Comments.

1. Pyramidal groups:

The pyramidal groups C_{nv} are generated by adding a vertical mirror plane to the original cyclic groups. Because of the existence of a principal C_n-axis, further mirror planes are generated, for a total of n-σ_v mirror planes. These symmetry groups of the n-pyramid are the two-dimensional dihedral groups of an n-polygon. The symmetry elements are

$$\{1, C_n^1, \ldots, C_n^{n-1}, \sigma_v, C_n\sigma_v, \ldots, C_n^{n-1}\sigma_v\}$$

The order of C_{nv} is $2n$. For example, CH_3CCl_3 has the C_{3v} point group (but a different version of the same molecule CH_3CCl_3 has only the C_3 symmetry. There are five crystallographic pyramidal groups, all obtained from the type-b algorithm.

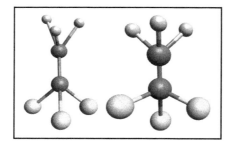

A molecule with C_3 (left) and C_{3v} (right) point groups.

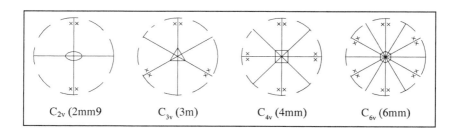

C_{2v} (2mm9)　　　C_{3v} (3m)　　　C_{4v} (4mm)　　　C_{6v} (6mm)

Stereograms of the four pyramidal groups.

2.

The type-b algorithm introduces two further reflection groups, C_{1h} and C_{3h}. As with a C_1 main axis, the horizontal and vertical planes are not distinguishable, and the group C_{1v} is identical to C_{1h}.

3. The type-b improper rotation group S_4

$$S_4^0, S_4^1, S_4^2, S_4^3$$

is generated by the an axis of rotatory reflection, consisting of a rotation by $90°$ around an axis and a reflection at a plane perpendicular to the main rotation axis.

4.

Further type-b groups are the prismatic group D_{3h} and the antiprismatic group D_{2d}.

5. The group T_d:

Finally, the type-b cubic group T_d contains all symmetry transformations of a regular tetrahedron.

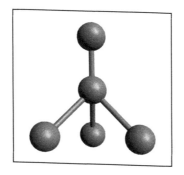

A unit with T_d-symmetry: Si, Ge, and GaAs crystals (three important materials in the semiconductor technology) consist of a network of such tetrahedral units. The local tetrahedral geometry is responsible for most physical properties of these materials and for their functioning.

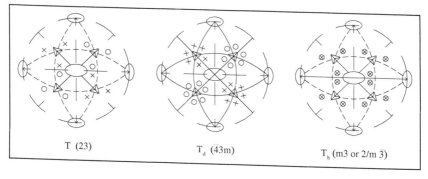

T (23) T_d ($\bar{4}$3m) T_h (m3 or 2/m $\bar{3}$)

Stereograms of the groups T, T_d, T_h.

Referring to the table describing the 48 O_h transformations of the (xyz)-coordinates, the elements of T_d are given by the sets of the type

$$C_1 \quad C_4^2 \quad C_3 \quad IC_4 \quad IC_2.$$

6. Summary:

The resulting 32 point groups represent the finite subgroups of O_3 which are encountered in crystallography: The proper ones are called enantiomorphic, the next class with I as the element are the centrosymmetric. The last 10 are non-enantiomorphic and non-centrosymmetric. In the final table, the 32 point groups relevant to 3D crystallography and their nomenclature are summarized.

Schönflies-symbol	HM-symbol	Schönflies-symbol	HM-symbol
C_1	1	C_3	3
C_i	$\bar{1}$	C_{3i}	$\bar{3}$
C_2	2	D_3	32
C_s	m	C_{3v}	$3m$
C_{2h}	$2/m$	D_{3d}	$3m$
D_2	222	C_6	$\bar{6}$
C_{2v}	$mm2$	C_{3h}	$\bar{6}$
D_{2h}	mmm	C_{6h}	$6/m$
C_4	4	D_6	$6mm$
S_4	$\bar{4}$	C_{6v}	$6mm$
C_{4h}	$4/m$	D_{3h}	$\bar{6}m2$
D_4	422	D_{6h}	$6/mmm$
C_{4v}	$4mm$	T	23
D_{2d}	$\bar{4}2m$	T_h	$m3$
D_{4h}	$4/mmm$	0	432
		T_d	$\bar{4}3m$
		O_h	$m3m$

5.1.8 *Non-crystallographic point groups*

The icosahedral point groups Y, Y_h contain fivefold rotations and are therefore excluded in crystallography. They, however, occur in some important finite systems, such as the C$_{60}$ molecule.

Besides the groups containing fivefold axis, there are further groups that are useful in chemistry.

Linear groups: These groups include $C_{\infty v}$ and $D_{\infty h}$, which are the symmetry groups of linear molecules. $C_{\infty v}$ has symmetry operations $\mathbb{1}$ and C_∞ and an infinite number of σ_v mirror planes that include the main rotation axis. Examples are CO, HCN, NO, and HCl. Linear molecules with a mirror plane perpendicular to the main rotation axis have the symmetry group $D_{\infty h}$, which consists of the symmetry operations $\{\mathbb{1}, C_\infty, \infty C_v, \sigma_h, I,$ and $\infty C_2\}$. Examples are CO_2, O_2, and N_2.

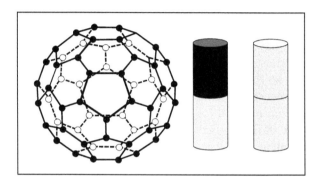

Left: Molecule with icosahedral symmetry. One recognizes the units with fivefold symmetry surrounded by sixfold symmetry units. Middle: Object with C_∞-symmetry (main rotation axis along the cylinder axis). Right: Object with $D_{\infty h}$-symmetry. A continuous circle marks the horizontal reflection plane.

5.2 Constructive crystallography

In this section, we explain and organize the symmetry groups of "crystals", i.e. the subset of isometries that transform a periodic pattern into itself. We focus on two-dimensional patterns (the so called "wallpaper patterns") with the understanding that the geometrical complexity of proofs increases when dealing with three-dimensional objects. Accordingly, we relax the accuracy of our reasoning when dealing with three-dimensional crystals. The approach in these lecture notes is a "constructive one": We construct patterns and symmetry elements, and in doing so, we get acquainted with analyzing a pattern according to its symmetry elements and obtain them. Readers who are interested in complete and specific information on crystallography and space groups should refer to the monumental work called "International Tables for Crystallography" (https://doi.org/10.1107/97809553602060000001).

There is probably an infinite number of patterns (crystals), but they can be classified into a finite number of so-called **space groups**. This means that all materials within a given space group have similar symmetry elements, despite the content of their unit cell being different. The number of space groups in a given space of dimension n is the subject of Hilbert's 18th problem (D. Hilbert, "Mathematical Problems". Bulletin of the American Mathematical Society. **8** 437–479, (1902) doi:10.1090/S0002-9904-1902-00923-3). It was solved by Bieberbach by proving that the number of space groups in n dimensions is finite. A still open question is, however, the exact number of the space groups. In one, two, and three dimensions, we have 2, 17, and 230 space groups, respectively. These numbers show that the complexity of constructing the space groups is growing with n, so that, in this lecture, we limit ourselves to construct the one- and two-dimensional space groups explicitly, while the results for three dimensions are summarized for the convenience of later applications. Now, we briefly review the history of crystallography and list some literature. The subject of crystallography was independently developed by J. S. Fjodorow and A. M. Schönflies in around 1890 and culminated with the classification of solids in terms of Bravais lattices (Auguste Bravais, 1811–1863, was a French physicist known for his work in crystallography and the conception of the Bravais lattices), crystal systems, crystal classes, and space groups. There are several books on the subject of crystallography, mostly reporting complete tables and extensive summaries of crystallographic data, which can be referred to for specific applications. The International Tables of Crystallography, for instance, is a monumental work that contains all knowledge on space groups available up to date in the shape of useful tables and diagrams. The one book which is not as comprehensive but is closer to our "educational" approach is the book by D. E. Sands, *Introduction to Crystallography*, Dover Publications, Inc., Mineola, New York, USA, 1993. You will find there an accurate account of simple and fundamental facts on crystallography, which are partly retold in this lecture.

5.2.1   *An introductory example*

To become acquainted with the subject of crystallography, we discuss a periodic pattern in two dimensions. The pattern consists of circles and "up" and "down" triangles that repeat in the plane. We define two vectors t_1 and t_2 and an origin "O" at which the two vectors have their starting points. "O" is chosen to be centered on one of the circular objects. The rectangle defined by $\mathbf{t}_1$ and $\mathbf{t}_2$ define what is known as a unit cell. The unit cell contains various objects. The most remarkable aspect is that the unit cell and its content repeat indefinitely in the plane, remaining identical. When repeated, the tip of the unit cell vectors define a set of points called the "lattice" points. The lattice points are described by integer multiples of the vectors $\mathbf{t}_1$ and $\mathbf{t}_2$, which are then said to be the basis vectors. Their integer linear combination is given by

$$\mathbf{t}_n = n_1 \cdot \mathbf{t}_1 + n_2 \cdot \mathbf{t}_2$$

$n_i \in \mathscr{Z} \cdot \mathbf{t}_n$ are the lattice vectors. The set of vectors $T \doteq \{\mathbf{t}_n\}$ constitutes a translation group. The pattern is invariant with respect to the elements of T.

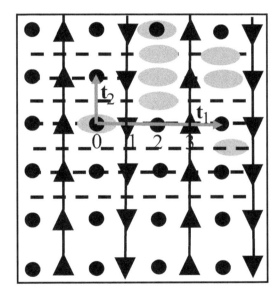

A wallpaper pattern: The basis vectors of the rectangular unit cell are indicated (in bold red). The origin 0 is chosen on top of a ∘-like object. The unit cell contains more than one type of object: The objects 1, 2, and 3 are the origin of further rectangular lattices displaced with respect to 0 by some fraction of $\mathbf{t}_1$. Some twofold rotation axes (indicated by an elliptical gray dot) perpendicular to the plane reside on top of the circular objects. Full vertical lines indicate reflection lines. Dashed horizontal lines indicate glide reflection lines.

The objects within the unit cell need not be on lattice points: Some of the circles and the up/down triangles are at some fractional positions.

However, one can note that similar objects are all on the same lattice, albeit displaced with respect to the unit cell chosen. This is an important rule for the construction of crystals: One can fill the unit cell with objects, but once filled, the objects must occupy the same lattice. To describe the pattern, we could have chosen a different origin, not necessarily on top of some of the objects. The lattice would be displaced, but its geometry would be identical — in this specific case, a "rectangular" two-dimensional lattice. The unit cell we have selected is the smallest unit that repeats, which is known as the "primitive unit cell". One can chose a larger one, containing more objects, that can also be repeated to build the entire pattern, which is called a "non-primitive" unit cell. There is an infinite number of primitive or non-primitive unit cells that can be used to build a pattern.

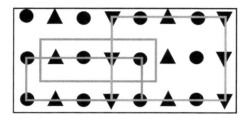

Possible unit cells for the wallpaper pattern shown in the previous figure. The red lines indicate possible primitive unit cells with different origins. The larger rectangle delimited by a blue line indicates a non-primitive unit cell.

Besides the translations $\{(E \mid \mathbf{t}_n)\}$, the pattern has further symmetry elements. Here, we summarize them in terms of the Seitz symbols:

- We can recognize one twofold rotation axis perpendicular to the plane of the pattern and passing through the origin (gray elliptical dot), defined by the transformation

$$\begin{pmatrix} x_1' \\ x_2' \end{pmatrix} = \begin{pmatrix} -1 & 0 \\ 0 & -1 \end{pmatrix} \begin{pmatrix} x_1 \\ x_2 \end{pmatrix}$$

  There is an infinite number of further twofold rotation axes, centered at each circular object and in the interstitial spaces (see the figure above). These rotational elements are given by the Seitz symbol

$$(C_2 \mid \mathbf{t}_n).$$

- We can recognize a reflection in the horizontal line followed by a translation along $\mathbf{t}_1$ by $\frac{1}{2} \mid \mathbf{t}_1 \mid$, given by the transformation

$$
\begin{pmatrix} x_1' \\ x_2' \end{pmatrix} = \begin{pmatrix} -1 & 0 \\ 0 & 1 \end{pmatrix} \begin{pmatrix} x_1 \\ x_2 \end{pmatrix} + \begin{pmatrix} \frac{1}{2} \\ 0 \end{pmatrix}
$$

This is known as the glide reflection line. The horizontal glide reflection line repeats along the vertical with a period of $\frac{|\mathbf{t}_2|}{2}$ and establishes a net of glide reflections described by the symbol

$$
\left(\sigma_v \mid \begin{pmatrix} \frac{1}{2} \\ 0 \end{pmatrix} + \mathbf{t}_n \right)
$$

- We also recognize that all lines joining the triangular objects along the vertical are reflection lines. They are described by

$$
\begin{pmatrix} x_1' \\ x_2' \end{pmatrix} = \begin{pmatrix} 1 & 0 \\ 0 & -1 \end{pmatrix} \begin{pmatrix} x_1 \\ x_2 \end{pmatrix} + \begin{pmatrix} \frac{1}{2} + n_1 \\ n_2 \end{pmatrix}
$$

and the Seitz symbol

$$
\left(\sigma_v' \mid \begin{pmatrix} \frac{1}{2} \\ 0 \end{pmatrix} + \mathbf{t}_n \right)
$$

These operations establish a net of vertical mirror lines.
- In summary, the set of symmetry transformations of the pattern contains the elements

$$
\left\{ (E \mid \mathbf{t}_n); (C_2 \mid \mathbf{t}_n); \left(\sigma_v \mid \begin{pmatrix} \frac{1}{2} \\ 0 \end{pmatrix} + \mathbf{t}_n \right); \left(\sigma_v' \mid \begin{pmatrix} \frac{1}{2} \\ 0 \end{pmatrix} + \mathbf{t}_n \right) \right\}
$$

Comments.

1. Observing these symmetry elements, we note that the rotational and reflectional components of the symmetry elements, taken together, builds the point group C_{2v}. This is an important feature of a crystal: The set of rotations and reflections that appear in the symmetry elements of a crystal builds the so-called "point group" of the crystal and establish its "crystal class". Note that these rotational and reflectional symmetry operations, taken alone, are not necessarily also the symmetry elements of the pattern: Some appear with fractional translations ("glides").

2. We have learned from this example that there are some elements that need to be specified for classifying a pattern:

- A: the type of lattice, defining the translations group;
- B: the point group of the pattern, sampling all rotational and reflectional elements;
- C: the fractional translations, which take into account the presence of objects within the unit cell.

5.2.2   *A. Bravais lattices and crystal systems*

The classification of the symmetry elements of a crystal starts with recognizing its Bravais lattice and its crystal system.

Definition. A lattice is a set of points that are described, with respect to an origin O, by the positional vectors

$$\mathbf{t}_n = n_1 \mathbf{t}_1 + n_2 \mathbf{t}_2 + n_3 \mathbf{t}_3$$

$\mathbf{t}_i$ are the basis vectors of the lattice, and n_i are integer numbers.

Comments:

1. The set $T_3 = \{\mathbf{t}_n\}$ builds a group, called the group of discrete translations. T_2 and T_1 in two respectively one dimensions are defined accordingly.
2. The parallelogram spanned by the vectors $\{\mathbf{t}_i\}$ is called the unit cell.
3. Any periodic pattern (crystal) is associated with one type of lattices, and finding and describing all lattice types is the starting point of the classification of crystals. The construction of the various Bravais lattices is based on two theorems that involve the symmetry elements of the lattice other than translations.

Theorem. *Given a lattice, the group of symmetry operations of the lattice that fixes the origin is finite, i.e. it is one of the point groups.*

Proof. Starting from a lattice point, a continuous rotation or a continuous set of reflection planes would produce an infinite set of lattice points in the vicinity of the origin, contradicting the very definition of lattice, which requires it to be discrete.   ◇

Theorem (crystallographic restriction). *Given a lattice, the order of rotation of a symmetry element of the lattice that fixes a point can only be $n = 1, 2, 3, 4, 6$.*

Proof. Set the origin of the unit cell at some point along the rotation axis and choose a set of mutually orthogonal basis vectors so that $\mathbf{t}_3$ is along the rotation axis. With respect to this orthogonal set of unit vectors, the symmetry element is then described by the rotation matrix

$$\begin{pmatrix} \cos\frac{2\pi}{n} & -\sin\frac{2\pi}{n} & 0 \\ \sin\frac{2\pi}{n} & \cos\frac{2\pi}{n} & 0 \\ 0 & 0 & 1 \end{pmatrix}$$

Because of the requirement of translational symmetry, the symmetry element rotates the primitive vectors by a matrix

$$\begin{pmatrix} a_{11} & a_{12} & a_{13} \\ a_{21} & a_{22} & a_{23} \\ a_{31} & a_{32} & a_{33} \end{pmatrix}$$

with the matrix elements being necessarily integers. Since the two matrices represent the same symmetry element with respect to two basis sets, the matrix must be related by a similarity transformation which leaves the trace invariant. This produces the restriction

$$2\cos\frac{2\pi}{n} \in \mathscr{Z} \Leftrightarrow \cos\frac{2\pi}{n} = \left\{-\frac{1}{2}, 0, \frac{1}{2}\right\} \Leftrightarrow \frac{2\pi}{n} \in \left\{\frac{2\pi}{6}, \frac{2\pi}{4}, \frac{2\pi}{3}, \frac{2\pi}{2}, \frac{2\pi}{1}\right\}$$

Comments.

1. In virtue of this theorem, a one-dimensional lattice can only have the point groups C_1, C_i.
2. A two-dimensional lattice can only have the point groups $C_1, C_2, C_3, C_4, C_6, D_1, D_2, D_3, D_4, D_6$.
3. A three-dimensional lattice can only have one of the 32 point groups of three-dimensional crystallography constructed in the previous section.

Definition. For a given lattice,[3] find the largest point group that transforms the lattice into itself: It is called the **holohedry** group.

Definition. Lattices within the same holohedry group are said to belong to the same **crystal system**.

Comment. A crystal system is characterized by the length of the basis vectors and the angle between them.

We proceed now to construct all possible Bravais lattices and organize them into crystal systems.

---

[3]The pattern or the crystal is not the subject of this definition, but only the lattice itself, onto which the crystal or pattern is defined.

One dimension: The most general lattice consists of equidistant points along one direction. Imposing the point groups C_1 and C_i does not produce a different type of lattice. For the classification of crystals in one dimensions, we ascertain that:

1. there is only one type of lattice, which is called the primitive lattice (P). Any crystal in one dimension has the P lattice.

2. The largest set of elements that transform the lattice into itself and fixes the origin — the so-called holohedry group of the lattice — is C_i. This defines one crystal system, the P-crystal system in one dimensions. The crystal system contains only one Bravais lattice, the P lattice. Any space group in one dimension belong to this crystal system and has the P lattice.

Two dimensions: We construct the Bravais lattice of wallpaper patterns by considering the action of the 10 point groups on $\mathbf{t}_1, \mathbf{t}_2$. We chose a coordinate system where $\mathbf{t}_1$ lies along the positive x-axis and $\mathbf{t}_2$ is in the first upper quadrant of the xy-plane.

1. The oblique crystal system: $|\mathbf{t}_2| \neq |\mathbf{t}_1|$ and the angle between $\mathbf{t}_2$ and $\mathbf{t}_2$ can be any value. The only orthogonal transformations which preserve the oblique lattice by fixing the origin are the identity and the rotation by π about the origin. The holohedry group is C_2. The oblique crystal system has only one Bravais lattice, the oblique Bravais lattice.

2. The rectangular crystal system: Starting from an oblique lattice, we impose the point group symmetry D_1, i.e. we impose a mirror line m.

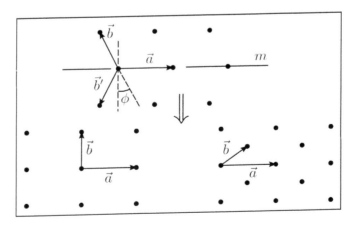

Construction of the two Bravais lattices of the rectangular crystal system: One starts with an oblique lattice defined by the vectors **a** (along the horizontal line) and **b**. Bottom left: The rectangular lattice (Solution I). Bottom right: The rhombic (or rectangular centered) lattice (Solution II).

Referring to the figure, if we mirror the basis vectors **a** and **b** at the horizontal line, we obtain

$$\mathbf{a}' = \mathbf{a}$$
$$\mathbf{b}' = -2 \cdot \sin \phi \cdot |\,\mathbf{b}\,| \cdot \frac{\mathbf{a}}{|\,\mathbf{a}\,|} - \mathbf{b}$$

There is a constraint: **b′ must** be a lattice point, i.e.

$$\mathbf{b}' = n_1 \mathbf{a} + n_2 \mathbf{b} \qquad n_1, n_2 \in \mathscr{L}$$

The equation

$$-2 \cdot \sin \phi \cdot |\,\mathbf{b}\,| \cdot \frac{\mathbf{a}}{|\,\mathbf{a}\,|} - \mathbf{b} = n_1 \mathbf{a} + n_2 \mathbf{b}$$

has two solutions:

Solution I:

$$n_2 = -1 \quad \text{and} \quad \phi = 0$$

This solution leads to a rectangular lattice.

Solution II:

$$n_2 = -1 \quad \sin \phi = \frac{\frac{|a|}{2}}{|\,b\,|}$$

leading to what is called the centered rectangular lattice (or "rhombic" lattice).

These two lattices can both be described by the same orthogonal basis vectors with different lengths. In the case of the rectangular lattice, these are primitive translations, whereas in the case of the centered rectangular lattice, these are non-primitive translations. The largest set of symmetry operations that fixes a point in both Bravais lattices — the holohedry group — is D_2. Accordingly, both belong to the same crystal system, which is called the rectangular crystal system.

3. The square crystal system: We start from an oblique lattice and impose the point group symmetry C_4. This produces the square lattice. The holohedry is D_4, and there is only one lattice in this crystal system.

4. The hexagonal crystal system: We start from an oblique lattice and impose the point group symmetry C_3. This produces the hexagonal lattice. The point groups C_6, D_3, D_6 do not produce any different lattice types. The hexagonal lattice provides one hexagonal crystal system (characterized by the D_6 holohedry group) with one hexagonal Bravais lattice.

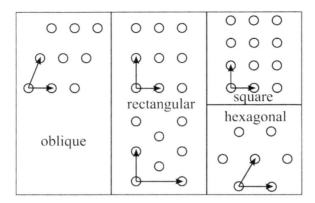

The four crystal systems and five Bravais lattices in two dimensions.

Three dimensions: The lattice types in three dimensions are constructed by starting with a general set of linearly independent vectors, t_1, t_2, t_3, and imposing successively the symmetry elements originating from the point groups. These elements will impose restrictions on the geometry of the lattice. One generates in this way 14 Bravais lattices (Bravais found all possible types of space lattices in 1850). These lattices are assigned to seven holohedry groups that determine seven crystal systems. A complete description of the 14 Bravais lattices can be found, for example, at https://en.wikipedia.org/wiki/Bravais_lattice.

- Triclinic system: The ratio between the length of the basis vectors and the angles between them can assume any value. The holohedry is C_i. There is one Bravais lattice in the so-called triclinic system.
- Monoclinic system: A twofold rotation axis is now imposed. If we assume that this axis coincides with the c axis of the unit cell, a and b can be chosen on the lattice plane normal to c. The angle between a and b is non-constrained. A close inspection of the effect of the symmetry elements on the monoclinic lattice shows that, besides the primitive monoclinic Bravais lattice, a second one is generated, which has lattice points in the ab-plane. There are therefore two lattices with this crystal system (the C_{2h} holohedry group).
- Orthorhombic system: Adding two more twofold axis produces a type of lattice with mutually orthogonal basis vectors and arbitrary reciprocal

length. A close inspection of the effect of these symmetry elements on the monoclinic lattice shows that, besides the primitive orthorhombic lattice, three further Bravais lattices are generated, with lattice points being on the face of the unit cell or at the center. The holohedry is $D_{2h} = mmm$.

- Tetragonal system: Imposing a fourfold axis (take it as the c-axis) constraints the a and b axes to be perpendicular to it and having the same length: The angles between the basis vectors is $90°$ and the ratio of a:b:c is 1:1:c. Careful consideration of the implications of a fourfold axis produces two Bravais lattices in the tetragonal system, one primitive and one centered, with holohedry D_{4h}.

- Trigonal system. Introducing a threefold axis produces a lattice with three basis vectors inclined by the same angle with respect to the threefold axis, having $a = b = c$ and encompassing equal angles. The crystal system generated is the trigonal one (holohedry: D_{3d}). There is only one unit cell belonging to the crystal system, called the rhobohedral unit cell.

- Hexagonal system. Imposing a sixfold axis generates one hexagonal lattice: If we assume that the hexagonal axis is the c-axis, a and b will be perpendicular to it and include an angle of $120°$. The ratio between a,b,c is 1:1:c. The crystal system generated by these groups is the hexagonal one, with one Bravais lattice and holohedry D_{6h}.

- Cubic system. Taking four threefold axes that can be thought as lying along the four diagonal of a cube and three two fold axis produces, after close consideration, three cubic Bravais lattices. The crystal system is characterized by unit cells having basis vectors with a = b = c and angle of $90°$ between them. The holohedry is O_h.

5.2.3 B. Crystal classes

The holohedry is the largest point group compatible with a given crystal system. Each holohedry, however, has a set of subgroups that are also compatible with a given crystal system.

Definition. Given a crystal system, those subgroups of the holohedry group that are not already formed in crystal systems with lower symmetry are the crystal classes of that given crystal system.

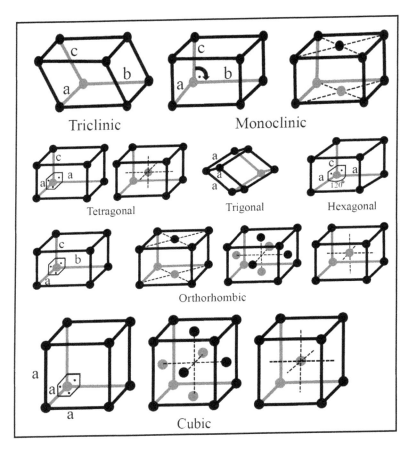

The seven crystal systems and 14 Bravais lattices of three-dimensional crystallography.

Comment.

All crystals can be classified as belonging to a crystal system, possessing a Bravais lattice and belonging to one of its crystal classes.

One-dimensional crystals: There are two crystal classes within the C_i holohedry: C_1 and C_i.

Two-dimensional systems: The following table provides insight into the crystal classes of 2D systems.

Three-dimensional systems: The following figure provides insight into the crystal classes of 3D systems.

Crystal classes in 2D: (Left) the crystal system, (Center) Schönflies notation, (Right) Hermann–Mauguin notation.

Crystal system	S	HM
Oblique	C_1	1
	C_2	$\bar{2}$
Rectangular	D_1	$1m$
	D_2	$2m$
Hexagonal	C_3	3
	D_3	$3m$
	C_6	6
	D_6	$6m$
Square	C_4	4
	D_4	$4m$

Crystal system	S	HM
Triclinic	C_1	1
	C_i	$\bar{1}$
Monoclinic	C_2	2
	C_{2s}	m
	C_{2h}	2/m
Orthorhombic	D_2	222
	C_{2v}	mm2
	D_{2h}	mmm
Tetragonal	C_4	4
	S_4	$\bar{4}$
	C_{4h}	4/m
	D_4	422
	C_{4v}	4mm
	D_{2d}	$\bar{4}2m$
	D_{4h}	4/mmm

Crystal system	S	HM
Trigonal	C_3	3
	C_{3i}	$\bar{3}$
	D_3	32
	C_{3v}	3m
	D_{3d}	$\bar{3}m$
Hexagonal	C_6	6
	C_{3h}	$\bar{6}$
	C_{6h}	6/m
	D_6	622
	C_{6v}	6mm
	D_{3h}	$\bar{6}m2$
	D_{6h}	6/mmm
Cubic	T	23
	T_h	m3
	O	432
	T_d	$\bar{4}3m$
	O_h	m3m

Crystal classes in 3D: (Left) the crystal system, (Center) Schönflies notation, (Right) Hermann–Mauguin notation.

5.2.4   C. Space groups

From the introductory example, we have learned that the pattern repeats with the periodicity given by the lattice, but the unit cell can be filled with more objects. Translational symmetry demands that these various objects belong to the same type of lattice: The various lattices hosting similar objects are interpenetrating, i.e. they are displaced by some fraction of the basis vectors with respect to each other. The extra degree of freedom represented by the possibility of constructing interpenetrating lattices of the same type introduces a further differentiation in the classification of crystals that goes beyond Bravais lattices, crystal systems, and crystal classes, and produces the so-called space groups (or wallpaper groups in two dimensions).

Definition. Let G be a space group with the point group G_0 and the translation group T. For each $g \in G_0$, there is a vector $\mathbf{t}_g$ such that $(g \mid \mathbf{t}_g) \in G$. $\mathbf{t}_g$ are the so-called point group vectors and are uniquely determined up to addition of an element of T.

Comments.

1. In a compact way, we write

$$G = \{(g \mid \mathbf{t}_g + \mathbf{t}_n) : g \in G_0, \mathbf{t}_n \in T\}$$

2. In order to describe G, we need to specify T (i.e. the lattice), G_0 (the crystal class), **and** the point group vectors $\mathbf{t}_g$: The rotational elements of G_0 become symmetry elements of a crystal only in connection with suitable non-primitive translations $\mathbf{t}_g$, which are appropriately called point group vectors. This is the case when the symmetry elements are, for example, glide reflections or screw rotations. The degree of freedom represented by the point group vectors leads, ultimately, to the diversification of the various crystal classes into space groups.

Proposition (see, for example, P. J. Morandi, The Classification of Wallpaper Patterns: From Group Cohomology to Escher's Tessellations, Department of Mathematical Sciences, New Mexico State University, Las Cruces, NM 88003, free on-line book, https://drive.google.com/file/d/12ft i6FKkwqlCyntRGAkXtjbrfUyw3p-g/view). *The point group vectors* $\mathbf{t}_g$ *are unique modulo addition of an element of* T *and must satisfy the relation*

$$\mathbf{t}_g + g(\mathbf{t}_h) - \mathbf{t}_{gh} \in T$$

Proof. The proof consists of suitably multiplying the elements of the group to generate the element $(\mathbb{1} \mid \mathbf{t}_g + g(\mathbf{t}_h) - \mathbf{t}_{gh})$. One can verify that, if $(g \mid \mathbf{t}_g)$ and $(h \mid \mathbf{t}_h)$ are any elements of the group, then

$$(\mathbb{1} \mid \mathbf{t}_g + g(\mathbf{t}_h) - \mathbf{t}_{gh}) = (g \mid \mathbf{t}_g)(h \mid \mathbf{t}_h)(gh \mid \mathbf{t}_{gh})^{-1} \qquad \diamond$$

Comments.

1. We use this relation for generating point group vectors.

2. Some useful notation: In one- and two-dimensional crystallography, one uses several internationally recognized characters, p, c, m, and g, and numbers, 1, 2, 3, 4, and 6. The letter "p" refers to the lattice and stands for the word "primitive". For the centered rectangular lattice, one uses a non-primitive cell together with its center as the building block of the lattice and indicate it with the letter "c". The character "m" denotes a reflection, and "g" denotes a glide reflection. The numbers 1, 2, 3, 4, and 6 indicate rotations of the corresponding order. There is also a set of (self-explanatory, sometimes "locally defined" symbols) used to identify symmetry axis and reflection planes, see the following figures. For instance, thin lines are used to represent the lattice, thick lines to represent reflection lines, and dashed lines to represent glide reflection lines. If colors are available, one can use them too. Three-dimensional crystallography requires a larger set of symbols and characters, as explained later. To designate a space group, one uses the international notation which specifies that the first letter gives the type of lattice. The first letter is followed by a set of characters identifying the symmetry elements.

3. **Definition.** A space group is said to be symmorphic (split) if, by a suitable coordinate transformation, all point group vectors can be rendered vanishing (the origin of the lattice can be located in such a way that some point group vectors turn out to be non-vanishing; however, these groups are still called symmorphic). Otherwise, the group is said to be non-symmorphic.

5.2.5 *One-dimensional space groups*

Crystal class C_1: For the crystal class C_1, the point group vector is vanishing. In fact, using the relation for point group vectors, we find

$$t_{\mathbb{1}} + t_{\mathbb{1}} - t_{\mathbb{1}} = t_{\mathbb{1}} \in \mathcal{Z}$$

i.e. $t_{\mathbb{1}}$ can be set to vanish. The space group contains the elements

$$p1 = \{(\mathbb{1} \mid n)\}$$

In one dimension, we have only a primitive lattice (p). The next symbol is "1", identifying the point group. This 1D space group is therefore called the group $p1$.

Crystal class C_i: Again, $t_\mathbb{1} = 0$. The desired point group vector t_I can assume arbitrary values, as any value of t_I fulfills the relation for point group vectors. However, the element $(I \mid t_I)$ is similar to $(I \mid 0)$:

$$\left(\mathbb{1} \mid -\frac{t_I}{2}\right)(I \mid t_I)\left(\mathbb{1} \mid -\frac{t_I}{2}\right)^{-1} = (I \mid 0)$$

so that the point group vector can be rendered vanishing by a suitable choice of the origin. Therefore, the crystal class C_i sustains one space group with the elements

$$p\bar{1} = \{(\mathbb{1} \mid n), (I \mid n)\}$$

$p1$ and $p\bar{1}$ are symmorphic space groups.

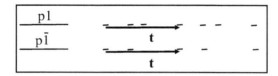

Graphical representation of the space groups $p1$ and $p\bar{1}$: The thin lines indicate the Bravais lattice. The primitive translation vector is indicated by **t**. The $p\bar{1}$ space group has two identical components in the unit cell and therefore has inversion centers located in the middle of the two nearest dashes.

5.2.6 *Two-dimensional space groups*

A large assembly of patterns with two-dimensional space groups can be found at https://mcescher.com/.

The oblique crystal system

In this crystal system, the lattice is oblique. There are two possible crystal classes: C_1 and C_2.

Crystal class C_1: Within this crystal class, we have only one space group: **p1**. Its elements are the primitive translations

$$p1 = \{(\mathbb{1} \mid m\mathbf{t}_1 + n\mathbf{t}_2)\}$$

m, n in $\mathscr{Z}$ (symmorphic).

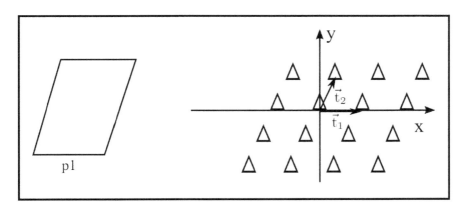

Left: Graphical representation of the space groups $p1$. Right: Wallpaper pattern with space group $p1$.

Crystal class C_2: The elements of C_2 are $\mathbb{1}$ and C_2. The question is now to establish whether the element C_2 has a non-zero point group vector, which we designate as $\mathbf{t}_2$. To answer this question, we apply the general relation to the desired point vector $\mathbf{t}_2$:

$$\mathbf{t}_2 + C_2(\mathbf{t}_2) \in T$$

As this relation can be is fulfilled for any $\mathbf{t}_2$, this rotational element can be assigned any point group vector. But the equation $C_2\mathbf{x} + \mathbf{t}_2 = \mathbf{x}$ has, as unique solution, a fixed point of a half-turn, and we may as well take the fixed point of this half-turn as the origin of the lattice, so that the point group vector can be chosen to be vanishing. The elements of the space group are

$$p2 = \{(\mathbb{1} \mid m\mathbf{t}_1 + n\mathbf{t}_2), (C_2 \mid m\mathbf{t}_1 + n\mathbf{t}_2)\}$$

(symmorphic). Note that an operation of the type $(C_2, m\mathbf{t}_1 + n\mathbf{t}_2)$ has a fixed point at $\frac{m\mathbf{t}_1 + n\mathbf{t}_2}{2}$. In other words, it is a half-turn at the points $\frac{m\mathbf{t}_1 + n\mathbf{t}_2}{2}$.

The rectangular crystal system

The rectangular system has two Bravais lattices, and there are four orthogonal transformations that preserve the lattice: the identity, half-turn, reflection in the horizontal axis, and reflection in the vertical axis. These

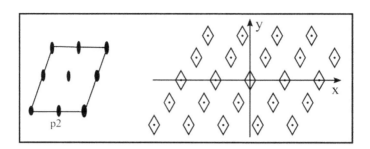

Left: Graphical representation of the space groups $p2$. Right: Wallpaper pattern with space group $p2$.

elements provide the holohedry group

$$D_2 = \{\mathbb{1}, C_2, \sigma_h, \sigma_v\}$$

(σ indicates a reflection at a line). The point groups of any wallpaper group of this crystal system is one of the subgroups of the holohedry group. The two crystal classes not encountered in the previous crystal system are D_1 and D_2.

Crystal class D_1: This crystal class sustains both Bravais lattices: primitive and centered rectangular. We start with the primitive rectangular one.

Primitive rectangular lattice: The point group D_1 contains the elements $\mathbb{1}$ and σ_h. The question is whether the point group vector $\mathbf{t}_h$ of σ_h is finite. Let us choose $\mathbf{t}_1$ along the horizontal axis and take $\mathbf{t}_h = \alpha\mathbf{t}_1 + \beta\mathbf{t}_2$ ($0 \leq \alpha, \beta \leq 1$). As $\mathbf{t}_{\mathbb{1}} = 0$, the relation for the putative point group vector $\mathbf{t}_h$ reads

$$\underbrace{\mathbf{t}_h}_{\alpha\mathbf{t}_1 + \beta\mathbf{t}_2} + \underbrace{\sigma_h(\mathbf{t}_h)}_{\alpha\mathbf{t}_1 - \beta\mathbf{t}_2} -\mathbf{0} \in T$$

i.e.

$$2\alpha \in \mathscr{Z} \quad \alpha < 1$$

For this equation to be satisfied, we have two alternatives: $\alpha = 0$ or $\alpha = \frac{1}{2}$, with β being arbitrary. There are therefore two wallpaper groups with the primitive lattice in the crystal class D_1:

$$pm = \{(\mathbb{1} \mid m\mathbf{t}_1 + n\mathbf{t}_2)), (\sigma_h \mid m\mathbf{t}_1 + n\mathbf{t}_2))$$

is symmorphic. pm contains, besides all translations, a reflection at the horizontal lines $y = \frac{n}{2}$, $n = 0, \pm1, \pm2, \dots$.

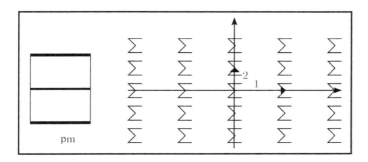

Left: Graphical representation of the space groups *pm*. Right: Wallpaper pattern with space group *pm*.

The wallpaper group

$$pg = \left\{ (\mathbb{1} \mid m\mathbf{t}_1 + n\mathbf{t}_2), \left(\sigma_h \mid \frac{1}{2}\mathbf{t}_1 + m\mathbf{t}_1 + n\mathbf{t}_2 \right) \right\}$$

contains, besides the translations, all glides along horizontal lines that run through the lattice points or midway between the lattice points. The length of each glide is an odd multiple of $\frac{1}{2}$. The group is non-symmorphic.

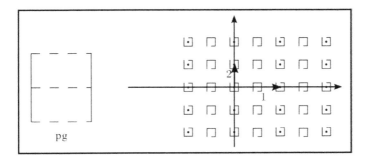

Left: Graphical representation of the space groups *pg*. Right: Wallpaper pattern with space group *pg*.

Centered rectangular lattice: To analyze the wallpaper group arising within this crystal class for the centered rectangular lattice, we choose the basis vectors **a** and **b** joining two centered lattice points lying one above the other below the horizontal line. The element σ_h exchanges **a** and **b**, so that a putative point group vector $\mathbf{t}_h = \alpha\mathbf{a} + \beta\mathbf{b}$ must satisfy the equation

$$\alpha\mathbf{a} + \beta\mathbf{b} + \alpha\mathbf{b} + \beta\mathbf{a} \in T$$

i.e.

$$(\alpha + \beta)(\mathbf{a} + \mathbf{b}) \in T$$

For this to be an element of T, we must have

$$\alpha + \beta \in \mathcal{Z}$$

We can choose $\beta = -\alpha$, so that

$$\mathbf{t}_h = \alpha(\mathbf{a} - \mathbf{b})$$

Groups with different α are related by a similarity operation $(\mathbb{1} \mid \alpha\mathbf{a})$ and are isomorphic. Thus, the wallpaper group can be chosen to have

$$\alpha = \beta = 0$$

and is therefore symmorphic. The group is called **cm** and has the elements

$$cm = \{(\mathbb{1} \mid m\mathbf{a} + n\mathbf{b})), (\sigma_h \mid m\mathbf{a} + n\mathbf{b}))\}$$

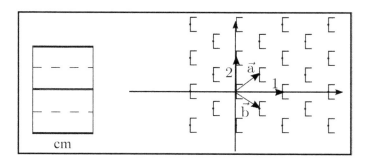

Left: Graphical representation of the space groups *cm*. Right: Wallpaper pattern with space group *cm*.

Referring to the non-primitive basis vectors of the centered lattice, we have $\mathbf{a} = \frac{1}{2}\mathbf{t}_1 + \frac{1}{2}\mathbf{t}_2$ and $\mathbf{b} = \frac{1}{2}\mathbf{t}_1 - \frac{1}{2}\mathbf{t}_2$, so that the elements of *cm* are written as

$$cm = \left\{ \left(\mathbb{1} \mid \frac{m+n}{2}\mathbf{t}_1 + \frac{m-n}{2}\mathbf{t}_2 \right), \left(\sigma_h \mid \frac{m+n}{2}\mathbf{t}_1 + \frac{m-n}{2}\mathbf{t}_2 \right) \right\}$$

Those lines that go through the lattice points and midway between the lattice points are reflection lines. The reflection at those lines that go through the quarter lengths are glide lines with translations along the horizontal axis by odd multiples of $\frac{1}{2}\mathbf{t}_1$.

Comment. This is an example of a symmorphic space group that appears with fractional translations when a non-suitable coordinate system is used.

Crystal class D_2: This crystal class contains four wallpaper groups. Three of them fall on the primitive rectangular lattice.

Primitive rectangular lattice: The point group elements are $\mathbb{1}, C_2, \sigma_h, \sigma_v$. We address first the question about the existence of point group vectors for the reflection elements. We assign a putative point group vector $\mathbf{u} = \alpha\mathbf{t}_1 + \beta\mathbf{t}_2$ to σ_h and test for its existence using the relation

$$\alpha\mathbf{t}_1 + \beta\mathbf{t}_2 + \sigma_h(\alpha\mathbf{t}_1 + \beta\mathbf{t}_2) \in T$$

We obtain the possible solutions:

$$\alpha = 0 \quad \text{or} \quad \alpha = \frac{1}{2}; \quad \text{any} \quad \beta$$

Setting $\beta = 0$, we obtain the possible point group vectors for σ_h:

$$\mathbf{0} \quad \text{or} \quad \frac{1}{2}\mathbf{t}_1$$

With the same *ansatz*, we obtain the possible point group vectors for σ_v:

$$\mathbf{0} \quad \frac{1}{2}\mathbf{t}_2$$

These solutions produce four possible wallpaper groups. The two with the elements

$$(\sigma_h, 0), \left(\sigma_v, \frac{1}{2}\mathbf{t}_2\right)$$

and

$$\left(\sigma_h, \frac{1}{2}\mathbf{t}_1\right), (\sigma_v, \mathbf{0})$$

are isomorphic, as their elements are conjugated by the similarity operation C_4. There are therefore three space groups with a primitive rectangular lattice and in the crystal class D_2.

- The first group is called *p2mm* and contains the elements

$$p2mm = \{(\mathbb{1} \mid m\mathbf{t}_1 + n\mathbf{t}_2), (C_2 \mid m\mathbf{t}_1 + n\mathbf{t}_2), (\sigma_h \mid m\mathbf{t}_1 + n\mathbf{t}_2), (\sigma_v \mid m\mathbf{t}_1 + n\mathbf{t}_2)$$

This group has half-turns centered at $\frac{m}{2}\mathbf{t}_1 + \frac{n}{2}\mathbf{t}_2$, horizontal reflection lines crossing the vertical coordinate at $\frac{n}{2}\mathbf{t}_2$, and vertical reflection lines crossing the horizontal coordinate at $\frac{n}{2}\mathbf{t}_1$.

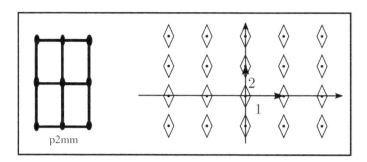

Left: Graphical representation of the space groups $p2mm$. Right: Wallpaper pattern with space group $p2mm$.

- The next group contains the elements $(\sigma_h \mid 0)$ and $(\sigma_v \mid \frac{1}{2}\mathbf{t}_2)$. The product

$$\left\{ \left(\sigma_v \mid \frac{1}{2}\mathbf{t}_2 \right)(\sigma_h \mid 0) \right\} = \left(C_2 \mid \frac{1}{2}\mathbf{t}_2 \right)$$

indicates that the orthogonal element C_2 appears in the space group with the fractional translation $\frac{1}{2}\mathbf{t}_2$. The group is called $p2mg$ and consists of the elements

$$p2mg = \left\{ (\mathbb{1} \mid m\mathbf{t}_1 + n\mathbf{t}_2), \left(C_2 \mid \frac{1}{2}\mathbf{t}_2 + m\mathbf{t}_1 + n\mathbf{t}_2 \right), \right.$$

$$\left. (\sigma_h \mid m\mathbf{t}_1 + n\mathbf{t}_2), \left(\sigma_v \mid \frac{1}{2}\mathbf{t}_2 + m\mathbf{t}_1 + n\mathbf{t}_2 \right) \right\}$$

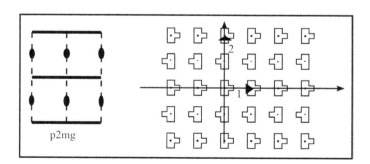

Left: Graphical representation of the space groups $p2mg$. Right: Wallpaper pattern with space group $p2mg$.

This groups has half-turns centered at the points $\frac{m}{2}\mathbf{t}_1 + \frac{1}{2}(n + \frac{1}{2})\mathbf{t}_2$. The horizontal reflection lines pass through the lattice points or midway between them. The vertical reflection lines are glides and pass through the lattice points or midway between them.

- The next group contains the elements $(\sigma_h \mid \frac{1}{2}\mathbf{t}_1))$ and $(\sigma_v \mid \frac{1}{2}\mathbf{t}_2)$. The product of these elements is $(C_2 \mid \frac{1}{2}(\mathbf{t}_1 - \mathbf{t}_2))$. The group is called $p2gg$ and contains the elements

$$\left\{ (\mathbb{1} \mid m\mathbf{t}_1 + n\mathbf{t}_2), \left(C_2 \mid \frac{1}{2}(\mathbf{t}_1 - \mathbf{t}_2) + m\mathbf{t}_1 + n\mathbf{t}_2 \right), \right.$$

$$\left. \left(\sigma_h \mid \frac{1}{2}\mathbf{t}_1 + m\mathbf{t}_1 + n\mathbf{t}_2 \right), \left(\sigma_v \mid \frac{1}{2}\mathbf{t}_2 + m\mathbf{t}_1 + n\mathbf{t}_2 \right) \right\}$$

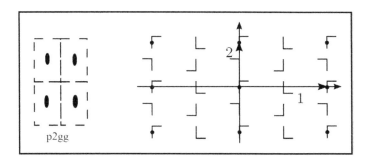

Left: Graphical representation of the space groups $p2gg$. Right: Wallpaper pattern with space group $p2gg$.

This group has vertical and horizontal glide lines passing through the lattice points or midway between them. The half-turns are centered at the points $\frac{1}{4}(\mathbf{t}_1 - \mathbf{t}_2) + \frac{1}{2}(m\mathbf{t}_1 + n\mathbf{t}_2)$.

Centered rectangular lattice: In the centered rectangular lattice, the crystal class D_2 have one wallpaper group: **c2mm**. Using the same arguments introduced for the crystal class D_1, we find that the point group vectors for the reflections can be chosen to be vanishing. This implies that the point group vector of C_2 is vanishing, so that the group **c2mm** contains the elements

$$c2mm = \{(\mathbb{1} \mid m\mathbf{a} + n\mathbf{b})), (C_2 \mid m\mathbf{a} + n\mathbf{b})),$$

$$(\sigma_h \mid m\mathbf{a} + n\mathbf{b}), (\sigma_v \mid m\mathbf{a} + n\mathbf{b}))$$

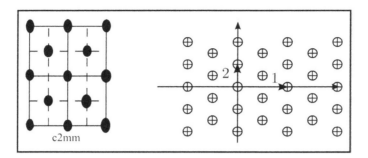

Left: Graphical representation of the space groups *c2mm*. Right: Wallpaper pattern with space group *c2mm*.

Translated to the non-primitive basis vectors, we have

$$\mathbf{a} = \frac{1}{2}\mathbf{t}_1 + \frac{1}{2}\mathbf{t}_2$$

$$\mathbf{b} = \frac{1}{2}\mathbf{t}_1 - \frac{1}{2}\mathbf{t}_2$$

The four elements of **c2mm** now read:

a. $(\mathbb{1} \mid \frac{m+n}{2}\mathbf{t}_1 + \frac{m-n}{2}\mathbf{t}_2)$. These are integer or half-integer translations in both vertical and horizontal directions.

b. $(C_2 \mid \frac{m+n}{2}\mathbf{t}_1 + \frac{m-n}{2}\mathbf{t}_2)$. These are half-turns centered at points within the non-primitive unit cell (see figure).

c. $(\sigma_h \mid \frac{m+n}{2}\mathbf{t}_1 + \frac{m-n}{2}\mathbf{t}_2)$. These are horizontal reflection lines that go through the lattice points and glide reflection lines in between them.

d. $(\sigma_v \mid \frac{m+n}{2}\mathbf{t}_1 + \frac{m-n}{2}\mathbf{t}_2)$. These are vertical reflection lines that go through the lattice points and glide reflection lines in between them.

The square crystal system

There are two crystal classes associated with the square Bravais lattice: C_4 and D_4.

Crystal class C_4: It has one wallpaper group called **p4**. We determine the point group vectors. Assume that **u** is the point group vector of the operation C_4. We have $(C_4, \mathbf{u})^2 = (C_2, C_4\mathbf{u} + \mathbf{u})$ and $(C_4, \mathbf{u})^3) = (C_4^3\mathbf{u} + C_4\mathbf{u} + C_2\mathbf{u})$. As $(C_4, \mathbf{u})$ fixes a point, we can set the origin of the lattice exactly at that point and consider $\mathbf{u} = 0$. This means that the wallpaper group **p4** is symmorphic and contains the operations

$$p4 = \{(\mathbb{1} \mid m\mathbf{t}_1 + n\mathbf{t}_2), (C_4 \mid m\mathbf{t}_1 + n\mathbf{t}_2), (C_2 \mid m\mathbf{t}_1 + n\mathbf{t}_2), (C_4^3 \mid m\mathbf{t}_1 + n\mathbf{t}_2)\}$$

Crystal class D_4: It contains the subgroup C_4 and four reflections. Let us set again the origin of the coordinate system so that we can assume that $\mathbf{t}_{C_i} \equiv \mathbf{0}$. We need to find possible point group vectors for the reflections. Assume the point group vector $\mathbf{u}$ for σ_h. One can use the relation for point group vectors with $g = (C_4\sigma_h)$ and $h = C_4$. Then, we have

$$\mathbf{t}_{C_4\sigma_h} + C_4\sigma_h(\mathbf{0}) - \mathbf{t}_{C_4\sigma_h C_4}$$

The point group vector of $C_4\sigma_h$ is just $C_4\mathbf{u}$ while $C_4\sigma_h C_4 = \sigma_h$, so that the relation for the point group vectors provides a constraint for $\mathbf{u}$:

$$C_4\mathbf{u} - \mathbf{u} \in T$$

Let us apply this constraint to $\alpha\mathbf{t}_1 + \beta\mathbf{t}_2$:

$$\alpha\mathbf{t}_2 - \beta\mathbf{t}_1 - \alpha\mathbf{t}_1 - \beta\mathbf{t}_2 = (-\alpha - \beta)\mathbf{t}_1 + (\alpha - \beta)\mathbf{t}_2 \in T$$

This means that

$$(\alpha \pm \beta) \in \mathscr{Z}$$

Because $0 \leq \alpha, \beta \leq 1$, the only possible solutions are

$$\alpha = \beta = 0 \quad \text{or} \quad \alpha = \beta = \frac{1}{2}$$

The point group vectors for the further reflection operations $C_4\sigma_h$, $C_2\sigma_h, C_4^3\sigma_h$ are obtained by rotating the point group vector $\mathbf{u}$ accordingly and are related to $\mathbf{u}$ by a primitive translation. The two space groups originating within the D_4 crystal class are

$$\begin{aligned} p4mm = \{&(\mathbb{1} \mid m\mathbf{t}_1 + n\mathbf{t}_2), (C_4 \mid m\mathbf{t}_1 + n\mathbf{t}_2), (C_2 \mid m\mathbf{t}_1 + n\mathbf{t}_2), (C_4^3 \mid m\mathbf{t}_1 + n\mathbf{t}_2) \\ &(\sigma_0 \mid m\mathbf{t}_1 + n\mathbf{t}_2), (\sigma_{\frac{\pi}{4}} \mid m\mathbf{t}_1 + n\mathbf{t}_2), (\sigma_{\frac{\pi}{2}} \mid m\mathbf{t}_1 + n\mathbf{t}_2), (\sigma_{\frac{3\pi}{4}} \mid m\mathbf{t}_1 + n\mathbf{t}_2)\} \end{aligned}$$

and

$$\begin{aligned} p4gm = &(\mathbb{1} \mid m\mathbf{t}_1 + n\mathbf{t}_2), (C_4 \mid m\mathbf{t}_1 + n\mathbf{t}_2), (C_2 \mid m\mathbf{t}_1 + n\mathbf{t}_2), (C_4^3 \mid m\mathbf{t}_1 + n\mathbf{t}_2) \\ &(\sigma_0 \mid \frac{1}{2}(\mathbf{t}_1 + \mathbf{t}_2) + m\mathbf{t}_1 + n\mathbf{t}_2), (\sigma_{\frac{\pi}{4}} \mid \mathbf{t}_1 + \mathbf{t}_2) + m\mathbf{t}_1 + n\mathbf{t}_2) \\ &(\sigma_{\frac{\pi}{2}} \mid \mathbf{t}_1 + \mathbf{t}_2) + m\mathbf{t}_1 + n\mathbf{t}_2), (\sigma_{\frac{3\pi}{4}} \mid \mathbf{t}_1 + \mathbf{t}_2) + m\mathbf{t}_1 + n\mathbf{t}_2) \end{aligned}$$

It is left to the reader to recognize the elements geometrically. Their summary is given in the following figure.

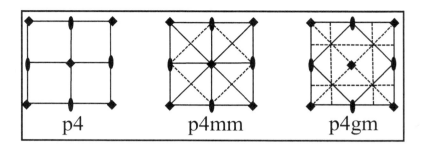

Crystallographic symbols for the 2D square wallpaper groups.

Hexagonal crystal system

This crystal system sustains the hexagonal lattice and has four crystal classes: C_3, C_6, D_3, D_6. The five space groups originating from them by addition of the point groups vectors to the point group operations are summarized in the following figure.

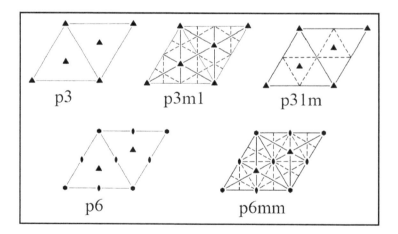

Graphical representation of the hexagonal 2D space groups.

5.2.7 *Three-dimensional space groups*

There is a total of 230 crystallographic space groups in the three-dimensional Euclidean space. They were constructed at the end of the

19th century by the mathematicians Fedorov (1890) and Schoenflies (1891) and are tabulated in the excellent book on crystallography by C. Giacovazzo (Carmelo Giacovazzo, *Symmetry in Crystals*, https://doi.org/10.1093/ac prof:oso/9780199573653.003.0001). Combining the 32 point groups with the 14 Bravais lattices, one obtains 73 (symmorphic) space groups. The remaining space groups are non-symmorphic: They arise when proper and improper rotation axes are replaced by screw axes of the same order and mirror planes by glide planes. This means that some of the operations of the point group have finite point group vectors. The notations and the information necessary to identify space groups are given, for example, in the book, by Carmelo Giacovazzo, or in the *International Tables for Crystallography, Volume A, Space-Group Symmetry* (5th edn.), Editor: Theo Hahn, Published for the International Union of Crystallography by Springer, 2005, DOI: https://doi.org/10.1107/97809553602060 000001. The international notation for labeling space groups is the Hermann–Mauguin one, although in the literature,, one also encounters the Schönflies notation (very often, actually). We provide, in the following, some hints on how to read the space group tables, such as the one by Giacovazzo.

The first symbol encountered in the notation for a space group identifies the type of lattice. P labels the primitive lattice with one lattice point per unit cell. The letter I labels the body-centered lattice with two lattice points per unit cell. The letters A, B, and C label the A, B, and C face- centered lattices, respectively, with two lattice points per unit cell. F is the all face-centered lattice (four lattice points).

The first letter is followed by a set of characters identifying the symmetry elements. The characters related to the point group operations have been discussed in relation to the two-dimensional space groups (the numbers 1, 2, 3, 4, and 6 for the n-fold rotation axes, and the letter m for a mirror plane). The symbol identifying screw axes is n_p, indicating a rotation by $\frac{2\pi}{n}$ followed by a translation of $\frac{p}{n}$ in the direction of the axis. For example, a 2_1 axis involves a rotation by $180°$ followed by a translation by one half of a unit cell parallel to the rotation axis. The combination of a reflection with a translation is called a glide plane. If the glide plane is parallel to the a, b, and c axes, then the symmetry element is a reflection in the plane and a translation by $\frac{1}{2}$ along a, b, or c, respectively. The symbol of the operation is a, b, or c, respectively. A diagonal glide is denoted by n and involves translations by $\frac{a+b}{2}, \frac{b+c}{2}$, or $\frac{a+c}{2}$, i.e. the direction of the glide is along the face diagonals. In the tetragonal, rhombohedral, and cubic systems one can have a glide reflection along the diagonal with the translation $\frac{a+b+c}{2}$. The diamond glide is denoted by d and has the translation of $\frac{a\pm b}{4}$, $\frac{a\pm c}{4}$, $\frac{b\pm c}{4}$, or $\frac{a\pm b\pm c}{4}$ for tetragonal and cubic, respectively.

Let us discuss some of the space groups (see the following table, compiled from Giacovazzo).

Excerpt of the table in the book of C. Giacovazzo on the classification of 3D space groups: (Left column) crystal system, (Central column), crystal class, and (Right column) Space group.

Crystal system	Point group	Space groups
Triclinic	1	$P1$
	$\bar{1}$	$P\bar{1}$
Monoclinic	2	$P2$, $P2_1$, $C2$
	m	Pm, Pc, Cm, Cc
	$2/m$	$P2/m$, $P2_1/m$, $C2/m$, $P2/c$, $P2_1/c$, $C2/c$
...	...	...
...	...	...
...	...	...
...	...	...
Cubic	...	...
	...	...
	...	...
	$m\bar{3}m$	..., ..., ..., ..., $Fm\bar{3}m(=O_h^5)$,..., $Fd\bar{3}m(=O_h^7)$,...

- The triclinic crystal system has only one lattice (P) and two point groups $(1,\bar{1})$. The only possible space groups are $P1$ and $P\bar{1}$ — no point groups vectors are allowed.
- The monoclinic system has P and C lattices and three possible crystal classes: 2, m, and $2/m$. Take, for example, the class $2/m$. We expect at least the two symmorphic space groups $P2/m$ and $C2/m$. Rotations and reflections can be replaced by the corresponding screw axes and glide planes. In the example of the crystal class $2/m$, this replacement gives the space groups $P2_1/m$, $P2/c$, $P2_1/c$, and $C2/c$. During this replacement, one must avoid duplications: For instance, $P2/c$ is the same as $P2/a$ or $P2/b$ except for the renaming of the axes. $C2_1/c$ and C_2/c are obtained by shifting the origin of the coordinate system and must be considered identical.
- A final look at the cubic crystal system: In the table, two space groups in the crystal class $m\bar{3}m$ are highlighted. On one side the symmorphic space group Fm$\bar{3}$m ($= O_h^5$ in the Schönflies notation), to which materials such

as Au, Cu, and Ag belong (F: face-centered cubic lattice with the point group symmetry m$\bar{3}$m). The Schönflies symbol for the space group also highlights the underlying point group (O_h): the superscript 5, instead, does not have a particular crystallographic significance. The second space group has a mirror plane replaced by a diamond glide and is the space group of the diamond crystal structure (C, Si, and Ge). Again, the superscript in the Schönflies notation is not of particular significance.

Lecture 6

APPLICATIONS TO SOLID-STATE PHYSICS

The aim of this lecture is to offer guidance on analyzing the electronic structure of a crystal in terms of the irreducible representations (irr. reps.) of the corresponding space group. In addition, we use symmetry-adapted wave functions to perform explicit computations of realistic band structures starting from a model potential (empty-lattice approximation). The lecture is accompanied by three projects that work out explicitly advanced applications of the symmetry analysis in solid-state physics. One, which is relatively more didactical, is titled "Empty-Lattice Band Structure". A second one, called "Symmetry Analysis of the Graphene Band Structure and the Dirac Cones", deals with the application of symmetry arguments to finding Dirac cones in the band structure of graphene and parent crystals. The third one, called "Topological Protection by Non-Symmorphic Degeneracy", explains the fundamental mechanism that produces topological protection in a so-called non-symmorphic crystal.

6.1 Irreducible Representations of the Translation Group T

We learned in Lecture 5 on crystallography that a space group G is an object with a large degree of complexity owing to the appearance of point group vectors. The general symmetry element is written as

$$(g \mid \mathbf{f}_g + \mathbf{t}_n)$$

g is some point group operation, and $\mathbf{f}_g$ is the corresponding point group vector — a fractional translation which is required in order that g becomes a symmetry element of the crystal. $\mathbf{t}_n$ is one of the primitive translations. The strategy to finding the irr. reps. of the space group uses the fact that the set of primitive translations T is an Abelian subgroup. As such, it offers a suitable *ansatz* for constructing the irr. reps. of the larger, less trivial space group. We have already used the relationship between an Abelian subgroup and group when we made the transition from SO_2 to SO_3: For an Abelian subgroup, we know that there exists a set of basis states that will diagonalize any matrix representation of all primitive translations simultaneously. All operations of the space group other than primitive translations will mix these basis states to provide a non-trivial, generally non-diagonal matrix representation of the larger space group. In other words, the invariant subspaces of the representations of the translation group constitute the fundamental units from which the possibly larger-dimensional representation of the space groups will emerge (analogous to the mixing of spherical harmonics Y_l^m by the non-SO_2 rotations of SO_3). There are, however, substantial differences between the couple (SO_2, SO_3) and (T, G): The primitive translations are not representatives of all conjugacy classes: Accordingly, the trace of their representing matrices has a limited application, as it does not provide a complete set of characters. Accordingly, the Herring[1] method[2] used to analyze the energy levels in terms of symmetry does not follow the standard path of organizing the group into conjugacy classes and working out their characters. Its starting point, however, remains the use of the basis states that diagonalize the translation group elements. We therefore proceed with finding these basis states.

6.1.1 *One-dimensional lattice*

A technical point we need to consider is that a crystal is an infinite set of objects. Accordingly, the group of discrete translations is not a continuous group but does not have an average, so that we do not expect its irr. reps. to be a countable set. To find the irr. reps. of $T = \{\mathbb{1} \mid n \cdot a\}; n = 0, 1, \ldots$:

1. We replace T with $T_N = \{(\mathbb{1} \mid n \cdot a); n = 0, 1, 2, \ldots, N - 1\}$ and assume $N \cdot a = 0$, i.e. we adopt periodic Born–von Karman boundary conditions. T_N has a finite number of elements and an average and is commutative and cyclic. Its character table is the following:

$$k = \frac{2\pi}{N \cdot a} \cdot m; \quad m \in 0, 1, \ldots, N{-}1 \quad \text{or} \quad m = -\frac{N-1}{2}, \ldots, 0, \ldots, \frac{N-1}{2}$$

---

[1]William Conyers Herring (1914–2009), American physicist, see e.g. https://en.wikip edia.org/wiki/Conyers_Herring.

[2]C. Herring, J. Franklin Inst., 233 (1942), p. 525; see also G.F. Koster, Space Groups and Their Representations, in Solid State Physics, edited by F. Seitz and D. Turnbull, Academic Press, Vol. 5, pp 173–256, 1957, DOI: https://doi.org/10.1016/S0081-1947(08) 60103-4) for an extended reference about the irreducible representations of space groups.

T_N	$t_0 = 0 \cdot a$	$t_1 = a$	...	$t_n = n \cdot a$	...	t_{N-1}
D^0	$e^{-i \cdot 0 \cdot t_0}$	$e^{-i \cdot 0 \cdot t_1}$	...	$e^{-i \cdot 0 \cdot t_n}$	...	$e^{-i \cdot 0 \cdot t_{N-1}}$
...	...	...	...	...	...	
...	...	...	...	...	...	
D^k	$e^{-i \cdot k \cdot t_0}$	$e^{-i \cdot k \cdot t_1}$	...	$e^{-i \cdot k \cdot t_n}$	...	$e^{-i \cdot k \cdot t_{N-1}}$
...	...	...	...	...	...	
...	...	...	...	...	...	
D^{N-1}	$e^{-i \cdot (N-1) \cdot t_0}$	$e^{-i \cdot (N-1) \cdot t_1}$	...	$e^{-i \cdot (N-1) \cdot t_n}$	...	$e^{-i \cdot (N-1) \cdot t_{N-1}}$

This means that the irr. reps. of T_N are labeled by a countable and finite set of k values distributed within the interval $[-\frac{\pi}{a}, \frac{\pi}{a}]$.

2. When $N \to \infty$, the distance between the k values decreases until the finite countable set is replaced by the continuous set of real values $k \in [-\frac{\pi}{a}, \frac{\pi}{a}]$. The segment $[-\frac{\pi}{a}, \frac{\pi}{a}]$ along the k-axis hosting the labels for the irr. reps. of T is called the **first Brillouin zone** (BZ) of the one-dimensional lattice. The k-axis is called the **reciprocal space**. Through this limiting procedure, one can construct the irr. reps. of the discrete translation group with an infinite number of elements. The irr. reps. turn out to constitute a continuous set labeled by k-values in the first BZ of a reciprocal lattice. These results can be rendered graphically as follows and generalized to two and three dimensions.

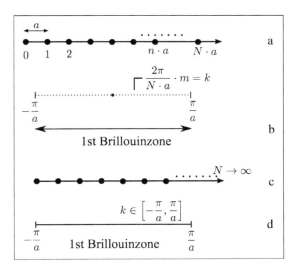

(a) The finite set of points represents a finite lattice with periodic boundary conditions. (b) The corresponding first BZ. (c) The infinite set of points represents a crystal in one dimension. (d) the first BZ is the segment $[-\frac{\pi}{a}, +\frac{\pi}{a}]$ of the reciprocal space.

Comments.

1. The one-dimensional lattice has a basis vector

$$\mathbf{a} = a \cdot \mathbf{e}$$

and its lattice is given by

$$n \cdot \mathbf{a} \quad n = 0, 1, 2, \ldots$$

2. In the reciprocal space, one defines a basis vector $\mathbf{b}$ through the relation

$$\mathbf{a} \cdot \mathbf{b} = 2\pi \Rightarrow \mathbf{b} = \frac{2\pi}{a}\mathbf{e}$$

3. The vectors

$$p \cdot \mathbf{b} \equiv \mathbf{h}; \quad p = 0, \pm 1, \ldots$$

define the reciprocal lattice vectors in the reciprocal space.
4. The first BZ cuts the two reciprocal lattice vectors $\pm\mathbf{b}$ in half.

6.1.2 *Two-dimensional lattice (square lattice for the sake of illustration)*

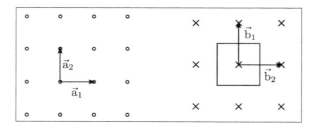

Left: The primitive square lattice with basis vectors. Right: The reciprocal lattice space with basis vectors and the first BZ of the square lattice.

Comments.

1. The lattice has two basis vectors

$$\mathbf{a}_1 = a \cdot (1, 0) \quad \mathbf{a}_2 = a \cdot (0, 1)$$

2. The basis vectors of the reciprocal space are defined by

$$\mathbf{a}_i \cdot \mathbf{b}_j = 2\pi \delta_{ij}$$

$i, j = 1, 2$. The solution of this equation is

$$\mathbf{b}_1 = \frac{2\pi}{a}(1,0) \quad \mathbf{b}_2 = \frac{2\pi}{a}(0,1)$$

3. The reciprocal space lattice vectors are given by

$$\mathbf{h} = p_1 \cdot \mathbf{b}_1 + p_2 \cdot \mathbf{b}_2$$

4. The first BZ containing the $\mathbf{k}$ vectors labeling the irr. reps. of the group $T = \{n_1\mathbf{a}_1 + n_2\mathbf{a}_2\}; n_1, n_2 = 0, \pm 1, \ldots$ has the geometry of a square with side lengths $\frac{2\pi}{a}$.

6.1.3   *Three-dimensional lattice (face-centered cubic lattice for the sake of illustration)*

The generalization to three-dimensional lattices requires that the crystal is terminated at the planes

$$\mathbf{r} \cdot \mathbf{a}_1 = N_1 \quad \mathbf{r} \cdot \mathbf{a}_2 = N_2 \quad \mathbf{r} \cdot \mathbf{a}_3 = N_3$$

with N_1, N_2, N_3 being large numbers. We then impose the Born–von Karman boundary conditions

$$(\mathbb{1} \mid \mathbf{t}_{(N_1,00)}) = (\mathbb{1} \mid \mathbf{t}_{(0,N_2,0)}) = (\mathbb{1} \mid \mathbf{t}_{(0,0,N_3)}) = (\mathbb{1} \mid \mathbf{0})$$

The collection of translational elements with such boundary conditions constitutes a finite, Abelian group T with as many elements as the number of unit cells in the crystal, i.e. $g = N_1 \cdot N_2 \cdot N_3$.

The reciprocal lattice of a crystal consists of integer number of linear combinations of the basis vectors $\mathbf{b}_j$, defined as those vectors which satisfy the equation

$$\mathbf{a}_i \cdot \mathbf{b}_j = 2\pi\delta_{ij}$$

$i, j = 1, 2, 3$. The solution of these equations are

$$\mathbf{b}_1 = 2\pi\frac{\mathbf{a}_2 \times \mathbf{a}_3}{\mathbf{a}_1 \cdot \mathbf{a}_2 \times \mathbf{a}_3} \quad \mathbf{b}_2 = 2\pi\frac{\mathbf{a}_3 \times \mathbf{a}_1}{\mathbf{a}_1 \cdot \mathbf{a}_2 \times \mathbf{a}_3} \quad \mathbf{b}_3 = 2\pi\frac{\mathbf{a}_1 \times \mathbf{a}_2}{\mathbf{a}_1 \cdot \mathbf{a}_2 \times \mathbf{a}_3}$$

The basis vectors of the primitive unit cell of the fcc lattice and those of its reciprocal space are written as

$$\mathbf{a}_1 = \tfrac{a}{2}(1,1,0) \quad \mathbf{b}_1 = \tfrac{2\pi}{a}(1,1,\bar{1})$$
$$\mathbf{a}_2 = \tfrac{a}{2}(0,1,1) \quad \mathbf{b}_2 = \tfrac{2\pi}{a}(\bar{1},1,1)$$
$$\mathbf{a}_3 = \tfrac{a}{2}(1,0,1) \quad \mathbf{b}_3 = \tfrac{2\pi}{a}(1,\bar{1},1)$$

Within the reciprocal lattice, one can define the first BZ, containing $N_1 \cdot N_2 \cdot N_3$ **k** vectors. It has the polyhedral geometry sketched for the fcc lattice in the figure. Some lines and points within the BZ are labeled: We will discuss their significance when the remaining symmetry elements of the space group are introduced.

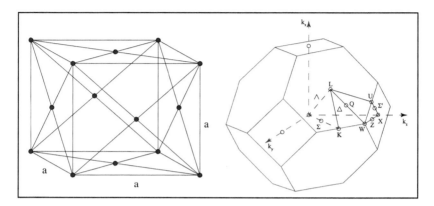

Left. The unit cell of the fcc lattice. Right: The 1st BZ of the fcc lattice. Special points and lines are indicated (see discussion in 6.2.2).

Proposition. *Each element of T is a class on its own and each vector* **k** *in the first BZ labels an irr. rep.* $D^{\mathbf{k}}$ *of T with the matrix element*

$$D^{\mathbf{k}}(\mathbb{1} \mid \mathbf{t}_n) = e^{-i\mathbf{k} \cdot \mathbf{t}_n}$$

Proof:

— The group is Abelian, so we have $N_1 \cdot N_2 \cdot N_3$ irr. reps., corresponding to the number of elements.

— $e^{-i\mathbf{k}\mathbf{t}_n}$ is a representation, as

$$\underbrace{e^{-i\mathbf{k}\cdot\mathbf{t}_n}}_{D^{\mathbf{k}}(E|\mathbf{t}_n)} \cdot \underbrace{e^{-i\mathbf{k}\mathbf{t}_m}}_{D^{\mathbf{k}}(E|\mathbf{t}_m)} = \underbrace{e^{-i\mathbf{k}(\mathbf{t}_n + \mathbf{t}_m)}}_{D^{\mathbf{k}}(E|\mathbf{t}_n + \mathbf{t}_m)}$$

— The representation is irreducible, as

$$\frac{1}{N_1 \cdot N_2 \cdot N_3} \sum \mid e^{-i\mathbf{k}\cdot\mathbf{t}_n} \mid^2 = 1$$

These are all the irr. reps. of T. ◇

Symmetry-adapted wave functions: We use the projector technique,

$$P^{\mathbf{k}}\psi(\mathbf{r}) \cong \sum_n e^{+i\mathbf{k}\mathbf{t}_n} \cdot \psi(\mathbf{r} - \mathbf{t}_n)$$

$$= e^{i\mathbf{k}\cdot\mathbf{r}} \cdot \sum_n e^{-i\mathbf{k}(\mathbf{r}-\mathbf{t}_n)} \cdot \psi(\mathbf{r} - \mathbf{t}_n)$$

$$= \underbrace{e^{i\mathbf{k}\cdot\mathbf{r}} \cdot u(\mathbf{k},\mathbf{r})}_{\text{Bloch functions}} \quad \text{and} \quad u(\mathbf{k}, \mathbf{r} - \mathbf{t}_n) = u(\mathbf{k},\mathbf{r})$$

to find that symmetry-adapted functions are *Bloch functions (Bloch theorem)*. We summarize the results for T in the following character table.

T	$\dots$	$\mathbf{t}_n$	$\dots$	Basis functions
$\dots$	$\dots$	$\dots$	$\dots$	$\dots$
$\dots$	$\dots$	$\dots$	$\dots$	
$\dots$	$\dots$	$\dots$	$\dots$	
$D^{\mathbf{k}}$	$\dots$	$e^{-i\mathbf{k}\cdot\mathbf{t}_n}$	$\dots$	$\sim e^{i\mathbf{k}\cdot\mathbf{r}}$
$\dots$	$\dots$	$\dots$	$\dots$	
$\dots$	$\dots$	$\dots$	$\dots$	
$\dots$	$\dots$	$\dots$	$\dots$	

6.2 Irreducible Representations of Space Groups (C. Herring)

6.2.1 *Introduction*

The translation group establishes a quantum number $\mathbf{k}$ which is specific to crystals and will be used to label the energy levels of a crystal. For the sake of illustrating what we expect for the energy levels of a crystal, we introduce here a graphical analogy between the energy levels we expect in systems with SO_3 symmetry and the energy levels we expect in a system with space group symmetry, see the following figure.

Roughly, the SO_3 symmetry introduces a specific quantum number l (the orbital quantum number, displayed along the horizontal axis on the left figure) that is used to label the energy levels. The discrete energy levels, displayed along the vertical in the left figure, are labeled by a further, non-symmetry-determined quantum number, the so-called main quantum number n. In a crystal (right), the energy levels are displayed as a function of the quantum number $k \cdot \mathbf{e}$, with $\mathbf{e}$ being some fixed direction in the

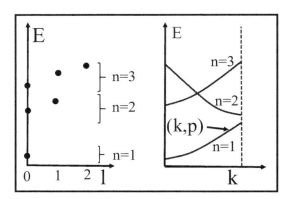

Left: Sketch of the energy levels in an atom. Right: Sketch of the energy levels in a solid.

BZ. k assumes continuous values, so that the discrete atomic energy levels broaden to **energy bands** in crystals. The bands accumulating along the vertical are identified by their "main quantum number", called the "band number". The presence of symmetry elements other then translations introduces, typically, a degeneracy of the bands, which are attributed to a further quantum number "p" labeling the irr. rep. $D^{k,p}$ of the space group. These further degeneracies play an important role in the physical properties of the materials; therefore, it is our aim, in the following sections, to find them.

A typical example of the electronic band structure in solids and of the method of displaying them is the band structure of GaAs, displayed in the following figure. One recognizes, along the horizontal axis, two selected directions in the first BZ of the crystal structure (the lattice is fcc). The energy bands contain further labels, which designate characteristic irr. reps. of the space group. These labels are the subject of the remaining part of this lecture.

6.2.2 *Irreducible representations of the group of* k

We consider one vector $\mathbf{k}$ within the first BZ and specifically those elements $R_\mathbf{k}$ of the point group of the crystal G_0 that transform $\mathbf{k}$ into itself (up to reciprocal lattice vectors):

$$R_\mathbf{k}\mathbf{k} = \mathbf{k} + \mathbf{h}$$

Definition (the small point group of k). The set

$$\{R_\mathbf{k}\} \doteq G_0(\mathbf{k})$$

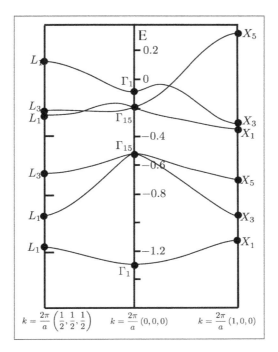

Band structure of GaAs, reproduced freely from F. Bassani and M. Yoshmine, *Phys. Rev.* **130**, 20 (1963).

constitutes a group $G_0(\mathbf{k})$, which is a subgroup of G_0 and is called the **small point group** of $\mathbf{k}$.

Comments.

1. At a general $\mathbf{k}$-point, $G_0(\mathbf{k}) = \{E\}$, but at some special $\mathbf{k}$-points or lines of the BZ — called the high-symmetry locations of the first BZ — $G_0(\mathbf{k})$ is larger than $\{E\}$.
2. Examples of high-symmetry locations are the Γ, L, and X-points and the Δ, Λ directions of an fcc lattice. For example,

$$G_0(\Gamma) = G_0 \quad G_0(\Delta) = C_{4v} \quad G_0(\Lambda) = C_{3v}$$

(Reference: L. P. Bouckaert, R. Smoluchowsky and E. Wigner, *Phys. Rev.* **50**, 58 (1936).)

Using the elements of the small point group, including their point group vectors, one can build the so-called group of $\mathbf{k}$ — $G(\mathbf{k})$ (sometimes also referred to as the "little" group of **k**). The group of **k** is relevant for classifying the energy bands at the given **k** according to their symmetry.

Definition (the group of k). The collection of elements of the space group

$$\{R_{\mathbf{k}} \mid \mathbf{t}_n + \mathbf{f}_{R_{\mathbf{k}}}\}$$

is called the little group of **k**, or simply the group of **k**, $G_{\mathbf{k}}$.

Comments.

1. At general **k**, $G_{\mathbf{k}} \equiv T$.
2. If **k** is a high-symmetry location, the group $G_{\mathbf{k}}$ contains more complex symmetry elements that include rotations and reflections with the corresponding point group vectors and their composition with all primitive translations.

Theorem. *Let* **k** *be inside the first BZ, i.e.*

$$R_{\mathbf{k}}\mathbf{k} = \mathbf{k} + \mathbf{0}$$

Then, from any irreducible matrix representation $D^p(R_{\mathbf{k}})$ of the elements of the point group $G_0(\mathbf{k})$, one can obtain an irr. rep. $D^{p,\mathbf{k}}$ of the group $G_{\mathbf{k}}$ by associating to every element $(R_{\mathbf{k}} \mid \mathbf{f}_{R_{\mathbf{k}}} + \mathbf{t}_n)$ of $G_{\mathbf{k}}$ the matrix

$$D^{p,\mathbf{k}}(R_{\mathbf{k}} \mid \mathbf{f}_R + \mathbf{t}_n) \doteq e^{-i\mathbf{k}(\mathbf{f}_{R_{\mathbf{k}}} + \mathbf{t}_n)} D^p(R_{\mathbf{k}})$$

Comments.

1. These irr. reps. are said to satisfy the Bloch theorem because their symmetry-adapted basis functions are, by construction, Bloch functions. They are therefore "physically relevant".
2. There are many irr. reps. of $G_{\mathbf{k}}$ which are not physically relevant, as translations are not represented by $e^{-i\mathbf{k}\mathbf{t}_n}$ and the symmetry-adapted basis functions are not-Bloch functions. One of these is certainly the identical representation, which — surprisingly — is not a physically relevant representation.
3. The general structure of the character table of $G_{\mathbf{k}}$ is given in the following typical table.
4. This theorem means that energy levels in solids at **k** points inside the BZ are classified according to the irr. reps. of a point group $G_0(\mathbf{k})$. Its character table, for the special case of points within the BZ, is often represented in a simplified way, as shown in the simplified table. Only

A typical character table of the group of **k**. The physically relevant representations are listed in the top section of the table.

$G_{\mathbf{k}}$	$\ldots$	$(R_i \mid \mathbf{f}_i)$	$\ldots$	$(R_i \mid \mathbf{f}_i + \mathbf{t}_n)$	$\ldots$
$\ldots$	$\ldots$	$\ldots$	$\ldots$	$\ldots$	$\ldots$
$\ldots$	$\ldots$	$\ldots$	$\ldots$	$\ldots$	$\ldots$
D^p	$\ldots$	$\chi_p(R_i) \cdot e^{-i\mathbf{k}(\mathbf{f}_i)}$	$\ldots$	$\chi_p(R_i) \cdot e^{-i\mathbf{k}(\mathbf{f}_i + \mathbf{t}_n)}$	$\ldots$
$\ldots$	$\ldots$	$\ldots$	$\ldots$	$\ldots$	$\ldots$
$\ldots$	$\ldots$	$\ldots$	$\ldots$	$\ldots$	$\ldots$
$\ldots$	$\ldots$	$\ldots$	$\ldots$	$\ldots$	$\ldots$
"1"	1	1	1	1	1
$\ldots$	$\ldots$	$\ldots$	$\ldots$	$\ldots$	$\ldots$
$\ldots$	$\ldots$	$\ldots$	$\ldots$	$\ldots$	$\ldots$

the elements with small point group operations are listed in the top row of the simplified character table and only point group irr. reps. are listed in the column on the right-hand side. The bulk of the table contains the corresponding characters. They are composed by multiplying the small point group characters (known from our knowledge of the point groups) with an exponential prefactor that takes into account possible fractional translations.

A simplified character table of the group of **k**.

$G_{\mathbf{k}}$	$\ldots$	$(R_i \mid \mathbf{f}_i)$	$\ldots$
$\ldots$	$\ldots$	$\ldots$	$\ldots$
$\ldots$	$\ldots$	$\ldots$	$\ldots$
D^p	$\ldots$	$\chi_p(R_i) \cdot e^{-i\mathbf{k}(\mathbf{f}_i)}$	$\ldots$
$\ldots$	$\ldots$	$\ldots$	$\ldots$
$\ldots$	$\ldots$	$\ldots$	$\ldots$

Proof of the Theorem: We show first that the matrices are indeed representations:

$$(R_i \mid \underbrace{\mathbf{f}_i + \mathbf{t}_n}_{\mathbf{a}_i}) \, (R_j \mid \underbrace{\mathbf{f}_j + \mathbf{t}_m}_{\mathbf{a}_j}) = \underbrace{(R_i R_j \mid R_i[\mathbf{f}_j + \mathbf{t}_m] + \mathbf{f}_i + \mathbf{t}_n)}_{D^p(R_i R_j) \cdot e^{-i\mathbf{k}\cdot(R_i \mathbf{a}_j + \mathbf{a}_i)}}$$

$$\underbrace{e^{-i\mathbf{k}\cdot\mathbf{a}_i} \cdot D^p(R_i)}_{} \; \underbrace{e^{-i\mathbf{k}\cdot\mathbf{a}_j} \cdot D^p(R_j)}_{}$$

The left-hand side can be written as

$$D^p(R_i R_j) e^{-i\mathbf{k}(\mathbf{a}_i + \mathbf{a}_j)}$$

The right-hand side can be written as

$$D^p(R_i R_j) e^{-i\mathbf{R}_i^{-1}\mathbf{k}\mathbf{a}_j + \mathbf{k}\mathbf{a}_i}$$

If $\mathbf{k}$ is inside the BZ, then $R_i^{-1}\mathbf{k} = \mathbf{k}$ and the two sides become identical, proving the homomorphism.

Comment (regarding k points on the surface of the BZ). The phase factor on the right-hand side contains $\mathbf{k} \cdot R_i\mathbf{a}_j$, which is necessarily equal to $\mathbf{k} \cdot \mathbf{a}_j$ *only* for $\mathbf{k}$ inside the BZ. At the surface of the BZ, however, we might have $R_i^{-1}\mathbf{k} = \mathbf{k} + \mathbf{h}$ and the phase factor becomes $\mathbf{k} \cdot \mathbf{t}_n + \mathbf{h} \cdot \mathbf{f}_j$. The presence of the extra term $\mathbf{h} \cdot \mathbf{f}_j$ corrupts, in general, the condition of homomorphism, unless $f_j \equiv \mathbf{0}$. Thus, the theorem can be extended to $\mathbf{k}$ at the surface of the BZ only for *symmorphic* crystals. For non-symmorphic crystals and for $\mathbf{k}$ points at the surface of the BZ, finding the irr. reps. becomes a more complex exercise (see later).

Next, we prove that the representations $D^{p,\mathbf{k}}$ are irreducible. In fact, if $g_0(\mathbf{k})$ is the number of elements in $G_0(\mathbf{k})$, we have

$$\frac{1}{g_0(\mathbf{k})N_1N_2N_3} \sum_{R_{\mathbf{k}}|\mathbf{f}+\mathbf{t}_n} \mid \chi_p(R_{\mathbf{k}} \mid \mathbf{f} + \mathbf{t}_n) \mid^2 = \frac{1}{g_0} \sum_{R_{\mathbf{k}}} \mid \chi_p(R_{\mathbf{k}}) \mid^2 = 1 \quad \diamond$$

Examples.

1. **Point Γ in O_h^5:** This is the point at the origin of the first BZ. The small point group of $\mathbf{k}$ is the entire point group O_h. Since we are inside the BZ, the characters of the physically relevant irr. reps. at Γ are simply given by the characters of the point group O_h without any modification.

2. **Line Λ in O_h^5:** $\mathbf{k} = \frac{\pi}{a}(\lambda, \lambda, \lambda)$, $\lambda < 1$. This is a point on the line joining the center of the BZ to the midpoint of a hexagonal face. The small point group of $\mathbf{k}$ is made up of the operations whose rotational parts interchange (x, y, z) among themselves because these are the only operations which leave $\mathbf{k}$ invariant. The small point group of $\mathbf{k}$ is thus C_{3v}. Since no fractional translation is associated with the elements of C_{3v}, the characters of the irr. reps. at Λ are the characters of the group C_{3v} without any modification.

3. **Line Δ in O_h^5 ($\mathbf{k} = \frac{2\pi}{a}(\delta, 0, 0)$, with $\delta < 1$):** This is a point on the line joining the center of the BZ to the midpoint of the square face. The small point group of $\mathbf{k}$ is given by the operations whose rotational part leave x unchanged. The small point group of $\mathbf{k}$ is thus C_{4v}. The simplified character table of the group G_Δ is given next.

4. **The space group 0_h^7 (diamond structure):**
   The diamond lattice is invariant with respect to primitive translations $(\mathbb{1} \mid \mathbf{t}_n)$ with $\mathbf{t}_n = \sum_i n_i\mathbf{a}_i$ and $\mathbf{a}_1 = \frac{a}{2}(0, 1, 1)$, $\mathbf{a}_2 = \frac{a}{2}(1, 0, 1)$,

The simplified character table of the group
G_Δ for the space group O_h^5.

G_Δ	E	C_{2x}	C_{4x}^{-1} C_{4x}	IC_{2x} IC_{2y}	IC_{2xy} $IC_{2y\bar{z}}$
Δ_1	1	1	1	1	1
Δ_1'	1	1	1	-1	-1
Δ_2	1	1	-1	1	-1
$\Delta_{2'}$	1	1	-1	-1	1
Δ_5	2	-2	0	0	0

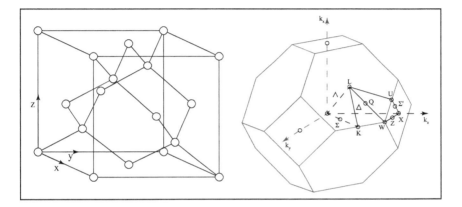

Left: The unit cell of the diamond structure. Right: Its 1st Brillouin zone.

$\mathbf{a}_3 = \frac{a}{2}(1,1,0)$. ($a$ is the length of the cube edge). The first BZ it the
one of the fcc lattice. There are two equal atoms in the unit cell in
the positions $\mathbf{d}_1 = (0,0,0); \mathbf{d}_2 = \frac{a}{4}(1,1,1)$. The diamond lattice can be
considered to consist of two interpenetrating fcc sublattices displaced
with respect to each other by the vector $\mathbf{d}_2 := \mathbf{f}$. The point group G_0
of the diamond lattice is the cubic group 0_h. Note that some of the
symmetry operations of 0_h appear in the space group associated with
the fractional translation

$$\mathbf{f} = \frac{a}{4}(1,1,1) = \frac{1}{4}(\mathbf{a}_1 + \mathbf{a}_2 + \mathbf{a}_3)$$

By choosing a lattice point as the origin of a cubic coordinate system,
we obtain the symmetry operations given in the following table.
The space group is $O_h^7 (Fd\bar{3}m)$ is non-symmorphic. The small point group
of Δ is C_{4v}. The character table of the group of Δ can be derived from
the character table of C_{4v} by multiplying the characters of the point

Symmetry elements of the diamond structure and the transformation properties of the (x, y, z) coordinates.

Type	Operation	Coordinate	Type	Operation	Coordinate
C_1	$(E \mid 0)$	xyz	I	$(I \mid \mathbf{f})$	$\bar{x} + \frac{a}{4}\,\bar{y} + \frac{a}{4}\,\bar{z} + \frac{a}{4}$
C_4^2	C_{2z}	$\bar{x}\bar{y}z$	IC_4^2	$(IC_{2z} \mid \mathbf{f})$	$x + \frac{a}{4}\,y + \frac{a}{4}\,\bar{z} + \frac{a}{4}$
	C_{2x}	$x\bar{y}\bar{z}$		$(IC_{2x} \mid \mathbf{f})$	$\bar{x} + \frac{a}{4}\,y + \frac{a}{4}\,z + \frac{a}{4}$
	C_{2y}	$\bar{x}y\bar{z}$		$(IC_{2y} \mid \mathbf{f})$	$x + \frac{a}{4}\,\bar{y} + \frac{a}{4}\,z + \frac{a}{4}$
C_4	$(C_{4z}^{-1} \mid \mathbf{f})$	$\bar{y} + \frac{a}{4}\,x + \frac{a}{4}\,z + \frac{a}{4}$	IC_4	IC_{4z}^{-1}	$y\bar{x}\bar{z}$
	$(C_{4z} \mid \mathbf{f})$	$y + \frac{a}{4}\,\bar{x} + \frac{a}{4}\,z + \frac{a}{4}$		IC_{4z}	$\bar{y}x\bar{z}$
	$(C_{4x}^{-1} \mid \mathbf{f})$	$x + \frac{a}{4}\,\bar{z} + \frac{a}{4}\,y + \frac{a}{4}$		IC_{4x}^{-1}	$\bar{x}z\bar{y}$
	$(C_{4x} \mid \mathbf{f})$	$x + \frac{a}{4}\,z + \frac{a}{4}\,\bar{y} + \frac{a}{4}$		IC_{4x}	$\bar{x}\bar{z}y$
	$(C_{4y}^{-1} \mid \mathbf{f})$	$z + \frac{a}{4}\,y + \frac{a}{4}\,\bar{x} + \frac{a}{4}$		IC_{4y}^{-1}	$\bar{z}\bar{y}x$
	$(C_{4y} \mid \mathbf{f})$	$\bar{z} + \frac{a}{4}\,y + \frac{a}{4}\,x + \frac{a}{4}$		IC_{4y}	$z\bar{y}\bar{x}$
C_2	$(C_{2xy} \mid \mathbf{f})$	$y + \frac{a}{4}\,x + \frac{a}{4}\,\bar{z} + \frac{a}{4}$	IC_2	IC_{2xy}	$\bar{y}\bar{x}z$
	$(C_{2xz} \mid \mathbf{f})$	$z + \frac{a}{4}\,\bar{y} + \frac{a}{4}\,x + \frac{a}{4}$		IC_{2xz}	$\bar{z}y\bar{x}$
	$(C_{2yz} \mid \mathbf{f})$	$\bar{x} + \frac{a}{4}\,z + \frac{a}{4}\,y + \frac{a}{4}$		IC_{2yz}	$x\bar{z}\bar{y}$
	$(C_{2x\bar{y}} \mid \mathbf{f})$	$\bar{y} + \frac{a}{4}\,\bar{x} + \frac{a}{4}\,\bar{z} + \frac{a}{4}$		$IC_{2x\bar{y}}$	yxz
	$(C_{2\bar{x}z} \mid \mathbf{f})$	$\bar{z} + \frac{a}{4}\,\bar{y} + \frac{a}{4}\,\bar{x} + \frac{a}{4}$		$IC_{2\bar{x}z}$	zyx
	$(C_{2y\bar{z}} \mid \mathbf{f})$	$\bar{x} + \frac{a}{4}\,\bar{z} + \frac{a}{4}\,\bar{y} + \frac{a}{4}$		$IC_{2y\bar{z}}$	xzy
C_3	C_{3xyz}^{-1}	zxy	IC_3	$(IC_{3xyz}^{-1} \mid \mathbf{f})$	$\bar{z} + \frac{a}{4}\,\bar{x} + \frac{a}{4}\,\bar{y} + \frac{a}{4}$
	C_{3xyz}	yzx		$(IC_{3xyz} \mid \mathbf{f})$	$\bar{y} + \frac{a}{4}\,\bar{z} + \frac{a}{4}\,\bar{x} + \frac{a}{4}$
	$C_{3x\bar{y}\bar{z}}^{-1}$	$z\bar{x}\bar{y}$		$(IC_{3x\bar{y}\bar{z}}^{-1} \mid \mathbf{f})$	$\bar{z} + \frac{a}{4}\,x + \frac{a}{4}\,y + \frac{a}{4}$
	$C_{3x\bar{y}\bar{z}}$	$\bar{y}\bar{z}x$		$(IC_{3x\bar{y}\bar{z}} \mid \mathbf{f})$	$y + \frac{a}{4}\,\bar{z} + \frac{a}{4}\,\bar{x} + \frac{a}{4}$
	$C_{3\bar{x}\bar{y}z}^{-1}$	$\bar{z}\bar{x}y$		$(IC_{3\bar{x}\bar{y}z}^{-1} \mid \mathbf{f})$	$z + \frac{a}{4}\,x + \frac{a}{4}\,\bar{y} + \frac{a}{4}$
	$C_{3\bar{x}\bar{y}z}$	$\bar{y}z\bar{x}$		$(IC_{3\bar{x}\bar{y}z} \mid \mathbf{f})$	$y + \frac{a}{4}\,\bar{z} + \frac{a}{4}\,x + \frac{a}{4}$
	$C_{3\bar{x}y\bar{z}}^{-1}$	$\bar{z}x\bar{y}$		$(IC_{3\bar{x}y\bar{z}}^{-1} \mid \mathbf{f})$	$z + \frac{a}{4}\,\bar{x} + \frac{a}{4}\,y + \frac{a}{4}$
	$C_{3\bar{x}y\bar{z}}$	$y\bar{z}\bar{x}$		$(IC_{3\bar{x}y\bar{z}} \mid \mathbf{f})$	$\bar{y} + \frac{a}{4}\,z + \frac{a}{4}\,x + \frac{a}{4}$

group with the suitable phase factor

$$e^{(-i\mathbf{k}\cdot\mathbf{f})} = e^{-i(\frac{\pi}{2}\cdot\delta)}$$

For example, the simplified character table for the group of the line Δ $((\mathbf{k} = \frac{2\pi}{a}(\delta, 0, 0), \delta < 1))$ is given as follows.

$G_\Delta(O_h^7)$	$(E \mid 0)$	$(C_{2x} \mid 0)$	$(C_{4x}^{-1} \mid \mathbf{f})$ $(C_{4x} \mid \mathbf{f})$	$(IC_{2x} \mid \mathbf{f})$ $(IC_{2y} \mid \mathbf{f})$	$(IC_{2xy} \mid 0)$ $(IC_{2y\bar{z}} \mid 0)$
Δ_1	1	1	$e^{-i(\frac{\pi}{2}\cdot\delta)}$	$e^{-i(\frac{\pi}{2}\cdot\delta)}$	1
Δ_1'	1	1	$e^{-i(\frac{\pi}{2}\cdot\delta)}$	$-e^{-i(\frac{\pi}{2}\cdot\delta)}$	-1
Δ_2	1	1	$-e^{-i(\frac{\pi}{2}\cdot\delta)}$	$e^{-i(\frac{\pi}{2}\cdot\delta)}$	-1
$\Delta_{2'}$	1	1	$-e^{-i(\frac{\pi}{2}\cdot\delta)}$	$-e^{-i(\frac{\pi}{2}\cdot\delta)}$	1
Δ_5	2	-2	0	0	0

6.2.3   *The star operation*

The final step in finding the irr. reps. of the space group requires taking into account those elements of the space group that send $\mathbf{k}$ into a set called the "star" of $\mathbf{k}$.

Definition. If $(R \mid \mathbf{0})$ is a member of G_0 but not of $G_0(\mathbf{k})$, then $R\mathbf{k}$ is not equivalent to $\mathbf{k}$. The effect of such operations on $\mathbf{k}$ is to send $\mathbf{k}$ into a number of inequivalent vectors. $\mathbf{k}_1 = \mathbf{k}, \mathbf{k}_2, \ldots$, constitute a set of vectors called "the star" of $\mathbf{k}$.

Comment. If g_0 is the number of elements of G_0 and $g_0(\mathbf{k})$ the number of elements of $G_0(\mathbf{k})$, then the number of vectors $S_\mathbf{k}$ in the star of $\mathbf{k}$ is given by $g_0(\mathbf{k}) \cdot S_k = g_0$

Example. For $\mathbf{k} = 0$ (Γ-point), we have only one member of the star of $\mathbf{k}$, which is $\mathbf{k} = 0$. For $\mathbf{k}$ along Δ in a fcc lattice, $\mid 0_h \mid = 48$ and $\mid C_{4v} \mid = 8$, so that we have six vectors in the BZ which are members of the star of $\mathbf{k}$: $\frac{2\pi}{a}(\pm\lambda, 0, 0), \frac{2\pi}{a}(0, \pm\lambda, 0), \frac{2\pi}{a}(0, 0, \pm\lambda)$.

Definition. Pick up one member of the set of vectors constituting a star as a representative of the star. The set of all representatives lying in the same sector of the BZ builds the reduced Brillouin zone. The volume of the reduced BZ is the volume of the BZ divided by g_0, the order of G_0.

Example. The following figure shows the BZ of a plane square lattice with a lattice constant of a. The triangle containing the special points and lines of symmetry is the reduced first BZ.

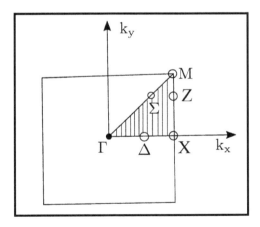

The BZ of the square lattice extends over the intervals $k_x \in [-\frac{\pi}{a}, \frac{\pi}{a}]$ and $k_y \in [-\frac{\pi}{a}, \frac{\pi}{a}]$, with a being the lattice constant. The reduced BZ of the two-dimensional primitive square lattice is the triangular zone indicated by vertical gray lines and showing the lines and points of high symmetry.

The following figure shows the stars of various high-symmetry **k** vectors for a plane square lattice.

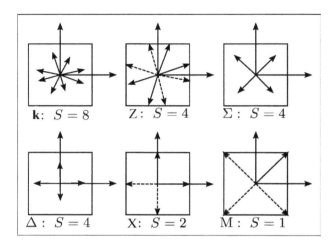

The star of some high-symmetry **k** for the square BZ. The members of the star are indicated by the solid arrows. Dashed vectors are related to one solid vector by a reciprocal lattice vector and are therefore not counted separately.

Theorem. *Let* $\{\varphi_k^1, \ldots, \varphi_k^{\ell p}\}$ *carry the* ℓp-*dimensional irr. rep.* $D^{p\mathbf{k}}$ *of* $G_{\mathbf{k}}$. *The* $S_{\mathbf{k}} \cdot \ell p$ *basis functions obtained by transforming the* ℓ_p-*basis functions with the "star" operations carry a* $S_{\mathbf{k}} \cdot \ell_p$-*dimensional irr. rep. of the entire space group* G. *These are all irr. reps. of the space group.*

Short Check: We recall that for irr. reps., the sum of their square dimensions is the number of elements of the group. We check this equations for the representations obtained with the star operation:

$$\sum_{\mathbf{k} \in \text{red. vol.}} (S_k \cdot l_p)^2 = \sum_{\mathbf{k} \in \text{red. vol.}} (S_k)^2 \cdot \underbrace{\sum_p (l_p)^2}_{g_0 \mathbf{k})}$$

$$= \sum_{\mathbf{k} \in \text{red. vol.}} S_k \cdot \underbrace{(S_k \cdot g_0(\mathbf{k}))}_{g_0} = g_0 \sum_{\mathbf{k} \in 1\text{BZ}} 1$$

$$= g_0 \cdot N_1 \cdot N_2 \cdot N_3$$

Fazit. The electronic structure obtained within the reduced BZ is the essential part of the electronic structure, while the remaining part of the

BZ increases degeneracy. To find the electronic band structure of a crystal, therefore, one must construct the band structure for $\mathbf{k}$ within the reduced BZ. A decisive factor is that the irr. reps. of the group $G(\mathbf{k})$ are obtained (in most cases) *from the representations of the small point group* $G_0(\mathbf{k})$ (times a phase factor $e^{-i\mathbf{k}\cdot\mathbf{t}}$) (we discuss later some exceptions in the non-symmorphic space groups).

6.3 Symmetry Analysis of the Band Structure of Solids

In the following examples, we observe that point group symmetries introduce special features, such as accidental/essential degeneracies (i.e. degeneracies which are imposed by symmetry) and avoided crossing exactly in those sectors of the band structure which are related to high-symmetry points and lines of the reduced BZ. However, these special features are not limited to the special points and lines of high symmetry. In fact, these special features are translated, by virtue of continuity, to a sizable volume of the BZ surrounding them.[3] Accordingly, these special features are crucial to the electronic and physical properties of the material, which, in general, are the result of an average over the entire BZ (or over large subsets of it).

6.3.1 *Example 1. Empty-lattice band structure of the one-dimensional primitive lattice: Accidental versus essential degeneracy*

The band structure of a simple one-dimensional lattice contains essential symmetry-induced elements that also appear in more complex solids. We therefore analyze this lattice using the symmetry arguments worked out in the previous section.

Symmetry elements

- The one-dimensional chain of atoms with a lattice constant of a has

$$T_N = \{m \cdot a\} \quad m = 0, \pm 1, \pm 2, \ldots$$

---

[3]See, for example, R. Allenspach, F. Meier and D. Pescia, "Experimental symmetry analysis of electronic states by spin-dependent photoemission", *Phys. Rev. Lett.* **51**, 2148 (1983), https://journals.aps.org/prl/abstract/10.1103/PhysRevLett.51.2148.

- The reciprocal lattice vectors are

$$h = \frac{2\pi}{a} \cdot n \quad n = 0, \pm 1, \pm 2, \ldots$$

- The first BZ is the segment $[-\frac{\pi}{a}, \frac{\pi}{a}]$
- The point group of the lattice is $G_0 = \{E, I\}$. The space group is

$$G = \{(E \mid na) \quad (I \mid na)\}$$

and it is *symmorphic*.

- $k \in]0, \frac{\pi}{a}[$ (the line Δ): $G_0(\Delta) = \{\mathbb{1}\}$. The stars are k and $-k$, and the character table is as follows.

G_Δ	E	...	...	$(E \mid ma)$	...	...	...
Δ_1	1	...	...	$e^{-i \cdot k \cdot m \cdot a}$	...	...	...

- $k = 0$ (Γ-point): $G_0(\Gamma) = \{E, I\}$. Star: $k = 0$.

G_Γ	E	I	...	$(E \mid na)$	...	$(I \mid na)$	...
Γ_1	1	1	...	1	...	1	...
$\Gamma_{\bar{1}}$	1	$\bar{1}$	...	1	...	$\bar{1}$	...

- $k = \frac{\pi}{a}$ (X-point): $G_0(X) = \{E, I\}$. Star: $\frac{\pi}{a}$

G_X	E	I	...	$(E \mid na)$	...	$(I \mid na)$	...
X_1	1	1	...	$e^{-i\pi n}$	...	$e^{-i\pi n}$	...
$X_{\bar{1}}$	1	$\bar{1}$	...	$e^{-i\pi n}$	...	$-e^{-i\pi n}$	...

Empty-lattice band structure

The Schrödinger equation for one electron in the lattice reads

$$\left[-\frac{\hbar^2}{2m} \triangle + V(x) \right] \psi(x) = E \cdot \psi(x)$$

In order to find wave functions adapted to translational symmetry, which are suitable for solving this equation for a general potential $V(x)$, we solve the "unperturbed" problem, i.e. the problem where the strength of $V(x)$ is set to 0 but its space group symmetry is retained. This is the so-called "empty"-lattice approximation, which generates the empty-lattice band structure. The empty-lattice energy levels has one index corresponding to an irr. rep. of the translation group: This is the k value (in the following, we use $\kappa = k/\frac{2\pi}{a}$). At some special points (Γ and X), a further quantum index, arising from the irr. reps. of the group of k might appear. There is a further index labeling the energy levels, the band index, provided by the reciprocal lattice vectors $h = \frac{2\pi}{a} \cdot n$. The solution to the empty-lattice problem reads

$$\frac{E_{\kappa,n}}{\frac{(2\pi\hbar)^2}{2ma^2}} = (\kappa + n)^2$$

The symmetry-adapted functions are

$$\sqrt{\frac{1}{Na}} e^{i\frac{2\pi}{a}(\kappa+n)\cdot x}$$

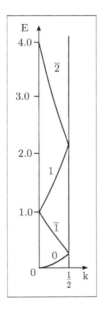

Empty-lattice band structure of an electron in a one-dimensional lattice. The bands are labeled by the reciprocal lattice numbers.

Symmetry analysis

→ I. $n = 0$ generates the lowest band. The plane wave with $(\kappa, n) = (0,0)$ carries the irr. rep. Γ_1 of $G(\Gamma)$.

For $\kappa \neq 0$ and $n = 0$, one has a plane wave with symmetry Δ_1 and a band with quantum numbers $(\kappa, \Delta_1, n = 0)$.

→ II. At X, the translational symmetry-adapted plane waves

$$\sqrt{\frac{1}{Na}}e^{i\frac{2\pi}{a}(\frac{x}{2})} \qquad \sqrt{\frac{1}{Na}}e^{i\frac{2\pi}{a}(\frac{x}{2}-x)}$$

have the same energy $\frac{1}{4}$. This level carries a representation $X_{\frac{1}{4}}$ of $G(X)$:

$$E = \begin{pmatrix} 1 & 0 \\ 0 & 1 \end{pmatrix} \qquad I = \begin{pmatrix} 0 & 1 \\ 1 & 0 \end{pmatrix}$$

constructed over the two plane waves.

Simplified character table of
the group $G(X)$.

G_X	E	I
X_1	1	1
$X_{\bar{1}}$	1	$\bar{1}$
$X_{\frac{1}{4}}$	2	0

According to the character table of $G(X)$, $X_{\frac{1}{4}}$ is reducible: $X_{\frac{1}{4}} = X_1 \oplus X_{\bar{1}}$. The degeneracy of these two irr. rep. is known as *accidental* degeneracy, i.e. it is not required by symmetry. In fact, $G(X)$ has only one-dimensional irr. reps. We expect therefore that modifying slightly the potential, such as by giving it a finite strength, will lift the accidental double degeneracy into two separate levels with symmetry X_1 and $X_{\bar{1}}$, as the decomposition theorem tells us that $X_{\frac{1}{4}} = X_1 \oplus X_{\bar{1}}$.

→ III. Using the projector technique, we are able to compute the symmetry-adapted linear combinations of

$$\sqrt{\frac{1}{Na}}e^{i\frac{2\pi}{a}(\frac{x}{2})} \qquad \sqrt{\frac{1}{Na}}e^{i\frac{2\pi}{a}(\frac{x}{2}-x)}$$

that transform according to X_1 and $X_{\bar{1}}$:

$$P_{X_1}\sqrt{\frac{1}{Na}}e^{i\frac{2\pi}{a}(\frac{x}{2})} = \sqrt{\frac{1}{Na}}e^{i\frac{2\pi}{a}(\frac{x}{2})} + \sqrt{\frac{1}{Na}}e^{i\frac{2\pi}{a}(\frac{-x}{2})}$$

$$\propto \sqrt{\frac{2}{Na}}\cdot\cos\left(\frac{2\pi}{a}\frac{x}{2}\right)$$

$$P_{X_{\bar{1}}}\sqrt{\frac{1}{Na}}e^{i\frac{2\pi}{a}(\frac{x}{2})} = \sqrt{\frac{1}{Na}}e^{i\frac{2\pi}{a}(\frac{x}{2})} - \sqrt{\frac{1}{Na}}e^{i\frac{2\pi}{a}(\frac{-x}{2})}$$

$$\propto \sqrt{\frac{2}{Na}}\cdot\sin\left(\frac{2\pi}{a}\frac{x}{2}\right)$$

→ IV. The general expression for the symmetry-adapted wave functions at $k = 0$ is

$$\sqrt{\frac{2}{Na}} \cos \frac{n \cdot 2\pi}{a} \cdot x \qquad \sqrt{\frac{2}{Na}} \sin \frac{n \cdot 2\pi}{a} \cdot x$$

$n = 0, 1, 2, \ldots$. From this expression, it is clear that the deepest lying level $(k = G = 0)$ is not degenerate, while the degeneracy of the higher lying levels is two.

→ V. The general expression for the wave function at $k = \frac{\pi}{a}$ is

$$\sqrt{\frac{2}{Na}} \cos \left[\frac{\pi}{a} + n\frac{2\pi}{a} \right] x \qquad \sqrt{\frac{2}{Na}} \sin \left[\frac{\pi}{a} + n\frac{2\pi}{a} \right] x$$

$n = 0, 1, \ldots$.

→ VI. When a crystal potential with finite strength is switched on, the accidental degeneracies at Γ and X are lifted and gaps open up. In the spirit of the perturbation theory, the symmetry-adapted functions can be used to compute the energy levels in the vicinity of the gaps when the perturbation $V(x)$ is switched on. For example, the band energies at the first gap at X reads, exactly,

$$E_{X_1} = E_{\frac{1}{4}} + \frac{2}{Na} \int_0^{Na} dx V(x) \cdot \cos^2 \left(\frac{2\pi}{a} \frac{x}{2} \right)$$

$$E_{X_{\bar{1}}} = E_{\frac{1}{4}} + \frac{2}{Na} \int_0^{Na} dx V(x) \cdot \sin^2 \left(\frac{2\pi}{a} \frac{x}{2} \right)$$

Comments.

1. The situation encountered in the one-dimensional band structure is generally that the degeneracy at points lying at the border of the BZ is an accidental one, i.e. it is not required by symmetry. In fact, the representation of the k vector realized at the border of the BZ is a two-dimensional one, while the point groups of Γ and X have only one-dimensional irr. reps. We expect therefore that modifying slightly the potential (e.g. giving it a finite strength) will lift the accidental double degeneracy at the center and at the edge of the BZ, thus introducing an essential and general element of the band structure of solids (also found in higher dimensionalities) called the **the energy gap**. Within the energy gap, no states of the system are allowed. Group theoretical arguments, therefore, predict the opening of gaps at the border of the

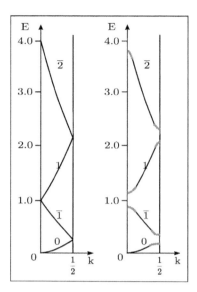

Left: Empty-lattice band structure of a one-dimensional crystal. Right: Opening of gaps at the edge of the BZ by a potential with a finite strength.

BZ and tell us how to deal with it. Note that, in one dimension, the energy gap extends over the entire BZ: This is, however, a feature which is specific to one dimension.

2. Within a perturbational approach, symmetry-adapted wave functions can be used to compute the first-order correction caused by a small, general crystal potential $V(x)$. The energy of the levels at the two singular points of the BZ can be calculated, using the first-order perturbation theory, to be

$$E_{\pm}(k=0) = \frac{\hbar^2}{2m}\left(\frac{2\pi}{a}\right)^2 \cdot n^2 + \frac{1}{4Na}\int_0^{Na} dx \cdot V(x)$$

$$\pm \frac{1}{4Na}\int_0^{Na} dx \cdot V(x) \cdot \cos\left(2n \cdot \frac{2\pi}{a}\right)x$$

(the negative-sign level is suppressed for $n = 0$) and

$$E_{\pm}(k=\frac{\pi}{a}) = \frac{\hbar^2}{2m}\left(\frac{2\pi}{a}\right)^2 \cdot \left(\frac{2n+1}{2}\right)^2 + \frac{1}{4Na}\int_0^{Na} dx \cdot V(x)$$

$$\pm \frac{1}{4Na}\int_0^{Na} dx \cdot V(x) \cdot \cos\left[(2n+1) \cdot \frac{2\pi}{a}\right]x$$

$(n = 0, 1, 2, \ldots)$. Thus, the gap is twice the Fourier transform of the crystal potential perturbing the empty-lattice band structure. This computational result (adapted to higher dimensions) has general applications and validity.

6.3.2 *Example 2. The two-dimensional primitive square lattice: Essential degeneracy and avoided crossing*

The band structure of the two-dimensional square lattice shows two types of band structure singularities: essential degeneracies and "Von Neumann–Wigner avoided crossing" (anticrossing of bands carrying the same irr. rep.), which is only possible in $d \geq 2$.

The symmetry analysis of the band structure proceeds according to the lines given for the one-dimensional case. We refer to the previous section for a figure containing the reduced BZ. The high-symmetry point and lines and their point groups are summarized by the following table.

The small point groups of selected points and lines of the reduced BZ of a square lattice.

$\mathbf{k}$	Symbol	$G_0(\mathbf{k})$
(k_x, k_y)	general	C_1
(k_x, k_x)	Σ	$C_{1,h}$
$(\frac{1}{2}, k_y)$	Z	$C_{1,h}$
$(k_x, 0)$	Δ	$C_{1,h}$
$(\frac{1}{2}, 0)$	X	$C_{2,v}$
$(\frac{1}{2}, \frac{1}{2})$	M	$C_{4,v}$
$(0, 0)$	Γ	$C_{4,v}$

The starting point for a symmetry analysis is again the *empty lattice model*, which keeps track of the translational symmetry of the lattice but sets the strength of the crystal potential to zero. A suitable set of basis functions, symmetry adapted to translational symmetry, are the plane waves

$$\phi_{\mathbf{kG}} = \frac{1}{\sqrt{Na^2}} \cdot e^{i(\mathbf{k}+\mathbf{G})\mathbf{r}}$$

with $\mathbf{G}$ being some reciprocal lattice vectors with coordinates $\frac{2\pi}{a}(m, n)$, $m, n \in \mathscr{Z}$.

Line Δ: As a way of illustration, we consider the bands along the Δ direction. The energy bands generated by the numbers (m, n) have the energy

$$E(m,n) = \frac{\hbar^2}{2m}\left(\frac{2\pi}{a}\right)^2\left[(\kappa+m)^2 + n^2\right]$$

and the corresponding plane waves are

$$e^{\frac{i\cdot2\pi}{a}[(\kappa+m)\cdot x+n\cdot y]}$$

The following figure plots the empty lattice $\mathscr{E} \equiv \frac{E}{\frac{\hbar^2}{2m}\left(\frac{2\pi}{a}\right)^2}$ versus κ.

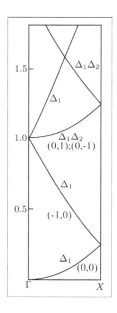

Empty-lattice band structure of the primitive square lattice along the Δ direction. $\mathscr{E}$ is along the vertical coordinate.

Symmetry analysis:

The small point group of Δ is $\{E, \sigma_x\} = C_{1,h}$ and thus has two irr. reps., which we call (Δ_1, Δ_2).

$\rightarrow$ I: The lowest lying band is the one with $m = n = 0$ and has Δ_1 symmetry. The next band is characterized by $m = -1, n = 0$ and has, again, Δ_1 symmetry. The degeneracy of these bands is 1.

$\rightarrow$ II: These two bands are followed by a doubly degenerate band corresponding to the indices $(m = 0, n = 1)$ and $(m = 0, n = \bar{1})$. The subspace determined by $(0, 1)$ and $(0, \bar{1})$ carries the representation Δ_x, which consists of 2×2 matrices:

$$\begin{pmatrix} (\psi_{01}, E\psi_{01}) & (\psi_{01}, E\psi_{0\bar{1}}) \\ (\psi_{0\bar{1}}, E\psi_{01}) & (\psi_{0\bar{1}}, E\psi_{0\bar{1}}) \end{pmatrix} = \begin{pmatrix} 1 & 0 \\ 0 & 1 \end{pmatrix}$$

$$\begin{pmatrix} (\psi_{01}, m_x\psi_{01}) & (\psi_{01}, m_x\psi_{0\bar{1}}) \\ (\psi_{0\bar{1}}, m_x\psi_{01}) & (\psi_{0\bar{1}}, m_x\psi_{0\bar{1}}) \end{pmatrix} = \begin{pmatrix} 0 & 1 \\ 1 & 0 \end{pmatrix}$$

where the matrix elements are between the corresponding plane waves $e^{i\frac{2\pi}{a}(\kappa \cdot x + y)}$ and $e^{i\frac{2\pi}{a}(\kappa \cdot x - y)}$. The resulting representation Δ_x is twice degenerate and reduces to $\Delta_x = \Delta_1 \oplus \Delta_2$. This means that this band will split when the potential is switched on.

Simplified character table of G_Δ.

G_Δ	E	m_x
Δ_1	1	1
Δ_2	1	-1
Δ_x	2	0

The symmetry-adapted functions, computed using the projector method, read

$$P_{\Delta_1} e^{i\frac{2\pi}{a}(\kappa \cdot x + y)} \sim e^{i\frac{2\pi}{a}(\kappa \cdot x + y)} + e^{i\frac{2\pi}{a}(\kappa \cdot x - y)}$$

$$\sim e^{i\frac{2\pi}{a}(\kappa \cdot x)} \cdot \cos \frac{2\pi}{a} y$$

$$P_{\Delta_1} e^{i\frac{2\pi}{a}(\kappa \cdot x + y)} \sim e^{i\frac{2\pi}{a}(\kappa \cdot x + y)} - e^{i\frac{2\pi}{a}(\kappa \cdot x - y)}$$

$$\sim e^{i\frac{2\pi}{a}(\kappa \cdot x)} \cdot \sin \frac{2\pi}{a} y$$

According to the first-order perturbation theory, the symmetry-adapted wave functions are also the eigenfunctions of $-\frac{\hbar^2}{2m}\Delta + V(x, y)$ (non-diagonal matrix elements between functions with different symmetry vanish in virtue of the Wigner–Eckart–Koster theorem), and the energy corrections with respect to the empty-lattice band structure amount to

$$\left\langle e^{i\frac{2\pi\kappa \cdot x}{a}} \cos \frac{2\pi}{a} y, V(x, y) e^{i\frac{2\pi\kappa \cdot x}{a}} \cos \frac{2\pi}{a} y \right\rangle = E_{\Delta_1}$$

$$\left\langle e^{i\frac{2\pi\kappa \cdot x}{a}} \sin \frac{2\pi}{a} y, V(x, y) e^{i\frac{2\pi\kappa \cdot x}{a}} \sin \frac{2\pi}{a} y \right\rangle = E_{\Delta_2}$$

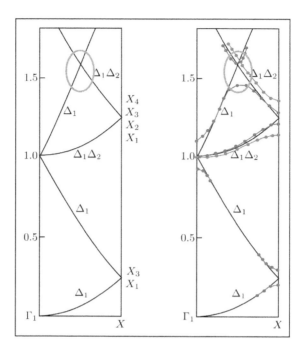

Left: Empty-lattice band structure of the primitive square lattice along Δ. Right: Realistic band structure including a potential with space group symmetry and finite strength. The Δ_1 bands are in blue, and the Δ_2 bands in red. The green circle indicates the presence of a singular point called the Wigner–von Neumann anticrossing (see the following point III for an explanation). The small dots used to draw the band structure in a finite potential are a guide to the eyes.

$\rightarrow$ III: Further up in the empty-lattice band structure, we encounter an additional typical situation for band structure computations: the *crossing* or the meeting of two bands carrying the same irr. rep., in this case Δ_1. Note that band crossing is only possible in $d \geq 2$, i.e. there is no band crossing for one-dimensional crystals. The question now is: What happens to bands of the same symmetry, crossing at some points of the BZ, when the crystal potential is switched on? At the crossing point, the representation contains more symmetry-adapted wave functions carrying the same irr. reps., i.e. those with the same symmetry. As non-diagonal matrix elements of the Hamilton operator between symmetry-adapted wave functions belonging to the same irreducible representatives are *not necessarily* vanishing, the eigenvalue problem left to be solved by explicit calculation is a non-trivial one, i.e. it involves computing explicitly the determinant of a non-diagonal matrix (second-order perturbation theory).

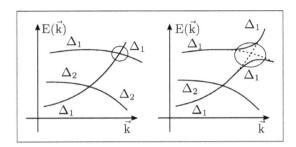

Formation of a hybridization gap at a singular point of the BZ.

The result of non-diagonal matrix elements is the formation of an *hybridization gap* or an **anticrossing** or a **Von Neumann–Wigner avoided crossing**, around which the wave function will be a linear combination of the original symmetry-adapted functions. We will discuss this important process of hybridization of wave functions with a concrete example in Lecture 7.

6.4 Non-symmorphic Space Groups

The strategy for finding the irr. rep. of the groups of **k** for **k** at the surface of the BZ in a non-symmorphic crystal is the one described by J. F. Cornwell, "Group theory and electronic energy bands in solids", in *Selected Topics in Solid State Physics*, Vol. 10, Editor: G. P. Wohlfarth, North-Holland, Amsterdam, 1969. This strategy can be generally applied, although it is really only useful for non-symmorphic space groups. It was invented by Herring. The starting point is that, in addition to the classes of adjunct elements, there is a further useful way of organizing a group.

6.4.1 *Cosets*

Definition. Let U be a subgroup of the order s of a group G of the order g, and suppose that the elements of U are $u_1, \ldots, u_s$. If a is an element of

G, then the set $\{au_1, \ldots, au_s\}$ — in short, aU — is a left coset of U with respect to G (Ua is a right coset).

Properties of cosets:

1. The right and left cosets do not contain necessarily the same elements. As an example, we show this for the group $C_{3v} = \{E, C_3, C_3^{-1}, 3\sigma\}$ and its subgroup $U = \{E, \sigma_3\}$. Recall the multiplication table of the group C_{3v}:

C_{3v}	E	C_3	C_3^{-1}	σ_1	σ_2	σ_3
E	E	C_3	C_3^{-1}	σ_1	σ_2	σ_3
C_3	C_3	C_3^{-1}	E	σ_3	σ_1	σ_2
C_3^{-1}	C_3^{-1}	E	C_3	σ_2	σ_3	σ_1
σ_1	σ_1	σ_2	σ_3	E	C_3	C_3^{-1}
σ_2	σ_2	σ_3	σ_1	C_3^{-1}	E	C_3
σ_3	σ_3	σ_1	σ_2	C_3	C_3^{-1}	E

   The left cosets are

   $$EU = \{U\}, \qquad \sigma_1 U = \{\sigma_1, C_3^{-1}\},$$
   $$C_3 U = \{C_3, \sigma_2\}, \qquad \sigma_2 U = \{\sigma_2, C_3\},$$
   $$C_3^{-1} U = \{C_3^{-1}, \sigma_1\}, \quad \sigma_3 U = \{\sigma_3, E\} = U.$$

   The right cosets are

   $$UE \; = \{U\}, \qquad U\sigma_1 = \{\sigma_1, C_3\},$$
   $$UC_3 \; = \{C_3, \sigma_1\}, \qquad U\sigma_2 = \{\sigma_2, C_3^{-1}\},$$
   $$UC_3^{-1} = \{C_3^{-1}, \sigma_2\}, \; U\sigma_3 = \{\sigma_3, E\} = U.$$

   Comparison shows that the left cosets of U are not equal to the right cosets of U with respect to $U = \{E, \sigma_3\}$.

2. Each coset contains s distinct elements.
   Proof: Assume aU is a coset, where $au_1 = au_2$ with $u_1 \neq u_2$. Left-multiplication by a^{-1} leads to $u_1 = u_2$, which is a contradiction.

3. If $a \in U$, then $aU = Ua = U$.
   Proof: U is a subgroup, implying that it is closed with respect to multiplication of any two of its elements (subgroup axiom). If now $a \in U$, we obtain the same set U by multiplication from left aU or right Ua (although probably in a different order).

4. If a is *not* a member of U, then aU is *not* a group.

   Proof: Assume aU is a group and $a \notin U$. Then, there exists $u_m \in U$ such that $au_m = e$, or, $a = u_m^{-1} \in U$, contradicting the assumption that a is not in U.

5. Two cosets are either identical or have no elements in common.

   Proof: Consider the two cosets aU and bU, and let $au_n = bu_m$, i.e. assume they have one element in common. Then, $b^{-1}a = u_m u_n^{-1} \in U$, i.e. $b^{-1}a$ belongs to U. According to property 3, this implies that $b^{-1}aU = U$. Left-multiplication with b leads to $aU = bU$, i.e. all elements are in common; therefore, the two cosets are identical.

6. If b is a member of aU, then $bU = aU$.

   Proof: If b is a member of aU, then $b = au_n$, i.e. $a^{-1}b \in U$ and $a^{-1}bU = U$, from which follows $bU = aU$.

7. Each group G of order g can be divided into a subgroup $U \subset G$ of order s and its cosets, i.e.

$$G = a_1 U \cup a_2 U \cup \ldots \cup a_i U = U \cup a_2 U \cup \ldots \cup a_i U \quad g = s \cdot i$$

   Proof: Multiplying U by every element of G produces g cosets, each containing s elements. Due to the closure of the group G, we cannot create new elements $x \notin G$, i.e. $a_1 U \cup a_2 U \cup \ldots \cup a_g U = G$. According to property 5, the cosets are either identical or disjoint. This leads to the desired property.

Example. Any space group G can be divided into cosets of the type

$$(R_1 \mid \mathbf{f}_1) \cdot T, (R_2 \mid \mathbf{f}_2) \cdot T, \ldots, (R_{g_0} \mid \mathbf{f}_{g_0}) \cdot T$$

Definition. Consider a subgroup U and a fixed element $x \in G$. Then, the set of elements xUx^{-1} forms a group (prove this). If $xUx^{-1} = U$, the group U is called an **invariant (normal) subgroup or normal divisor**.

Comments.

1. If U is an invariant subgroup, then $Ux = xU$.
2. T is an invariant subgroup of a space group.

Definition (product of cosets). Let U be an invariant subgroup, and consider now the i-cosets $a_1 U, \ldots, a_i U$ as single entities, the content of the cosets being disregarded for the moment. One can define a product between two such entities as the coset corresponding to the product of the representative elements, i.e.

$$a_i U \circ a_j U = a_i \cdot a_j U$$

Comments.

1. In order for this definition to be meaningful, one must prove that the result of the multiplication $\circ$ is independent of the choice of coset representatives.
2. This is an important property: The product of cosets reduces to the product of their representatives producing a representative of a coset.

Definition. Once this multiplication is defined, one can define a new, abstract group, the **factor group** G/U, consisting of "elements" $a_1 U = U, a_2 U, \ldots, a_i U$.

Comment.

Factor groups are typically smaller in size than the original groups and therefore more manageable. Let us consider a concrete, simple example.

We consider a one-dimensional chain with basis, i.e. with an extra atom being, for example, displaced by $\frac{1}{4} \cdot a$ with respect to the atoms occupying the lattice sites.

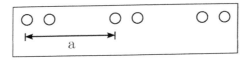

A one-dimensional lattice with a second atom at a fractional position of the unit segment.

We take the origin of the horizontal axis on top of one of the lattice atoms. We observe

- The space group writes

$$G = (E \mid 0), \left(I\frac{1}{4}a \mid 0\right), (E \mid \pm a), (E \mid \pm 2a), \ldots, \left(I\frac{1}{4}a \mid \pm a\right), \ldots$$

- Its multiplication table is as follows.

G	$(E \mid n)$	$(I \mid m + \frac{1}{4})$	$\ldots$
$(E \mid n')$	$(E \mid n + n')$	$(I \mid m + n' + \frac{1}{4})$	$\ldots$
$(I \mid m' + \frac{1}{4})$	$(I \mid -n + m' + \frac{1}{4})$	$(E \mid -m + m')$	$\ldots$

- The small point group is $G_0 = \{E, I\}$.
- The translation group is $U = 0 \cdot a, \pm a, \pm 2a, \ldots$.
- Further,

$$G/U = \left\{ EU, I\frac{1}{4}aU \right\}$$

$G/U \sim \{E, I\}$	EU	$I\frac{1}{4}U$
EU	EU	$I\frac{1}{4}U$
$I\frac{1}{4}U$	$I\frac{1}{4}U$	EU

We conclude that the factor group G/U contains two elements. Its multiplication table shows that the group is isomorphic to the point group C_i.

6.4.2 *Irreducible representations of the factor group of k*

Definition. For any $\mathbf{k}$, which might or might not be at the surface of a BZ, we define the group $T(\mathbf{k})$ consisting of all *primitive* translations $\mathbf{t}_n$ such that

$$e^{-i\mathbf{k}\cdot\mathbf{t}_n} = 1$$

Comments.

1. $T(\mathbf{k})$ is an invariant subgroup of $G_{\mathbf{k}}$. It is therefore possible to define the factor group

$$G_{\mathbf{k}}/T_{\mathbf{k}} = \{(E \mid \mathbf{0})T(\mathbf{k}), (R \mid \mathbf{f}_R + \mathbf{t}_m \neq \mathbf{t}_n)T(\mathbf{k})\}$$

Example. A linear chain at $k = \frac{\pi}{a}$.

— $T(\frac{\pi}{a}) = \{0, \pm 2a, \pm 4a, \ldots\}$
— $G(\frac{\pi}{a}) = \{(E \mid 0), (E \mid \pm a), (E \mid \pm 2a), \ldots, (I \mid \frac{a}{4}), (I \mid \frac{a}{4} \pm a) \ldots\}$
— $G(\frac{\pi}{a})/T(\frac{\pi}{a}) = \{(E \mid 0)T(\frac{\pi}{a}), (E \mid +a)T(\frac{\pi}{a}), (I \mid \frac{a}{4})T(\frac{\pi}{a}), (I \mid \frac{a}{4} + a)T(\frac{\pi}{a})\}$

2. For $\mathbf{k}$ residing inside the BZ, $T(\mathbf{k})$ contains only the identity element $\mathbf{t}_n = \mathbf{0}$. The factor group $G_{\mathbf{k}}/T_{\mathbf{k}}$ is identical with the group $G_{\mathbf{k}}$.

Theorem. *Let D be an irr. rep. of $G_{\mathbf{k}}/T_{\mathbf{k}}$. If D has the property (Bloch)*

$$D[(E \mid \mathbf{t}_m)T(\mathbf{k})] = e^{-i\mathbf{k}\cdot\mathbf{t}_m} D[(E \mid 0)T(\mathbf{k})]$$

for every coset $(E \mid \mathbf{t}_m)T(\mathbf{k})$ formed from primitive translations, then the set of matrices defined by

$$D(R \mid \mathbf{t}) := D[(R \mid \mathbf{t})T(\mathbf{k})]$$

constitutes an irr. rep. of $G_{\mathbf{k}}$, which satisfies the Bloch condition and is thus physically relevant. Furthermore, all physically relevant irr. reps. of $G_{\mathbf{k}}$ can be constructed in this way, starting from those of $G_{\mathbf{k}}/T(\mathbf{t}_{\mathbf{k}})$ (the proof of this last sentence is given in the book of J.F. Cornwell, quoted previously).

Comments.

1. This theorem is obvious for $\mathbf{k}$ inside the BZ. It states that only representations of $G_{\mathbf{k}}$ are physically relevant if they obey the Bloch condition.
2. Note that not all irr. reps. of $G_{\mathbf{k}}/T_{\mathbf{k}}$ satisfy the Bloch condition. In particular, for $\mathbf{k} \neq 0$, the identical representation "1" is an important example of physically irrelevant representations.
3. $G_{\mathbf{k}}/T_{\mathbf{k}}$ is not necessarily isomorphic to a crystallographic point group, so that the problem of finding the irr. rep. of $G_{\mathbf{k}}$ is considerably simplified by this theorem but cannot always be reduced to a point group problem. We will see an explicit example of this fact.

Proof:

- First, we have to prove that the matrix defined above is indeed a representation of $G(\mathbf{k})$:

$$D((R \mid \mathbf{t}))D((R' \mid \mathbf{t}')) = D((R \mid \mathbf{t})T(\mathbf{k}))D((R' \mid \mathbf{t}')T(\mathbf{k}))$$

$$= D((R \mid \mathbf{t})T(\mathbf{k})(R' \mid \mathbf{t}')T(\mathbf{k}))$$

$$= D((R \mid \mathbf{t})(R' \mid \mathbf{t}')T(\mathbf{k}))$$

$$= D((R \mid \mathbf{t})(R' \mid \mathbf{t}'))$$

- Next, we have to establish the irreducibility of the representations constructed in this way. We write

$$\sum_{G(\mathbf{k})} \mid \chi(R \mid \mathbf{t}) \mid^2 = t(\mathbf{k}) \cdot \sum_{G(\mathbf{k})/T(\mathbf{k})} \mid \chi((R \mid \mathbf{t})T(\mathbf{k})) \mid^2$$

where $t(\mathbf{k})$ is the order of the group $T(\mathbf{k})$. As $D((R \mid \mathbf{t})T(\mathbf{k}))$ is an irr. rep., we have

$$\sum_{G(\mathbf{k})/T(\mathbf{k})} \mid \chi((R \mid \mathbf{t})T(\mathbf{k})) \mid^2 = g(\mathbf{k})/t(\mathbf{k})$$

which immediately proves the irreducibility of the representation for $G(\mathbf{k})$.
- Finally, the representation constructed in this way obeys the Bloch condition, i.e.

$$D(E \mid \mathbf{t}_n) = D((E \mid \mathbf{t}_n)T(\mathbf{k})) = e^{-i\mathbf{k}\mathbf{t}_n} D((E \mid \mathbf{0})T(\mathbf{k}))$$
$$= e^{-i\mathbf{k}\mathbf{t}_n} D((E \mid \mathbf{0}))$$

showing that the elements containing primitive translations are represented by $e^{-i\mathbf{k}\mathbf{t}_n} \cdot E$, as it should be in order to obtain the Bloch functions as symmetry-adapted wave functions.

Example 1. Atomic chain with basis

a. Multiplication table: One must first find the multiplication table of the group

$$G\left(\frac{\pi}{a}\right) \Big/ T\left(\frac{\pi}{a}\right) = \left\{ \underbrace{(E \mid 0)T\left(\frac{\pi}{a}\right)}_{a \equiv e}, \underbrace{(E \mid +a)T\left(\frac{\pi}{a}\right)}_{b}, \right.$$

$$\left. \underbrace{\left(I \mid \frac{a}{4}\right) T\left(\frac{\pi}{a}\right)}_{c}, \underbrace{\left(I \mid \frac{a}{4} + a\right) T\left(\frac{\pi}{a}\right)}_{d} \right\}$$

Multiplication table for the
group $G\left(\frac{\pi}{a}\right)/T\left(\frac{\pi}{a}\right)$.

	e	b	c	d
e	e	b	c	d
b	b	e	d	c
c	c	d	e	b
d	d	c	b	e

Accordingly, the group will be either isomorphous to C_4 or C_{2v}.
The multiplication table tells us that the group is isomorphic to C_{2v}, so
that its character table is constructed as follows.

Character table for the factor group of the point $X = \frac{\pi}{a}$
of the first BZ of the linear chain with basis

$G_{\frac{\pi}{a}}/T_{\frac{\pi}{a}}$	$(E \mid 0)$	$(E \mid +a)$	$(I\frac{1}{4}a)$	$(I\frac{1}{4}a + a)$
X_1	1	1	1	1
X_2	1	1	$\bar{1}$	$\bar{1}$
X_3	1	$\bar{1}$	1	$\bar{1}$
X_4	1	$\bar{1}$	$\bar{1}$	1

b. Physically relevant representations: X_1 and X_2 are not physically
relevant, as, for example, $(E \mid 2n + 1)$ is represented by 1 and not by
$e^{-i\frac{\pi}{a}(2n+1)\cdot a} = -1$, as it should be. X_3 and X_4 are the only physically
relevant representations.

c. Symmetry-adapted functions: Starting from

$$e^{\pm i\frac{\pi}{a}\cdot x}$$

we construct the symmetry-adapted basis functions to X_3 and X_4 using
the projector method:

$$P_{X_3}e^{i\frac{\pi}{a}\cdot x} \simeq e^{i\frac{\pi}{a}\cdot x} - e^{i\frac{\pi}{a}\cdot(x-a)} + e^{i\frac{\pi}{a}\cdot(-x+\frac{a}{4})} - e^{i\frac{\pi}{a}\cdot(-x+\frac{a}{4}+a)}$$

$$\approx e^{i\frac{\pi}{a}\cdot x} + e^{i\frac{\pi}{4}}e^{-i\frac{\pi}{a}\cdot x}$$

$$P_{X_4}e^{i\frac{\pi}{a}\cdot x} \simeq e^{i\frac{\pi}{a}\cdot x} - e^{i\frac{\pi}{4}}e^{-i\frac{\pi}{a}\cdot x}$$

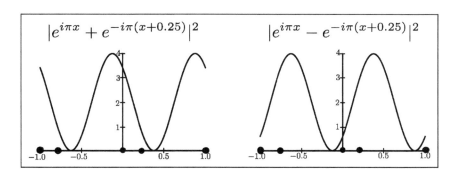

Left: X_3 symmetry-adapted plane wave. Right: X_4 symmetry-adapted plane wave. The black dots indicate the atomic positions.

Example 2. The X-point of O_h^7

The small point group of $\mathbf{k}$ is D_{4h} and has 16 elements; it is given by $C_{4v} \cup IC_{4v}$. In fact,

$$I\mathbf{k} = \frac{2\pi}{a}(-1,0,0) = \mathbf{k} - (\mathbf{h}_2 + \mathbf{h}_3)$$

where $\mathbf{h}_{2,3}$ are reciprocal lattice vectors. As I is associated with fractional translations, the irr. rep. of G_X cannot be derived from D_{4h}. Rather, we have to use the factor group G_X/T_X, which contains 32 elements and is not a point group. The character table of this group (see F. Bassani and G. Pastori Parravicini, *Electronic States and Optical Transitions in Solids*, Pergamon, Oxford, 1975), has only four irr. reps. that are compatible with the Bloch theorem, called in the literature (X_1, X_2, X_3, X_4). A characteristic of these four irr. reps. is that they are all two-dimensional — a fact that is referred to in the literature as "non-symmorphic degeneracy". Non-symmorphic degeneracy is an example of what is know in the modern literature as a "topologically protected" feature of the band structure. The band structure of Si, which has a diamond structure, shows explicitly this topological degeneracy, which is lifted when the Si atom at fractional positions is substituted by a C atom in SiC, which has a symmorphic space group.

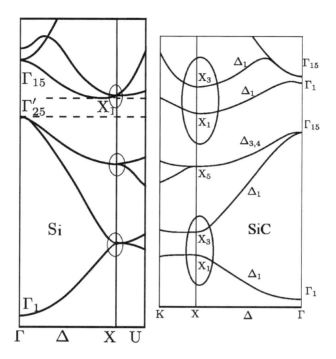

Left: Schematic band structure of Si along the Δ and U lines. The locations of the non-symmorphic degeneracies at X are encircled. The dashed horizontal lines indicate the gap. Reproduced freely from J. R. Chelikowsky and M. L. Cohen, *Phys. Rev. B* 10, 5095 (1974), https://journals.aps.org/prb/abstract/10.1103/PhysRevB.10.5095. Right: Schematic band structure of SiC. The origin of the vertical, energy axis has been shifted to align, approximately, the position of the lifted degenergies at X in SiC (encircled) with the corresponding degeneracies in Si. Reproduced freely from L. A. Hemstreet and C. Y. Fong, Silicon Carbide — 1973, Editors: R. C. Marshall, J. W. Faust and C. E. Ryan, University of South Carolina Press, Columbia, S.C. 1974, p. 284.

Lecture 7

THE RELATIVISTIC ELECTRONIC STRUCTURE OF ATOMS AND SOLIDS

In this lecture, we explain how a symmetry group is expanded to implement new degrees of freedom using the technique called the "direct product of groups" and "Kronecker product" of representations. We use this technique to implement, for instance, the spin of the electrons within the symmetry-based electronic structure calculations of atoms and solids. Finally, we introduce the generalized Wigner–Eckart–Koster theorem for symmetry-based computing of matrix elements. This lecture is accompanied by two projects. The project "Topological Aspects of Continuous Groups: Universal Covering Groups" deals with finding a mathematically rigorous proof of SU_2, which is the symmetry group of electrons with spin degree of freedom. The second project, "Optical Spin Orientation in Atoms and Solids", computes some symmetry-related quantities in optical transitions, produced by the entanglement of spin and orbital states in atoms and solids.

7.1 The Spin of the Electron and the Bethe Hypothesis

7.1.1 *Introduction*

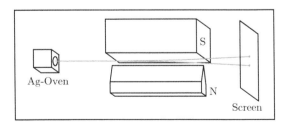

The Stern–Gerlach apparatus: The z-direction is vertical in the figure.

The algebra of the angular momentum operator is a direct consequence of the Schrödinger quantum mechanics and is a precise confirmation of Bohr–Sommerfeld quantization rules for the orbital angular momentum. In 1921, O. Stern and W. Gerlach performed a very remarkable experiment. They placed at the entrance of an instrument, which is nowadays known as the Stern–Gerlach apparatus, a Ag furnace. If the furnace is heated, Ag atoms evaporate. At that time, the Ag atom was known to have a single electron circulating around a set of shells — to be indented as Bohr circular orbits — containing the remaining electrons. According to the quantization rule by Bohr, such an electron has an angular momentum quantum number of 1. In such a state, the z component of the quantized angular momentum L_z was expected to be either $\hbar$, 0 or $-\hbar$ — the Cartan weights of the irreducible representation D_1 of SO_3. In classical physics, on the other hand, all values in between would also be possible. This means that the Ag atoms were expected to leave the furnace, being equally distributed in some state of the z-component of the angular momentum: one of the three quantized states (in the Bohr–Sommerfeld quantized model), or a continuous value in the interval $[-\hbar, \hbar]$ in the classical model. Such an experiment was therefore designed to discriminate between the two models. After leaving the furnace, the beam was directed to an area where there is a strong magnetic field gradient: $\mathbf{B}$ is highly inhomogeneous along the z-axis. Such inhomogeneity can be achieved by properly designing the poles of the magnet. It is known (from classical physics, actually) that the magnetic field gradient along z sends atoms with different L_z into different paths along the z-axis. In other words, in the absence of B-inhomogeneity, the Ag atoms passes unperturbed and arrive at the center of the screen. When the B-inhomogeneity is present, two scenarios were expected: In the Bohr-quantized scenario, three separate beams would form and produce three separate spots along z on the screen. In the scenario based on classical physics, the particles would distribute themselves along a continuous line in the z-direction. Stern and Gerlach observed neither of these two expected scenarios: They observed two separate spots. This was taken as a partial proof of quantization of angular momentum but led to speculations, such as those by A. Sommerfeld, regarding the absence of the third spot. Could it be possible that, for some reason, one of the three beams was somehow "suppressed"? The correct interpretation of this result was formulated as a hypothesis by Goudsmit and Uhlenbeck in 1925: The electron is actually in a state of vanishing orbital angular momentum (s state, as we know from the

Schrödinger quantum mechanics) but must have had some internal degree of freedom with angular momentum character and quantum number $\frac{1}{2}$ with the Cartan weights $+\frac{1}{2}$ and $-\frac{1}{2}$ in order to explain the existence of only two z components. This internal degree of freedom was called the "spin" of the electron.

The hypothesis by Goudsmit and Uhlenbeck was based on further observations that pointed to a new double-valued quantum number: In the same year, Pauli suggested his principle of double occupancy of the Bohr circular orbits. Meanwhile, there were spectroscopic observations of the Na vapor. The transition shown in the following figure is the well-known Na D-line ($\lambda \approx 589$ nm), which causes a yellow color of the Na flame. At a sufficiently high resolution of the spectrometer, it can be seen that it consists of two closely spaced lines, called a doublet (fine structure of the spectral lines, a term introduced by A. Sommerfeld). Again, this observation pointed to a new quantum number for the angular momentum with two Cartan weights.

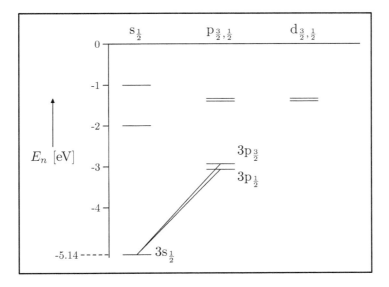

Term scheme of the Na atom indicating the transitions producing the "Na doublet". The energy levels are labeled with the main quantum number 3, the orbital quantum number $l = 1$ (p), and a third quantum number referring to the quantum number of the total angular momentum — orbital plus spin quantum number. This third quantum number will be explained later.

Comments.

1. The existence of half-integer spin quantum numbers raises the question of the existence of $S = \frac{1}{3}, S = \frac{1}{4}$, and so on. The charge, for example, occurs in integer multiples of $\mid e \mid$, but $\frac{1}{3} \mid e \mid$ exists too! We will find the answer to this question in the project on "Topological Aspects".
2. The Schrödinger equation predicts that quantum numbers for orbital angular momenta are integer numbers $l = 0, 1, 2, 3, \ldots$. Half-integer numbers can only be assigned to the intrinsic spin state of particles. Of course, there are particles that have integers values for the intrinsic spin state quantum numbers, such as photons (spin 1) and gravitons (spin 2).
3. Orbital angular momentum quantum numbers are often indicated by the small Latin letter l (or the capital letter L). Intrinsic spin angular momentum quantum numbers are often given with the small Latin letter s or the capital letter S. Quantum numbers resulting from both (see later for how to add angular momenta) are given with the small Latin letter j or the capital letter J. For more details on all these quantum mechanical aspects see e.g. D. Pescia, "Lectures on Quantum Mechanics for Material Scientists", online available at https://doi.org/10.3929/ethz-b-000621567.

7.1.2 The $D^{\frac{1}{2}}$ representation of SO_3

The postulate of Cartan weights $\pm\frac{1}{2}$ introduces, technically, a novel matrix representation of SO_3, which we can call $D^{\frac{1}{2}}$. This representation is constructed by generalizing the matrix elements for the infinitesimal generators, taken from E. U. Condon and G. Shortley, *The Theory of Atomic Spectra*, Cambridge University Press, Cambridge, England, 1951) to half-integer spins. Using the "Z" symbol to label states with any value of the angular momentum (integer or half-integer), one has ($m = p, p-1, \ldots, -p$)

$$(Z_p^m, I_z Z_p^m) = m$$
$$(Z_p^{m\pm1}, I_x Z_p^m) = \frac{1}{2}\sqrt{(p \mp m)(p \pm m + 1)}$$
$$(Z_p^{m\pm1}, I_y Z_p^m) = \mp\frac{i}{2}\sqrt{(p \mp m)(p \pm m + 1)}$$

One also gives usually the commutation relations

$$[Z_x, Z_y] = iI_z \quad [Z_y, Z_z] = i \cdot Z_x \quad [Z_z, Z_x] = i \cdot Z_y$$

Accordingly, for $p = \frac{1}{2}$ and $m = \pm\frac{1}{2}$, one obtains

$$
S_z = \frac{1}{2} \cdot \underbrace{\begin{pmatrix} 1 & 0 \\ 0 & -1 \end{pmatrix}}_{\sigma_z} \qquad S_x = \frac{1}{2} \cdot \underbrace{\begin{pmatrix} 0 & 1 \\ 1 & 0 \end{pmatrix}}_{\sigma_x} \qquad S_y = \frac{1}{2} \cdot \underbrace{\begin{pmatrix} 0 & -i \\ i & 0 \end{pmatrix}}_{\sigma_y}
$$

where the Pauli matrices $(\sigma_x, \sigma_y, \sigma_z)$ have been introduced.

We proceed now "*a la Rodriguez*", i.e. we use the exponential map algorithm to obtain the matrix representation $D^{\frac{1}{2}}$. Using $\mathbf{n} = (l, m, n)$ and the rotation angle $\varphi \in [0, 2\pi]$[1] we can write the rotation matrix as

$$
D^{\frac{1}{2}}(\mathbf{n}, \varphi) = e^{-i\frac{\varphi}{2}\begin{pmatrix} n & l - i \cdot m \\ l + i \cdot m & -n \end{pmatrix}}
$$

As $l^2 + n^2 + m^2 = 1$, we have

$$
\begin{pmatrix} n & l - i \cdot m \\ l + i \cdot m & -n \end{pmatrix}^2 = \begin{pmatrix} 1 & 0 \\ 0 & 1 \end{pmatrix}
$$

so that

$$
D^{\frac{1}{2}}(\mathbf{n}, \varphi) = \sum_{k_{\text{even}}} \frac{(-i\varphi/2)^k}{k!} \begin{pmatrix} 1 & 0 \\ 0 & 1 \end{pmatrix}
$$

$$
+ \sum_{k_{\text{odd}}} \frac{(-i\varphi/2)^k}{k!} \begin{pmatrix} n & l - i \cdot m \\ l + i \cdot m & -n \end{pmatrix}
$$

$$
= \begin{pmatrix} \cos\frac{\varphi}{2} - i \cdot n \sin\frac{\varphi}{2} & (-i \cdot l - m)\sin\frac{\varphi}{2} \\ (-i \cdot l + m)\sin\frac{\varphi}{2} & \cos\frac{\varphi}{2} + i \cdot n \sin\frac{\varphi}{2} \end{pmatrix}
$$

---

[1] When the range $[0, 2\pi]$ is used, it is enough to choose positive l, m, n to cover the entire group.

Comments.

1. Consider now the matrix representation of rotations about the z-axis by an angle φ:

$$D^{\frac{1}{2}}(z,\varphi) = \begin{pmatrix} e^{-i\frac{\varphi}{2}} & 0 \\ 0 & e^{+i\frac{\varphi}{2}} \end{pmatrix}$$

In particular,

$$D^{\frac{1}{2}}(z,\varphi = 0) = \begin{pmatrix} 1 & 0 \\ 0 & 1 \end{pmatrix}$$

is the identity matrix, which correctly represents the identity element of SO_3. A strange situation arises when the matrix with rotation by 2π is written:

$$D^{\frac{1}{2}}(z,\varphi = 2\pi) = \begin{pmatrix} \bar{1} & 0 \\ 0 & \bar{1} \end{pmatrix}$$

We have the situation that, in the $D^{\frac{1}{2}}$ representation, a rotation by 0 and 2π, which are the same element of SO_3, are represented by two different matrices. As a matter of fact, we find that this is systematic of $D^{\frac{1}{2}}$:

$$D^{\frac{1}{2}}(\mathbf{n},\varphi + 2\pi) = \bar{E} \cdot D^{\frac{1}{2}}(\mathbf{n},\varphi)$$

with

$$\bar{E} \doteq D^{\frac{1}{2}}(\mathbf{n},2\pi) = \begin{pmatrix} \bar{1} & 0 \\ 0 & \bar{1} \end{pmatrix}$$

We conclude that $D^{\frac{1}{2}}$ is a two-valued representation of SO_3, i.e. it is not a representation in the strictest sense. None of the theorems we have derived regarding representations can be applied to $D^{\frac{1}{2}}$.

2. The way out of this problem was suggested by H. Bethe. He noticed that rotating by 4π the identity is again obtained. He therefore suggested that SO_3 cannot be the physically relevant symmetry group for describing angular momentum in quantum mechanics. He suggested replacing SO_3 with a "double group" SO_3^D, where the rotation by 2π, say $R(\mathbf{n},2\pi)$, is a new element distinct from $R(\mathbf{n},0)$ and the number of elements is

doubled with respect to SO_3, according to

$$SO_3^D = SO_3 \cup R(2\pi)SO_3$$

with the set on the right-hand side describing the elements with rotation angles larger than 2π.

3. The representation theory of this new group entails two types of irr. reps.: the so-called "single group" representations, obtained by representing $R(2\pi)$ with the identity matrix E. For these representations, the characters of the elements $\varphi \geq 2\pi$ are exactly the same as the characters of the elements $\varphi \leq 2\pi$. The "double (or extra) group" representations are obtained by representing the element $R(2\pi)$ with $\bar{E} = -E$. Accordingly, the character table of SO_3^D is constructed.

SO_3^D	R_0	$R(\varphi)$	$R_{2\pi}$	$R_{2\pi}R(\varphi)$
$D^l(l = 0, 1, 2, \ldots)$	$2l + 1$	$\frac{\sin((2l+1)\frac{\varphi}{2})}{\sin\frac{\varphi}{2}}$	$2l + 1$	$\frac{\sin((2l+1)\frac{\varphi}{2})}{\sin\frac{\varphi}{2}}$
$D^s(s = \frac{1}{2}, \frac{3}{2}, \ldots)$	$2s + 1$	$\frac{\sin((2s+1)\frac{\varphi}{2})}{\sin\frac{\varphi}{2}}$	$-(2s + 1)$	$-\frac{\sin((2s+1)\frac{\varphi}{2})}{\sin\frac{\varphi}{2}}$

4. A question arises regarding the representation of the half-integer basis states Z_s^m using spherical harmonics on the unit sphere in 3D. "Half-integer spherical harmonics" (the basis functions carrying the D^s representations) cannot be defined on the sphere: They exist only in a space which is different from the conventional Euclidean (or Einstein-like) space. In this particular space, which we call $\Theta_s \equiv \mathscr{C}^{2S+1}$, the periodicity is $2\pi \cdot (2S+1)$ and not the periodicity of 2π implied by the set of conventional spherical harmonics on the unit sphere. The "spin" space only exists at the very small spatial scales corresponding to the world of subatomic physics, where half-integer particles reside. At macroscopic length scales, the 2π periodicity must emerge.

5. In the following, we take the convention of writing X_l^m for those states that are related to the "conventional" spherical harmonics $Y_l^m(\vartheta, \varphi)$. Those states that represent "spinors" are then written as Y_s^m or simply Y^+, Y^- for spin-$\frac{1}{2}$ states.

6. One can generally obtain the description of a spinor state Y_s^m as a $\mathscr{C}$-valued scalar field "a la Landau–Lifshitz", i.e. by introducing a configuration space $\{+s, \ldots, -s\}$ for the variable m and defining the wave function $Y_s(m)$ as the amplitude for the spin to assume the value m

of the z-component of the spin angular momentum. In this notation, the z-component of the spin angular momentum works as an independent, discrete-valued variable over which the wave function is described. For the basis states $Y_s^{m_s}$, $m_s = +s, \ldots, -s$, we have, for example,

$$Y_s^{m_s}(m) = \delta_{m,m_s}$$

7. A question arises regarding which matrices represent I (the inversion) and $\bar{I}$ in the $D^{\frac{1}{2}}$ representation. As the inversion commute with any other element, we have $I = +E$ or $I = -E$ and vice versa for $\bar{I}$. The conventional choice is

$$D^{\frac{1}{2}}(I) = E \quad D^{\frac{1}{2}}(\bar{I}) = -E$$

With this choice, one can extend the $D^{\frac{1}{2}}$ to become an irr. rep. of O_3^D.

7.2   Symmetry of Composite Systems

Introduction: The spin of the electron is an example of a **new degree of freedom** for a single electron moving in a potential landscape. We must learn to understand the rather technical, mathematical "devices" that allow expanding the Hilbert space to accommodate the new degrees of freedom and the inevitably associated extra symmetry elements. This technique goes under the name of "direct product of groups" and "tensor (or Kronecker) product of vector spaces". The same technique that allows for accommodating extra degrees of freedom is also used to built many-particle states (we will dedicate Lectures 8 and 9 to this topic).

As a way of illustration, we consider, for example, a molecule with atoms arranged at the corners of a triangle within a plane and an atom sticking out perpendicular to the plane along the $+z$-direction in order to build a regular pyramid. The symmetry group of this molecule is C_{3v}. Suppose we add a further atom in the specular position along the $-z$-direction. One introduces, with this move, a new degree of freedom and, simultaneously, a further symmetry element — the horizontal plane as a mirror plane. The symmetry group of the initial pyramid must be expanded to include the symmetry element σ_h properly. We anticipate the solution to this problem: The extra atom produce the existence of a symmetry group $\{E, \sigma_h\}$, and the symmetry group of the entire system consisting of the five vertices is called the direct (or Kronecker) product $C_{3v} \times \{E, \sigma_h\}$.

7.2.1 The direct product of groups

Definition. Let $H = e, h_1, h_2, \ldots$ and $K = e, k_1, k_2, \ldots$ be two groups such that all the elements h_j commute with k_j. If we multiply each element of H with each of K, we obtain a new set of elements, called the direct product $G \equiv H \times K$.

Proposition. $H \times K$ *is a group.*

Proof. We prove that the product of two elements in G is also an element of G. Given $h_i h_m = h_p$, $k_j k_n = k_q$, and $G = e, g_{11}, g_{12}, \ldots, g_{1k}, g_{21}, g_{22}, \ldots, g_{ij} \equiv h_i k_j, \ldots$, we compute

$$g_{ij} g_{mn} = (h_i k_j)(h_m k_n) = h_i h_m k_j k_n = h_p k_q = g_{pq}$$

◇

Comments.

1. In our previous example, we have $H = C_{3v} = \{e, 2C_3, 3\sigma_v\}$, $K = C_{1h} = \{e, \sigma_h\}$ and

$$G = C_{3v} \times C_{1h} = \{e, 2C_3, 3\sigma_v, \sigma_h, 2C_3\sigma_h, 3\sigma_v\sigma_h\}$$

2. The inversion I commute with any other symmetry element and can be used to produce new groups from existing ones. O_2, for instance, can be obtained from SO_2 by the direct product with $\bar{1} = \{e, i\} \equiv C_i$:

$$O_2 = SO_2 \times C_i$$

3. Similarly,

$$O_3 = SO_3 \times C_i = \{e, R(\varphi), I, IR(\varphi)\}$$

4. In three-dimensional crystallography, starting from the 11 proper rotation point groups, one obtains 11 new improper crystallographic point groups by the operation of direct product with $\{e, I\}$ (type "a" improper point groups, see Chapter 5). For example,

$$432(= O) \times \bar{1}(C_i) = \frac{4}{m}\bar{3}\frac{2}{m}(O_h)$$

(on the left is the Hermann–Mauguin labeling, and in brackets is the Schönflies nomenclature)

5. A question arises regarding whether and how the irr. reps. of the direct product group can be constructed knowing the irr. reps. of the components.

7.2.2　*The tensor (or Kronecker) product of vectors and matrices*

To answer this question, we need to introduce a new mathematical tool, known as the Kronecker (or tensor) product of vectors and matrices. We start with a set of basis vectors $\{e_1, \ldots, e_a\}$ in a vector space V_a and a set of basis vectors $\{f_1, \ldots, f_b\}$ in a vector space V_b (the vector spaces can be infinite-dimensional, in which case the vector spaces must be Hilbert spaces). The vectors in these vector spaces are identified by their components:

$$\mathbf{x} = (x_1, x_2, \ldots, x_a) \quad \mathbf{y} = (y_1, \ldots, y_b)$$

Definition. The products

$$\{x_1 \cdot y_1, x_1 \cdot y_2, \ldots, x_1 \cdot y_b, x_2 \cdot y_2, \ldots, x_2 \cdot y_b, \ldots, x_a \cdot y_b\} \equiv \mathbf{x} \otimes \mathbf{y}$$

are a new set of components of the Kronecker (tensor) product $\mathbf{x} \otimes \mathbf{y}$ with respect to the Kronecker (tensor) product of the basis states:

$$\{e_1 \otimes f_1, e_1 \otimes f_2, \ldots, e_2 \otimes f_1, e_2 \otimes f_2, \ldots\}$$

Comments.

1. The set of products $\mathbf{x} \otimes \mathbf{y}$ constitute a vector space, i.e. the operations of sum and multiplication by a scalar are well defined:

$$\mathbf{a} \otimes \mathbf{b} + \mathbf{c} \otimes \mathbf{d} = (a_1 \cdot b_1 + c_1 \cdot d_1, \ldots, a_i \cdot b_i + c_i \cdot d_i, \ldots)$$

$$\lambda(\mathbf{a} \otimes \mathbf{b}) = (\lambda \mathbf{a}) \otimes \mathbf{b} = \mathbf{a} \otimes \lambda \mathbf{b}$$

The vector space is called the tensor or Kronecker product space.

2. The tensor product is linear in each slot in the sense that for any complex numbers α and β, we have

$$(\alpha \mathbf{a}_1 + \beta \mathbf{a}_2) \otimes \mathbf{b} = \alpha \mathbf{a}_1 \otimes \mathbf{b} + \beta \mathbf{a}_2 \otimes \mathbf{b}$$

and

$$\mathbf{a} \otimes (\alpha \mathbf{b}_1 + \beta \mathbf{b}_2) = \mathbf{a} \otimes (\alpha \mathbf{b}_1) + \mathbf{a} \otimes (\beta \mathbf{b}_2)$$

3. In the tensor product space, a Hermitic metric can be obtained from the definition

$$(\mathbf{a}_1 \otimes \mathbf{b}_1, \mathbf{a}_2 \otimes \mathbf{b}_2) = (\mathbf{a}_1, \mathbf{a}_2) \cdot (\mathbf{b}_1, \mathbf{b}_2)$$

4. Provided the basis vectors $\{\mathbf{e}_i\}$ and $\{\mathbf{f}_j\}$ are complete orthononormal basis vectors in their respective vectors spaces, the basis vectors

$$\{\mathbf{e}_i \otimes \mathbf{f}_j\}$$

constitute a complete orthonormal set in the vector space $V_a \otimes V_b$, i.e. any vector in $V_a \otimes V_b$ can be expressed as a linear combination of the basis vectors:

$$\mathbf{a} = \sum_{i,j} c_{ij} \mathbf{e}_i \otimes \mathbf{f}_j$$

5. **Entanglement:** An interesting situation arises from the fact that, generally,

$$\sum_{i,j} c_{ij} \mathbf{e}_i \otimes \mathbf{f}_j \neq \mathbf{a} \otimes \mathbf{b}$$

Definition. Vectors in $V_a \otimes V_b$ that cannot be expressed as a Kronecker product of a vector in V_a and a vector in V_b are called **entangled**.

6. Given a $a \times a$ matrix A with

$$x_i' = \sum_{\lambda=1}^{a} A_{i\lambda} x_\lambda$$

and a $b \times b$ matrix B with

$$y_k' = \sum_{\mu=1}^{b} B_{k\mu} y_\mu$$

the product $x_i' \cdot y_k'$ transforms as

$$x_i' y_k' = \sum_{\lambda=1}^{a} \sum_{\mu=1}^{b} A_{i\lambda} B_{k\mu} x_\lambda y_\mu$$

If in the product space, we adopt the sequence of basis vectors

$$(1,1), (1,2), (1,3), \ldots, (1,b), (2,1), (2,2), \ldots, (2,b), \ldots, (n,1), \ldots$$

this transformation property can be written in vectorial style as

$$
\begin{pmatrix} x_1'y_1' \\ x_1'y_2' \\ \vdots \\ \vdots \\ x_1'y_b' \\ \vdots \\ \vdots \\ x_a'y_b' \end{pmatrix} = \begin{pmatrix} A_{11}\cdot B_{11} & A_{11}\cdot B_{12} & \ldots & \ldots & A_{11}\cdot B_{1b} & \ldots & \ldots & \ldots & A_{1a}\cdot B_{1b} \\ A_{11}\cdot B_{21} & A_{11}\cdot B_{22} & \ldots & \ldots & A_{11}\cdot B_{2b} & \ldots & \ldots & \ldots & A_{1a}\cdot B_{2b} \\ \vdots & \vdots & & & \vdots & & & & \vdots \\ \vdots & \vdots & & & \vdots & & & & \vdots \\ A_{11}\cdot B_{b1} & A_{11}\cdot B_{b2} & \ldots & \ldots & A_{11}\cdot B_{bb} & \ldots & \ldots & \ldots & A_{1a}\cdot B_{bb} \\ \vdots & \vdots & & & \vdots & & & & \vdots \\ \vdots & \vdots & & & \vdots & & & & \vdots \\ A_{a1}B_{b1} & A_{a1}B_{b2} & \ldots & \ldots & A_{a1}B_{bb} & \ldots & \ldots & \ldots & A_{aa}B_{bb} \end{pmatrix}
$$

$$
\times \begin{pmatrix} x_1y_1 \\ x_1y_2 \\ \vdots \\ \vdots \\ x_1y_b \\ \vdots \\ \vdots \\ x_ay_b \end{pmatrix}
$$

Definition. The transformation matrix

$$
\begin{pmatrix} a_{11}\cdot B & a_{12}\cdot B & \ldots & a_{1n}\cdot B \\ a_{21}\cdot B & a_{22}\cdot B & \ldots & a_{2n}\cdot B \\ \vdots & \vdots & & \vdots \\ a_{n1}\cdot B & a_{n2}\cdot B & \ldots & a_{nn}\cdot B \end{pmatrix}
$$

is called the Kronecker product of A and B, $A \otimes B$.

Comments.

1. For the trace of the Kronecker product, we have then the following important relation:

$$
\mathrm{tr}(A \otimes B) = \mathrm{tr}(A)\cdot\mathrm{tr}(B)
$$

2. Some computation rules:

a. The Kronecker product is linear in both arguments, i.e.

$$(A \otimes B_1 + B_2) = A_1 \otimes B_1 + A \otimes B_2)$$

$$(A_1 + A_2) \otimes B = A_1 \otimes B + A_2 \otimes B$$

$$\lambda(A \otimes B) = \lambda A \otimes B = A \otimes \lambda B$$

b.

$$(A_1 \otimes B_1) \cdot (A_2 \otimes B_2) = (A_1 \cdot A_2) \otimes (B_1 \cdot B_2)$$

and

c.

$$(A \otimes B)^k = A^{\otimes k} \cdot B^{\otimes k}$$
$$(A \otimes B)^{-1} = A^{-1} \otimes B^{-1}$$
$$(A \otimes B)^{\dagger} = A^{\dagger} \otimes B^{\dagger}$$

where the "·"-operation means multiplication of two operators.

The Kronecker product of vector spaces and matrices has many applications in science. One of them is the following proposition:

Proposition. *Let T_h be a representation of H and T_k a representation of K, i.e.*

$$T_h(h_i)T_h(h_m) = T_h(h_p)$$

$$T_k(k_j)T_k(k_n) = T_k(k_q)$$

Let $H \times K$ be the Kronecker product of the groups, i.e.

$$h_i k_j = g_{ij} \quad h_m k_n = g_{mn} \quad h_p k_q = g_{pq}$$

The Kronecker product $T_h \otimes T_k$ of the representations of two commuting groups is a representation of the direct product group.

Proof.

$$(T_h \otimes T_k)(g_{ij})(T_h \otimes T_k)(g_{mn})$$

$$= (T_h(h_i) \otimes T_k(k_j))(T_h(h_m) \otimes T_k(k_n))$$

$$= T_h(h_p) \otimes T_k(k_q)$$

$$= T_h \otimes T_k(g_{pq}) \qquad\qquad \diamond$$

Proposition. *If T_h and T_k are irr. reps. of H and K, then $T_h \otimes T_k \equiv T_g$ is an irreducible representation of $G \equiv H \times K$.*

Proof. T_h and T_k are irr. reps., i.e.

$$\mathcal{M}_{h_i \in H} \chi_h^*(h_i) \chi_h(h_i) = 1$$

$$\mathcal{M}_{k_i \in K} \chi_k^*(k_i) \chi_k(k_i) = 1$$

Taking the product of both sides of these equations leads to

$$1 = \mathcal{M}_{h_i, k_j} \chi_h^*(h_i) \chi_h(h_i) \chi_k^*(k_j) \chi_k(k_j) = \mathcal{M}_{g_{ij}} \chi_g^*(g_{ij}) \chi_g(g_{ij})$$

which proves that T_g is indeed an irr. rep. of the product group.   ◇

Proposition. All irr. reps. of G are the direct product of an irr. rep. of H and one of K.

Plausibility (for finite groups): Let the number of irr. reps. of H be n_h, and their dimensions be l_i^h, the number of irr. reps. of K be n_k, and their dimensions l_j^k:

$$\sum_i^{n_h} (l_i^h)^2 = h$$

$$\sum_j^{n_g} (l_j^k)^2 = k$$

Taking the product of both sides gives

$$h \cdot k = g = \sum_{i,j} (l_i^h)^2 (l_j^k)^2$$

The irr. reps. of G obtained by the direct product will have the dimensions $l_i^h \cdot l_j^k = l_{ij}^g$. The sum over all such irr. reps. gives

$$\sum_{i,j} (l_{ij}^g)^2 = \sum_{i,j} (l_i^h)^2 (l_j^k)^2 = g$$

so that the direct product of all irr. reps. exhausts all the irr. reps. of G. The number of such representations is $n_g = n_h \cdot n_k$.   ◇

Example 1. The character table of $C_{3v} \times C_{1h}$ can be constructed from the characters of the irr. reps. of C_{3v} and C_{1h}.

C_{3v}	e	$2C_3$	$3\sigma_v$
Λ_1	1	1	1
Λ_2	1	1	$\bar{1}$
Λ_3	2	$\bar{1}$	0

C_h	e	σ_h
$''g''$	1	1
$''u''$	1	$\bar{1}$

$C_{3v} \times C_h$	e	$2C_3$	$3\sigma_v$	σ_h	$2C_3\sigma_h$	$3\sigma_v\sigma_h$
Λ_1^+	1	1	1	1	1	1
Λ_2^+	1	1	$\bar{1}$	1	1	$\bar{1}$
Λ_3^+	2	$\bar{1}$	0	2	$\bar{1}$	0
Λ_1^-	1	1	1	$\bar{1}$	$\bar{1}$	$\bar{1}$
Λ_2^-	1	1	$\bar{1}$	$\bar{1}$	$\bar{1}$	1
Λ_3^-	2	$\bar{1}$	0	$\bar{2}$	1	0

Comment. The character table of the product group consists of four blocks. The characters of C_{3v} are repeated in the top blocks with the same positive sign (originating from σ_h being represented by $+1$). The bottom-left block contains the characters of C_{3v} with a positive sign. In the bottom-right block, their sign is reversed. This originates from σ_h being represented by -1.

Example 2. The character table of the 11 type-a improper point groups can be obtained from the character table of the parent proper point groups through the same block procedure used in example 1.

Example 3. The rotation inversion group O_3 is the set containing all proper and improper rotations in 3D. It is isomorphic to the set of 3×3 orthogonal matrices with determinants of $+1$ (proper rotations) and -1 (improper rotations). The elements of O_3 are those of SO_3, and the elements of SO_3

are multiplied by the inversion I. As I commutes with all other elements of the group, we can write

$$O_3 = SO_3 \times \{E, I\}$$

Regarding the character table, the irr. reps. of O_3 are obtained by multiplying the characters of the irr. rep. of SO_3 for the elements containing I with $+1$ or -1. The character table of O_3 is as follows.

O_3	E	$R(\varphi)$	I	$IR(\varphi)$	Basis functions
D_0^+	1	1	1	1	$f(r)Y_0^0$
D_0^-	1	1	-1	-1	
D_1^+	3	$1 + 2\cos\varphi$	3	$1 + 2\cos\varphi$	$\mathbf{L} = (L_x, L_y, L_z)$
D_1^-	3	$1 + 2\cos\varphi$	-3	$-1 - 2\cos\varphi$	$g(r) \cdot \{Y_1^1, Y_1^0, Y_1^{-1}\}$
D_2^+	5	$\frac{\sin(5\frac{\varphi}{2})}{\sin\frac{\varphi}{2}}$	5	$\frac{\sin(5\frac{\varphi}{2})}{\sin\frac{\varphi}{2}}$	$h(r) \cdot \{Y_2^2, Y_2^1, Y_2^0, Y_2^{-1}, Y_2^{-2}\}$
D_2^-	5	$\frac{\sin(5\frac{\varphi}{2})}{\sin\frac{\varphi}{2}}$	-5	$-\frac{\sin(5\frac{\varphi}{2})}{\sin\frac{\varphi}{2}}$	
...	...	...	...	...	...
...	...	...	...	...	...
D_l^+	$2l+1$	$\frac{\sin((2l+1)\frac{\varphi}{2})}{\sin\frac{\varphi}{2}}$	$2l+1$	$\frac{\sin((2l+1)\frac{\varphi}{2})}{\sin\frac{\varphi}{2}}$	...
D_l^-	$2l+1$	$\frac{\sin((2l+1)\frac{\varphi}{2})}{\sin\frac{\varphi}{2}}$	$-2l-1$	$-\frac{\sin((2l+1)\frac{\varphi}{2})}{\sin\frac{\varphi}{2}}$	...
...	...	...	...	...	...
...	...	...	...	...	...

Comment.

We show that $\mathbf{L} = \mathbf{r} \times \frac{d}{dt}\mathbf{r}$ (and the corresponding quantum mechanical operator) transform as D_1^+, which are called "pseudovectors".

- We first prove that $\mathbf{L}$ transforms as a vector under the action of an SO_3 elements R, i.e.

$$\mathbf{r}' = R\mathbf{r} \quad \dot{\mathbf{r}}' = R\dot{\mathbf{r}} \stackrel{?}{\Rightarrow} \mathbf{L}' = R\mathbf{L}$$

We limit ourselves to prove this for small rotational angles φ about any axis $\mathbf{n}$:

$$\mathbf{L}' = \left(\mathbf{r} + \varphi \cdot \mathbf{n} \times \mathbf{r}\right) \times \left(\dot{\mathbf{r}} + \varphi \cdot \mathbf{n} \times \dot{\mathbf{r}}\right)$$

$$\underbrace{=}_{\text{Jacobi identity}} \mathbf{r} \times \dot{\mathbf{r}} + \varphi \cdot \mathbf{n} \times (\mathbf{r} \times \dot{\mathbf{r}})$$

$$= R\mathbf{L}$$

- For the transformation under inversion, we obtain

$$\mathbf{L}' = \mathbf{r}' \times \dot{\mathbf{r}}' = -\mathbf{r} \times (-1) \cdot \dot{\mathbf{r}} = \mathbf{L}$$

i.e. L is invariant with respect to inversion. We conclude that the vector of angular momentum is indeed a pseudovector.

7.3 The Coupling of Degrees of Freedoms

Introduction: The most important applications in technical, biological, and social sciences arise from the coupling of degrees of freedom. Think of traffic jams: Each participant is, in principle, able to move freely along one street when moving alone, but their interaction with other participants produces often mysteriously long queues. A more physical example: A spin in a magnetic material is able to rotate and assume all possible values for its z component. Yet, when immersed into an ensemble of identical spins, their interaction produces the macroscopic alignment that we sense when we deal with magnetic systems. An excellent and very precise review of the simulations of coupling in social problems is given, for example, in D. Stauffer, "Social applications of two-dimensional Ising models", *Am. J. Phys.* **76**, 470 (2008), https://doi.org/10.1119/1.2779882. An application to neural networks is highlighted in the paper by E. Schneidman, M. Berry, R. Segev, *et al.* "Weak pairwise correlations imply strongly correlated network states in a neural population," *Nature* **440**, 1007–1012 (2006), https://doi .org/10.1038/nature04701.

Almost all approaches to describing the coupling between degrees of freedom start from a similar, fundamental "two-particles" interaction Hamiltonian that was found originally in physical problems (W. Heisenberg, "Über den Bau der Atomkerne. I," *Z. Physik* **77**, 1–11 (1932), https://doi. org/10.1007/BF01342433 and P. A. M. Dirac, *The Principles of Quantum Mechanics* (1st edn.), Oxford at the Clarendon Press, 1930). The most general model of this two-particle coupling operator involves the Kronecker product of two angular momentum vectors:

$$\mathbf{I}_p \otimes \mathbf{I}_q$$

with $\mathbf{I}_p, \mathbf{I}_q$ being the infinitesimal generators of the group $SO_3^D \times SO_3^D$.

7.3.1 *The symmetry group of the coupling operator* $\mathbf{I}_p \otimes \mathbf{I}_q$

To determine the transformation properties of this operator under rotations, we compute, as a first step, the transformation properties of the single particle operator $\mathbf{I}$ under the operations of SO_3. We must compute, for example,

$$\left(e^{-(\mathbf{I}\cdot\mathbf{n})\varphi}, \mathbf{I}_x e^{(\mathbf{I}\cdot\mathbf{n})\varphi}\right)$$

To simplify the algebra, we compute this expression in the limit of small rotation angles, i.e. we compute

$$(1 - i(\mathbf{I}\cdot\mathbf{n})\cdot\varphi)I_x(1 + i(\mathbf{I}\cdot\mathbf{n})\cdot\varphi) \cong I_x$$

$$+ i\cdot\varphi\cdot\left(n_x\underbrace{[I_x, I_x]}_{0} + n_y\cdot\underbrace{[I_x, I_y]}_{iI_z} + n_z\cdot\underbrace{[I_x, I_z]}_{-iI_y}\right)$$

$$= I_x + \varphi\cdot(n_z I_y - n_y I_z)$$

After computing I_z and I_y, we find that the quantum mechanical vector operator $\mathbf{I}$ transforms as

$$\mathbf{I} - i\varphi\mathbf{n}\times\mathbf{I}$$

i.e. it is rotated like a vector by an infinitesimal rotation around any axis. In precise terms, the vector operator $\mathbf{I}$ transforms according to the representation D_1 of SO_3.

We now compute the transformation properties of the coupled operator $\mathbf{I}_p \otimes \mathbf{I}_q$ under rotations about the axes $\mathbf{n}_p$ and $\mathbf{n}_q$ by the rotation angles φ_p and φ_q. We proceed again by assuming that the rotation angles are infinitesimally small and find

$$(1 - i(\mathbf{I}_p \cdot \mathbf{n}_p)\cdot\varphi_p) \otimes (1 - i(\mathbf{I}_q \cdot \mathbf{n}_q)\cdot\varphi_q)\mathbf{I}_p$$

$$\otimes\mathbf{I}_q(1 + i(\mathbf{I}_p \cdot \mathbf{n}_p)\cdot\varphi_p) \otimes (1 + i(\mathbf{I}_q \cdot \mathbf{n}_q)\cdot\varphi_q)$$

$$= \underbrace{(1 - i(\mathbf{I}_p \cdot \mathbf{n}_p)\cdot\varphi_p)\mathbf{I}_p(1 + i(\mathbf{I}_p \cdot \mathbf{n}_p)\cdot\varphi_p)}_{\mathbf{I}_p - \varphi_p\mathbf{n}_p\times\mathbf{I}_p}$$

$$\otimes\underbrace{(1 - i(\mathbf{I}_q \cdot \mathbf{n}_q)\cdot\varphi_q)\mathbf{I}_q(1 + i(\mathbf{I}_q \cdot \mathbf{n}_q)\cdot\varphi_q)}_{\mathbf{I}_q - \varphi_q\mathbf{n}_q\times\mathbf{I}_q}$$

i.e. the operator $\mathbf{I}_p \otimes \mathbf{I}_q$ is transformed to

$$\mathbf{I}_p \otimes \mathbf{I}_q - \varphi_q\mathbf{I}_p \otimes \mathbf{n}_q \times \mathbf{I}_q - \varphi_p\mathbf{n}_p \times \mathbf{I}_p \otimes \mathbf{I}_q$$

We conclude that general rotations in the p and q states do not leave the coupling operator invariant, i.e. the group $SO_3^D \times SO_3^D$ is *not* the symmetry group of a Hamiltonian that contains the coupling $\mathbf{I}_p \otimes \mathbf{I}_q$!

A different situation arises when only those elements that consist of simultaneous rotations in p and q spaces are considered, i.e. we can set in the last equation $\varphi_p = \varphi_q = \varphi$ and $\mathbf{n}_p = \mathbf{n}_q = \mathbf{n}$. In this new situation, we obtain that the coupling operator transforms to

$$\mathbf{I}_p \otimes \mathbf{I}_q - \varphi \cdot \underbrace{\mathbf{I}_p \otimes \mathbf{n} \times \mathbf{I}_q}_{-\mathbf{n} \times \mathbf{I}_p \otimes \mathbf{I}_q} - \varphi \cdot \mathbf{n} \times \mathbf{I}_p \otimes \mathbf{I}_q$$

The last two terms in the previous equation cancel out. This produces the invariance of the coupling operator with respect to the subgroup of $SO_3^D \times SO_3^D$ that consists of simultaneous rotations in the p and q spaces. The symmetry group of the Hamiltonian including the $\mathbf{I}_p \otimes \mathbf{I}_q$- operator is therefore isomorphous to SO_3^D.

7.3.2 *Clebsch–Gordan series*

The question that arises is: What happens to the representation $D^p \otimes D^q$ of $SO_3^D \times SO_3^D$ when it is restricted to the subgroup SO_3^D of simultaneous rotations?

In general, given two representations T_i and T_j of a group, the Kronecker product representation is reducible. For the reduction of the direct product of irr. reps. T_i and T_j, we expect an expansion as a sum of irr. reps.

Definition. An expansion of the type

$$T_i \otimes T_j = \oplus_k n_k^{i,j} T_k$$

where $n_k^{i,j}$ are non-negative integers, is known as the **Clebsch–Gordan series**.

We now proceed to find the Clebsch–Gordan series for $D^p \otimes D^q$: $D^p \otimes D^q$ as a representation of SO_3^D, is *reducible* and decomposes into the Clebsch–Gordan series containing the irr. reps. of SO_3^D.

Consider the representations $D^p(a, b, c)$ and $D^q(a, b, c)$ of SO_3^D (take $p \geq q$ for convenience). (a, b, c) indicates the three parameters required to identify a given operation of SO_3^D. The task is to find the irreducible components of $D^p \otimes D^q(a, b, c)$.

Proposition.

$$D^p \otimes D^q(a,b,c) = D^{p+q}(a,b,c) \oplus D^{p+q-1}(a,b,c) \oplus \cdots \oplus D^{p-q}(a,b,c)$$

Plausibility (Cartan): The "trick" is to prove this proposition not for the representations of a general element (a,b,c) but only for rotations about the z-axis by a small angle: $(a,b,c) = (0,0,\epsilon)$. In this situation, we can write

$$D^p \otimes D^q(0,0,\epsilon) = (\mathbb{1} - iI_{z,p} \cdot \epsilon) \otimes (\mathbb{1} - iI_{z,q} \cdot \epsilon)$$

We then compute (to the lowest order in ϵ)

$$(\mathbb{1} - iI_{z,p} \cdot \epsilon) \otimes (\mathbb{1} - iI_{z,q} \cdot \epsilon) \simeq \mathbb{1} \otimes \mathbb{1} - i\left(I_{z,p} \otimes \mathbb{1} + \mathbb{1} \otimes I_{z,q}\right) \cdot \epsilon$$

i.e.

$$I_{z,p\otimes q} = \left(I_{z,p} \otimes \mathbb{1} + \mathbb{1} \otimes I_{z,q}\right)$$

• We have just found, as a byproduct of this proof, the relation

$$i \cdot \frac{dD^p \otimes D^q(\epsilon)}{d\epsilon} = \left(I_{z,p} \otimes \mathbb{1} + \mathbb{1} \otimes I_{z,q}\right)$$

i.e. the infinitesimal generator of the representation $D^p \otimes D^q$ is the vector operator

$$\mathbf{I}_p \otimes \mathbb{1} + \mathbb{1} \otimes \mathbf{I}_q$$

(in short, $\mathbf{I}_p + \mathbf{I}_q$). This vector operator is called the **total angular momentum operator** of the coupled systems "p" and "q".

To prove the proposition, we need therefore to find all Cartan weights of $I_{p,z} \otimes \mathbb{1} + \mathbb{1} \otimes I_{q,z}$. A graphical way of finding the Cartan weights is a Cartan diagram: The Cartan weights $-p,\ldots,p$ are on the horizontal axis, the Cartan weights $-q,\ldots,q$ on the vertical, and the general lattice point (m_p, m_q) is occupied with the weight $m_p + m_q$. The set

$$\{m_p + m_q\}$$

provides *all* possible Cartan weights of $I_{z,p\otimes q}$. For instance, the Cartan diagram for finding all possible eigenvalues of $I_{z,3\otimes 2}$ is given in the following figure.

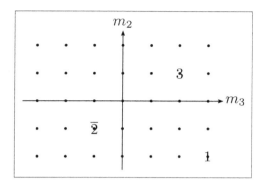

Cartan diagram for the representation $D^3 \otimes D^2$. Some weights are indicated.

Next, we have to organize the Cartan weights into sets that we can assign to an irr. rep. For this purpose, one picks up the highest Cartan weight $m_p + m_q$ in the Cartan diagram and sample all weights along a rectangular path, see the following figure.

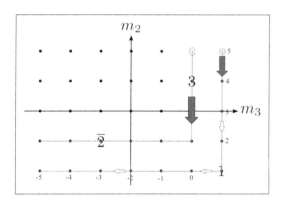

Illustration of the strategy to find the possible values appearing in the Clebsch–Gordan expansion of $D_3 \otimes D_2$.

Along this path, one can sample exactly $2(m_p + m_q) + 1$ weights which are then assigned to the irr. rep. $D^{m_p + m_q}$: This representation appears in the reduction of $D^p \otimes D^q$ at least once. The Cartan weight selected so far are then "eliminated," and the remaining are sampled using the same algorithm until all are "consumed". By inspection of the Cartan diagram, the proposition is proved.   ◇

Proof. The Clebsch–Gordan series reducing the direct product can also be found using the theorems on characters. The coefficients n_j in $D^p \otimes D^q = \oplus n_j D^j$ are given by

$$
\begin{aligned}
n_j &= \frac{2}{\pi} \int_0^\pi d\varphi \sin\left[(2j+1)\frac{\varphi}{2}\right] \chi_p(\varphi)\chi_q(\varphi) \sin\frac{\varphi}{2} \\
&= \frac{4}{\pi} \int_0^\pi d\varphi \sin\left[(2j+1)\frac{\varphi}{2}\right] \sin\left[(2p+1)\frac{\varphi}{2}\right] \cdot \sum_{k=0}^{q} \cos k\varphi \\
&= \frac{8}{\pi} \int_0^{\frac{\pi}{2}} dx \sin[(2j+1)x] \cdot \frac{1}{2}\sum_k [\sin[(2p+1+2k)x] \\
&\quad - \sin[(2p+1-2k)x]] \\
&= \frac{4}{\pi} \int_0^{\frac{\pi}{2}} dx \sin[(2j+1)x] \cdot \sum_k [\sin[(2(p+k)+1)x] \\
&\quad - \sin[(2(p-k)+1))x]]
\end{aligned}
$$

The expression on the right is $= 1$ if $j = p+k$ or $p-k$, with $k = 0, 1, \ldots, q$ and 0 otherwise. Hence, we have obtained an exact proof of the Clebsch–Gordan series for the direct product $D^p \otimes D^q$.   ◇

7.3.3   *An application: The fine structure of atomic levels*

Some more atomic physics: We have learned in Lecture 4 that the configuration Hamiltonian establishes a set of single-particle energy levels (shells) with quantum number n, l. In a further step of building the electronic structure of an atom, the single-particle energy levels are filled with the Z electrons forming an atom. Starting from the lowest energy, one continues filling according to a well-defined recipe, called the "shell" model, based on the Pauli principle: No more than two electrons can be placed in an atomic orbital (n, l, m). The final configuration will then be of the type $(1s)^x, (2s)^x, (2p)^x, \ldots$, until all Z electrons are accommodated. Note that the shell model provides a good explanation of the Mendeleev periodic table of elements and, ultimately, of the stability of atoms.

The eigenspaces of the configuration Hamiltonian, filled with the Z electrons, have an unrealistically large many-particle wave function degeneracy. Let us analyze, as an example, a p^2 configuration. If the spin state is considered, it contains 36 product states with the same configuration energy. Taking the Pauli principle into account, the technique of Slater determinants (see Lecture 9) eliminates some of them, leaving 15 non-vanishing Slater determinants with the same configuration energy. When the Coulomb interaction between the electrons is taken into account more precisely than within the mean field model, the degeneracy of the 15 Slater states is lifted, and only a few will belong to the state with the lowest energy.

In general, we expect that the high degeneracy of the eigenstates of the configuration Hamiltonian, when supplemented with the requirement of antisymmetry with respect to coordinate exchange *and* the Coulomb interaction, is lifted. The split energy states formed are called "multiplets". The word "multiplet" refer to the orbital part of the multi-particle wave function transforming according to a representation D^L of SO_3 and the spin part according to a representation D^S.[2] S and L are, respectively, the total spin and the total orbital angular momentum quantum numbers of the many-electron wave function constituting the eigenstates of the mean field Hamiltonian augmented by the Coulomb interaction (this Hamiltonian is pertinently called the "multiplet Hamiltonian"). There is a technique for finding the quantum numbers S and L, called "Hund's rules" (see also Lecture 9). The symbol for a multiplet is, conventionally, (^{2S+1}L).

The spin–orbit coupling operator: The mean field configuration Hamiltonian containing a spherically symmetric potential provides the key to computing atomic energy levels in the mean field approximation. It does not have spin-dependent terms, so the spin component just increases the degeneracy of the (n, L) energy level. When the relativistic **Dirac equation** is worked out at the lowest order,[3] the atomic Hamiltonian is corrected to contain terms that depend on the spin operator. The most important of them is the so-called spin–orbit coupling. The correction to the mean field Hamiltonian, originating from spin–orbit coupling, is written as[4]

$$H_{LS} = A_{LS} \cdot \mathbf{L} \otimes \mathbf{S}$$

---

[2] In the literature, one uses, when speaking of many-electron multiplets, the uppercase characters S and L.

[3] See, for example, T. Itoh, "Derivation of Nonrelativistic Hamiltonian for Electrons from Quantum Electrodynamics", *Rev. Mod. Phys.* **37**, 159 (1965).

[4] Spectroscopic observations relate to the average over the radial charge density of the coupling constant $A_{LS}(r)$, indicated as A_{LS}. For H-like atoms, we have

$$A_{LS} = \int_0^\infty r^2 dr f_{nL}^2(r) \cdot \frac{Z \cdot \hbar^2 e^2}{4\pi\epsilon_0 2m^2c^2r^3} = \frac{\alpha^2 \cdot Z^4}{2n^3(l+1)(l+\frac{1}{2}) \cdot l} \cdot \overbrace{\text{Ha}}^{\text{Hartree}}$$

Plugging in the universal constant

$$\alpha = \frac{e^2}{\hbar c} \approx \frac{1}{137}$$

(called the fine-structure constant), we obtain $A_{ls} \sim 10^{-4} - 10^{-5}$ Ha.

It is a convention that A_{LS} has the units of energy, while **L** and **S** are assumed to be dimensionless. The spin–orbit coupling operator is of the type we have discussed in the section 7.3.1.

Terms: The symmetry analysis of the coupling operator provides us with all instruments required to estimate the role of spin–orbit coupling in the energy level of a given multiplet ^{2S+1}L. The energy level carrying the multiplet ^{2S+1}L splits, when spin–orbit coupling is accounted for, into fine structure levels (called **terms**). A term carries the quantum numbers L, S *and* one of the total angular momentum quantum numbers

$$L + S \quad L + S - 1 \ \ldots \ L - S$$

originating from the Clebsch–Gordan series. The symbol for a term is, accordingly, $^{2S+1}L_J$.

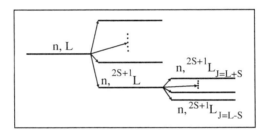

Schematic energy diagram summarizing the atomic levels. Left: Single-electron energy level hosting the (n, L) shell. This level is computed typically by solving, numerically, the eigenvalue problem of the single-electron configuration Hamiltonian. Middle: The Hamiltonian is supplemented by the Coulomb repulsion between the electrons (multiplet Hamiltonian). When the Pauli principle and the Hund's rule are also taken into account, the (n, L) shell splits into multiplet levels (the ordering of which is given by Hund's first and second rules). Right: When the spin–orbit coupling is added, each multiplet level splits into a fine structure of terms, each carrying a value for the total angular momentum quantum number. The values are computed using the Clebsch–Gordan series. The ordering of the levels is empirically given by Hund's third rule.

Computation of the energy of terms with symmetry arguments: We use a "trick" to solve, through perturbation, the eigenvalue problem of the spin–orbit operator. Within the Ritz space of the $(2L + 1) \cdot (2S + 1)$ product functions constituting the state with the multiplet ^{2S+1}L, we write the operator identity

$$\left(\underbrace{\mathbf{L} \otimes \mathbb{1} + \mathbb{1} \otimes \mathbf{S}}_{\mathbf{J}} \right)^2 = (\mathbf{L} \otimes \mathbb{1})^2 + (\mathbb{1} \otimes \mathbf{S})^2 + 2 \cdot \mathbf{L} \otimes \mathbf{S}$$

or, equivalently (and written in a simplified way),

$$\mathbf{L} \cdot \mathbf{S} = \frac{\mathbf{J}^2 - \mathbf{L}^2 - \mathbf{S}^2}{2}$$

From this identity, we can compute the eigenvalues of the spin–orbit coupling operator (left) using the eigenvalues of the operators appearing on the right-hand side of the operator identity. The advantage is that we know the eigenvalues of the operators on the right! Let us analyze them:

- $\mathbf{L}^2$ has the quantum number L corresponding to the eigenvalue $L \cdot (L+1)$.
- $\mathbf{S}^2$ has the quantum number S corresponding to the eigenvalue $S \cdot (S+1)$.
- $\mathbf{J}^2$ is the most interesting operator. We know, from the Clebsch–Gordan series, the it has many possible quantum numbers $L + S, \dots, L - S$. Accordingly, within the space of the mulitplet ^{2S+1}L, this operator will have the possible eigenvalues

$$(L + S) \cdot (L + S + 1), \dots, (L - S) \cdot (L - S + 1)$$

The many-eigenvalued operator $\mathbf{J}^2$ produces a many-eigenvalue situation for the spin–orbit coupling operator as well. The multiplet ^{2S+1}L splits into separate eigenvalues, each corresponding to a term with the total angular momentum quantum number $J \in \{L+S, \dots, L-S\}$. The eigenvalues of the spin–orbit coupling operator computed with this trick confirm an empirical interval rule found by Landé in 1923:

$$E_J - E_{J-1} = A_{LS} \cdot J$$

7.4 Clebsch–Gordan Coefficients

Basis states: The wave function of two degrees of freedom with wave function $\psi(q_a)$ and $\psi(q_b)$ is obtained by using the von Neumann[5] rendering of the tensor product of states ψ_a and ψ_b, respectively:

$$(\psi_a \otimes \psi_b)(q_a, q_b) = \psi_a(q_a) \cdot \psi_b(q_b)$$

The operation of $\cdot$ on the right-hand side is the common multiplication between complex numbers. This rendering is the generalization of the Kronecker product of vectors to functions of variables[6] Note that, in the

---

[5] J. von Neumann and N. A. Wheeler, *Mathematical Foundations of Quantum Mechanics* (New Edition), Princeton University Press, Princeton, 2018, https://doi.org/10.15 15/9781400889921.

[6] This rendering is in line with Born's postulate, stating that the probability amplitudes of composite systems must be **multiplied** to obtain the total probability amplitude.

following, the "$\otimes$" symbol is often omitted and the two states are simply juxtaposed.

Examples.

1. Basis states for an electron with spin in the state of the configuration Hamiltonian specified by the quantum numbers (n, l):

$$f_{nl} \otimes X_l^m \otimes Y^\sigma(r, \vartheta, \varphi, m_s) = f_{nl}(r) \cdot X_l^m(\vartheta, \varphi) \cdot Y^\sigma(m_s)$$

n is the main quantum number, varying from 1 to ∞ and accounting for the radial degree of freedom. $l = 0, 1, \ldots$ and $m = l, \ldots, -l$ are the quantum numbers assigned to the orbital motion. We assign the symbols X_l^m and $X_l^m(\vartheta, \varphi)$ for those spherical harmonics that account for the orbital motion. m_s and σ assume the values $\pm\frac{1}{2}$ (or simply $\pm$) of the z-component of the spin. We use the symbols $Y^\pm$ and $Y^\pm(m_s)$ to identify those "spherical harmonics" that account for the spin degree of freedom. In the symbolism of Dirac, the Kronecker product of states is written as

$$f_{nl} \otimes X_l^m \otimes Y^\sigma \equiv | n, l > | l, m > | \frac{1}{2}, \sigma >$$

2. The basis states carrying the multiplet ^{2S+1}L are multi-particle states. The orbital component transform as the set $\{X_L^{M_L}\}$ and the spin component as the set $\{Y_S^{M_S}\}$. The basis states carrying the multiplet ^{2S+1}L are the $(2L+1) \cdot (2S+1)$ tensor products

$$\{X_L^{M_L} \otimes Y_S^{M_S}\}$$

3. In general, for the basis states describing a subsystem with basis states $Z_p^{m_p}$ transforming according to D^p, coupled to a subsystem with basis states $Z_q^{m_q}$, we write the tensor products as

$$\{Z_p^{m_p} \otimes Z_q^{m_q}\} \quad \text{or simply} \quad \{Z_p^{m_p} Z_q^{m_q}\}$$

Entangled states: The Clebsch–Gordan series

$$D^p \otimes D^q \left.\right|_{SU_2 \subset SU_2 \times SU_2} = D^{p+q} \oplus D^{p+q-1} \oplus \cdots \oplus D^{p-q}$$

provides us with the irreducible components appearing in the decomposition of $D^p \otimes D^q$. Accordingly, in the space of the product functions

$$\{Z_p^{m_p} \otimes Z_q^{m_q}\}$$

there will be symmetry-adapted linear combinations that block-diagonalize the reducible representation $D^p \otimes D^q$. These symmetry-adapted linear combinations will also diagonalize both $\mathbf{I}_p \otimes \mathbf{I}_q$ and $(\mathbf{I}_p + \mathbf{I}_q)^2$.

Definition. The coefficients of the linear combinations of the product functions that block-diagonalize $D^p \otimes D^q \mid_{SU_2 \subset SU_2 \times SU_2}$ are known as the Clebsch–Gordan coefficients[7] (CGC) for SO_3^D.

Comment. We can understand the occurrence of linear combinations (i.e. the mixing of states) graphically by looking at the Cartan diagram, which we have already discussed previously, from a different point of view, see the following figure. Along a diagonal path, one observes a Cartan weight (3) that appears in three different irr. reps. D^5, D^4, and D^3 of the Clebsch–Gordan series of $D^3 \otimes D^2$. There are three product states that carry this Cartan weight, which can be obtained from the coordinates of the Cartan lattice points: $Z_3^3 Z_2^0$, $Z_3^2 Z_2^1$, and $Z_3^1 Z_2^2$, corresponding to the lattice points $(3,0)$, $(2,1)$, and $(1,2)$. These three product functions can all potentially appear as members of the linear combinations that transform according to D^5, D^4, and D^3. This mixing of product states is a situation which is also referred to as an example of the mechanism of **entanglement**.

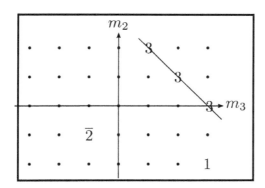

A path in the Cartan diagram of $D^3 \otimes D^2$ showing that the same Cartan weight (3) appears in the irr. reps. D^5, D^4, and D^3.

We now work out a specific example, from which we can learn the basic properties of CGCs.

---

[7]A. Clebsch (1833–1872) and P. Gordan (1837–1912). For details on the Clebsch–Gordan coefficients and their history see e.g. https://en.wikipedia.org/wiki/Clebsch-Gordan_coefficients.

7.4.1 Computation of Clebsch–Gordan coefficients for SO_3^D

To compute symmetry-adapted linear combinations for $D^p \otimes D^q$, using the projector operator technique is impractical: It entails performing complicated group integrals. Rather, one uses an algorithm proposed by Cartan. This algorithm uses the matrix elements of the infinitesimal generators, which are summarized again here:

$$(Z_{j,m}, I_z Z_{j,m}) = m$$

$$(Z_{j,m\pm 1}, I_x Z_{j,m}) = \frac{1}{2}\sqrt{(j \mp m)(j \pm m + 1)}$$

$$(Z_{j,m\pm 1}, I_y Z_{j,m}) = \mp \frac{i}{2}\sqrt{(j \mp m)(j \pm m + 1)}$$

The Cartan algorithm reads:

1. Within the $D^p \otimes D^q$ space, find all wave functions carrying the same Cartan weight k. This space V_k might have, in general, more than one dimension. For example, in $D^L \otimes D^{\frac{1}{2}}$ the wave functions

$$X_L^m Y^+, X_L^{m+1} Y^-$$

   carry the same Cartan weight $k = m + \frac{1}{2}$.

2. Define the auxiliary operators

$$I_\pm = \sqrt{\frac{1}{2}}[I_x \pm i \cdot I_y]$$

   with the property

$$I_\pm Z_j^m = \frac{1}{\sqrt{2}}\sqrt{j(j+1) - m(m \pm 1)} Z_j^{m\pm 1}$$

   and then, solve the equation

$$I_+ Z = 0$$

   within the space V_k. The vector Z that solves this equation transforms, by construction, according to the largest weight of the representation D^k, i.e. when $I_3(D^k)$ is applied to Z, Z gets multiplied by k.

3. The remaining vectors transforming according to D^k are found by applying the lowering operator I_- to the newly found Z.

4. **Computation rule for operating with the infinitesimal operators:**

$$I_\pm(Z_1 \otimes Z_2) = I_\pm Z_1 \otimes Z_2 + Z_1 \otimes I_\pm Z_2$$

Example:

$$D^1 \otimes D^{\frac{1}{2}} = D^{\frac{3}{2}} \oplus D^{\frac{1}{2}}$$

We consider $k = \frac{1}{2}$ and the space spanned by the product functions

$$X^1 Y^- \quad X^0 Y^+$$

We compute

$$
\begin{aligned}
I_+(\alpha X^1 \otimes Y^- + \beta X^0 \otimes Y^+) &= \alpha[(I_+ X^1) \otimes Y^- + X^1 \otimes (I_+ Y^-)] \\
&\quad + \beta[(I_+ X^0) \otimes Y^+ + X^0 \otimes (I_+ Y^+)] \\
&= \left[\frac{1}{\sqrt{2}}\alpha + \beta\right] X^1 \otimes Y^+ \\
&= {!}0
\end{aligned}
$$

This equation is solved by, for example,

$$\alpha = \sqrt{\frac{2}{3}}; \quad \beta = -\sqrt{\frac{1}{3}}$$

This means that the function

$$\sqrt{\frac{2}{3}} X^1_1 \otimes Y^- - \sqrt{\frac{1}{3}} X^0_1 \otimes Y^+ \doteq Z^{\frac{1}{2}}_{\frac{1}{2}}$$

belongs to the invariant subspace of $D^{\frac{1}{2}}$ with $k = \frac{1}{2}$

Comments.

1. A table for the CGCs for some "popular" direct product representations can be found on p. 564 of M. Tanabashi *et al.* "Particle data group", *Phys. Rev. D* **98**, 030001 (2018), https://journals.aps.org/prd/abstract/10.1103/PhysRevD.98.030001. The tables use the convention of Condon and Shortley.
2. We discuss now how to read these tables in the example of the $D^1 \otimes D^{\frac{1}{2}}$ representation:

```
 1 ⊗ 1/2  | 3/2 |
          | +3/2| 3/2  | 1/2 |
 +1 +1/2  |  1  | +1/2 | +1/2|
          | +1 -1/2 | 1/3 | 2/3 | 3/2  | 1/2 |
          |  0 +1/2 | 2/3 |-1/3 | -1/2 | -1/2|
                    | 0 -1/2 | 2/3 | 1/3 | 3/2 |
                    |-1 +1/2 | 1/3 |-2/3 | -3/2|
                              | -1 -1/2 | -1 |
```

Table of CGCs for the reduction of $D^1 \otimes D^{\frac{1}{2}}$. The significance of the dashed rectangles is explained in the text.

- The top entries provide the quantum numbers j, m of the symmetry-adapted wave functions. For instance,

$$1/2$$
$$+1/2$$

is the symmetry-adapted function $Z^{\frac{1}{2}}_{\frac{1}{2}}$.

- The left entries are the product functions entering the symmetry-adapted functions. For instance,

$$+1 \quad -1/2$$

represents the product function $X_1^{+1}Y^-$ and

$$0 \quad +1/2$$

the product function $X_1^0 Y^+$.

- The numbers appearing at the intersection between vertical and horizontal lines are the actual CGCs (for simplicity of writing the square-root sign has been omitted in the tables). Accordingly, we obtain from the table that

$$Z^{\frac{1}{2}}_{\frac{1}{2}} = \sqrt{\frac{2}{3}} X_1^{+1} Y^- - \sqrt{\frac{1}{3}} X_1^0 Y^+$$

as we did compute explicitly.

3. At https://www.wolframalpha.com/, the computation of CGCs $C(j1, j2, m1, m2, j, m)$ in

$$\psi(j, m) = \sum_{m1+m2=m} C(j1, j2, m1, m2, j, m) \cdot \psi(j1, m1) \otimes \psi(j2, m2)$$

is implemented by entering the line "Clebsch–Gordan calculator"

4. **Proposition.** Only basis functions with $m_1 + m_2 = m$ entangle to build the symmetry-adapted function Z_j^m.

Proof. We apply I_z on both sides of

$$\psi(j, m) = \sum_{m1, m2} C(j1, j2, m1, m2, j, m)\psi(j1, m1) \otimes \psi(j2, m2)$$

From

$$I_z \psi(j, m) = m \cdot \psi(j, m)$$

and

$$I_z \sum_{m1, m2} C(j1, j2, m1, m2, j, m)\psi(j1, m1) \otimes \psi(j2, m2)$$

$$= \sum_{m1, m2} C(j1, j2, m1, m2, j, m)(I_z \psi(j1, m1)) \otimes \psi(j2, m2)$$

$$+ \sum_{m1, m2} C(j1, j2, m1, m2, j, m)(\psi(j1, m1)) \otimes (I_z \psi(j2, m2))$$

$$= (m_1 + m_2) \cdot \sum_{m1, m2} C(j1, j2, m1, m2, j, m)(I_z \psi(j1, m1))$$

$$\otimes \psi(j2, m2)$$

we obtain the desired equation $m = m_1 + m_2$.   ◇

7.5 Double Point Groups and the Relativistic Band Structure

Introduction: When expanding single point groups to include the operation $\bar{E}$ and its product with the other operations of the single point group, one must consider that the number of elements is exactly doubled; however, in general, the number of classes does not necessarily double. Let us study, as a specific example, the transition from the point group C_{4v} to the double group C_{4v}^D. In C_{4v}, the element C_{2z} builds a class on its own. We show that $\bar{C}_{2z} = \bar{E}C_{2z}$ is conjugate to C_{2z}, so that, in C_{4v}^D, the two elements are in the same class. This originates from the fact that, when an element involves a rotation by 2π, it is multiplied by $\bar{E}$. For instance, the correct solution of the equation

$$X \circ C_2 = E$$

is not $X = C_2$ because $C_2 \circ C_2 = C_{2\pi} = \bar{E}$. The correct solution is $X = C_2 \circ \bar{E}$, so that $C_2^{-1} = C_2 \bar{E}$! The algebra of the double point groups is therefore a particular one. In the specific case of the operation C_{2z}, we have $(IC_{2x})^{-1} = I\bar{C}_{2x}$, so that

$$(I\bar{C}_{2x})C_{2z}(IC_{2x}) = \bar{C}_{2z}$$

In C_{4v}^D, C_{2z} and $\bar{C}_{2z}$ belong to the same conjugate class. Accordingly, in general, the number of irr. reps. of the double group is not necessarily twice that of the single group. Having said these words of caution, let us work out some practical examples of taking the spin into account.

The group C_1^D: The single group C_1 contains the identity E. The double group contains E and $\bar{E}$. They form two distinct classes, and there are two one-dimensional irr. reps. One, "1", is a single group representation, and the next one, 1^D, is the double group representation.

C_1^D	E	$\bar{E}$
1	1	1
1^D	1	-1
$D^{\frac{1}{2}}$	2	$\bar{2}$

The $D^{\frac{1}{2}}$-representation of E and $\bar{E}$ is

$$E = \begin{pmatrix} 1 & 0 \\ 0 & 1 \end{pmatrix} \quad \bar{E} = \begin{pmatrix} -1 & 0 \\ 0 & -1 \end{pmatrix}$$

It is clearly reducible to $1^D \oplus 1^D$.

Comments.

1. The extra (or double) group representations of a double group are the only physically relevant representations if the spin is considered, as in these representations, $\bar{E}$ is properly represented.
2. In order to find what happens to an energy level at a general point of the BZ (carrying the representation "1") when the spin–orbit coupling is introduced, one must compute the product

$$"1" \otimes D^{\frac{1}{2}} |_{C_1^D} = 1^D \oplus 1^D$$

This shows that, at a general point of the BZ, the introduction of spin–orbit coupling produces generally a splitting of the originally double degenerate single group energy level (time reversal symmetry, however, removes the splitting and restores degeneracy).[8]

The group C_i^D: The single group C_i contains the identity and the inversion I. The double group C_i^D contains the four elements $E, \bar{E}, I, \bar{I}$. Because all these elements commute, there are four classes and four one-dimensional irr. reps. given in the table. The two-dimensional representation $D^{1/2}$, when limited to C_i^D, has the matrices

$$E = \begin{pmatrix} 1 & 0 \\ 0 & 1 \end{pmatrix} \quad \bar{E} = \begin{pmatrix} -1 & 0 \\ 0 & -1 \end{pmatrix}$$

The $D^{\frac{1}{2}}$ matrix of I is considered to coincide with the identity, so that $\bar{I}$ is given by $\bar{E}$.

C_i^D	E	$\bar{E}$	I	$\bar{I}$
g	1	1	1	1
u	1	1	-1	-1
g^D	1	-1	1	-1
u^D	1	-1	-1	1
$D^{\frac{1}{2}}$	2	$\bar{2}$	2	$\bar{2}$

To find what happens to a $g(u)$ band when spin–orbit coupling is switched on, we have to consider that

$$g \otimes D^{\frac{1}{2}} = g^D \oplus g^D$$

$$u \otimes D^{\frac{1}{2}} = u^D \oplus u^D$$

The degeneracy of the two resulting energy levels $2g^D$ and $2u^D$ is restored by time reversal symmetry.

7.5.1 *Symmetry analysis along the Λ-direction*

The single group C_{3v} contains six elements:

$$E, C_{3xyz}, C_{3xyz}^{-1}, IC_{2x\bar{y}}, IC_{2\bar{x}z}, IC_{2y\bar{z}}$$

---

[8]For a discussion of time reversal symmetry in the electronic band structure of crystals see e.g. G. F. Koster, "Space groups and their representations", in *Solid State Physics* Vol. 5, Editors: F. Seitz and D. Turnbull, Academic Press, New York, 1957, pp. 173–256, https://doi.org/10.1016/S0081-1947(08)60103-4.

all transforming the **k**-vector along the Λ-direction into itself ($\mathbf{k}_\Lambda = \pi/a(\lambda, \lambda, \lambda)$ with $0 < \lambda < 1$). The symbols C_{3xyz} and C_{3xyz}^{-1}, for instance, means rotations by an angle $2\pi/3$ around an axis with director cosines on the ratio 1:1:1. There are three classes: the identity element, the two threefold rotations, and the three reflections.

C_{3v}	E	$C_{3xyz}\,C_{3xyz}^{-1}$	$IC_{2x\bar{y}}\,IC_{2\bar{x}z}\,IC_{2y\bar{z}}$
		$2C_3$	3σ
Λ_1	1	1	1
Λ_2	1	1	-1
Λ_3	2	-1	0

These irr. reps. are used to label the non-relativistic band structure along the Λ-direction of fcc cubic materials, such as GaAs. Non-relativistic means that the spin–orbit coupling is neglected, and the spin merely provides a double degeneracy predicted by the Pauli principle.

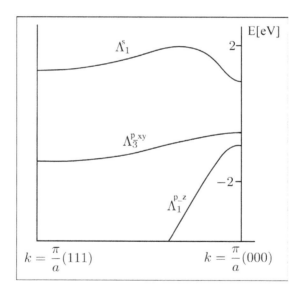

Sketch of the non-relativistic band structure of GaAs along the Λ-direction.

In the figure, the lower index denotes the symmetry of the bands according to the group representations of the group of Λ. These labels are typically encountered in the band structure calculation of solids. The upper letters are less "precise", as they indicate the atomic levels the bands originate from: s and p levels for the specific example of GaAs. Note that the upper Λ_1^s band is not occupied and separated by a gap from the occupied bands $\Lambda_3^{p_{xy}}$ and lower $\Lambda_1^{p_z}$. Not shown in the figure is also the lowest Λ_1^s band. The lower Λ_1^s band hosts one s-like electron, the $\Lambda_1^{p_z}$ band hosts a p_z electron (z along the Λ-direction), and the doubly degenerate $\Lambda_3^{p_{xy}}$ hosts two p-like electrons (xy perpendicular to the Λ-direction). This provides a total of four electrons per unit cell. Considering the double occupancy predicted by the Pauli principle, the occupied bands along Λ provide space for a total of eight electrons, in agreement with the outermost orbitals of Ga ($4s^2 4p$) and As ($4s^2 4p^3$).

The double group C_{3v}^D has twice the number of elements and twice the number of classes as C_{3v}. The 12 elements are divided into six classes, and the equation $\sum_\alpha (l_\alpha)^2 = 12$ has the solution $1^2 + 1^2 + 2^2 + 1^2 + 1^2 + 2^2 = 12$. Thus, there are a total of four one-dimensional representations and two two-dimensional representations. The three irr. reps. of C_{3v} can be extended as irr. reps. of C_{3v}^D by representing $\bar{E}$ with E and the $\bar{C}$ elements as the non-$\bar{C}$ elements. These are the so-called single-group representations of C_{3v}^D. There are three extra (double) additional irr. reps. to be found. We explore, for this purpose, the representation $D^{\frac{1}{2}}$, restricted to the elements of C_{3v}^D. In general, we have

$$D^{\frac{1}{2}}(l, m, n, \varphi) = \begin{pmatrix} \cos\frac{\varphi}{2} - i \cdot n \sin\frac{\varphi}{2} & (-i \cdot l - m) \sin\frac{\varphi}{2} \\ (-i \cdot l + m) \sin\frac{\varphi}{2} & \cos\frac{\varphi}{2} + i \cdot n \sin\frac{\varphi}{2} \end{pmatrix}$$

Inserting, for instance,

$$l = m = n = \sqrt{\frac{1}{3}} \quad \varphi = \frac{2\pi}{3}$$

we obtain

$$C_{3xyz} = \begin{pmatrix} \frac{1-i}{2} & -\frac{1+i}{2} \\ \frac{1-i}{2} & +\frac{1+i}{2} \end{pmatrix}$$

Similarly, one can obtain the matrices $IC_{2x\bar{y}}, IC_{2\bar{x}z}, IC_{2y\bar{z}}$ by considering that the inversion is followed by the rotation of $2\pi/2$ around an axis with director cosines in the ratio of $1:\bar{1}:0$, i.e.

$$\varphi = \pi \quad l = /\sqrt{2} \quad m = -1/\sqrt{2} \quad n = 0$$

Thus, for instance,

$$IC_{2x\bar{y}} = I \cdot \begin{pmatrix} 0 & \frac{1-i}{\sqrt{2}} \\ -\frac{1+i}{\sqrt{2}} & 0 \end{pmatrix}$$

The remaining matrices can be constructed in a similar way. In summary,

$$E = \begin{pmatrix} 1 & 0 \\ 0 & 1 \end{pmatrix}; \quad \bar{E} = \begin{pmatrix} \bar{1} & 0 \\ 0 & \bar{1} \end{pmatrix}$$

$$C_{3xyz} = 1/2 \begin{pmatrix} 1-i & -1-i \\ 1-i & 1+i \end{pmatrix}; \quad \bar{C}_{3xyz} = 1/2 \begin{pmatrix} -1+i & 1+i \\ -1+i & -1-i \end{pmatrix}$$

$$C_{3xyz}^{-1} = 1/2 \begin{pmatrix} 1+i & 1+i \\ -1+i & 1-i \end{pmatrix}; \quad \bar{C}_{3xyz}^{-1} = 1/2 \begin{pmatrix} -1-i & -1-i \\ 1-i & -1+i \end{pmatrix}$$

$$IC_{2x\bar{y}} = \sqrt{1/2}I \begin{pmatrix} 0 & 1-i \\ -1-i & 0 \end{pmatrix}; \quad I\bar{C}_{2x\bar{y}} = \sqrt{1/2}I \begin{pmatrix} 0 & -1+i \\ 1+i & 0 \end{pmatrix}$$

$$IC_{2\bar{x}z} = \sqrt{1/2}I \begin{pmatrix} -i & i \\ i & i \end{pmatrix}; \quad I\bar{C}_{2\bar{x}z} = \sqrt{1/2}I \begin{pmatrix} i & -i \\ -i & -i \end{pmatrix}$$

$$IC_{2y\bar{z}} = \sqrt{1/2}I \begin{pmatrix} i & -1 \\ 1 & -i \end{pmatrix}; \quad I\bar{C}_{2y\bar{z}} = \sqrt{1/2}I \begin{pmatrix} -i & 1 \\ -1 & i \end{pmatrix}$$

We note that the two-dimensional $D^{\frac{1}{2}}$ representation is irreducible: In fact, with the characters

$$\chi(E) = 2 \quad \chi(\bar{E}) = -2 \quad \chi(C_3) = 1 \quad \chi\bar{E}C_3 = -1 \quad \chi(\sigma_v) = 0$$

we have $\sum_A \chi(A)^2 = 12$, which is a necessary and sufficient condition for an irr. rep. The remaining two extra one-dimensional representations are found by applying the unitarity conditions. Λ_6 is the name of $D^{\frac{1}{2}}$ when restricted to this point group.

Character table of the double point group C_{3v}^D.

Irr. Rep.	E	$\bar{E}$	$2C_3$	$2\bar{C}_3$	$3\sigma_v$	$3\bar{\sigma}_v$
Λ_1	1	1	1	1	1	1
Λ_2	1	1	1	1	-1	-1
Λ_3	2	2	-1	-1	0	0
Λ_4	1	-1	-1	1	i	$-i$
Λ_5	1	-1	-1	1	$-i$	i
Λ_6	2	-2	1	-1	0	0

In order to find the "destiny" of the non-relativistic bands with symmetry Λ_1, Λ_2, and Λ_3, we need to find $\Lambda_1 \otimes D^{\frac{1}{2}}$, $\Lambda_2 \otimes D^{\frac{1}{2}}$, and $\Lambda_3 \otimes D^{\frac{1}{2}}$, as follows.

Character of representations of the double point group C_{3v}^D.

Irr. Rep.	E	$\bar{E}$	$2C_3^2$	$2\bar{C}_3^2$	$3\sigma_v$	$3\bar{\sigma}_v$
Λ_1	1	1	1	1	1	1
Λ_2	1	1	1	1	-1	-1
Λ_3	2	2	-1	-1	0	0
Λ_4	1	-1	-1	1	i	$-i$
Λ_5	1	-1	-1	1	$-i$	i
Λ_6	2	-2	1	-1	0	0
$D^{\frac{1}{2}}\big|_{C_{3v}^D}$	2	-2	1	-1	0	0
$\Lambda_1 \otimes D^{\frac{1}{2}}$	2	-2	1	-1	0	0
$\Lambda_2 \otimes D^{\frac{1}{2}}$	2	-2	1	-1	0	0
$\Lambda_3 \otimes D^{\frac{1}{2}}$	4	-4	-1	1	0	0

By computing $\Lambda_i \otimes D^{\frac{1}{2}}$ in terms of the double group representations, we can find out what happens to the single group energy bands when spin–orbit coupling is switched on:

$$\Lambda_1 \otimes D^{\frac{1}{2}} = \Lambda_6$$

$$\Lambda_2 \otimes D^{\frac{1}{2}} = \Lambda_6$$

$$\Lambda_3 \otimes D^{\frac{1}{2}} = \Lambda_6 + \Lambda_4 + \Lambda_5$$

Thus, the Λ_3 band splits into three sub-bands, each carrying one irr. rep. of the double group. The band Λ_1 remains degenerate and carries the irr. rep. Λ_6 of the double group. In the next figure, the superscripts indicate the single group irr. rep. from which the relativistic bands originate.

Remark: Λ_4 and Λ_5 carry complex conjugate wave functions as symmetry-adapted wave functions. Because of time reversal symmetry, they are degenerate even if the SO_3^D symmetry does not impose their degeneracy.

7.5.2   *The group C_{4v}^D*

- The character table for the double point group C_{4v}^D according to table 33 in G. F. Koster, J. O. Dimmock, R. G. Wheeler and H. Statz, *Properties of the Thirty-Two Point Groups*, MIT Press, Cambridge, Mass., 1963, https://babel.hathitrust.org/cgi/pt?id=mdp.39015022422318 (KDWS) is as follows.

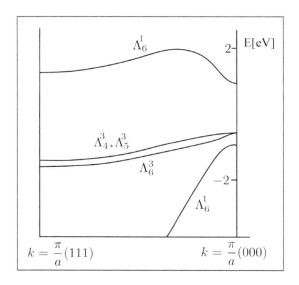

Sketch of the relativistic band structure along the Λ-direction of GaAs. The convention is that the subscripts indicate the label of the double group representation. The superscripts indicate the single group representation underlying the non-relativistic, parent energy bands.

Character table of C_{4v}^D (Table 33, p. 45, KWDS).

C_{4v}	E	$\bar{E}$	$C_{2z}, \bar{C}_{2z}$	$2C_4 = C_{4z}C_{4z}^{-1}$	$2\bar{C}_4$	$2\sigma_v = IC_{2x}, IC_{2z}, 2\bar{\sigma}_v$	$2\sigma_d = IC_{2xy}, IC_{2x\bar{y}}$
Δ_1	1	1	1	1	1	1	1
Δ_1'	1	1	1	1	1	-1	-1
Δ_2	1	1	1	-1	-1	1	-1
Δ_2'	1	1	1	-1	-1	-1	1
Δ_5	2	2	-2	0	0	0	0
Δ_6	2	-2	0	$\sqrt{2}$	$-\sqrt{2}$	0	0
Δ_7	2	-2	0	$-\sqrt{2}$	$\sqrt{2}$	0	0

- Some symmetry-adapted polynomial basis states are listed here:

$$\Delta_1 \rightarrow 1, z, 2z^2 - x^2 - y^2$$
$$\Delta_1' \rightarrow xy(x^2 - y^2)$$
$$\Delta_2 \rightarrow x^2 - y^2$$
$$\Delta_2' \rightarrow xy$$
$$\Delta_5 \rightarrow (x, y), (xz, yz)$$

- $D^{\frac{1}{2}}$ restricted to C_{4v}^D (some elements):

$$D^{\frac{1}{2}}_{0,0,1}(\pi) = \begin{pmatrix} -i & 0 \\ 0 & i \end{pmatrix} \quad D^{\frac{1}{2}}_{0,0,1}(\pi/2) = \begin{pmatrix} \frac{1-i}{\sqrt{2}} & 0 \\ 0 & \frac{1+i}{\sqrt{2}} \end{pmatrix}$$

$$D^{\frac{1}{2}}_{1,0,0}(\pi) = \begin{pmatrix} 0 & -i \\ i & 0 \end{pmatrix} \quad D^{\frac{1}{2}}_{1/\sqrt{2},1/\sqrt{2},0}(\pi) = \begin{pmatrix} 0 & -\frac{i+1}{\sqrt{2}} \\ \frac{-i+1}{\sqrt{2}} & 0 \end{pmatrix}$$

In summary, the characters are written as

	E	$\bar{E}$	$C_2, \bar{C}_2$	$2C_4$	$2\bar{C}_4$	$2\sigma_v, 2\bar{\sigma}_v$	$2\sigma_d, 2\bar{\sigma}_d$
$D^{\frac{1}{2}}\big\vert_{C_{4v}^D}$	2	-2	0	$\sqrt{2}$	$-\sqrt{2}$	0	0

By comparison, we find that

$$D^{\frac{1}{2}}\big\vert_{C_{4v}^D} \equiv \Delta_6$$

7.5.3 The Clebsch–Gordan coefficients for double point groups

Given an invariant subspace

$$\{X_1, \ldots, X_k\}$$

transforming according to a single-group representation $^S\Gamma^k$, we would like to compute the symmetry-adapted linear combinations of

$$\{X_1 Y^+, X_1 Y^-, \ldots, X_k Y^+, X_k Y^-\}$$

transforming according to $_D\Gamma_J$, where $_D\Gamma_J$ is the double-group representation occurring in

$$^S\Gamma^k \otimes D^{\frac{1}{2}} = \oplus_J n_J \cdot {}_D\Gamma_J$$

Remark. When double-group representations are involved, the superscript indicates the parent single-group representation $^S\Gamma^k$.

Example: The Λ-direction:

We construct, as example, the symmetry-adapted wave functions for

$$\Lambda^3 \otimes D^{\frac{1}{2}} = \Lambda_6^3 + \Lambda_4^3 + \Lambda_5^3$$

The space upon which $\Lambda^3 \otimes D^{\frac{1}{2}}$ acts is constructed by the the four product functions

$$\{X_3^1 Y^+, X_3^1 Y^-, X_3^{-1} Y^+, X_3^{-1} Y^-\}$$

(when functions are used, the upper index labels the partner function in the irr. rep. indicated by the lower index).

To find the symmetry- adapted linear combinations transforming according to Λ_6^3, Λ_4^3, and Λ_5^3, we use the projector technique, i.e. *we must indeed rotate the wave functions (no infinitesimal generators are available).* We set the z-axis to be along the Λ direction, so that we can rotate $X_3^{\pm 1}$ by an angle φ by simply multiplying them with $exp(\mp i\varphi)$. Similarly, we can rotate $Y^{\pm}$ by multiplying them with $exp(\mp i\frac{\varphi}{2})$. Reflections are irrelevant to the formation of the projector operator, as they have vanishing trace. Take, for example, $X_3^+ Y^+$:

$$E(X_3^+ Y^+) = X_3^+ Y^+$$

$$\bar{E}(X_3^+ Y^+) = -X_3^+ Y^+$$

$$C_3(X_3^+ Y^+) = e^{-i\varphi} X_3^+ e^{-\frac{\varphi}{2}} Y^+$$

$$C_3^{-1}(X_3^+ Y^+) = e^{+i\varphi} X_3^+ e^{\frac{\varphi}{2}} Y^+$$

$$\bar{C}_3(X_3^+ Y^+) = -e^{-i\varphi} X_3^+ e^{-\frac{\varphi}{2}} Y^+$$

$$\bar{C}_3^{-1}(X_3^+ Y^+) = -e^{i\varphi} X_3^+ e^{\frac{\varphi}{2}} Y^+$$

Referring to the character table of C_{3v}^D, we build the projectors:

$$P_{\Lambda 6} X_3^+ Y^+ \propto \underbrace{2 \cdot X_3^+ Y^+ + 2 \cdot X_3^+ Y^+}_{4 X_3^+ Y^+}$$

$$+ \underbrace{2 \cdot e^{-i\frac{3\varphi}{2}} X_3^+ Y^+ + 2 \cdot e^{i\frac{3\varphi}{2}} X_3^+ Y^+}_{4 \cdot \cos(\frac{3}{2} \frac{2\pi}{3}) X_3^+ Y^+}$$

$$= 4 X_3^+ Y^+ + 4 \cos \pi X_3^+ Y^+$$

$$= 0$$

Accordingly, the states $X_3^{\pm 1} Y^{\pm}$ must reside within the subspace belonging to $\Lambda_{4,5}^3$. We also find that

$$P_{\Lambda 6} X_3^{\pm 1} Y^{\mp} \propto 4 \cdot X_3^{\pm 1} Y^{\mp} + 4 \cos \frac{\varphi}{2} X_3^{\pm 1} Y^{\mp} \propto X_3^{\pm 1} Y^{\mp}$$

i.e.

$$\{ X_3^1 Y^-, X_3^{-1} Y^+ \}$$

are the basis states of Λ_6^3. Following these results, we are able to assign states to the various bands along Λ (time reversal restores the degeneracy of Λ_4 and Λ_5):

$X_0^0 Y^+$	$X_0^0 Y^-$	Λ_6^1	upper s-type band, unoccupied
$X_3^1 Y^+$	$X_3^{-1} Y^-$	Λ_{4+5}^3	upper occupied band
$X_3^1 Y^-$	$X_3^{-1} Y^+$	Λ_6^3	spin–orbit split occupied band

The complete list of CGCs for the 32 point groups is given in the tables of KDWS. We will learn how to use these tables in the project "Relativistic Band Structures and Spin-Polarized Transitions". CGCs are of general relevance, as they enter the computation of matrix elements of operators via the WEK theorem. As an example, consider (Table 34 on p. 45 in KDWS)

$$\Delta^1 \otimes \Delta_6 = \Delta_6^1 \quad \text{and} \quad \Delta^5 \otimes \Delta_6 = \Delta_6^5 \oplus \Delta_7^5$$

which describes the destiny of the single-group representations Δ^1 and Δ^5 when spin is taken into account. The CGCs describing the symmetry-adapted states in the product space of $\Delta^1 \otimes \Delta_6$ and $\Delta^5 \otimes \Delta_6$ can be found in the suitable tables of KWDS. In the case of Δ_6^1, the symmetry-adapted functions are directly given by the direct product of the basis functions and are not listed as a separate table (KWDS systematically use superscripts to the states to label the irr. reps. and subscripts to label the various partners):

$$\psi_{-1/2}^6 = u_1 v_{-1/2}^6,$$

$$\psi_{1/2}^6 = u_1 v_{1/2}^6.$$

where u_1 is some basis state transforming according to Δ_1. The Clebsch–Gordan (or coupling) coefficients in the case of $\Delta^5 \otimes \Delta^6 = \Delta_6 \oplus \Delta_7$ are listed in the table second from bottom on p. 46:

$$\psi^6_{-1/2} = +\frac{i}{\sqrt{2}}u^5_x v^6_{1/2} - \frac{i}{\sqrt{2}}u^5_y v^6_{1/2},$$

$$\psi^6_{1/2} = +\frac{i}{\sqrt{2}}u^5_x v^6_{-1/2} + \frac{i}{\sqrt{2}}u^5_y v^6_{-1/2},$$

$$\psi^7_{-1/2} = +\frac{i}{\sqrt{2}}u^5_x v^6_{1/2} + \frac{i}{\sqrt{2}}u^5_y v^6_{1/2},$$

$$\psi^7_{1/2} = +\frac{i}{\sqrt{2}}u^5_x v^6_{-1/2} - \frac{i}{\sqrt{2}}u^5_y v^6_{-1/2}.$$

Here, the functions $u_{x,y}$ and $v_{-1/2,1/2}$ transform according to Δ_5 and Δ_6, respectively.

7.6 The Wigner–Eckart–Koster Theorem

(G. F. Koster, *Phys. Rev.* **109**, 227 (1958), https://journals.aps.org/pr/abstract/10.1103/PhysRev.109.227). We have introduced a simplified version of this theorem when we treated the matrix elements of the Hamilton operator. Let us recall. Let H commute with a representation $U(g)$ of a group within some vector space V. We can divide the space V into an orthogonal sum of symmetry-adapted spaces $\{f_{n,l} \otimes Y^m_l\}$. l labels the irreducible representation of G appearing in $\Gamma(g) = \oplus n_l \Gamma_l$. Y^m_l labels the symmetry-adapted wave function transforming according to the mth column of the lth irr. rep.[9] $f_{n,l}$ is a part of the function taking care of the degrees of freedom not concerned by the symmetry group. In the case of $SO(3)$, for example, functions which have a certain symmetry under rotation might have a different radial dependence, i.e. they are described by $f_{n,l}(r)Y^m_l(\vartheta, \varphi)$, with n labeling the main quantum number.

The most simple version of the Wigner–Eckart–Koster (WEK) theorem, which holds for *any* operator H invariant with respect to G (including $\mathbb{1}$), states that the matrix elements of H are exactly vanishing if the initial (right, i) and final (left, f) states do not transform according to the *same*

---

[9]The basis functions Y^m_l have a further index m^p that takes into account the fact that, for a given U_l, there might be more orthogonal functions transforming according to the same column m of the representation U_l. This situation does not occur in atomic physics, but one encounters it, for example, in solid-state physics or particle physics. For simplicity of writing, we include this index tacitly in the index n and render it explicit only when needed.

column of the *same* irr. rep.:

$$\left(f_{n,f}Y_f^{m_f}, f_{m,i}Y_i^{m_i}\right) = a(n,m,i)\cdot\delta_{if}\cdot\delta_{m_im_f} \quad \text{WEK 1}$$

and

$$\left(f_{n,f}Y_f^{m_f}, H[f_{m,i}Y_i^{m_i}]\right) = b(n,m,i)\cdot\delta_{if}\cdot\delta_{m_im_f} \quad \text{WEK 2}$$

One uses WEK 2 to block-diagonalize the Hamilton operator of a system with the symmetry group $U(g)$.

The generalization of the WEK theorems involves the same symmetry-adapted initial and final states, but the operators involved are more general.

Definition. A set of operators $Q_o^{m_o}$ transform according to the irr. rep. Γ_o of a group G if

$$\Gamma_o^{-1}(g)Q_o^{m_o}\Gamma_o(g) = \sum_{m_p}\Gamma_o(g)_{m_pm_o}Q_o^{m_p}$$

Examples:

– With R being any proper or improper rotation and Γ its representation in a function space, we have

$$\Gamma(R^{-1})\mathbf{r}\Gamma(R) = R\mathbf{r}$$

This means that $\mathbf{r}$ transforms as the irr. rep. D^1 of SO_3 or $D^{1,-}$ of O_3.

– With R being any proper or improper rotation, we have

$$\Gamma(R^{-1})\mathbf{p}\Gamma(R) = R\mathbf{p}$$

This means that $\mathbf{p}$ transforms as the irr. rep. D^1 of SO_3 or $D^{1,-}$ of O_3.

– With R being any proper or improper rotation, we have

$$\Gamma(R^{-1})\mathbf{L}\Gamma(R) = \det(R)\cdot R\mathbf{L}$$

This means that $\mathbf{L}$ transforms as the irr. rep. D^1 of SO_3 or $D^{1,+}$ of O_3.

– With $\mathbf{t}$ being any translation, we have

$$\Gamma(-\mathbf{t})\mathbf{p}\Gamma(\mathbf{t}) = \mathbf{p}$$

This means that $\mathbf{p}$ is left invariant by any translation. In particular, $\mathbf{p}$ does not transform according to what we have called the physical representation of the group of $\mathbf{k}$ in Lecture 6 because it is invariant with respect to translations and therefore does not acquire the multiplicative factor $e^{-i\mathbf{k}\cdot\mathbf{t}_n}$ upon translations. From this example, we understand that our wording "physical representation" did refer to the states of a system. $\mathbf{p}$ is certainly a physically relevant operator, although it does not transform according to "physical representations"!

Theorem (selection rule theorem). *Given that* $Q_o^{m_o}(f_{m,i}Y_i^{m_i})$ *is in the subspace invariant with respect to* $\Gamma_o \otimes \Gamma_i$

$$\left(f_{n,f}Y_f^{m_f}, Q_o^{m_o}(f_{m,i}Y_i^{m_i}) \right) = 0$$

if

$$\Gamma_f \notin \Gamma_o \otimes \Gamma_i$$

Proof. Use WEK 1.   ◇

Comments.

1. In words, in order for the matrix element not to vanish by symmetry, the final state irr. rep. must be contained in the Clebsch–Gordan series of $\Gamma^{op} \otimes \Gamma^i$.
2. If $\Gamma_f \in \Gamma_o \otimes \Gamma_i$, the matrix element is expected to be different from zero, but can vanish by accident.

This selection rule theorem is, however, the weakest form of the WEK. One can make a more precise statement about the values of the matrix elements by looking at a table for CGCs from a slightly different point of view.

General scheme for a table of CGCs. We use the scheme adapted for the tables on tensor products of the representations D^j of SO_3^D, see, for example, the "Table of Clebsch–Gordan coefficients for the reduction of $D^1 \otimes D^{\frac{1}{2}}$". The roman superscripts refer to the fact that the final representation Γ^f can appear more than once in the decomposition of $\Gamma^o \otimes \Gamma^i$.

$\Gamma^o \otimes \Gamma^i$	... ...	$Y_f^{1,I}$	$Y_f^{2,I}$	...	$Y_f^{1,II}$	...
...	... ...	...	...	...	...	...
...	... ...	...	...	...	...	...
$Q_0^{m_o}Y_i^{m_i}$	... ...	$c(I,o,i,f,m_o,m_i,1)$	$c(I,o,i,f,m_o,m_i,2)$	...	$c(II,o,i,f,m_o,m_i,1)$	...
...	... ...	...	...	...	...	...
...	... ...	...	...	...	...	...
...	... ...	...	...	...	...	...

Typically, one reads this table along vertical lines. Along these lines, one finds the CGCs required to express the symmetry-adapted state on the top (say $Y_f^{\cdot,m_f}$) as a function of the product functions $Q_o^{m_o}Y_i^{m_i}$:

$$Y_f^{\cdot,m_f} = \sum_{m_o,m_i} c(.,o,i,f,m_o,m_i,m_f)Q_o^{m_o}Y_i^{m_i}$$

One could also attempt to express the product functions on the left as a linear combination of the symmetry-adapted functions on the top (this is not the usual way of reading the matrix since, typically, one is interested in the symmetry-adapted vectors): Consider the bulk of the table as a matrix and compute its inverse. However, the bulk of the table is a unitary matrix, and its inverse is the complex conjugate transpose. In other words, one can express the product functions on the left just by reading the matrix along horizontal lines *and* taking the complex conjugate of the matrix elements:

$$Q_o^{m_o} Y_i^{m_i} = \sum_{\alpha=I}^{n_f} \sum_f \overline{c(\alpha, o, i, f, m_o, m_i, m_f = (m_o + m_i))} Y_f^{\alpha, m_f}$$

This relation allows us to write the full version of the WEK theorem:

Theorem (WEK).

$$\left(f_{n,f} Y_f^{m_f}, Q_o^{m_o} (f_{m,i} Y_i^{m_i}) \right)$$

$$= \sum_{\alpha=I}^{n_f} a(n, m, f, \alpha) \cdot \overline{c(\alpha, o, i, f, m_o, m_i, m_f = (m_o + m_i))}$$

Proof. Use WEK 1. $\diamond$

The parameters $a(n, m, f, \alpha)$ are not related to symmetry and *must* be calculated through analytic methods or numerically.

Example 1: Dipole selection rules for spherically symmetric potential

The matrix element governing the optical transition of an electron between an initial state ψ_i and a final state ψ_f reads

$$M_{i \to f} = -q \cdot E_0 \cdot (\psi_f, (e_0 \cdot \mathbf{r} \otimes \mathbb{1})_{op} \psi_i)$$

where E_0 is the strength of the light electric field, e_0 is the polarization vector of the radiation, and q is the charge of the electron. The radiation does not interact with the spin (in dipole approximation), and this is highlighted by $\mathbb{1}$ in the spin sector of the operator.

$e_0 \cdot \mathbf{r}$ transforms according to the representation D^{1-} of the group O_3. If we consider the initial state to transform according to a representation $D^{i,\sigma}$, then the transition is only allowed if $D^{f,\sigma'}$ has the symmetry of one of the states

$$D^{1-} \otimes D^{i,\sigma} = D^{i+1,-\sigma} \oplus D^{i,-\sigma} \oplus D^{i-1,-\sigma}$$

Comments:

1. A final state $i, -\sigma$ by the given initial state i, σ does not exist so that the $i \to i$ transition is forbidden.
2. By an optical transition, the spin angular momentum remains unchanged and the orbital angular momentum can only change by ± 1.

   Let us evaluate the matrix elements in detail for some special but important situations.

   - We choose monochromatic, linearly polarized light propagating along some direction **k**. Take, for example, $\mathbf{e}_0 = (0, 0, z)$ (linear polarization along z). The operator $\mathbf{e}_0 \cdot \mathbf{r}$ transforms according to the $m = 0$ column of the representation D^1 of SO_3, i.e. as X_1^0. Take, for example, an initial state that transforms according to X_1^1 and look in the table for the CGCs $D^1 \otimes D^1$ for the function $X_1^0 \cdot X_1^1$.

Table of CGCs for the reduction of $D^1 \otimes D^1$. The top row gives the quantum numbers of the symmetry-adapted linear combinations. The left column contains the product states. Adapted from p. 564 of M. Tanabashi *et al.* (Particle Data Group), Phys. Rev. D **98**, 030001 (2018), https://journals.aps.org/prd/abstract/10.1103/PhysRevD.98.030001.

$D^1 \otimes D^1$	(2,2)	(2,1)	(1,1)	(2,0)	(1,0)	(0,0)	(2,−1)	(1,−1)	(2,−2)
$X^1 X^1$	1	0	0	0	0	0	0	0	0
$X^1 X^0$	0	$\sqrt{\frac{1}{2}}$	$\sqrt{\frac{1}{2}}$	0	0	0	0	0	0
$X^0 X^1$	0	$\sqrt{\frac{1}{2}}$	$-\sqrt{\frac{1}{2}}$	0	0	0	0	0	0
$X^1 X^{\bar{1}}$	0	0	0	$\sqrt{\frac{1}{6}}$	$\sqrt{\frac{1}{2}}$	$\sqrt{\frac{1}{3}}$	0	0	0
$X^0 X^0$	0	0	0	$\sqrt{\frac{2}{3}}$	0	$-\sqrt{\frac{1}{3}}$	0	0	0
$X^{\bar{1}} X^1$	0	0	0	$\sqrt{\frac{1}{6}}$	$-\sqrt{\frac{1}{2}}$	$\sqrt{\frac{1}{3}}$	0	0	0
$X^0 X^{\bar{1}}$	0	0	0	0	0	0	$\sqrt{\frac{1}{2}}$	$\sqrt{\frac{1}{2}}$	0
$X^{\bar{1}} X^0$	0	0	0	0	0	0	$\sqrt{\frac{1}{2}}$	$-\sqrt{\frac{1}{2}}$	0
$X^{\bar{1}} X^{\bar{1}}$	0	0	0	0	0	0	0	0	1

From the forth line of the table, we find that

$$X_1^0 \cdot X_1^1 = \sqrt{\frac{1}{2}} X_2^1 - \sqrt{\frac{1}{2}} X_1^1$$

This means that an electron in the state X_1^1 will make, by excitation with light that is linearly polarized along z, a transition to a state with symmetry X_2^1 with an amplitude of $1/\sqrt{2}$ and to a state with symmetry X_1^1 with an amplitude of $-1/\sqrt{2}$. In general, CGCs predict that $0 + m_i = m_f$, i.e. the selection rule for linearly polarized light is $m_f = m_i$.

- We now assume circularly polarized light propagating along z. Then, $\mathbf{e}_0 = \frac{1}{\sqrt{2}}(\mathbf{e}_x \pm i\mathbf{e}_y)$ (depending on the handedness of the electric field vector), and the matrix element is proportional to

$$(\psi_f, [x \pm i \cdot y]\psi_i)$$

The operator transforms according to the representation D^1, so that the selection rules for the orbital momentum are $\Delta_l = \pm 1$. The operator transforms as $X_1^{\pm 1}$. Take, for example, the "plus" sign and an initial state with symmetry X_1^{-1}. From the fifth line of the previous table, we find that

$$X_1^1 \cdot X_1^{-1} = \sqrt{\frac{1}{6}}X_2^0 + \sqrt{\frac{1}{2}}X_1^0 + \sqrt{\frac{1}{3}}X_0^0$$

A transition to the three states X_2^0, X_1^0, X_0^0 is then possible, with the corresponding amplitudes. The selection rule for circularly polarized light is $\Delta_m = \pm 1$. When parity is taken into account, the transition to the state with $l_i = l_f$ is prohibited.

Example 2: Spin polarization of $p \to s$ optical transitions in spherically symmetric potential with spin–orbit coupling

A possible sequence of energy levels in atoms, including spin–orbit splitting, is given in the following:

$$\underbrace{X_0^0 Y^+}_{\frac{1}{2}} \quad \underbrace{X_0^0 Y^-}_{-\frac{1}{2}} \quad {}^2S_{\frac{1}{2}}$$

$$\underbrace{X_1^1 Y^+}_{\frac{3}{2}} \quad \underbrace{\sqrt{\frac{1}{3}}X^1Y^- + \sqrt{\frac{2}{3}}X^0Y^+}_{\frac{1}{2}} \quad \underbrace{\sqrt{\frac{2}{3}}X^1Y^- - \sqrt{\frac{1}{3}}X^0Y^+}_{-\frac{1}{2}} \quad \underbrace{X_{\bar{1}}^1 Y^-}_{-\frac{3}{2}} \quad {}^2P_{\frac{3}{2}}$$

$$\underbrace{\sqrt{\frac{2}{3}}X^1Y^- - \sqrt{\frac{1}{3}}X^0Y^+}_{\frac{1}{2}} \quad \underbrace{\sqrt{\frac{1}{3}}X^1Y^- - \sqrt{\frac{2}{3}}X^0Y^+}_{-\frac{1}{2}} \quad {}^2P_{\frac{1}{2}}$$

The highest level is the final state we expect the electrons to be excited to. It originates from an atomic s level. Spin–orbit coupling just shifts the level, which is twice degenerate. The quantum number of the z component of the total angular momentum is given underneath the level. The two lower levels are possible initial states. They originate, through spin–orbit splitting, from the p levels. The $^2P_{\frac{3}{2}}$ level is fourfold degenerate, while the $^2P_{\frac{1}{2}}$ is twice degenerate. The basis states are taken from the following table of CGCs.

$1 \otimes \frac{1}{2}$		3/2	3/2	1/2				
		+3/2	+1/2	+1/2				
+1	+1/2	1						
		+1 -1/2	1/3	2/3	3/2	1/2		
		0 +1/2	2/3	-1/3	-1/2	-1/2		
				0 -1/2	2/3	1/3	3/2	
				-1 +1/2	1/3	-2/3	-3/2	
						-1 -1/2	-1	

Table of CGCs for the reduction of $D^1 \otimes D^{\frac{1}{2}}$.

We investigate light excitation by, for example, left-hand circularly polarized light propagating along $+z$, i.e. the operator is $\sim X_1^{-1}$. We consider, for instance, the $P_{\frac{3}{2}} \to S_{\frac{1}{2}}$ transition. We need to compute the eight matrix elements

$$\left(S_{\frac{1}{2}}^{m_f}, [X_1^{-1} \otimes \mathbb{1}]_{op} P_{\frac{3}{2}}^{m_i} \right)$$

with $m_i = \frac{3}{2}, \frac{1}{2}, -\frac{1}{2}, -\frac{3}{2}$, $m_f = \frac{1}{2}, -\frac{1}{2}$. Take, for instance,

$$\left(X_0^0 Y^-, [X_1^{-1} \otimes \mathbb{1}]_{op} \left[\sqrt{\frac{1}{3}} X_1^1 Y^- + \sqrt{\frac{2}{3}} X_1^0 Y^+ \right] \right)$$

The orthogonality of the spin functions reduces the matrix element to

$$\left(X_0^0, X_1^{-1} \left[\sqrt{\frac{1}{3}} X_1^1 \right] \right) = \sqrt{\frac{1}{3}} \cdot (X_0^0, X_1^{-1} X_1^1)$$

On the right-hand side, a CGC appears, originating from the entanglement between orbital and spin state. This entanglement is produced by spin–orbit splitting. To evaluate the remaining matrix element, we go back to the table of CGCs for the tensor product $D^1 \otimes D^1$. There, we find that

$$X_1^{-1}X_1^1 = \sqrt{\frac{1}{6}}X_2^0 - \sqrt{\frac{1}{2}}X_1^0 + \sqrt{\frac{1}{3}}X_0^0$$

Accordingly, we have

$$\left(X_0^0, X_1^{-1}X_1^1\right) = c \cdot \frac{1}{\sqrt{3}}$$

Finally, we have

$$\left(X_0^0 Y^-, X_1^{-1} \otimes \mathbb{1}\left[\sqrt{\frac{1}{3}}X_1^1 Y^- + \sqrt{\frac{2}{3}}X_1^0 Y^+\right]\right) = c \cdot \sqrt{\frac{1}{3}} \cdot \sqrt{\frac{1}{3}}$$

i.e. the product of two CGCs. Evaluating all other matrix elements using the same method, we find that the optical transition $P_{\frac{3}{2}} \to S_{\frac{1}{2}}$ can occur via two channels. The first one is the transition

$$P_{\frac{3}{2}}^{\frac{3}{2}} \to S_{\frac{1}{2}}^{\frac{1}{2}} \quad \text{matrix element} \quad c \cdot 1 \cdot \sqrt{\frac{1}{3}}$$

Interestingly, this transition produces electrons with spin "up" (Y^+). The second channel involves the transition

$$P_{\frac{3}{2}}^{\frac{3}{2}} \to S_{\frac{1}{2}}^{-\frac{1}{2}} \quad \text{matrix element} \quad c \cdot 1 \cdot \sqrt{\frac{1}{3}} \cdot \sqrt{\frac{1}{3}}$$

This transition produces electrons with spin "down" (Y^-).

• Suppose we want to sample the electrons excited in an ensemble of atoms during this transition. One could think of measuring their spin polarization

$$P_z \doteq \frac{N_\uparrow - N_\downarrow}{N_\uparrow + N_\downarrow}$$

with $N_\uparrow(N_\downarrow)$ being the number of electrons with spin "up" (spin "down", respectively). In the case of our transition, we obtain

$$N_\uparrow \sim \left|c \cdot 1 \cdot \sqrt{\frac{1}{3}}\right|^2 \qquad N_\downarrow \sim \left|c \cdot 1 \cdot \sqrt{\frac{1}{3}} \cdot \sqrt{\frac{1}{3}}\right|^2$$

resulting in a spin-polarized ensemble with

$$P_z = 50\%$$

One can argue (see the project on relativistic band structures) that the optical transition across the gap of Ge or GaAs is, from the point of view

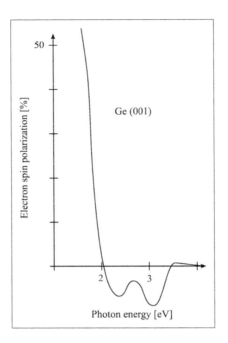

The spin polarization of electrons photoemitted from Ge as a function of the photon energy: At the threshold photon energy (indicated by an arrow), Ge single crystals have a transition which can be described by the $P_{\frac{3}{2}} \rightarrow S_{\frac{1}{2}}$ transition. Reproduced from D. Pescia, Diss. ETH Zurich Nr: 7326, https://doi.org/10.3929/ethz-a-000300178

of symmetry, very much of the type $P_{\frac{3}{2}} \rightarrow S_{\frac{1}{2}}$, so that it could be used to produce a beam of spin-polarized electrons. In fact, measurements of the spin polarization of electrons excited across the energy gap of GaAs or Ge provide an exquisite confirmation of the 50% spin polarization result, which we have predicted purely on the basis of symmetry arguments.

Such beams of highly spin-polarized electrons are routinely produced at SLAC to test, for example, the parity violation in weak interactions. See, for example, M. Woods, "Polarization at SLAC", *AIP Conf. Proc.* **343**, 230 (1995), https://doi.org/10.1063/1.48861.

D. T. Pierce, R. J. Celotta, G.-C. Wang, W. N. Unertl, A. Galejs, C. E. Kuyatt and S. R. Mielczarek, "The GaAs spin polarized electron source", *Rev. Sci. Instrum.* **51**, 478 (1980), https://doi.org/10.1063/1.1136250

A. V. Subashiev, Yu. A. Mamaev, Yu. P. Yashinand A. N. Ambrazhei, J. E. Clendenin, T. Maruyama, G. A. Mulhollan, A. Yu. Egorov, V. M. Ustinov and A. E. Zhukov, SLACPUB7922 PPRC-TN-98-3, August 1998, https://www.slac.stanford.edu/pubs/slacpubs/7750/slac-pub-7922.pdf.

Example 3: Optical orientation along Λ

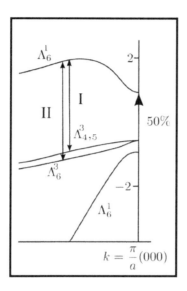

Relativistic band structure of GaAs along the Λ-direction. The optical transition at the Γ point across the energy gap is of the type $P_{\frac{3}{2}} \to S_{\frac{1}{2}}$ and produces 50% spin-polarized electrons. I indicates the optical transition $\Lambda_{4,5} \to \Lambda_6$, and II the transition $\Lambda_6 \to \Lambda_6$.

The level scheme along Λ, as obtained by symmetry analysis, predicts a final state band of Λ_6-label, originating from a single-group representation Λ^1 and containing two basis states with the symmetry $X_0^0 Y+, X_0^0 Y^-$. The basis states of the initial state bands that we have obtained by symmetry analysis are summarized in the following level scheme:

$$\underline{X_0^0 Y^+ \quad X_0^0 Y^-} \quad \Lambda_6^1$$

$$\underline{X_3^1 Y^+ \quad X_3^{-1} Y^-} \quad \Lambda_{4,5}^3$$

$$\underline{X_3^1 Y^- \quad X_3^{-1} Y^+} \quad \Lambda_6^3$$

The optical transition with circularly polarized light (X_1^{-1}) produces polarized electrons: The $\Lambda_{4,5}^3 \to \Lambda_6^1$ transition (I) produces 100% spin-up electrons, and the $\Lambda_6^3 \to \Lambda_6^1$ (II) transition produces 100% spin-down electrons. Taken together, we expect that the result of the optical transition from the top two bands along Λ is a vanishing spin polarization. Toward the Γ point, the two spin–orbit split bands come infinitely close to each other, so that, toward Γ, we expect the simultaneous excitation of both bands and thus vanishing spin polarization. The optical transition at Γ, however, produces 50% spin-polarized electrons. There is, accordingly, something very unphysical in this level scheme: Upon approaching the Γ point, we expect vanishing spin polarization for transitions along Λ, while the correct value should be 50%. Experimentally, all electrons at Γ and close to Γ are actually sampled by the photoemission experiment at 1.4 eV used to excite the electrons in GaAs or Ge, so that we need to find a mechanism that restores a 50% polarization along Λ when Γ is approached.

The desired mechanism is the **von-Neumann–Wigner hybridization** of orbital components originating from different single-group representations (in this case, Λ^1 and Λ^3) into the same double group irr. rep. (in this case, Λ_6). As we approach the Γ point, the two Λ_6 bands originating from the single-group irr. rep. Λ^1 (lowest band) and Λ^3 (upper valence band) come indeed close together, so that we might hybridize their single-group components. The degree of hybridization varies upon approaching Γ, but close to Γ, we can estimate it so that the resulting spin polarization is 50%:

$$
\begin{array}{ccc}
\underline{X_0^0 Y^+} & \underline{X_0^0 Y^-} & \Lambda_6^1 \\[4pt]
\underline{X_3^1 Y^+} & \underline{X_3^{-1} Y^-} & \Lambda_{4+5}^3 \\[4pt]
\underline{\tfrac{1}{\sqrt{3}} X_3^1 Y^- + \tfrac{\sqrt{2}}{\sqrt{3}} X_1^0 Y^+} & \underline{\tfrac{1}{\sqrt{3}} X_3^{-1} Y^+ + \tfrac{\sqrt{2}}{\sqrt{3}} X_1^0 Y^-} & \Lambda_6^{1,3}
\end{array}
$$

According to this mechanism of hybridization, the spin polarization at Γ and around Γ are similar, as required by experiment and physical continuity.

Lecture 8

YOUNG DIAGRAMS AND
PARTICLE PHYSICS

In this lecture, we introduce the technique of Young (after A. Young, (1873–1940), British mathematician) diagrams (YDs) and apply it to particle physics.

8.1 Irreducible Representations of $GL(n, \mathscr{C})$ and Some Important Subgroups

Introduction: The classical group $GL(n, \mathscr{C})$ consists of the set of all complex, non-singular $n \times n$ matrices. They describe formally the mapping of the n-coordinates $x_1, \ldots, x_n$ of a point X onto the n-coordinates $x'_1, \cdots, x'_n$ of a point X' in an n-dimensional space:

$$
\begin{pmatrix} x'_1 \\ x'_2 \\ x'_3 \\ \vdots \\ x'_n \end{pmatrix} = \begin{pmatrix} a_{11} & a_{12} & \cdots & \cdots & a_{1n} \\ a_{11} & a_{12} & \cdots & \cdots & a_{1n} \\ a_{21} & a_{22} & \cdots & \cdots & a_{2n} \\ \vdots & \vdots & & & \vdots \\ a_{n1} & a_{n2} & \cdots & \cdots & a_{nn} \end{pmatrix} \begin{pmatrix} x_1 \\ x_2 \\ x_3 \\ \vdots \\ x_n \end{pmatrix}
$$

The matrix elements are complex numbers. The determinant of the matrix is non-vanishing, so the inverse matrix exists. Each matrix element assumes values within some set of the complex plane. If this set is connected, we refer to it as a continuous, connected $GL(n, \mathscr{C})$ subgroup. If it is compact, we refer to it as a compact $GL(n, \mathscr{C})$ subgroup. The $n \times n$ matrix describing the transformation of the coordinates is the natural representation of the group $GL(n, \mathscr{C})$. We would like to find all irreducible representations of $GL(n, \mathscr{C})$ and its subgroups $SL(n, \mathscr{C})$, U_n. and SU_n. The technique we adopt is that of transforming the tensor product of the coordinates. This technique uses Young diagrams and Young tableaux (YTs) for displaying the tensor products.

8.1.1   *Scalars, vectors, and tensors*

We start with a linear transformation that maps a point X with coordinates $(x_1, \ldots, x_n)$ onto a point X' with coordinates $(x'_1, \ldots, x'_n)$ of an n-dimensional space:

$$
\begin{pmatrix} x'_1 \\ x'_2 \\ x'_3 \\ \vdots \\ x'_n \end{pmatrix} = \begin{pmatrix} a_{11} & a_{12} & \cdots & \cdots & a_{1n} \\ a_{21} & a_{22} & \cdots & \cdots & a_{2n} \\ a_{31} & a_{32} & \cdots & \cdots & a_{3n} \\ \vdots & \vdots & & & \vdots \\ a_{n1} & a_{n2} & \cdots & \cdots & a_{nn} \end{pmatrix} \begin{pmatrix} x_1 \\ x_2 \\ x_3 \\ \vdots \\ x_n \end{pmatrix}
$$

Let us describe this mapping by

$$
X' = AX
$$

where A is the transformation matrix.

Scalars:

Definition. A scalar is a number that transforms into itself by the transformation A.

Example: We compute the product of X and Y defined by

$$
(X, Y) \equiv \sum_i \overline{x_i} \cdot y_i
$$

We find that

$$
(X', Y') = \sum_i \sum_j \overline{a_{ij} x_j} \sum_m a_{im} y_m = \sum_j \left[\sum_{i,m} \overline{a_{ij}} a_{im} \right] \overline{x_j} x_m
$$

$$
= \sum_j \left[\sum_{i,m} \overline{a_{ji}^T} a_{im} \right] \overline{x_j} x_m
$$

a_{ji}^T are the matrix elements of the transposed matrix, i.e. of A^T. The defined product of the coordinates is a scalar if and only if $\sum_i \overline{a_{ji}^T} a_{im} = \delta_{j,m}$, i.e. if the matrix A is a unitary matrix $(\overline{A^T} A = \mathbb{1})$.

Vectors.

Definition. A set of n numbers $(u_1, \ldots, u_n)$ that transforms like the coordinates upon the coordinate transformation $X' = AX$ is called a vector and symbolically noted as **u**.

Examples:

1. The set $(x_1, \ldots, x_n)$ is a vector, and a short notation for it is $\mathbf{x}$.
2. We find now the transformation properties of the gradient of a function $\frac{\partial f}{\partial x_1}, \frac{\partial f}{\partial x_2}, \ldots, \frac{\partial f}{\partial x_n}$:

$$\frac{\partial f}{\partial x_i} = \sum_j \frac{\partial f}{\partial x'_j} \cdot \frac{\partial x'_j}{\partial x_i} = \sum_j \frac{\partial f}{\partial x'_j} a_{ji}$$

The transformation property of the gradient is written accordingly as

$$\begin{pmatrix} \frac{\partial f}{\partial x_1} \\ \frac{\partial f}{\partial x_2} \\ \vdots \\ \vdots \\ \frac{\partial f}{\partial x_n} \end{pmatrix} = \begin{pmatrix} a_{11} & a_{21} & \cdots & \cdots & a_{n1} \\ a_{12} & a_{22} & \cdots & \cdots & a_{n2} \\ a_{13} & a_{23} & \cdots & \cdots & a_{n3} \\ \vdots & \vdots & & & \vdots \\ a_{1n} & a_{2n} & \cdots & \cdots & a_{nn} \end{pmatrix} \begin{pmatrix} \frac{\partial f}{\partial x'_1} \\ \frac{\partial f}{\partial x'_2} \\ \vdots \\ \vdots \\ \frac{\partial f}{\partial x'_n} \end{pmatrix}$$

The set of numbers $(\frac{\partial f}{\partial x_1}, \frac{\partial f}{\partial x_2}, \ldots, \frac{\partial f}{\partial x_n})$ consisting of the partial derivative transform according to the matrix $(A^T)^{-1}$. This matrix is called the contra-gradient matrix to A.

Definition. A set of numbers that transforms according to A is said to be a contra-variant vector.

Definition. A set of numbers that transforms according to $(A^T)^{-1}$ is said to be a covariant vector.

Comments.

1. In the following, we refer to vectors and use the $\mathbf{x}$ symbol for them, regardless of whether they are contra-variant or covariant. If necessary, we label the components of contra-variant vectors with subscripts, whiles those of covariant vectors with superscripts.
2. Although not relevant for the current purpose, which deals with groups of linear transformations, this framework can be generalized to nonlinear variable transformations

$$x'_i = x'_i(x_1, x_2, \ldots, x_n)$$

where $x'_i(x_1, \ldots, x_n)$ are typically continuous functions. For linear transformations, $x'_i(x_1, x_2, \ldots, x_n) = \sum_j a_{ij} x_j$. The definition of vectors in the case of nonlinear transformations occurs by means of the Jacobi matrix J. Construct

$$
\begin{pmatrix} dx'_1 \\ dx'_2 \\ \vdots \\ dx'_n \end{pmatrix} = \underbrace{\begin{pmatrix} \frac{\partial x'_1}{\partial x_1} & \frac{\partial x'_1}{\partial x_2} & \cdots & \frac{\partial x'_1}{\partial x_n} \\ \frac{\partial x'_2}{\partial x_1} & \frac{\partial x'_2}{\partial x_2} & \cdots & \frac{\partial x'_2}{\partial x_n} \\ \vdots & \vdots & & \vdots \\ \frac{\partial x'_2}{\partial x_1} & \frac{\partial x'_2}{\partial x_2} & \cdots & \frac{\partial x'_2}{\partial x_n} \end{pmatrix}}_{\doteq J} \begin{pmatrix} dx_1 \\ dx_2 \\ \vdots \\ dx_n \end{pmatrix}
$$

Definition. A set of numbers $(u_1, u_2, \ldots, u_n)$ is called a (contra-variant) vector if it transforms according to

$$
\begin{pmatrix} u'_1 \\ u'_2 \\ u'_3 \\ \vdots \\ u'_n \end{pmatrix} = \begin{pmatrix} J_{11} & J_{12} & \cdots & \cdots & J_{1n} \\ J_{21} & J_{22} & \cdots & \cdots & J_{2n} \\ J_{31} & J_{32} & \cdots & \cdots & J_{3n} \\ \vdots & \vdots & & & \vdots \\ J_{n1} & J_{n2} & \cdots & \cdots & J_{nn} \end{pmatrix} \begin{pmatrix} u_1 \\ u_2 \\ u_3 \\ \vdots \\ u_n \end{pmatrix}
$$

A covariant vector is defined accordingly.

Tensors: The concept of tensor is a generalization of the concept of a vector. Consider two contra-variant vectors $(u_1, u_2, \ldots, u_n)$ and $(v_1, v_2, \ldots, v_n)$ in a vector space V_n.

Definition. The n^2 numbers

$$u_1 \cdot v_1, u_1 \cdot v_2, \ldots, u_1 \cdot v_n, u_2 \cdot v_1, \ldots, u_n \cdot v_1, \ldots, u_n \cdot v_n$$

are said to be the components of the tensor product of the two vectors. As the original vectors are contra-variant, the resulting tensor is also said to be contra-variant.

Comments.

1. Tensors can be visualized by placing the product $u_i \cdot v_j$ at site ij of a square lattice with n^2 lattice points. This is called a rank 2 tensor. A vector aligns the components along one dimension and is called, accordingly, a rank 1 tensor. A scalar is, by generalization, a rank 0 tensor.

2. As an alternative to occupying a lattice, one can consider the tensor products as the components of a vector in a V_{n+n}-dimensional space called the tensor product space $V_n \otimes V_n$. The transformation properties of a contra-variant tensor are found through

$$u'_i \cdot v'_j = \sum_{l,m} a_{il} a_{jm} u_l \cdot v_m$$

and, respectively,

$$u'_i \cdot v'_j = \sum_{l,m} J_{il} J_{jm} u_l \cdot v_m$$

for nonlinear transformations. If the components are ordered to form a vector the way we have done, $a_{il} \cdot a_{jm}$ are the matrix elements of a $n^2 \times n^2$ matrix, called the Kronecker (Leopold Kronecker (1823–1891), German mathematician) product of A with itself:

$$A \otimes A = \begin{pmatrix} a_{11} \cdot A & a_{12} \cdot A & \cdots & a_{1n} \cdot A \\ a_{21} \cdot A & a_{12} \cdot A & \cdots & a_{2n} \cdot A \\ \vdots & \vdots & & \vdots \\ \vdots & \vdots & & \vdots \\ a_{n1} \cdot A & a_{n2} \cdot A & \cdots & a_{nn} \cdot A \end{pmatrix}$$

3. As a way of generalization, we can start with vectors $(u_1, u_2, \ldots, u_n) \in V_n$ and $(v_1, v_2, \ldots, v_m) \in V_m$ that transform as

$$\begin{pmatrix} u'_1 \\ u'_2 \\ u'_3 \\ \vdots \\ u'_n \end{pmatrix} = \begin{pmatrix} a_{11} & a_{12} & \cdots & \cdots & a_{1n} \\ a_{21} & a_{22} & \cdots & \cdots & a_{2n} \\ a_{31} & a_{32} & \cdots & \cdots & a_{3n} \\ \vdots & \vdots & & & \vdots \\ a_{n1} & a_{n2} & \cdots & \cdots & a_{nn} \end{pmatrix} \begin{pmatrix} u_1 \\ u_2 \\ u_3 \\ \vdots \\ u_n \end{pmatrix}$$

and

$$\begin{pmatrix} v'_1 \\ v'_2 \\ v'_3 \\ \vdots \\ v'_m \end{pmatrix} = \begin{pmatrix} b_{11} & b_{12} & \cdots & \cdots & b_{1m} \\ b_{21} & b_{22} & \cdots & \cdots & b_{2m} \\ b_{31} & b_{32} & \cdots & \cdots & b_{3m} \\ \vdots & \vdots & & & \vdots \\ b_{m1} & b_{m2} & \cdots & \cdots & b_{mm} \end{pmatrix} \begin{pmatrix} v_1 \\ v_2 \\ v_3 \\ \vdots \\ v_n \end{pmatrix}$$

respectively and constitute the vector products $\{u_i \cdot v_j\}$, which transform according to the Kronecker product

$$A \otimes B = \begin{pmatrix} a_{11} \cdot B & a_{12} \cdot B & \cdots & a_{1n} \cdot B \\ a_{21} \cdot B & a_{12} \cdot B & \cdots & a_{2n} \cdot B \\ \vdots & \vdots & & \vdots \\ \vdots & \vdots & & \vdots \\ a_{n1} \cdot B & a_{n2} \cdot B & \cdots & a_{nn} \cdot B \end{pmatrix}$$

4. For nonlinear transformations, A and B are the corresponding Jacobi matrices.

5. Covariant tensors of rank 2 are obtained by the tensor product of two covariant vectors $\{u^i \cdot v^j\}$. They transforms according to $(A^T)^{-1} \otimes (B^T)^{-1}$.

6. A mixed rank 2 tensor can be obtained by the tensor product of a contra-variant vector with a covariant one. The transformation property of the element $u_i \cdot v^j$ ($u_i, v_j \in V_n$) reads

$$u_i' \cdot v'^j = \sum_{m,l} A_{il} u_l \cdot A_{mj}^{-1} v^m = \sum_{m,l} A_{il} (u_l \cdot v^m) A_{mj}^{-1}$$

This transformation property, describing the transformation of the tensor element $T_{ij} \equiv u_i \cdot v^j$, can be summarized as

$$T' = A T A^{-1}$$

which is the transformation property of matrix T under the mapping A: Mixed tensors are therefore conventional $n \times n$ matrices.

7. Tensors with higher ranks: The tensor product of three contra-variant vectors u_i, v_j, z_k produces a tensor of rank 3 with components T_{ijk}. One can picture these components as components of a vector in the vector space $V_n \otimes V_m \otimes V_p$ or locate them on a three-dimensional lattice. The transformation properties read

$$T_{ijk}' = u_i \cdot v_j \cdot z_k \sum_{lmn} A_{il} A_{jm} A_{kn} \underbrace{u_l \cdot v_m \cdot z_n}_{T_{lmn}}$$

Arranged as a vector in the space $V_n \otimes V_m \otimes V_p$, this transformation property can be summarized by the Kronecker product of three matrices

$$A \otimes (A \otimes A) = (A \otimes A) \otimes A$$

8. Higher-ranking contra-variant, covariant and mixed tensors can be defined accordingly.

8.1.2 *An application: Constructing representations of $GL(2, \mathscr{C})$*

We illustrate the use of tensor products for constructing representations and the associated diagrammatic technology for the case of the classical group $GL(2, \mathscr{C})$, within which we have learned that the subgroup SU_2 plays an important role in spin physics. The group $GL(2, \mathscr{C})$ is defined by the transformation matrix

$$\begin{pmatrix} x_1' \\ x_2' \end{pmatrix} = \begin{pmatrix} a_{11} & a_{12} \\ a_{21} & a_{22} \end{pmatrix} \begin{pmatrix} x_1 \\ x_2 \end{pmatrix}$$

The matrix elements have the property that $a_{11} \cdot a_{22} - a_{12} \cdot a_{21} \neq 0$. This is known as the natural representation of $GL(2, \mathscr{C})$.

One can build the representations of $GL(2, \mathscr{C})$ using the diagrammatic technique of the YTs to classify the tensor products constituting the representation space.

Rank 1 tensor representation of $GL(2, \mathscr{C})$: We associate to the coordinates x_1' the YT (Young Tableau) $\boxed{1}$ and to the coordinate x_2' the YT $\boxed{2}$. The YD (Young Diagram) $\square$ identifies therefore the two YTs $\boxed{1}$ and $\boxed{2}$, which, taken together, define a representation space built by a vector with the two components (x_1', x_2'): $\square$ is the YD used to label a two-dimensional representation of the group $GL(2, \mathscr{C})$. To find the matrix representation within the representation space given by $\square$, we write the basis functions (x_1', x_2') in terms of the coordinates (x_1, x_2):

$$\begin{pmatrix} x_1' \\ x_2' \end{pmatrix} = \begin{pmatrix} a_{11} & a_{12} \\ a_{21} & a_{22} \end{pmatrix} \begin{pmatrix} x_1 \\ x_2 \end{pmatrix}$$

We observe that the matrix obtained is just the natural representation of $GL(2, \mathscr{C})$. Note that the pair $\{x_1, x_2\}$ transforms exactly as the coordinates (x_1, x_2). The pair $\{x_1, x_2\}$ is a vector (rank 1 tensor). So, $\square$ is the rank 1 tensor representation of $GL(2, \mathscr{C})$ (the natural representation). It is irreducible. $\square$ also provides a two-dimensional irreducible representation of the subgroups $SL(2, \mathscr{C}), U_2, SU_2$.

Comment. We have used the convention of filling the YD with the primed variables, so that we can read out the transformation properties of the basis functions directly using the original transformation matrix, which transforms unprimed variables into primed variables. We keep this convention throughout, but drop, for obvious practical purposes, the prime sign within the YT itself or when referring to it.

Rank 2 tensor representation of $GL(2,\mathscr{C})$**:** We study now the representation identified by $\boxed{}$. The basis functions associated with this diagram are given by the YTs

$$\boxed{1\,1}\quad\boxed{1\,2}\quad\boxed{2\,1}\quad\boxed{2\,2}$$

In terms of the coordinates. one could think of these tableaux as identifying the following tensor products of coordinates:

$$\boxed{1\,1}=x_1\cdot y_1\quad\boxed{1\,2}=x_1\cdot y_2\quad\boxed{2\,1}=x_2\cdot y_1\quad\boxed{2\,2}=x_2\cdot y_2$$

Let us construct the transformation matrix on these four basis functions:

$$\begin{pmatrix}x_1'\cdot y_1'\\x_1'\cdot y_2'\\x_2'\cdot y_1'\\x_2'\cdot y_2'\end{pmatrix}=\begin{pmatrix}a_{11}a_{11}&a_{11}a_{12}&a_{12}a_{11}&a_{12}a_{12}\\a_{11}a_{21}&a_{11}a_{22}&a_{12}a_{21}&a_{12}a_{22}\\a_{21}a_{11}&a_{21}a_{12}&a_{22}a_{11}&a_{22}a_{12}\\a_{21}a_{21}&a_{21}a_{22}&a_{22}a_{21}&a_{22}a_{22}\end{pmatrix}\begin{pmatrix}x_1\cdot y_1\\x_1\cdot y_2\\x_2\cdot y_1\\x_2\cdot y_2\end{pmatrix}$$

The four products of the coordinates $(x_1y_1,x_1y_2,x_2y_1,x_2y_2)$ have a special transformation property: They transform according to a 4×4 matrix that can be written as the Kronecker product of the matrix

$$\begin{pmatrix}a_{11}&a_{12}\\a_{21}&a_{22}\end{pmatrix}$$

with itself:

$$\begin{pmatrix}a_{11}a_{11}&a_{11}a_{12}&a_{12}a_{11}&a_{12}a_{12}\\a_{11}a_{21}&a_{11}a_{22}&a_{12}a_{21}&a_{12}a_{22}\\a_{21}a_{11}&a_{21}a_{12}&a_{22}a_{11}&a_{22}a_{12}\\a_{21}a_{21}&a_{21}a_{22}&a_{22}a_{21}&a_{22}a_{22}\end{pmatrix}=\begin{pmatrix}a_{11}&a_{12}\\a_{21}&a_{22}\end{pmatrix}\otimes\begin{pmatrix}a_{11}&a_{12}\\a_{21}&a_{22}\end{pmatrix}$$

The four basis functions are said to build a tensor of rank 2, i.e. they transform according to the tensor product of the two vectors (x_1,x_2) and (y_1,y_2). They are the coordinates of a point in the tensor product space $\mathscr{C}\otimes\mathscr{C}$.

Comments.

1. The matrix we have just constructed is a four-dimensional representation of the group $GL(2,\mathscr{C})$ and is identified by the YD $\boxed{}$.
2. This Kronecker product representation is reducible. For SU_2, we can immediately recognize that $\boxed{}$ is the representation $D^{\frac{1}{2}}\otimes D^{\frac{1}{2}}$ which decomposes into $D^1\oplus D^0$.

3. Our aim, however, is to construct irreducible representations and, if possible, all of them. Here is where a refined application of the YD technique provides us with the suitable way to construct the irreducible representation associated with the YD ⬚⬚. This entails choosing those basis functions that are made of the tensor product of the vector (x_1, x_2) with itself. For the specific case, we can construct three independent products

$$x_1^2 \quad x_1 \cdot x_2 \quad x_2^2$$

which can also be recognized as the monomials in the variables x_1, x_2 of second degree. We associate to these monomials the YTs

$$\boxed{1\,1} \quad \boxed{1\,2} \quad \boxed{2\,2}$$

with the tableau $\boxed{2\,1}$ leading to the same monomials as $\boxed{1\,2}$ and thus dropped. The representation sustained by these three basis functions is a three-dimensional representation of $GL(2, \mathscr{C})$, and it is irreducible. For example, starting from the subgroup SU_2 consisting of the matrices

$$\begin{pmatrix} a & b \\ -\bar{b} & \bar{a} \end{pmatrix}$$

we construct explicitly the matrix representation associated with the diagram ⬚⬚ by writing

$$\begin{pmatrix} x_1'^2 \\ x_1' x_2' \\ x_2'^2 \end{pmatrix} = \begin{pmatrix} (ax_1 + bx_2)^2 \\ (ax_1 + bx_2)(-\bar{b}x_1 + \bar{a}x_2) \\ (-\bar{b}x_1 + \bar{a}x_2)^2 \end{pmatrix}$$

$$= \begin{pmatrix} a^2 & 2ab & b^2 \\ -a\bar{b} & (|a|^2 - |b|^2) & b\bar{a} \\ \bar{b}^2 & -2\bar{a}\bar{b} & \bar{a}^2 \end{pmatrix} \begin{pmatrix} x_1^2 \\ x_1 x_2 \\ x_2^2 \end{pmatrix}$$

Rank 0 representation of $GL(2, \mathscr{C})$: We study now the representation defined by the YD ⬚ (vertical). By locating the boxes along the vertical, one defines a new type of product of coordinates, namely an antisymmetric tensor product. For instance, $\begin{smallmatrix}1\\2\end{smallmatrix}$ designates the product function

$$x_1 y_2 - x_2 y_1 = \left| \begin{pmatrix} x_1 & y_1 \\ x_2 & y_2 \end{pmatrix} \right|$$

A second product function $\begin{smallmatrix}2\\1\end{smallmatrix}$ is identical with $\begin{smallmatrix}1\\2\end{smallmatrix}$ up to a change in sign and is therefore dropped. The product functions $\begin{smallmatrix}1\\1\end{smallmatrix}$ and $\begin{smallmatrix}2\\2\end{smallmatrix}$ are vanishing.

This explain the rule that only YTs with increasing numbers along the vertical need to be considered. The transformation property of $\begin{array}{|c|} \hline 1 \\ \hline 2 \\ \hline \end{array}$ under a matrix of $GL(2, \mathscr{C})$ reads

$$(a_{11}x_1 + a_{12}x_2)(a_{21}y_1 + a_{22}y_2) - (a_{21}x_1 + a_{22}x_2)(a_{11}y_1 + a_{12}y_2)$$

$$= (a_{11}a_{22} - a_{21}a_{12}) \cdot (x_1y_2 - x_2y_1)$$

The representation identified by $\begin{array}{|c|} \hline \\ \hline \end{array}$ is a one-dimensional representation associating to each matrix its determinant (a scalar, therefore the name "rank 0" tensor).

Mixed YDs: One can use YDs that have both horizontal and vertical boxes. For example, we discuss the representation $\begin{array}{|c|c|} \hline & \\ \hline & \\ \hline \end{array}$ of the group $GL(2, \mathscr{C})$. The corresponding basis functions are obtained by inserting the index of the variables x_1, x_2 according to the rules for horizontal and vertical tableaux. Accordingly, we have the two basis functions $\begin{array}{|c|c|} \hline 1 & 1 \\ \hline 2 & \\ \hline \end{array}$ and $\begin{array}{|c|c|} \hline 1 & 2 \\ \hline 2 & \\ \hline \end{array}$. The corresponding tensor products are

$$\left| \begin{pmatrix} x_1 & y_1 \\ x_2 & y_2 \end{pmatrix} \right| \cdot x_1$$

and

$$\left| \begin{pmatrix} x_1 & y_1 \\ x_2 & y_2 \end{pmatrix} \right| \cdot x_2$$

The two-dimensional irreducible representation matrix is obtained by analyzing the transformation properties of these product functions:

$$\begin{pmatrix} \begin{array}{|c|c|} \hline 1 & 1 \\ \hline 2 & \\ \hline \end{array} \\[6pt] \begin{array}{|c|c|} \hline 1 & 2 \\ \hline 2 & \\ \hline \end{array} \end{pmatrix} = \left| \begin{pmatrix} a_{11} & a_{12} \\ a_{21} & a_{22} \end{pmatrix} \right| \cdot \begin{pmatrix} a_{11} & a_{12} \\ a_{21} & a_{22} \end{pmatrix} \begin{pmatrix} \begin{array}{|c|c|} \hline 1 & 1 \\ \hline 2 & \\ \hline \end{array} \\[6pt] \begin{array}{|c|c|} \hline 1 & 2 \\ \hline 2 & \\ \hline \end{array} \end{pmatrix}$$

The matrix representation constructed over the diagram $\begin{array}{|c|c|} \hline & \\ \hline & \\ \hline \end{array}$ associates to every element of $GL(2, \mathscr{C})$ its natural matrix representation multiplied by its determinant. When restricted to the subgroup SU_2, the determinant of the matrices is 1; therefore, $\begin{array}{|c|c|} \hline & \\ \hline & \\ \hline \end{array}$ and $\begin{array}{|c|} \hline \\ \hline \end{array}$ produces the same (irreducible) representations of SU_2.

Computation with YD: On can compute, for example, the Kronecker product $\square \otimes \square$ by finding all possible locations that the second diagram can be placed with respect to the first one without violating the rules for horizontal and vertical boxing. In this specific case, we have

$$\square \otimes \square = \boxed{\ \ } \oplus \begin{array}{c}\square\\\square\end{array}$$

For the subgroup SU_2, this computation reads

$$D^{\frac{1}{2}} \otimes D^{\frac{1}{2}} = D^1 \oplus D^0$$

As a further example, we compute

$$\begin{array}{c}\square\square\\\square\end{array} \otimes \begin{array}{c}\square\square\\\square\end{array} = \boxed{\ \ \ \ } \oplus \begin{array}{c}\square\square\square\\\square\square\square\end{array}$$

When restricted to SU_2, this operation reads

$$D^{\frac{1}{2}} \otimes D^{\frac{1}{2}} = D^1 \oplus D^0$$

owing to the identities of the representations $\square\square$ and $\begin{array}{c}\square\\\square\end{array}$ with $\begin{array}{c}\square\square\square\\\square\square\end{array}$ and $\begin{array}{c}\square\square\square\\\square\square\square\end{array}$, respectively (when restricted to SU_2).

8.1.3 Generalizations and general theorems

The search for the irreducible representations of $GL(n,\mathscr{C})$ required us to learn the YD and YT techniques. These techniques are based on some exact theorems of discrete mathematics. Rather than providing detailed proofs, we present arguments supporting their validity and physical relevance.

Definition. A YD is a scheme with r lines, each containing $m_1, m_2, \ldots, m_r$ boxes with $m_1 \geq m_2 \geq \cdots \geq m_r$. The total number of boxes, $m_1 + m_2 + \cdots + m_r = f$, is the rank of the tensor used for finding the representation of the group. To each ordered partition $m_1 + m_2 + \cdots + m_r = f$, there is exactly one diagram associated.

Comment. the number of boxes along one line cannot increase when descending along the lines, as it is our convention that the variables (when the diagram is filled with numbers) along the vertical are combined to form antisymmetric products of the corresponding variables.

Definition. A YT is a YD filled with integers $1, \ldots, n$.

Comments.

1. When the YT is used to produce basis monomials of degree f to represent the group $GL(n, \mathscr{C})$ or one of its subgroups, the integers label the n variables that are transformed by the elements of the group.
2. The rule for building the tensor product basis monomials is to perform the **product** of the variables appearing along horizontal lines and construct an **antisymmetric product** of the variables positioned along the vertical lines. Because of the requirement of antisymmetry, no two numbers along a vertical column can be equal. This means that the maximum number of lines allowed for a YD with a group that transforms n variables is just n, i.e $r \le n$.

Examples:

1. Take $n = 3$ and $f = 1$. We will have only one type of YD, namely ▢, representing the rank 1 tensor. There are three YTs, 1 , 2 , and 3 , that correspond to the monomials x_1, x_2, and x_3. The representation created over these YTs is the natural representation of $GL(3, \mathscr{C})$.
2. Take, for example, $n = 3$ and $f = 4$. The number of horizontal lines must be $\le n$, and the ordered partitions of 4 that are relevant for producing all possible YDs are

$$4 = 4 + 0 + 0 \quad 4 = 3 + 1 + 0 \quad 4 = 2 + 2 + 0 \quad 4 = 2 + 1 + 1$$

The partition $4 = 1 + 1 + 1 + 1$ must be dropped, as it would require a fourth variable which we do not have when dealing with $GL(3, \mathscr{C})$. From these partitions, we can obtain the YD that can be constructed for $GL(3, \mathscr{C})$ with power 4 monomials of the variables:

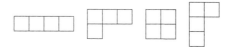

Each of these YTs symbolizes a representation of $GL(3, \mathscr{C})$.

We now formulate two important theorems regarding the representations of the group $GL(n, \mathscr{C})$ with YDs (for their proof see H. Boerner, *Darstellungen von Gruppen*, Springer Verlag, 1967, https://link.springer.com/book/10.1007/978-3-642-86032-4).

Theorem 1. *Each YD with $r \le n$ symbolizes an irreducible representation of the group $GL(n, \mathscr{C})$.*

Theorem 2. *The representations by YDs are all irreducible representations of the group $GL(n, \mathscr{C})$.*

Comments.

1. The YD representations remain irreducible if limited to the following subgroups:

$$U_n \quad SU_n \quad GL(n, \mathscr{R}) \quad SL(n, \mathscr{C}) \quad SL(n, \mathscr{R})$$

2. One obtains in this way all irreducible representations of the subgroups as well, although some YDs, when limited to the subgroups, produce the same irreducible representation (we have seen some examples above in this lecture).

In the specialized literature, one can find theorems from discrete mathematics that allow for the computation of the dimensionality of the irreducible representations originating from YDs and their character. However, one should not overestimate the usefulness of these theorems for concrete applications. We limit ourselves to illustrating the background leading to these theorems.

Dimensionality of a YD: The dimensionality of a representation depends on the number of tensor products one can generate on the base of a YD, i.e. on the number of possible YTs. The filling rules for the group $GL(n, \mathscr{C})$ determine those YT that are called the standard YT:

1. When moving along horizontal lines from left to right, the numbers filling the boxes do not decrease.
2. When moving along a vertical line from top to bottom, the numbers are always increasing.

We consider, as a way of giving an example ($f = 4$ and $n = 3$),

We have 15 (standard) YTs:

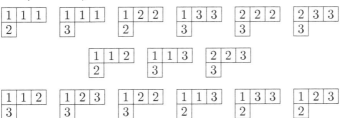

Character of a representation: To each non-singular matrix A, there exists a non-singular matrix S such that SAS^{-1} is diagonal, i.e.

$$\begin{pmatrix} \lambda_1 & 0 & & \cdots & 0 \\ 0 & \lambda_2 & 0 & \cdots & 0 \\ & & & \cdots & 0 \\ \vdots & \vdots & \vdots & \cdots & 0 \\ 0 & & & \cdots & \lambda_n \end{pmatrix}$$

The original matrix A is therefore conjugate to a diagonal matrix. The conjugacy classes have therefore diagonal matrices as representatives. The diagonal matrices that are obtained by the similarity operation might have the same eigenvalues but ordered differently. They belong to the same class, as the character is identifying the classes. To compute the character of any matrix in a representation given by a YD, one must find the transformation properties of each YT under the operation of such a diagonal matrix. This operation produces, for each YT, a monomial:

$$\lambda_1^{p_1} \cdot \lambda_2^{p_2} \cdot \cdots \cdot \lambda_n^{p_n}$$

The character is the sum of such monomials.

Definition. The vector

$$\mathbf{p} = (p_1, \ldots, p_n)$$

$(p_1 + p_2 + \cdots p_n = f)$ is called the **weight** of a YT. Each component of the vector gives the number of times a given coordinate appears in the YT.

Comment.

1. As an example, the weight of the YT $\begin{array}{|c|c|c|} \hline 1 & 1 & 1 \\ \hline 2 \\ \cline{1-1} \end{array}$ is $(3, 1, 0)$, and this basis state contributes a monomial $\lambda_1^3 \cdot \lambda_2^1$ to the character of the representations $\begin{array}{ccc} \hline & & \\ \hline & \end{array}$ of the group $GL(3, \mathscr{C})$.

2. Let us work out a complete example of the applications of these concepts. We consider the YD

of an irreducible representation of $GL(2, \mathscr{C})$ $(n = 2, f = 3)$. The possible YTs are

$\begin{array}{|c|c|} \hline 1 & 1 \\ \hline 2 \\ \cline{1-1} \end{array}$ weight $(2, 1)$ monomial $\lambda_1^2 \cdot \lambda_2$

and

$$\boxed{\begin{array}{|c|c|} \hline 1 & 2 \\ \hline 2 \\ \cline{1-1} \end{array}} \quad \text{weight} \quad (1,2) \quad \text{monomial} \quad \lambda_1 \cdot \lambda_2^2.$$

The character is therefore

$$\lambda_1^2 \cdot \lambda_2 + \lambda_1 \cdot \lambda_2^2$$

8.1.4   *The group* SU_2 *and its irreducible representations*

The matrices

$$D^{\frac{1}{2}} = \begin{pmatrix} \cos\frac{\varphi}{2} - i \cdot n \sin\frac{\varphi}{2} & (-i \cdot l - m)\sin\frac{\varphi}{2} \\ (-i \cdot l + m)\sin\frac{\varphi}{2} & \cos\frac{\varphi}{2} + i \cdot n \sin\frac{\varphi}{2} \end{pmatrix}$$

that we have constructed in Lecture 7 have an important property: They are unitary matrices with a determinant of 1 and the set of them obtained by running over the parameter space $l^2 + m^2 + n^2 = 1$ ($l, m, n \geq 0$) and $\varphi \in [0, 4\pi]$ builds the group called $SU(2)$. This group is known as the special unitary group of 2×2 matrices. It is a three-parameter continuous Lie group with an average. By construction, the elements of SO_3 are uniquely mapped onto the elements of SU_2 with $\varphi \leq 2\pi$. The elements $R(2\pi) \cup SO_3$ are uniquely mapped onto the elements of SU_2 with $\varphi \in [2\pi, 4\pi]$. Because of this isomorphism, the task of finding the irreducible representations of Bethe's double group SO_3^D can be solved by finding the irreducible representations of SU_2.

The Weyl average of SU_2:

In order to prove the irreducibility of the representations constructed using the YD technique, we need to define the average of the group. To find the average, we define first the parameters (we adopt the convention that φ is in the range $[0, 2\pi]$ and allow negative values for l, n, m)

$$\cos\frac{\varphi}{2} \doteq \alpha_1 \quad -n \cdot \sin\frac{\varphi}{2} \doteq \alpha_2 \quad -m \cdot \sin\frac{\varphi}{2} \doteq \alpha_3 \quad -l \cdot \sin\frac{\varphi}{2} \doteq \alpha_4$$

to obtain the matrix

$$\begin{pmatrix} \alpha_1 + i\alpha_2 & \alpha_3 + i\alpha_4 \\ -\alpha_3 + i\alpha_4 & \alpha_1 - i\alpha_2 \end{pmatrix}; \quad \alpha_1^2 + \alpha_2^2 + \alpha_3^2 + \alpha_4^2 = 1$$

with α_i being real numbers in the interval $[-1, 1]$. We now compute the density of symmetry elements in the parameter space defined by α_i.

We multiply the matrices

$$\begin{pmatrix} \alpha_1 + i\alpha_2 & \alpha_3 + i\alpha_4 \\ -\alpha_3 + i\alpha_4 & \alpha_1 - i\alpha_2 \end{pmatrix} \quad \text{and} \quad \begin{pmatrix} 1 + \epsilon_1 + i\epsilon_2 & \epsilon_3 + i\epsilon_4 \\ -\epsilon_3 + i\epsilon_4 & 1 + \epsilon_1 - i\epsilon_2 \end{pmatrix}$$

to find the composition law in the vicinity of the identity:

$$\beta_1 = \alpha_1 + \epsilon_1 \cdot \alpha_1 - \epsilon_2 \cdot \alpha_2 - \epsilon_3 \cdot \alpha_3 - \epsilon_4 \cdot \alpha_4$$

$$\beta_2 = \alpha_2 + \epsilon_1 \cdot \alpha_2 + \epsilon_2 \cdot \alpha_1 - \epsilon_3 \cdot \alpha_4 + \epsilon_4 \cdot \alpha_3$$

$$\beta_3 = \alpha_3 + \epsilon_1 \cdot \alpha_3 + \epsilon_2 \cdot \alpha_4 + \epsilon_3 \cdot \alpha_1 - \epsilon_4 \cdot \alpha_2$$

$$\beta_4 = \alpha_4 + \epsilon_1 \cdot \alpha_4 - \epsilon_2 \cdot \alpha_3 + \epsilon_3 \cdot \alpha_2 + \epsilon_4 \cdot \alpha_1$$

The Jacobi matrix for this transformation law at $\epsilon = 0$ reads

$$\begin{pmatrix} \alpha_1 & -\alpha_2 & -\alpha_3 & -\alpha_4 \\ \alpha_2 & \alpha_1 & -\alpha_4 & \alpha_3 \\ \alpha_3 & \alpha_4 & \alpha_1 & -\alpha_2 \\ \alpha_4 & -\alpha_3 & \alpha_2 & \alpha_1 \end{pmatrix}$$

The determinant of this matrix is 1, and the integral in α space is over the infinitesimal element

$$1 \cdot d\alpha_1 \cdot d\alpha_2 \cdot d\alpha_3 \cdot d\alpha_4$$

The requirement of $\alpha_1^2 + \alpha_2^2 + \alpha_3^2 + \alpha_4^2 = 1$ means that the integral must be performed over the unit sphere in four dimensions. We implement this requirement by choosing spherical coordinates in four dimensions:

$$\alpha_1 \doteq \cos\phi; \quad \phi \in [0, \pi]$$

$$\alpha_2 \doteq \cos\vartheta \sin\phi; \quad \vartheta \in [0, \pi]$$

$$\alpha_3 \doteq \sin\vartheta \sin\psi \sin\phi; \quad \psi \in [0, 2\pi]$$

$$\alpha_4 \doteq \sin\vartheta \cos\psi \sin\phi$$

When normalized to "1", the average over the group SU_2 reads

$$\frac{1}{2\pi^2} \int_0^\pi d\phi \cdot \sin^2\phi \cdot \int_0^\pi d\vartheta \cdot \sin\vartheta \cdot \int_0^{2\pi} d\psi \cdot f(\phi, \vartheta, \varphi)$$

By comparing the spherical coordinates to the original variables (l, m, n, φ), we can identify

$$\phi = -\frac{\varphi}{2}$$

$$n = \cos \vartheta$$
$$m = \sin \vartheta \sin \psi$$
$$l = \sin \vartheta \cos \psi$$

i.e. ϕ is (the negative of) half the rotation angle (the negative sign indicates the sense of the rotation and arises because of our particular choice of parameters). The angles ϑ, ψ are the spherical angles identifying the direction of the rotation axis in the three-dimensional space. The group SU_2 of all 2×2 unitary matrices with a determinant of 1 is important in problems involving spin. According to the theorem on $GL(n, \mathscr{C})$, the YD with two lines at the most (corresponding to the two variables transformed by SU_2) of length $l_1 \geq l_2$ and $l_1 \geq 1$, $l_2 \geq 0$ carry all irreducible representations of SU_2.

The YD of SU_2**:** Let us analyze the YD technique as applied to SU_2. Vertically aligned boxes carry a tensor product that transforms according to the determinant $(= 1)$, i.e. they are left invariant. This mechanism allows us to produce YDs that carry the same representation, for instance

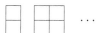

All these YDs produce the same, identical representation, which is called D^0 in the "Cartan" terminology used in Lecture 7. When creating a further representation, one must consider that only those boxes in the YD which do not have underlying boxes are "active" ones, i.e. can be filled with basis monomials that produce a representation beyond D_0. There are $l_1 - l_2$ active boxes which can host $l_1 - l_2 + 1$ monomials of the variables (x_1, x_2), so that the dimension of the representation carried by a YD with parameters (l_1, l_2) is $l_1 - l_2 + 1$.

Regarding the notation, in the literature, one finds often that a representation of SU_2 with dimension $l_1 - l_2 + 1$ is denoted by a symbol containing its dimension, i.e. $[d = l_1 - l_2 + 1]$. Accordingly, the identical representation is denoted by $[1]$, the natural representation of SU_2 is denoted by $[2]$, and so forth.

Proposition. *The representations* $[d]$ *are:* (i) *irreducible and* (ii) *provide all irreducible representations of* SU_2.

Proof of (i): Any unitary 2×2 matrix with a determinant of 1 can be diagonalized to

$$\begin{pmatrix} e^{i\phi} & 0 \\ 0 & e^{-i\phi} \end{pmatrix}$$

The diagonalized matrices can be taken as representatives of the conjugacy classes. Under the transformation

$$x_1' = e^{i\phi} x_1$$
$$x_2' = e^{-i\phi} x_2$$

the active monomials of the representation $[d] = [l_1 - l_2 + 1]$, i.e.

$$\underbrace{x_1' \cdot x_1' \ldots \cdot x_1'}_{\text{d-1-times}} \quad x_1' \cdot x_1' \ldots \cdot x_2' \quad x_1' \cdot x_1' \cdot \ldots \cdot x_2' \cdot x_2' \quad \ldots$$

transform as

$$\begin{pmatrix} e^{i(d-1)\phi} & 0 & & \cdots & \\ 0 & e^{i(d-3)\phi} & 0 & \cdots & \\ \vdots & \vdots & & \cdots & \cdots \\ 0 & 0 & 0 & \cdots & e^{-i(d-1)\phi} \end{pmatrix}$$

The character of this matrix amounts to[1,2]

$$\chi(d) = \frac{\sin(d \cdot \phi)}{\sin \phi} \quad d = 1, 2, 3, \ldots.$$

As the integral

$$\mathcal{M} \frac{\sin(d \cdot \phi)}{\sin \phi}$$

is equal to 1, the representations $[d]$ are irreducible.

---

[1] To compute the sum, multiply the desired sum $\chi(d)$ with $e^{i \cdot 2 \cdot \phi}$ and subtract them:

$$e^{i \cdot 2 \cdot \phi} \chi(d) - \chi(d) = e^{i\phi(d+1)} - e^{-i\phi(d-1)}$$

Multiply both sides with $e^{-i\phi}$ and solve for $\chi(d)$.

[2] We generalize the character formula to $d = 1$, which corresponds to the irreducible representation "1" (▢).

Proof of (ii): Assume there exists an irreducible representation $\Gamma \neq [d]$ which is also irreducible. Then,

$$\mathscr{M} \frac{\sin(d \cdot \phi)}{\sin \phi} = \frac{2}{\pi} \int_0^\pi d\phi \cdot \sin \phi \cdot \sin(d \cdot \phi) \Gamma(\phi) = 0 \quad \forall d$$

But the set $\sin(\phi), \sin 2\pi, \ldots$ constitutes a complete set in the space of continuous functions on the interval $0, \pi$ with $f(\pi) = f(0)$, so that $\Gamma(\phi) = 0$. As the character of an irreducible representation cannot vanish identically, Γ is not irreducible, contrary to the assumptions.

Comment. One can also use, as a parameter for the group SU_2, the rotation angle $\varphi = 2\phi$. The Cartan weights of the representation $[d]$ are written, accordingly, as

$$\frac{d-1}{2} \quad \frac{d-3}{2} \quad \cdots \quad \frac{-(d-1)}{2}$$

One recognizes that

$$[d] \quad \leftrightarrow \quad D^{\frac{d-1}{2}}$$

with $D^{\frac{d-1}{2}}$ being the notation we used for the group SO_3^D.

The relation between SU_2 and SO_3. We have found in Chapter 7 that, in the $D^{\frac{1}{2}}$ representation, one element of SO_3 was associated with two distinct elements of SU_2, thus producing a double-valued representation of SO_3. This double-valuedness was amended by Bethe with his double-group construction. An alternative way to finding the relation between SU_2 and SO_3 is the so-called Cayley[3] parametrization of SO_3. For the purpose of demonstrating this parametrization, we construct explicitly the representation [3] of SU_2 and show that it is the natural representation of SO_3. Within the space of the monomials $x_1^2, x_2^2, x_1 \cdot x_2$, we choose the basis polynomials

$$X_1(x_1, x_2) \doteq \frac{x_1^2 - x_2^2}{2} \quad X_2(x_1, x_2) \doteq \frac{x_1^2 + x_2^2}{2i} \quad X_3(x_1, x_2) \doteq x_1 \cdot x_2$$

and transform them with

$$\begin{pmatrix} x_1 \\ x_2 \end{pmatrix} = \underbrace{\begin{pmatrix} \overline{a} & -b \\ \overline{b} & a \end{pmatrix}}_{D_{\frac{1}{2}}^{-1}} \begin{pmatrix} x_1' \\ x_2' \end{pmatrix}$$

[3] A. Cayley, "Sur quelques propriétés des déterminants gauches." **32**, 119–123 (1846); https://doi.org/10.1515/crll.1846.32.119.

to

$$X_1 = \frac{1}{2}\left[x_1'^2(\bar{a}^2 - \bar{b}^2) - x_2'^2(a^2 - b^2) + 2x_1'x_2'(\bar{a}b + a\bar{b})\right]$$

$$X_2 = \frac{1}{2i}\left[x_1'^2(\bar{a}^2 + \bar{b}^2) + x_2'^2(a^2 + b^2) + 2x_1'x_2'(a\bar{b} - \bar{a}b)\right]$$

$$X_3 = \left[x_1'^2(\bar{a}\bar{b}) - x_2'^2(ab) + x_1'x_2'(a\bar{a} - \bar{b}b)\right]$$

The expressions on the right-hand side must be arranged to show their linearity in X_1', X_2', X_3'. After some algebra, we find

$$X_1 = \frac{1}{2}[(\bar{a}^2 - \bar{b}^2) + (a^2 - b^2)]X_1'$$
$$+ \frac{i}{2}[(\bar{a}^2 - \bar{b}^2) - (a^2 - b^2)]X_2' + (\bar{a}b + a\bar{b})X_3'$$

$$X_2 = \frac{1}{2i}[(\bar{a}^2 + \bar{b}^2) - (a^2 + b^2)]X_1'$$
$$+ \frac{1}{2}[(\bar{a}^2 + \bar{b}^2) + (a^2 + b^2)]X_2' + \frac{1}{i}(-\bar{a}b + a\bar{b})X_3'$$

$$X_3 = [\bar{a}\bar{b} + ab]X_1' + i[\bar{a}\bar{b} - ab]X_2' + (\bar{a}a - b\bar{b})X_3'$$

or, using

$$a = \alpha_1 + i \cdot \alpha_2 \quad b = \alpha_3 + i \cdot \alpha_4$$

$$\begin{pmatrix} X_1 \\ X_2 \\ X_3 \end{pmatrix} = \underbrace{\begin{pmatrix} \alpha_1^2 - \alpha_2^2 - \alpha_3^2 + \alpha_4^2 & 2(-\alpha_1\alpha_2 - \alpha_3\alpha_4) & 2(\alpha_1\alpha_3 - \alpha_2\alpha_4) \\ 2(\alpha_1\alpha_2 - \alpha_3\alpha_4) & \alpha_1^2 - \alpha_2^2 + \alpha_3^2 - \alpha_4^2 & 2(\alpha_2\alpha_3 + \alpha_1\alpha_4) \\ 2(\alpha_1\alpha_3 + \alpha_2\alpha_4) & 2(\alpha_2\alpha_3 - \alpha_1\alpha_4) & \alpha_1^2 + \alpha_2^2 - \alpha_3^2 - \alpha_4^2 \end{pmatrix}}_{D_1^{-1}}$$

$$\times \begin{pmatrix} X_1' \\ X_2' \\ X_3' \end{pmatrix}$$

Comments.

1. The transformation matrix is an orthogonal 3×3 matrix with a determinant of 1 and depends on three parameters. By definition, the set of such matrices is SO_3.

2. The D^1 matrices depend quadratically on the parameters $\alpha_1, \alpha_2, \alpha_3, \alpha_4$, while the matrices $D^{\frac{1}{2}}$ depend linearly on them. Accordingly, the map associates to one matrix D^1 the two matrices $\pm D^{\frac{1}{2}}$.

3. The diagonal matrix

$$\begin{pmatrix} e^{-i\phi} & 0 \\ 0 & e^{i\phi} \end{pmatrix}$$

representing the elements of the conjugacy classes is mapped by the Cayley parametrization into

$$\begin{pmatrix} \cos(2\phi) & -\sin(2\phi) & 0 \\ \sin(2\phi) & \cos(2\phi) & 0 \\ 0 & 0 & 1 \end{pmatrix}$$

which describes a rotation about the z-axis by $\varphi = 2\phi$.

8.1.5 *Irreducible representations of SU_3*

The group SU_3 of all 3×3 unitary matrices with a determinant of 1 is important in problems of particle physics. The irreducible representations are described by a YD with three lines at the most, $l_1 \geq l_2 \geq l_3$, with $l_1 \geq 1$, $l_2 \geq 0$, $l_3 \geq 0$. Vertically aligned boxes carry a tensor product that transforms according to the determinant $(= 1)$, i.e. are left invariant. Accordingly, the identical representation is

The representations are often denoted by mentioning their dimension.

Examples of irreducible representations and computations with them are given in the following:

1.

$$[1] \equiv \begin{array}{|c|} \hline \\ \hline \\ \hline \end{array}$$

is the trivial representation.
2. $[3] \equiv \boxed{}$ has three YTs: $\boxed{1}, \boxed{2}, \boxed{3}$. It is the three-dimensional natural representation of SU_3.
3. There is a further three-dimensional representation of SU_3, defined by the YD $\begin{array}{|c|}\hline\\\hline\\\hline\end{array}$. This YD has the three basis states $\begin{array}{|c|}\hline 1\\\hline 2\\\hline\end{array}, \begin{array}{|c|}\hline 1\\\hline 3\\\hline\end{array}, \begin{array}{|c|}\hline 2\\\hline 3\\\hline\end{array}$. The difference

with respect to [3] is apparent when one computes their character. Any unitary matrix with a determinant of 1 can be diagonalized to

$$\begin{pmatrix} e^{i\theta_1} & 0 & 0 \\ 0 & e^{i\theta_2} & 0 \\ 0 & 0 & e^{i\theta_3} \end{pmatrix}$$

with $\theta_1 + \theta_2 + \theta_3 = 0$. The trace in the representation [3] is, accordingly,

$$e^{i\theta_1} + e^{i\theta_2} + e^{i\theta_3}$$

The matrix representing the diagonalized matrix in the basis set $\boxed{\frac{1}{2}}, \boxed{\frac{1}{3}}, \boxed{\frac{2}{3}}$ is

$$\begin{pmatrix} e^{i(\theta_1+\theta_2)} & 0 & 0 \\ 0 & e^{i(\theta_1+\theta_3)} & 0 \\ 0 & 0 & e^{i(\theta_2+\theta_3)} \end{pmatrix}$$

and the character is, accordingly,

$$e^{i(\theta_1+\theta_2)} + e^{i(\theta_1+\theta_3)} + e^{i(\theta_2+\theta_3)}$$

This character can be rewritten, taking the relation $\theta_1 + \theta_2 + \theta_3 = 0$ into account, as

$$e^{-i\theta_3} + e^{-i\theta_1} + e^{-i\theta_2}$$

and is the complex conjugate of the character in the representation [3]. The representation $\boxminus$ is therefore denoted as $[\bar{3}]$.

4. Further representations to remember are

$$\text{(tableau)} \equiv [8] \qquad \text{(tableau)} \equiv [10]$$

8.2 Applications to Particle Physics

(Parts of this section are a personal, tutorial rendering of the particle physics topics treated in the (unpublished) lecture notes of Prof. G. M. Graf on the subject of symmetry.)

8.2.1 *Introduction*

In 1932, Heisenberg[4] introduced an important principle that has since governed the development of fundamental physics (we will later see, for example, that the quark hypothesis by Murray Gell-Mann (1964) originates within this principle): Particles must be regarded as basis states transforming according to some irreducible representation of the symmetry group of the Hamiltonian (or the S matrix). As such, it must be possible to assign them a "Cartan" weight, i.e. to group them into so-called "multiplets" (a "multiplet" is a word used to designate the set of basis states transforming according to a certain irreducible representation). The known particles at that time and until about 1950 (electron, positron, neutron, proton, and π-mesons) were characterized by the charge Q. If we were to attempt using this principle to classify these particles, we would consider Q as the "Cartan weight" $Q = 0, \pm 1, \pm 2, \ldots$ designating the irreducible representation Γ^Q of the group $U(1)$ and set e, p, and π^+ as transforming according to Γ^{+1}, n and π^0 as transforming according to Γ^0, and π^- as transforming according to Γ^{-1}. A principle is, of course, only useful if it "predicts" something. In the specific case, the symmetry $U(1)$ underlying our Cartan weight assignment can be used to compute matrix elements of the type

$$\langle \psi_f | S | \psi_i \rangle$$

ψ_i is the initial state wave function consisting of the tensor product of the two particles that undergo a scattering event. Each particle transforms according to an irreducible representation Γ^Q, so that the initial state transforms according to $\Gamma^{Q_1} \otimes \Gamma^{Q_2} = \Gamma^{Q_1 + Q_2}$ The operator S is required to be $U(1)$-invariant, i.e. to transform according to the irreducible representation Γ^0. In virtue of the Wigner–Seitz theorem, a selection rule for this matrix element to be non-vanishing is that

$$\Gamma^{Q_f} = \Gamma^{Q_1 + Q_2} \quad \text{i.e.} \quad Q_f = Q_1 + Q_2$$

Using the principle of assigning particles to some irreducible representations, we have obtained the charge conservation law, which is certainly

---

[4]W. Heisenberg, "Über den Bau der Atomkerne. I", *Z. Phys.* **77**, 1–11 (1932), https://doi.org/10.1007/BF01342433.

a sound scientific result. As an example, we are able to predict that the scattering process

$$\pi^+ + p \to \pi^+ + p$$

is certainly possible, but when one scatters π^- and p, two scattering channels open:

$$\pi^- + p \to \pi^- + p \quad \text{and} \quad \pi^- + p \to \pi^0 + n$$

as experimentally observed.

Among the hadrons (which are particles interacting through the strong interaction), there are particles that have similar masses. For example, the neutron (n) and proton (p) have masses of around 940 Mev, and the three pions π^-, π^0, π^+, have masses of around 140 MeV. This symmetry was used by Heisenberg to conjecture that n, p, and π^-, π^0, π^+ are multiplets of some irreducible representations with dimensionalities higher than one (two and three, respectively). This required the introduction of a more complex symmetry group that contains $U(1)$ as a subgroup, allowing the desired irreducible representations to decompose when the more complex group is limited to the subgroup $U(1)$. This relationship is required by the fact that charge conservation is a sound scientific result that must remain valid even if a more complex symmetry group is introduced. An educated guess about this more complex symmetry group was formulated by Heisenberg: $SU(2)$. Accordingly, particles should be classified as being basis states $|I, I_3\rangle$ $(I_3 = I, I - 1, \ldots, -I)$ of a isospin multiplet transforming according to the representation D_I of $SU(2)$. Specifically, the two nucleons are the states

$$n = Y_{\frac{1}{2}}^{-\frac{1}{2}} \equiv \left| \frac{1}{2}, \frac{\bar{1}}{2} \right\rangle \quad p = Y_{\frac{1}{2}}^{\frac{1}{2}} \equiv \left| \frac{1}{2}, \frac{1}{2} \right\rangle$$

transforming according to the representation $D_{\frac{1}{2}}$ of the group $SU(2)$ and the Cartan weights $I_3 = -\frac{1}{2}$ and $+\frac{1}{2}$, respectively. The three mesons are the states

$$\pi^- = |1, \bar{1}\rangle \quad \pi^0 = |1, 0\rangle \quad \pi^1 = |1, 1\rangle$$

transforming according to D_1. Regarding particles as basis states (symmetry-adapted vectors) of a multiplet of a certain group is equivalent to regarding the Hamilton operator (or, more generally, the S matrix (entering the matrix elements between these states) as invariant with respect to the group. The isospin hypothesis by Heisenberg therefore entails that the strong interaction is $SU(2)$-invariant.

The group $SU(2)$ has an attractive property: It contains the set of diagonal matrices

$$\begin{pmatrix} e^{i\frac{\varphi}{2}} & 0 \\ 0 & e^{-i\frac{\varphi}{2}} \end{pmatrix}$$

as a subgroup ($\varphi \in [0, 4\pi]$). This set of matrices is isomorphous to $U(1)$, which allows the embedding of $U(1)$ in the larger group. The link between the components Γ^{I_3} appearing in the decomposition $D_I | U(1) = \oplus_{I_3} \Gamma^{I_3}$ and Γ^Q is realized if one defines the relation

$$Q = I_3 + \frac{1}{2}$$

between the Cartan weights Q and I_3. This recalibration of Cartan weights is equivalent to using a different parameter set for describing $U(1)$. Within the framework of using symmetry groups to classify particles, the "charge" is regarded as a "symmetry-breaking field" that reduces the original symmetry group to a subgroup.

8.2.2 *Experimental proof of SU_2-symmetry*

The question now is whether the isospin hypothesis, based on the observation of mass degeneracy, provides useful information (beyond charge conservation) that can be verified experimentally. An important prediction of the isospin hypothesis relates to the total cross-section for pion–proton scattering. In a π–p scattering experiment, a pion beam is directed toward a sample consisting of hydrogen gas, with protons acting as scattering centers. The scattering cross-section of a process that leads to an initial state "in" $(p + \pi^{\bar{1},0,1})$ to a final state "out" is, in general, related to the S matrix by

$$\sigma(\text{in} \to \text{out}) \propto \left| \langle \text{out} | S | \text{in} \rangle \right|^2$$

If the kinetic energy of the pions is taken to be large enough so that scattering by the Coulomb interaction can be neglected (i.e. the charge is not important), the S-matrix operator will involve only the strong interaction between pions and protons. A simple model of the S-matrix operator that explicitly shows its SU_2-invariance is

$$S \propto \mathbf{I}_1 \cdot \mathbf{I}_2$$

where $\mathbf{I}_1$ is the isospin operator of the incoming pion and $\mathbf{I}_2$ is the isospin operator of the p target. We now show that the matrix elements of S can be computed using arguments based solely on the SU_2-symmetry of the S matrix.

We make use of the SU_2-invariance of the S matrix within the framework of the Wigner–Eckart theorem, which simplifies the computation of the elements of an SU_2-invariant operator if the states involved are symmetry adapted. In the specific case, the desired states, symmetry adapted to the operator S within the set of product states

$$\{|\pi^-, p\rangle, |\pi^-, n\rangle, |\pi^0, p\rangle, |\pi^0, n\rangle, |\pi^+, p\rangle, |\pi^+, n\rangle\}$$

are those that block-diagonalize the tensor product representation

$$D^1 \otimes D^{\frac{1}{2}} = D^{\frac{3}{2}} \oplus D^{\frac{1}{2}}$$

From a table of Clebsch–Gordan coefficients (see end of Lecture 7 for the appropriate table), we can obtain the necessary relations (experimentally, mainly the π^+ and π^- beams are used; the lifetime of the π^0 beam is too short to be organized into a useful beam):

$$|\pi^+ p\rangle = \left|\frac{3}{2}, \frac{3}{2}\right\rangle$$

$$|\pi^- p\rangle = \sqrt{\frac{1}{3}}\left|\frac{3}{2}, -\frac{1}{2}\right\rangle + \sqrt{\frac{2}{3}}\left|\frac{1}{2}, -\frac{1}{2}\right\rangle$$

$$|\pi^0 n\rangle = \sqrt{\frac{2}{3}}\left|\frac{3}{2}, -\frac{1}{2}\right\rangle - \sqrt{\frac{1}{3}}\left|\frac{1}{2}, -\frac{1}{2}\right\rangle$$

We compute now the matrix elements of S using the development of the $|\pi, p>$-states in terms of the symmetry-adapted states:

$$\sigma(\pi^+ p \to \pi^+ p) \propto \left| \left\langle \frac{3}{2}, \frac{3}{2} \right| S \left| \frac{3}{2}, \frac{3}{2} \right\rangle \right|^2$$

$$\doteq |a_{\frac{3}{2}}|^2$$

$$\sigma(\pi^- p \to \pi^- p) \propto \left| \left\langle \frac{3}{2}, -\frac{1}{2} \right| \sqrt{\frac{1}{3}} + \left\langle \frac{1}{2}, -\frac{1}{2} \right| \sqrt{\frac{2}{3}} |S| \sqrt{\frac{1}{3}} \left| \frac{3}{2}, -\frac{1}{2} \right\rangle \right.$$

$$\left. + \sqrt{\frac{2}{3}} \left| \frac{1}{2}, -\frac{1}{2} \right\rangle \right|^2$$

$$= \left| \frac{1}{3} a_{\frac{3}{2}} + \frac{2}{3} a_{\frac{1}{2}} \right|^2$$

$$\sigma(\pi^- p \to \pi^0 n) \propto \left| \left\langle \frac{3}{2}, -\frac{1}{2} \right| \sqrt{\frac{2}{3}} \right.$$

$$-\left\langle \frac{1}{2}, -\frac{1}{2} \middle| \sqrt{\frac{1}{3}} \middle| S \middle| \frac{3}{2}, -\frac{1}{2} \right\rangle \sqrt{\frac{1}{3}} + \middle| \frac{1}{2}, -\frac{1}{2} \right\rangle \sqrt{\frac{2}{3}} \right\rangle \Big|^2$$

$$= \left| \frac{\sqrt{2}}{3} a_{\frac{3}{2}} - \frac{\sqrt{2}}{3} a_{\frac{1}{2}} \right|^2$$

The scattering process can proceed along two channels, governed by the Wigner–Eckart constants $a_{\frac{3}{2}}$ and $a_{\frac{1}{2}}$. These constants contain details of the scattering matrix elements that cannot be ascertained using symmetry arguments. So, in principle, one cannot compute the matrix elements without numerical work and without knowing about the details of the interaction. However, we might be interested at computing

$$\frac{\sigma(\pi^+ + p \to \text{anything})}{\sigma(\pi^- + p \to \text{anything})}$$

at such an energy where one channel (e.g. the $I = \frac{3}{2}$ channel) dominates, and in this case, the Wigner–Eckart constants cancel out, leading to a precise prediction of the ratio of the cross-sections from symmetry arguments alone (precisely, from the knowledge of the Clebsch–Gordan coefficients). For the specific case considered, at an energy of 1236 Mev, one observes indeed a resonant behavior of the total cross-section. Assuming that the scattering process is dominated by the $I = \frac{3}{2}$ channel ($a_{\frac{3}{2}} \gg a_{\frac{1}{2}}$)

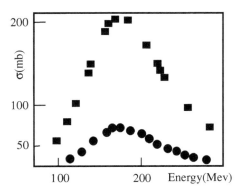

The total cross-sections σ for scattering processes involving pions, as a function of the pion kinetic energy, reproduced freely from A. A. Carter *et al.*, *Nucl. Phys. B* **26** 445–460 (1971): Black squares denote the $\pi^+ p$ total cross-section, and black circles denote the $\pi^- p$ total cross-section. At resonance, the predicted factor of 3 on the base of the isospin model is observed.

we compute

$$\frac{\sigma(\pi^+ + p \to \pi^+ + p)}{\sigma(\pi^- + p \to \pi^- + p) + \sigma(\pi^- + p \to \pi^0 + n)} = \frac{1}{\frac{1}{9} + \frac{2}{9}} = 3$$

Experimentally, at resonance, one observes indeed the expected ratio (see, for example, A. A. Carter *et al.*, *Nucl. Phys. B* **26** 445–460 (1971)).

The short-lived particle produced at resonance is a hadron called Δ with $I = \frac{3}{2}$. The weights of this isospin multiplet correspond to the particles $\Delta^-, \Delta^0, \Delta^+, \Delta^{++}$.

8.2.3 *The hadron particle multiplets: The early 1960s*

After the discovery of proton, neutron, and π-mesons, further hadrons where observed in laboratory experiments and in cosmic radiation. By the early 1060s, the known particles could be grouped together into systematic schemes that could not be explained as multiplets of the group SU_2. In the first place, the baryon number B could be introduced as a further quantum label (besides I and I_3) that was conserved during all observed scattering or particle generation processes. B gives the number of protons that appear when a particle decays. For instance, $B = 0$ for the π-mesons, and $B = 1$ for p and n (remember that the neutron has a finite lifetime and decays into a proton, an electron, and a neutrino). The Δ particles have a baryon number of $B = 1$. The addition of a further quantum number, according to group theory, indicates the existence of a further commuting infinitesimal generator and calls for the expansion of the symmetry group beyond SU_2. However, it turned out that B is not describing the new degree of freedom completely. Further hadrons were discovered in cosmic radiation, such as the Λ^0 baryon and the K^0-meson. The Λ^0, for instance, decays into a π^- and a proton, i.e. it has $B = 1$. However, its lifetime was about 10^{-11} s, i.e. 10^{11} times larger than the lifetime of a Δ particle, which has $B = 1$ and a similar mass. Pais, Gell-Mann, and Nishijima, around 1953, called them "strange particles" and introduced a new quantum number (besides B) — the strangeness number S (conserved by the strong interaction) — to explain their long lifetimes. The "strangeness" hypothesis predicts that processes involving the strong interaction can generate strange particles, such as $\pi^- + p \to K^0 + \Lambda^0$. K^0 is assigned $S = -1$, and Λ^0 is assigned $S = +1$. Their decay into non-strange $(S = 0)$ particles $(K^0 \to \pi^+ + \pi^-, \Lambda^0 \to \pi^- + p)$ is so slow because it can only happen via a non-S-conserving interaction (called "weak" interaction). It turned out that despite the introduction of the new quantum number S, one parameter beyond (I, I_3) is enough to organize most of the known particles: the hypercharge $Y = B + S$. Accordingly, the known particles were grouped into sets which could be described quite accurately in a $(I_3 - Y)$-diagram, for a given spin quantum number, parity, and baryon quantum number (the charge Q is given by $I_3 + \frac{1}{2}Y$).

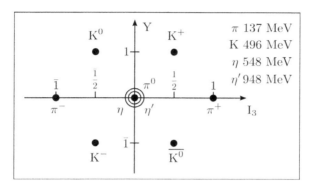

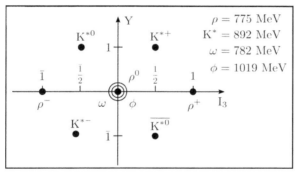

$Y - I_3$ diagrams containing the spin-0 mesons (top) and the spin-1 mesons (bottom).

The top diagram contains the nine pseudo-scalar mesons (with a Baryon number of 0, parity of -1, and spin of 0). The similarity of the masses along horizontal lines allows us to recognize one SU_2 multiplet $D^{\frac{1}{2}}$ at $Y = 1$, the π-D^1 multiplet at $Y = 0$, and a further SU_2 multiplet $D^{\frac{1}{2}}$ at $Y = -1$. One also suspects two D^0 singlets residing at $Y = 0$. This assignment suggests that SU_2 continues to be an approximate symmetry, but it must be certainly expanded to a larger group to include the hypercharge. A possibility is the group $SU_2 \times U(1)$. In this scenario, the hypercharge quantum number is generated by an irreducible representation of $U(1)$. Note that the masses of the most recently discovered particles are different from the mass of the π-mesons. Looking at this diagram, one concludes that either the expanded group does not need to have an irreducible representation containing all nine particles or that, at these lower masses, the putative irreducible representation of the putative "new" group is reduced into smaller components, provided by the irreducible representations of $SU_2 \times U(1)$. The bottom diagram are the vector mesons (with a spin of 1, positive parity, and baryon number of 0). Again, one recognizes isospin as an approximate symmetry. In addition, eight of the nine particles have a similar mass. These eight particles in this diagram might therefore be described by a eight-dimensional irreducible representation of a group that contains $SU_2 \times U(1)$ as a subgroup. At these larger energies, this representation does not appear to decompose into the irreducible components of the subgroup.

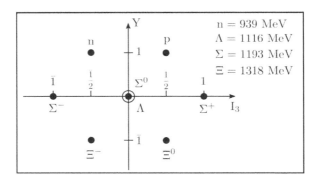

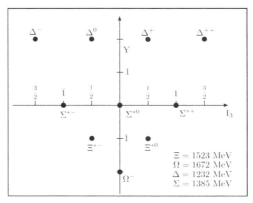

$Y - I_3$ diagrams containing the spin-$\frac{1}{2}$ baryons (top) and the spin-$\frac{3}{2}$ baryons (bottom).

The two diagrams in the above figure show the grouping of particles with $B = 1$. The top one in the figure hosts the eight (lighter) baryons with a spin of $\frac{1}{2}$ and parity of $+1$. The bottom one contains 10 particles with a spin of $\frac{3}{2}$ and parity of $+1$. Within the two groupings, the masses are similar, suggesting the scenario of a new group with eight- and ten-dimensional irreducible representations and with $SU_2 \times U(1)$ as subgroup. Note that each meson diagram contains both particles and antiparticles. The anti-baryons are described by two separate diagrams (not shown). There is an important point regarding the fourth diagram: The Ω^--particle is included here, but, at the time of constructing these (on the base of phenomenological observations), the Ω^--particle was not observed, and the heavy-hadron diagram contained only the nine known particles. Precisely, the "story" of this particle is very relevant to the development of modern particle physics, as explained in the following.

8.2.4 *The eightfold way*

The irreducible representations of the group $SU_2 \times U(1)$ host all particles observed in the four diagrams, but the idea is that these are components resulting from the breaking of the irreducible representations of a larger group that includes $SU_2 \times U(1)$ as a subgroup. Gell-Mann and Ne'eman $(1961)^5$ suggested that this larger group describing the symmetry of the Hamiltonian involving the strong interaction is the group SU_3.

Why SU_3? This group has a mathematical structure that would explain most of the observed particles.

- First, it has the eight- and ten-dimensional irreducible representations (called [8] and [10] in the previous section) that can (almost) explain the observed hadrons diagrams with eight and nine particles.
- Second, SU_3 contains the subgroup

$$\begin{pmatrix} SU_2 \cdot e^{i\theta} & 0 \\ 0 & e^{-2i\theta} \end{pmatrix} \quad \theta \in [0, 2\pi]$$

which is isomorphic to $SU_2 \times U(1)$. This subgroup plays in SU_3 the same role played by $U(1)$ in the isospin group SU_2.

---

[5]M. Gell-Mann, *The Eightfold Way: A Theory of Strong Interaction Symmetry*, Synchrotron Laboratory, California Institute of Technology, Pasadena, CA, 1961, https://www.osti.gov/biblio/4008239; Y. Ne'eman, "Derivation of strong interactions from a gauge invariance", *Nucl. Phys.* (North-Holland Publishing Co., Amsterdam) **26** 222–229, https://www.sciencedirect.com/science/article/pii/0029558261901341.

- Third, the irreducible representations of the group $SU_2 \times U(1)$ are written as

$$D^j \otimes e^{i \cdot n \cdot \theta} \equiv D_j^n$$

with $j = 0, \frac{1}{2}, \ldots$ and $n = 0, \pm 1, \pm 2, \ldots$. The irreducible representations of SU_3 are reducible when limited to the subgroup $SU_2 \times U(1)$. For instance, we have

$$[3] = D_{\frac{1}{2}}^1 \oplus D_0^{-2} \quad [\bar{3}] = D_{\frac{1}{2}}^{-1} \oplus D_0^2$$

$$[3] \otimes [3] = (D_{\frac{1}{2}}^1 \oplus D_0^{-2}) \otimes (D_{\frac{1}{2}}^1 \oplus D_0^{-2}) = D_1^2 \oplus 2D_{\frac{1}{2}}^{-1} \oplus D_0^2 \oplus D_0^{-4}$$

$$[3] \otimes [\bar{3}] = (D_{\frac{1}{2}}^1 \oplus D_0^{-2}) \otimes (D_{\frac{1}{2}}^{-1} \oplus D_0^2) = D_1^0 \oplus D_{\frac{1}{2}}^3 \oplus D_{\frac{1}{2}}^{-3} \oplus 2D_0^0$$

so that

$$[6] = D_1^2 \oplus D_{\frac{1}{2}}^{-1} \oplus D_0^{-4} \quad [8] = D_1^0 \oplus D_{\frac{1}{2}}^3 \oplus D_{\frac{1}{2}}^{-3} \oplus D_0^0$$

From $[3] \otimes [3] \otimes [3]$, we obtain

$$[10] = D_{\frac{3}{2}}^6 \oplus D_1^0 \oplus D_{\frac{1}{2}}^{-3} \oplus D_0^{-6}$$

In view of these decomposition results, one can assign precisely the irreducible components D_j^n to the particles building the four diagrams above by setting $Y = \frac{n}{3}$.

The possibility of assigning the particle to the representations D_j^{3Y} might indicate that the group $SU_2 \times U(1)$ is enough to describe the symmetry of the strong interaction. The larger irreducible representations are not necessarily compelling, particularly in view of the fact that the masses within the larger multiplets are similar but not identical. However, the scenario with $SU_2 \times U(1)$ and D_j^n has a drawback: Based on symmetry alone, there is no need for further components to appear. For instance, the second baryon diagram could be regarded, at the time of the SU_3-hypothesis proposed by Gell-Mann, as an ensemble of nine D_j^{3Y} components: The particle Ω^- was yet to be discovered. The decuplet hypothesis inherent to SU_3 led Gell-Mann to postulate its existence. It was finally discovered in 1964, thus confirming that SU_3 is the exact symmetry group.

The scenario described so far explains the four hadron diagrams almost completely and predicted the existence of a particle which was later discovered. However, there remained few unanswered questions. The most crucial one was: Why are other irreducible representations of SU_3 not observed (in particular, the natural representations $[3]$)?

8.2.5 *The quark hypothesis*

To answer this question, Gell-Mann (M. Gell-Mann, "A schematic model of baryons and mesons", *Physics Letters* **8** 214 (1964); https://www.science direct.com/science/article/pii/S0031916364920013) proposed, in 1964, the quark hypothesis: spin-$\frac{1}{2}$ particles in the representation [3] (and antiquarks in [$\bar{3}$]) of SU_3. As

$$[3] = D^{\frac{1}{3}}_{\frac{1}{2}} \oplus D^{-\frac{2}{3}}_0$$

we must assign $Y = \frac{1}{3}$ and $Y = -\frac{2}{3}$, respectively, to the basis states of [3], so that the $(Y - I_3)$ diagrams representing the three basis states of [3] (and [$\bar{3}$]) are constructed as follows.

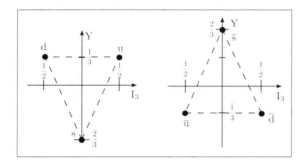

$Y - I_3$ diagrams containing the quarks and antiquarks constituting the [3] and [$\bar{3}$] irreducible representations of SU_3.

The following table summarizes the properties of quarks and antiquarks.

Name	Symbol	J	I	I_3	B	S	Y	Q
up	u	$\frac{1}{2}$	$\frac{1}{2}$	$\frac{1}{2}$	$\frac{1}{3}$	0	$\frac{1}{3}$	$\frac{2}{3}$
down	d	$\frac{1}{2}$	$\frac{1}{2}$	$\frac{1}{2}$	$\frac{1}{3}$	0	$\frac{1}{3}$	$\frac{1}{3}$
strange	s	$\frac{1}{2}$	0	0	$\frac{1}{3}$	$\bar{1}$	$\frac{2}{3}$	$\frac{1}{3}$
antiup	$\bar{u}$	$\frac{1}{2}$	$\frac{1}{2}$	$\frac{1}{2}$	$\frac{1}{3}$	0	$\frac{1}{3}$	$\frac{2}{3}$
antidown	$\bar{d}$	$\frac{1}{2}$	$\frac{1}{2}$	$\frac{1}{2}$	$\frac{1}{3}$	0	$\frac{1}{3}$	$\frac{1}{3}$
antistrange	$\bar{s}$	$\frac{1}{2}$	0	0	$\frac{1}{3}$	1	$\frac{2}{3}$	$\frac{1}{3}$

The basis labels u, d, and s are referred to as the "flavors" of the quarks. Quark with the properties entailed by the [3]-irreducible representations were finally inferred from experimental observations.

There is a further aspect of the quark hypothesis that is particularly intriguing: Because the mathematical structure of $SU(3)$ allows for performing the tensor product of [3]-irreducible representations, it is possible that baryons are composed of three quarks (with each $B = \frac{1}{3}$) and mesons of a quark and antiquark (giving $B = 0$). Let us see how the quark hypothesis relates to the observed diagrams. First, we consider the mesons. The tensor product

$$[3] \otimes [\overline{3}] = [1] \oplus [8]$$

explains the two observed nine-particle diagrams, which accordingly consist of an octet and a singlet (having either the spin quantum numbers of 0 or 1). There is a member of the representation [8] with the same quantum numbers $(Y = 0, I = 0, I_3 = 0)$ as the [1] meson. These two states are orthogonal, just as the $(l = 0, s = 0)$-state of the D^0 representation of SU_2 is orthogonal to the $(l = 0, s = 0)$-state of the D^1 representation. The observed physical particles (η, η') and (in the spin-1 diagram) (ω, ϕ) are some linear combination of these basis states.

Now, consider the baryons. From

$$[3] \otimes [3] \otimes [3] = [10] \oplus [8] \oplus [8] \oplus [1]$$

one obtains the required baryon octets and decuplets.

The SU_3 hypothesis predicts quarks, and explains how mesons and baryons can be constructed from $q\bar{q}$ and qqq states, respectively. This hypothesis has a lengthy history since its formulation. One important aspect is that physicists were confronted with apparent contradictions, and their solutions did animate to new ideas. For example, it was soon realized (after the quark hypothesis was formulated) that the Δ^{++}-baryon is the state $|u \uparrow\rangle \, |u \uparrow\rangle \, |u \uparrow\rangle$. This is the completely symmetric $(\frac{3}{2}, +\frac{3}{2})$-spin state with vanishing orbital momentum (s-orbital state). As such, this particle violates the Pauli principle. This contradiction was solved in 1973[6] by introducing a further degree of freedom of the quarks other than their flavors, namely their "color": r for "red", g for "green", and b for "blue".

The Cartan weights of the [3] representation of SU_3

The quark states providing the basis state of the [3] representation of SU_3 are described by a set of quantum numbers (Y, I_3), which we have inferred from the decomposition of representations of SU_3 that describe the mesons

---

[6]A number of authors contributed to this development. Among them are O. Greenberg, Moo-Young Han, Yoichiro Nambu, William Bardeen, Harald Fritzsch, and Murray Gell-Mann.

and the hadrons when these representations are restricted to the subgroup $SU_2 \times U(1)$. Using this translation key, we have found, for example, the fractional values of Y for the quark states. It would be interesting to obtain the quantum numbers describing the quark states from the Cartan weights of the representation [3].

What do we mean by "Cartan weights"? We recall the concept of Cartan weights (or Cartan multiplicators) that was introduced when seeking the irreducible representations of a similar continuous group SO_3. Given an irreducible representation D^l of SO_3, one can find the Cartan weights by taking the representation matrices of the Abelian subgroup SO_2 and expanding them for small angles as

$$D^l(d\varphi) \approx \mathbb{1} - i \cdot I_3 \cdot d\varphi$$

The Cartan weights are along the diagonal of the infinitesimal generator I_3 and amount to $l, l-1, \ldots, -l$. They are used to specify the basis functions $|l, m >$ carrying the irreducible representation D^l. When the alternative technique of Lie algebra is used, the Cartan weights are found directly using suitable means (algebraic or analytic, for instance), and the matrix representations at finite angles are then constructed via exponential maps. We now search for the quark quantum numbers (Y, I_3) using the Abelian subgroup strategy.

The Abelian subgroups of SU_3 we have in mind is that of the diagonal unitary matrices with a determinant of 1. The immediate choice is

$$\begin{pmatrix} e^{i\theta_1} & 0 & 0 \\ 0 & e^{i\theta_2} & 0 \\ 0 & 0 & e^{i\theta_3} \end{pmatrix}$$

The condition of having the determinant as 1 is realized by taking $\theta_3 = -\theta_1 - \theta_2$, which reduces the subgroup to a continuous group with two parameters, θ_1 and θ_2. We, however, have some freedom in choosing the parameters describing the group, and we use this freedom while keeping the final result in mind, i.e.

$$\left[Y = \frac{1}{3}, I_3 = \pm\frac{1}{2} \right] \quad \left[Y = -\frac{2}{3}, I_3 = 0 \right]$$

With a suitable choice of parameters, the diagonal matrix can be written, without loss of generality, as

$$\begin{pmatrix} e^{-i\left(\frac{\theta_1}{2} + \frac{\theta_2}{3}\right)} & 0 & 0 \\ 0 & e^{-i\left(\frac{-\theta_1}{2} + \frac{\theta_2}{3}\right)} & 0 \\ 0 & 0 & e^{-i\frac{2\cdot\theta_2}{3}} \end{pmatrix}$$

We now rewrite this matrix for small angles, with the aim of determining the Cartan weights. We have two parameters, so we have two generators, and the Cartan weights will be in pairs (in SO_3, the subgroup had only one parameter, one infinitesimal generator, and one set of Cartan weights):

$$\mathbb{1} - i \cdot \begin{pmatrix} \frac{1}{2} & 0 & 0 \\ 0 & -\frac{1}{2} & 0 \\ 0 & 0 & 0 \end{pmatrix} d\theta_1 - i \cdot \begin{pmatrix} \frac{1}{3} & 0 & 0 \\ 0 & \frac{1}{3} & 0 \\ 0 & 0 & -\frac{2}{3} \end{pmatrix} d\theta_2$$

From these matrices, we obtain the following Cartan weights:

$$\left(\frac{1}{2}, \frac{1}{3} \right) \rightarrow \text{"u"-quark}$$

$$\left(-\frac{1}{2}, \frac{1}{3} \right) \rightarrow \text{"d"-quark}$$

$$\left(0, -\frac{2}{3} \right) \rightarrow \text{"s"-quark}$$

With this result, the quark quantum numbers are obtained from purely algebraic arguments. This result therefore confirms SU_3 as the fundamental symmetry group for the strong interaction.

THE PERMUTATION GROUP AND ITS APPLICATIONS TO MANY-BODY PROBLEMS

This lecture shows how the symmetry of the permutation group is applied to many-body electron computations. The lecture is accompanied by a project on "Quantum Chemistry". This project applies the content of this lecture to practical computations.

9.1 Definitions and Notations

• We consider a finite set of distinct elements which we number with integers, i.e. $\{1, 2, \ldots, N\}$.

• The set consisting of all permutations of N elements is denoted by S_N. A permutation is a bijective function from N into itself that interchanges the objects $1, \ldots, N$. We use σ to denote a permutation operation: σ transforms each element of N into an element of N and can be represented by a two-line notation. The convention is that the elements $1, 2, 3, \ldots, N$ occupy the first row, and their images under σ form the second row:

$$
\sigma = \begin{pmatrix} 1 & 2 & 3 & \ldots & N \\ \sigma(1) & \sigma(2) & \sigma(3) & \ldots & \sigma(N) \end{pmatrix}
$$

The order of the elements in the first row is, however, irrelevant.

• **Example:** S_3: The elements of S_3 are:

$$
\begin{pmatrix} 1 & 2 & 3 \\ 1 & 2 & 3 \end{pmatrix}
\begin{pmatrix} 1 & 2 & 3 \\ 3 & 1 & 2 \end{pmatrix}
\begin{pmatrix} 1 & 2 & 3 \\ 2 & 3 & 1 \end{pmatrix}
\begin{pmatrix} 1 & 2 & 3 \\ 1 & 3 & 2 \end{pmatrix}
\begin{pmatrix} 1 & 2 & 3 \\ 3 & 2 & 1 \end{pmatrix}
\begin{pmatrix} 1 & 2 & 3 \\ 2 & 1 & 3 \end{pmatrix}
$$

• The set S_N builds a group. The group operation is the composition of permutations $\sigma_2 \circ \sigma_1$, where, by convention, the permutation on the right-hand side is performed first:

$$\begin{pmatrix} 1 & 2 & 3 & \ldots & N \\ \sigma_2(1) & \sigma_2(2) & \sigma_2(3) & \ldots & \sigma_2(N) \end{pmatrix} \circ \begin{pmatrix} 1 & 2 & 3 & \cdots & N \\ \sigma_1(1) & \sigma_1(2) & \sigma_1(3) & \ldots & \sigma_1(n) \end{pmatrix}$$

$$= \begin{pmatrix} 1 & 2 & 3 & \ldots & n \\ \sigma_2(\sigma_1(1)) & \sigma_2(\sigma_1(2)) & \sigma_2(\sigma_1(3)) & \ldots & \sigma_2(\sigma_1(n)) \end{pmatrix}$$

The sign of operation $\circ$ is often omitted. σ^{-1} is obtained from

$$\sigma = \begin{pmatrix} 1 & 2 & 3 & \ldots & N \\ \sigma(1) & \sigma(2) & \sigma(3) & \ldots & \sigma(N) \end{pmatrix}$$

as

$$\sigma^{-1} = \begin{pmatrix} \sigma_1 & \sigma_2 & \sigma_3 & \ldots & \sigma_N \\ 1 & 2 & 3 & \ldots & N \end{pmatrix}$$

and satisfies $\sigma\sigma^{-1} = \mathbb{1}$. The property of associativity follows from the law of composition. The number of elements in S_N is $N!$.
• Instead of using the two-line notation, we can use the cyclic notation: In a cycle, the convention is that the elements between two pairs of brackets are cycled from left to right:

$$(l_1, l_2, l_3, \ldots, l_m) \equiv \begin{pmatrix} l_1 & l_2 & l_3 & \ldots & l_m \\ l_2 & l_3 & l_4 & \ldots & l_1 \end{pmatrix}$$

• Permuting cyclically, the elements of a cycle produces the same permutations: $(l_1, l_2, l_3, \ldots, l_m) \equiv (l_m, l_1, l_2, \ldots, l_{m-1}) \equiv (l_{m-1}, l_m, l_1, \ldots, l_{m-2}) \ldots$.
• $m = 1$ is the trivial cycle. The cycle composed of two numbers (l_1, l_2) is called a transposition.
• Every permutation σ can be represented as a product of cycles with no common elements. In fact, take the element number 1, and consider it as the first element of a cycle. As a second element, take the map of 1 under the permutation σ. Let l_2 be this second element. The third element is the map of l_2 under the permutation σ. Continue until the element that transforms to 1 under the permutation σ. This is the last element of the cycle. The cycle constructed in this way contains strictly different elements, but not

all N elements. Among the remaining elements, take one to start a new cycle.

• As an example, we write the cycle building S_3:

$$(1)(2)(3) \quad (12)(3) \quad (13)(2) \quad (1)(23) \quad (123) \quad (132)$$

• A permutation σ can be endowed with a sign. Associate to the numbers $1, \ldots, N$ the Vandermonde polynomial

$$h(x_1, \ldots, x_N) = \Pi_{i<j}(x_i - x_j)$$

and transforms it with σ to

$$h(x_{\sigma_1}, x_{\sigma_2}, \ldots, x_{\sigma_n}) = \Pi_{i<j}(x_{\sigma_1} - x_{\sigma_j})$$

The transformed polynomial differs from the original only by the sequence of the factors and, possibly, its sign. If the transformed polynomial is multiplied by $+1$, σ is said to be an even permutation with sgn $= 1$, otherwise it is an odd one with sgn $= -1$.

• Any cycle $(l_1, l_2, \ldots, l_m)$ can be represented as a product of transpositions:

$$(l_1, l_2, \ldots, l_m) = (l_1 l_2)(l_2 l_3) \ldots (l_{m-1} l_m)$$

so that any permutation is some product of transpositions. As one transposition is odd, a permutation consisting of an even number of transpositions is itself even, otherwise it is odd.

• Let $a_1, a_2, \ldots, a_h$ be the length of the cycles in the permutation σ. Let us order the cycles such that

$$a_1 \geq a_2 \geq \cdots \geq a_h$$

This produces a cycle representation in the normal form. As

$$\sum_{l=1}^{h} a_l = N$$

each cycle sequence corresponds to a partition of the number N. For example, in S_3, we have the cycles of length $3 = 3 + 0$, $3 = 2 + 1$, and $3 = 1 + 1 + 1$.

• Let us consider two permutations P and Q:

$$P = \begin{pmatrix} 1 & 2 & 3 & \ldots & N \\ p(1) & p(2) & p(3) & \ldots & p(N) \end{pmatrix}$$

and

$$Q = \begin{pmatrix} 1 & 2 & 3 & \ldots & N \\ q(1) & q(2) & q(3) & \ldots & q(N) \end{pmatrix} = \begin{pmatrix} p(1) & p(2) & p(3) & \ldots & p(N) \\ s(1) & s(2) & s(3) & \ldots & s(N) \end{pmatrix}$$

We have

$$PQ^{-1} = \begin{pmatrix} 1 & 2 & 3 & \ldots & n \\ p(1) & p(2) & p(3) & \ldots & p(N) \end{pmatrix} \begin{pmatrix} q(1) & q(2) & q(3) & \ldots & q(N) \\ 1 & 2 & 3 & \ldots & n \end{pmatrix}$$

$$= \begin{pmatrix} q(1) & q(2) & q(3) & \ldots & q(N) \\ p(1) & p(2) & p(3) & \ldots & p(N) \end{pmatrix}$$

Accordingly,

$$QPQ^{-1} = \begin{pmatrix} p(1) & p(2) & p(3) & \ldots & p(N) \\ s(1) & s(2) & s(3) & \ldots & s(N) \end{pmatrix} \begin{pmatrix} q(1) & q(2) & q(3) & \ldots & q(N) \\ p(1) & p(2) & p(3) & \ldots & p(N) \end{pmatrix}$$

$$= \begin{pmatrix} q(1) & q(2) & q(3) & \ldots & q(N) \\ s(1) & s(2) & s(3) & \ldots & s(N) \end{pmatrix}$$

Accordingly, the conjugate elements belonging to the same conjugacy class to a permutation

$$P = \begin{pmatrix} 1 & 2 & 3 & \ldots & N \\ p(1) & p(2) & p(3) & \ldots & p(N) \end{pmatrix}$$

are obtained by applying any permutation Q to both rows of P. This operation does not change the cycle structure of the permutation, so that we have proved the following theorem.

Theorem. *Elements in the same conjugacy class have the same cycle segmentation.*

Comments.

1. Each conjugacy class is univocally associated with some partition of the number N.
2. Each conjugacy class is represented by a Young diagram (YD).
3. The permutations associated with each YD can be represented with the convention that the Young boxes are filled with the numbers to be permuted and the numbers along horizontal lines in a YD are cycled.

Example:

$$3 = 1 + 1 + 1 \Rightarrow \;\begin{array}{c}\square\\\square\\\square\end{array}\; \leftrightarrow [1^3] \leftrightarrow \{(1)(2)(3)\}$$

$$3 = 2 + 1 \Rightarrow \;\begin{array}{c}\square\square\\\square\end{array}\; \leftrightarrow [2,1] \leftrightarrow \{(12)(3); (13)(2); (1)(23)\}$$

$$3 = 3 \Rightarrow \;\square\square\square\; \leftrightarrow [3] \leftrightarrow \{(123); (132)\}$$

The following table summarizes the number of conjugacy classes for S_1–S_{10}.

n	1	1	3	4	5	6	7	8	9	10
$\#(n)$	1	2	3	5	7	11	15	22	30	42

9.2   The Irreducible Representations of S_N

According to the Burnside theorem, the number of irreducible representations of S_N equals the number of classes. When using a YD to produce the representations of S_N, one must consider that the permutation group acts to change the index of the variables filling the boxes, so that the number of boxes of a YD equals N. The filling of the boxes with monomials is governed by an important theorem.

Theorem. (H. Boerner, *Darstellungen von Gruppen*, Springer Verlag, 1967, https://link.springer.com/book/10.1007/978-3-642-86032-4). *The basis states required to generate all irreducible representations of S_N are obtained by filling both lines and columns with strictly increasing numbers.*

Examples:

1. $N = 3$:

$$[3] \doteq \square\square\square \;\rightarrow\; \boxed{1\,|\,2\,|\,3}$$

There is only one basis monomial, $x_1 \cdot x_2 \cdot x_3$. The monomial is transformed into itself by any of the permutations of S_3. This YD is a graphical rendering of the one-dimensional, trivial representation of S_3.

$$[2,1] \doteq \;\begin{array}{c}\square\square\\\square\end{array}\; \rightarrow \left\{ \begin{array}{|c|c|}\hline 1 & 2\\\hline 3\\\cline{1-1}\end{array}, \begin{array}{|c|c|}\hline 1 & 3\\\hline 2\\\cline{1-1}\end{array} \right\}$$

is a two-dimensional representation of S_3.

$$[1^3] \doteq \begin{array}{c} \Box \\ \Box \\ \Box \end{array} \rightarrow \begin{array}{|c|} \hline 1 \\ \hline 2 \\ \hline 3 \\ \hline \end{array}$$

is a one-dimensional representation associating to each class its sign.

2. $N = 4$: We have the irreducible representations

$$[4] \doteq \boxed{\ \ \ \ } \qquad [3,1] \doteq \begin{array}{c} \boxed{\ \ \ } \\ \boxed{\ } \end{array} \qquad [2^2] \doteq \begin{array}{c} \boxed{\ \ } \\ \boxed{\ \ } \end{array} \qquad [1^4] \doteq \begin{array}{c} \Box \\ \Box \\ \Box \\ \Box \end{array}$$

Comments.

1. The dimensionality of the representations is the number of Young tableaux (YTs) one can construct from a YD using the filling rule for the permutation group.

2. There are results from discrete mathematics that provide general formulas for computing dimensionality and characters. It is probably a better way to compute them for any specific case one would like to treat. Accordingly, we give here a concrete example of character computations for S_3.

• $\boxed{\ \ \ }$: There is one basis monomial symbolized by $\boxed{1|2|3}$, which is simply $x_1 \cdot x_2 \cdot x_3$. Any permutation transforms this tensor product into itself, and the character is just 1 for all classes.

• $\begin{array}{c}\Box\\\Box\\\Box\end{array}$: There is one basis function

$$\left| \begin{pmatrix} x_1 & y_1 & z_1 \\ x_2 & y_2 & z_2 \\ x_3 & y_3 & z_3 \end{pmatrix} \right|$$

The character of even permutations transforming this basis function is $+1$, and the character of odd permutations is -1.

• $\begin{array}{c}\boxed{\ \ }\\\boxed{\ }\end{array}$: The basis tensor products symbolized by $\begin{array}{|c|c|}\hline 1 & 2 \\ \hline 3 \\ \cline{1-1}\end{array}$ and $\begin{array}{|c|c|}\hline 1 & 3 \\ \hline 2 \\ \cline{1-1}\end{array}$ are

$$\left| \begin{pmatrix} x_1 & y_1 \\ x_3 & y_3 \end{pmatrix} \right| \cdot x_2 \qquad \left| \begin{pmatrix} x_1 & y_1 \\ x_2 & y_2 \end{pmatrix} \right| \cdot x_3$$

The character of the identity permutation is 2. To find the character of the representative $(12)3$ of the class $[2,1]$, we operate with this permutation onto the two basis tensor products:

$$P_{(12)3}\begin{vmatrix} x_1 & y_1 \\ x_3 & y_3 \end{vmatrix}\cdot x_2 = \begin{vmatrix} x_2 & y_2 \\ x_3 & y_3 \end{vmatrix}\cdot x_1 \quad P_{(12)3}\begin{vmatrix} x_1 & y_1 \\ x_2 & y_2 \end{vmatrix}\cdot x_3 = \begin{vmatrix} x_2 & y_2 \\ x_1 & y_1 \end{vmatrix}\cdot x_3$$

We observe that

$$P_{(12)3}\;\boxed{\begin{array}{c}1\;2\\3\end{array}} = \boxed{\begin{array}{c}1\;2\\3\end{array}} - \boxed{\begin{array}{c}1\;3\\2\end{array}} \qquad P_{(12)3}\;\boxed{\begin{array}{c}1\;3\\2\end{array}} = -\boxed{\begin{array}{c}1\;3\\2\end{array}}$$

The matrix representation of this element in the space $\left\{\boxed{\begin{array}{c}1\;2\\3\end{array}}, \boxed{\begin{array}{c}1\;3\\2\end{array}}\right\}$ is therefore

$$\begin{pmatrix} 1 & 0 \\ \bar{1} & \bar{1} \end{pmatrix}$$

and has a character of 0. Further characters can be found similarly (the character of the irreducible representation $[1^N]$, for example, is given by its sign).

3. Character tables: The character tables of S_2, S_3, and S_4 (characters in red) are here summarized.

S_2	$[1^2]$	$[2]$	S_3	$[1^3]$	$[2,1]$	$[3]$
$[2]$	1	1	$[3]$	1	1	1
$[1^2]$	1	$\bar{1}$	$[2,1]$	2	0	$\bar{1}$
			$[1^3]$	1	$\bar{1}$	1

S_4	$[1^4]$	$[2,1^2]$	$[2^2]$	$[3,1]$	$[4]$	
$[4]$	1	1	1	1	1	
$[3,1]$	3	1	$\bar{1}$	0	$\bar{1}$	
$[2^2]$	2	0	2	$\bar{1}$	0	
$[2,1^2]$	3	$\bar{1}$	$\bar{1}$	0	1	
$[1^4]$	1	$\bar{1}$	1	1	$\bar{1}$	

9.3 Applications to Physical Problems

Introduction: Physical problems concerned with the permutation groups are those involving, for example N identical particles. In order to use the permutation group, we need to define how permutations are represented in the Hilbert space of the N particles. Let the configuration space of the N particles be specified by the spatial coordinates $\mathbf{r}_1, \ldots, \mathbf{r}_N$ and by the spin coordinates $m_1, \ldots, m_N$ (spin coordinates are taken to assume discrete values within some range $(+S, \ldots, -S)$).

Definition. The permutation operator $\Pi(\sigma)$ representing the permutation σ into the space of wave functions $\psi(\mathbf{r}_1, m_1, \ldots, \mathbf{r}_N, m_N)$ is defined by

$$\Pi(\sigma)\psi\left(\mathbf{r}_1, m_1, \ldots, \mathbf{r}_N, m_n\right) \doteq \psi\left(\mathbf{r}_{\sigma(1)}, m_{\sigma(1)}, \ldots, \mathbf{r}_{\sigma(N)}, m_{\sigma(N)}\right)$$

9.3.1 *Example I: Two spin-$\frac{1}{2}$ particles*

We consider a Hamilton operator that contains scalars made up of the spin operators $S_x(m_1), S_y(m_1), S_z(m_1)$ acting in the Hilbert space of particle "1". For example,

$$- |\alpha| S_z^2(m_1)$$

is an uniaxial anisotropy favoring the alignment of spin "1" along the z-axis. If a second, identical spin is added, the Hamilton operator must be augmented by the same scalars made up of the operators $S_x(m_2), S_y(m_2), S_z(m_2)$ acting in the Hilbert space of particle "2", so that it is invariant with respect to the permutations $(1)(2)$ and (12). In this specific case,

$$H(1, 2) = - |\alpha| \cdot \left(S_z^2(m_1) + S_z^2(m_2)\right)$$

An interaction between the spins might occur, e.g. the exchange interaction:

$$H(1, 2) = - |\alpha| \cdot \left(S_z^2(m_1) + S_z^2(m_2)\right) - J \cdot S_z(m_1) \cdot S_z(m_2)$$

This coupling must also be symmetric with respect to particle interchange. Note that the invariance with respect to particle interchange is not at all restricted to Hamilton operators and quantum mechanics: Any many-particle Hamilton function of classical physics has the property that it does not change its functional dependence on the particle coordinates when the particle coordinates are permuted. This property is inherent to the particles being identical. In quantum mechanics, this invariance implies that the permutation operators constitute a symmetry group of the Hamilton operator, and thus, the eigenfunctions of the Hamilton operators can be classified according to the symmetry-adapted functions transforming with respect to the irreducible representations of the symmetry group — in this specific case, S_2.

Let us work out, for example the symmetry-adapted functions to the Hamilton operator $H(1,2)$ specified above. The Hilbert space carrying this Hamiltonian is made up of the four basis functions

$$Y^+(m_1) \cdot Y^+(m_2) \quad Y^+(m_1) \cdot Y^-(m_2) \quad Y^-(m_1) \cdot Y^+(m_2) \quad Y^-(m_1) \cdot Y^-(m_2)$$

To find the symmetry-adapted wave functions. we use the projector technique:

$$P_{[2]} Y^+(m_1) \cdot Y^+(m_2) = 1 \cdot \underbrace{\Pi_{(1)(2)}[Y^+(m_1) \cdot Y^+(m_2)]}_{Y^+(m_1) \cdot Y^+(m_2)}$$

$$+ 1 \cdot \underbrace{\Pi_{(12)}[Y^+(m_1) \cdot Y^+(m_2)]}_{Y^+(m_1) \cdot Y^+(m_2)}$$

$$\propto Y^+(m_1) \cdot Y^+(m_2)$$

$$P_{[2]} Y^+(m_1) \cdot Y^-(m_2) = 1 \cdot \underbrace{\Pi_{(1)(2)}[Y^+(m_1) \cdot Y^-(m_2)]}_{Y^+(m_1) \cdot Y^-(m_2)}$$

$$+ 1 \cdot \underbrace{\Pi_{(12)}[Y^+(m_1) \cdot Y^-(m_2)]}_{Y^-(m_1) \cdot Y^+(m_2)}$$

$$\propto \frac{1}{\sqrt{2}} \cdot [Y^+(m_1) \cdot Y^-(m_2) + Y^-(m_1) \cdot Y^+(m_2)]$$

$$P_{[2]} Y^-(m_1) \cdot Y^-(m_2) = 1 \cdot \underbrace{\Pi_{(1)(2)}[Y^-(m_1) \cdot Y^-(m_2)]}_{Y^+(m_1) \cdot Y^+(m_2)}$$

$$+ 1 \cdot \underbrace{\Pi_{(12)}[Y^-(m_1) \cdot Y^-(m_2)]}_{Y^-(m_1) \cdot Y^-(m_2)}$$

$$\propto Y^-(m_1) \cdot Y^-(m_2)$$

The projector technique delivers three symmetry-adapted spin wave functions transforming according to the irreducible representation [2] of the group S_2. There is one wave function transforming according to the irreducible representation $[1^2]$:

$$P_{[1^2]}Y^+(m_1) \cdot Y^-(m_2) = 1 \cdot \underbrace{\Pi_{(1)(2)}[Y^+(m_1) \cdot Y^-(m_2)]}_{Y^+(m_1) \cdot Y^-(m_2)}$$

$$+ \bar{1} \cdot \underbrace{\Pi_{(12)}[Y^+(m_1) \cdot Y^-(m_2)]}_{Y^-(m_1) \cdot Y^+(m_2)}$$

$$\propto \frac{1}{\sqrt{2}} \cdot [Y^+(m_1) \cdot Y^-(m_2) - Y^-(m_1) \cdot Y^+(m_2)]$$

We can summarize the results in terms of symmetry-adapted spin states in a table, as follows.

S_2	⊟	⬚⬚	Basis states		
[2]	1	1	$\underbrace{Y^+Y^+}_{\uparrow\uparrow}$	$\underbrace{\frac{1}{\sqrt{2}} \cdot (Y^+Y^- + Y^-Y^+)}_{\frac{1}{\sqrt{2}}(\uparrow\downarrow+\downarrow\uparrow)}$	$\underbrace{Y^-Y^-}_{\downarrow\downarrow}$
$[1^2]$	1	$\bar{1}$	$\underbrace{\frac{1}{\sqrt{2}} \cdot (Y^+Y^- - Y^-Y^+)}_{\frac{1}{\sqrt{2}}(\uparrow\downarrow-\downarrow\uparrow)}$		

9.3.2   *Example II: Three spin-$\frac{1}{2}$ particles*

- We list the basis states ($Y^+ \rightarrow \uparrow$, $Y^- \rightarrow \downarrow$):

$$\uparrow\uparrow\uparrow \quad \uparrow\uparrow\downarrow \quad \uparrow\downarrow\uparrow \quad \uparrow\downarrow\downarrow \quad \downarrow\uparrow\uparrow \quad \downarrow\uparrow\downarrow \quad \downarrow\downarrow\uparrow \quad \downarrow\downarrow\downarrow$$

- We project out symmetry-adapted states and find four states that each transform according to the one-dimensional irreducible representation [3]:

$$P_{[3]} \uparrow\uparrow\uparrow = \uparrow\uparrow\uparrow$$

$$P_{[3]} \uparrow\uparrow\downarrow \propto \uparrow\uparrow\downarrow + \uparrow\downarrow\uparrow + \downarrow\uparrow\uparrow + \uparrow\uparrow\downarrow + \downarrow\uparrow\uparrow + \uparrow\downarrow\uparrow$$

$$= \sqrt{\frac{1}{3}} (\uparrow\uparrow\downarrow + \uparrow\downarrow\uparrow + \downarrow\uparrow\uparrow)$$

$$P_{[3]} \uparrow\downarrow\downarrow = \sqrt{\frac{1}{3}} \left(\uparrow\downarrow\downarrow + \downarrow\downarrow\uparrow + \downarrow\uparrow\downarrow \right)$$

$$P_{[3]} \downarrow\downarrow\downarrow = \downarrow\downarrow\downarrow$$

- To find the states transforming according to the two-dimensional representation $[2, 1]$, we use the projector based on the character:

$$P_{[2,1]} \uparrow\uparrow\downarrow = \sqrt{\frac{1}{6}} \left(2\uparrow\uparrow\downarrow - \downarrow\uparrow\uparrow - \uparrow\downarrow\uparrow \right) = [2,1]_1$$

This produces one basis vector. We find the second one by applying one permutation onto this basis vector, e.g. the permutation $(1)(23)$:

$$[2,1]_2 = \sqrt{\frac{1}{6}} \left(2\uparrow\downarrow\uparrow - \uparrow\uparrow\downarrow - \downarrow\uparrow\uparrow \right)$$

There is a second pair of basis vectors obtained using a different trial vector:

$$P_{[2,1]} \uparrow\downarrow\downarrow = \sqrt{\frac{1}{6}} \left(2\uparrow\downarrow\downarrow - \downarrow\uparrow\downarrow - \downarrow\downarrow\uparrow \right) = [2,1]_3$$

and

$$[2,1]_4 = \sqrt{\frac{1}{6}} \left(2\downarrow\uparrow\downarrow - \downarrow\downarrow\uparrow - \uparrow\downarrow\downarrow \right)$$

- There are no states transforming according to $[1^3]$.
- We summarize the symmetry adapted states in the following table.

S_3	$[1^3]$	$[2,1]$	$[3]$	Basis states
$[3]$	1	1	1	$[3]_1 = \uparrow\uparrow\uparrow$ $[3]_2 = \sqrt{\frac{1}{3}} \left(\uparrow\uparrow\downarrow + \uparrow\downarrow\uparrow + \downarrow\uparrow\uparrow \right)$ $[3]_3 = \sqrt{\frac{1}{3}} \left(\uparrow\downarrow\downarrow + \downarrow\downarrow\uparrow + \downarrow\uparrow\downarrow \right)$ $[3]_4 = \downarrow\downarrow\downarrow$
$[2,1]$	2	0	$\bar{1}$	$[2,1]_1 = \sqrt{\frac{1}{6}} \left(2\uparrow\uparrow\downarrow - \downarrow\uparrow\uparrow - \uparrow\downarrow\uparrow \right) \quad [2,1]_2 = \sqrt{\frac{1}{6}} \left(2\uparrow\downarrow\uparrow - \uparrow\uparrow\downarrow - \downarrow\uparrow\uparrow \right)$ $[2,1]_3 = \sqrt{\frac{1}{6}} \left(2\uparrow\downarrow\downarrow - \downarrow\uparrow\downarrow - \downarrow\downarrow\uparrow \right) \quad [2,1]_4 = \sqrt{\frac{1}{6}} \left(2\downarrow\uparrow\downarrow - \downarrow\downarrow\uparrow - \uparrow\downarrow\downarrow \right)$
$[1^3]$	1	$\bar{1}$	1	

9.4   The Schur–Weyl Duality

9.4.1   *An introduction to the problem*

The symmetry-adapted states we have just computed have an interesting property. Consider the operator for the square of the total spin vector

$$(\mathbf{S})^2 \doteq (\mathbf{S}_1 + \mathbf{S}_2 + \mathbf{S}_3)^2$$

The eigenvalues of this operator in the space $\{\uparrow\uparrow\uparrow, \uparrow\uparrow\downarrow, \uparrow\downarrow\uparrow, \uparrow\downarrow\downarrow, \downarrow\uparrow\uparrow, \downarrow\uparrow\downarrow, \downarrow\downarrow\uparrow, \downarrow\downarrow\downarrow\}$ can be found by the writing the Clebsch–Gordan series:

$$D_{\frac{1}{2}} \otimes D_{\frac{1}{2}} \otimes D_{\frac{1}{2}} = D_{\frac{3}{2}} \oplus D_{\frac{1}{2}} \oplus D_{\frac{1}{2}}$$

This result tells us that $(\mathbf{S})^2$, in this space, has quantum numbers $\frac{3}{2}$ and $\frac{1}{2}$, from which we read out the corresponding eigenvalues $\hbar^2 \cdot \frac{3}{2}(\frac{3}{2} + 1)$ and $\hbar^2 \cdot \frac{1}{2}(\frac{1}{2} + 1)$ (both fourfold degenerate).

Now, the Clebsch–Gordan method tells us how to construct basis functions (out off the original product functions spanning the Hilbert space of the three-spin system) that diagonalize $\mathbf{S}^2$. One computes first the symmetry-adapted function for a two-spin system. These are given by the Clebsch–Gordan coefficients for the sum of two spin-$\frac{1}{2}$ particles to build a system with a total spin quantum number of 1 or 0. We can obtain them from suitable tables to be

$$S = 1 \rightarrow Y_1^1 = \uparrow\uparrow$$

$$Y_1^0 = \frac{1}{\sqrt{2}}(\uparrow\downarrow + \downarrow\uparrow)$$

$$Y_1^{-1} = \downarrow\downarrow$$

$$S = 0 \rightarrow Y_0^0 = \frac{1}{\sqrt{2}}(\uparrow\downarrow - \downarrow\uparrow)$$

We couple the third spin from the left and build the symmetry-adapted states to $S = \frac{3}{2}$ from the Clebsch–Gordan coefficients appearing in the tensor product $D^{\frac{1}{2}} \otimes D^1$:

$$Y_{\frac{3}{2}}^{\frac{3}{2}} = Y^+ Y_1^1 = \uparrow\uparrow\uparrow$$

$$Y_{\frac{3}{2}}^{\frac{1}{2}} = \sqrt{\frac{2}{3}} Y^+ Y_1^0 + \sqrt{\frac{1}{3}} Y^- Y_1^1 = \sqrt{\frac{1}{3}}\,(\uparrow\uparrow\downarrow + \uparrow\downarrow\uparrow + \downarrow\uparrow\uparrow)$$

$$Y^{\frac{1}{2}}_{\frac{1}{2}} = \sqrt{\frac{1}{3}}Y^+Y_1^{-1} + \sqrt{\frac{2}{3}}Y^-Y_1^0 = \sqrt{\frac{1}{3}}(\uparrow\downarrow\downarrow + \downarrow\uparrow\downarrow + \downarrow\downarrow\uparrow)$$

$$Y^{\frac{3}{2}}_{\frac{3}{2}} = Y^-Y_1^{-1}1 = \downarrow\downarrow\downarrow$$

$$Y^{\frac{1}{2}}_{I,\frac{1}{2}} = \sqrt{\frac{1}{3}}Y^+Y_1^0 - \sqrt{\frac{2}{3}}Y^-Y_1^1 = \sqrt{\frac{1}{6}}(\uparrow\uparrow\downarrow + \uparrow\downarrow\uparrow - 2\downarrow\uparrow\uparrow)$$

$$Y^{\frac{1}{2}}_{I,\frac{1}{2}} = \sqrt{\frac{2}{3}}Y^+Y_1^{-1} - \sqrt{\frac{1}{3}}Y^-Y_1^0 = \sqrt{\frac{1}{6}}(2\uparrow\downarrow\downarrow - \downarrow\uparrow\downarrow - \downarrow\downarrow\uparrow)$$

There are states resulting from the tensor product $D^{\frac{1}{2}} \otimes D^0$:

$$Y^{\frac{1}{2}}_{II,\frac{1}{2}} = Y^+Y_0^0 = \sqrt{\frac{1}{2}}(\uparrow\uparrow\downarrow - \uparrow\downarrow\uparrow)$$

$$Y^{\frac{1}{2}}_{II,\frac{1}{2}} = Y^-Y_0^0 = \sqrt{\frac{1}{2}}(\downarrow\uparrow\downarrow - \downarrow\downarrow\uparrow)$$

Comments.

1. Compare the four $S = \frac{3}{2}$ states with the four [3]-states: They are identical, i.e. It appears as if all eigenvectors transforming according to [3] are also exactly the eigenvectors of $(\mathbf{S})^2$ to the same eigenvalue $\hbar^2 \cdot \frac{3}{2}(\frac{3}{2}+1)$.

2. The situation with the symmetry-adapted states transforming according to [2, 1] is more complicated. The $D^{\frac{1}{2}}$ representation appears twice in the decomposition of $D^{\frac{1}{2}} \otimes D^{\frac{1}{2}} \otimes D^{\frac{1}{2}}$, so that the space carrying the $S = \frac{1}{2}$ value has a degeneracy of four. On the other hand, the space carrying the [2, 1] representation also has a degeneracy of four. This means that one has the possibility of doing linear combinations while staying within the [2, 1] space and produce the symmetry-adapted functions transforming according to the $\frac{1}{2}$-value of $\mathbf{S}^2$. In fact, we observe that

$$Y^{\frac{1}{2}}_{I,\frac{1}{2}} = [2,1]_1 + [2,1]_2 \quad Y^{\frac{1}{2}}_{II,\frac{1}{2}} = \sqrt{\frac{1}{3}}([2,1]_1 - [2,1]_2)$$

This means that two linear combinations of the two basis states $[2,1]_1, [2,1]_2$ produces states that transform according to the $+\frac{1}{2}$-Cartan weight of the irreducible representation $D^{\frac{1}{2}}$. Similarly, two linear combinations of the two basis states $[2,1]_3, [2,1]_4$ produces states that transform according to the $-\frac{1}{2}$-Cartan weight of the irreducible representation $D^{\frac{1}{2}}$.

3. In this example, coupling the third spin from the right through the Clebsch-Gordan technique generates a further set of states symmetry-adapted to $\mathbf{S}^2$, which are not necessarily identical to the states obtained by the left coupling. The two sets of basis states are transformed into each other by the Racah coefficients.[1] Their behavior under permutations does not depend on whether the third spin is coupled from the left or from the right.

4. From the two examples worked out above, one could conclude that there is some kind of relation between the value of S, the value of S_z, and the type of symmetry under permutation. This conjecture is supported by an exact theorem proved by Schur and Weyl[2] for an ensemble of N spin-$\frac{1}{2}$ particles.

9.4.2   *The Schur–Weyl duality and the Young diagram technique*

We start from the irreducible representation $D^{\frac{1}{2}}$ within a vector space with basis states $\uparrow, \downarrow$ and establish a basis set for N such spin-$\frac{1}{2}$ particles by performing all possible products

$$\underbrace{\uparrow\uparrow\downarrow \cdots \uparrow\downarrow\uparrow\downarrow}_{N\,\text{times}}$$

Within this product space, the Clebsch–Gordan series

$$D^{\frac{1}{2}} \otimes D^{\frac{1}{2}} \cdots \otimes D^{\frac{1}{2}} = [D^{\frac{1}{2}}]^{\otimes N} = \oplus_{S=0(\frac{1}{2})}^{\frac{N}{2}} d_S \cdot D^S$$

establishes symmetry-adapted states that transform according to the irreducible representations D^S. S are the possible total spin quantum numbers that a set N of spin-$\frac{1}{2}$ particles can assume. The representation D^S appears d_S times in the Clebsch–Gordan expansion: d_s is called the multiplicity of the representation D^S. To each quantum number (S, m_S), there are exactly n_S symmetry-adapted linear combinations of the basis spin products. They can be found, in principle, using the Clebsch–Gordan coefficients for the group SU_2.

---

[1]L. C. Biedenharn, J. M. Blatt and M. E. Rose, "Some properties of the Racah and associated coefficients", *Rev. Mod. Phys.* **24**, 249 (1952), https://journals.aps.org/rmp/abstract/10.1103/RevModPhys.24.249.

[2]H. Weyl, *The Classical Groups: Their Invariants and Their Representations*, Princeton University Press, 1939.

In the proof of the following theorem, we must compute the numbers n_S since they play an important role. For this purpose, we analyze the following scheme and try to catch a systematicity.

	D^0	$D^{\frac{1}{2}}$	D^1	$D^{\frac{3}{2}}$	D^2	$D^{\frac{5}{2}}$
$D^{\frac{1}{2}\otimes 2}$	1	0	1	0	0	0
$\otimes 3$	0	2	0	1	0	0
$\otimes 4$	2	0	3	0	1	0
$\otimes 5$	0	5	0	4	0	1
$\vdots$	$\vdots$	$\vdots$	$\vdots$	$\vdots$	$\vdots$	$\vdots$

In this table, one recognizes a tree structure that leads to the combinatorial result

$$n_S = \binom{N}{S} - \binom{N}{(S-1)}$$

which we rewrite, for convenience, as

$$\binom{N}{(\frac{N-2S}{2})} \cdot \frac{2S+1}{\frac{N+2S+2}{2}}$$

Suppose now that we have collected the n_S symmetry-adapted states to a given S and to any of its Cartan weights m_S.

Theorem (Schur and Weyl). *The d_S product states belonging to any given (S, m_S) constitute an invariant space transforming according to the same unique d_S-dimensional representation $\Delta(S)$ of the group S_N.*[3]

Comment. To each S, one can assign a unique $\Delta(S)$ but not all irreducible representations of S_N can be assigned a total spin S.

Proof. We provide a proof of this theorem using the YD method and well-known results from mathematics. We first consider that the tensor product

---

[3]The Schur–Weyl duality has important applications in modern quantum information theory, see e.g. A. Botero, Revista Colombiana de Matematicas **50**, 191–209 (2016) DOI:10.15446/recolma.v50n2.62210 for a tutorial introduction.

of N-$D^{\frac{1}{2}}$ representations can be obtained by computing

Only those diagrams that consist of two lines of length $l_1 \geq l_2$ are taken into account, as they represent the irreducible representations $D^{\frac{l_1-l_2}{2}}$ of SU_2, according to an exact theorem (see "Boerner"). One important property of these diagrams is that

$$l_1 + l_2 = N$$

This property is only valid for the product of YD $\square$ of SU_2. Because of this property, one can associate to each YD, resulting from the tensor product of $D^{\frac{1}{2}}$ and identifying a value of S, a unique irreducible representation of S_N.

We have therefore the situation that two exact theorems of the YD technique allow us to associate the SAME YD to both the total spin S of an ensemble of N particles (the irreducible representation D^S) and to an irreducible representation of S_N, which we call $\Delta(S)$. We now find a set of basis states among the product states that can be used on one side as basis states for $\Delta(S)$ and on the other as basis states for D^S.

- Take a general YD with lengths (l_1, l_2).
- Write down all YTs for a given YD using the variables x_i (first horizontal line) and y_j (second line). They provide (Boerner) basis monomials for the irreducible representation Δ_S in N variables.
- Let the index identify a site in the string $(1, \ldots, N)$.
- Set all x variables at a given site to assume the value $Y^+ \equiv\uparrow$ and all the y variables at a given site to assume the value $Y^- \equiv\downarrow$.
- Using this mapping, one generates a number δ_S of many-spin states that can be equally used, as alternatives to the variables $x_1, \ldots, x_N$ and $y_1, \ldots, y_N$, to build a δ_S-dimensional representation Δ_S.
- As $\mathbf{S}^2$ and S_z commute with the irreducible representation $\Delta(S)$, the basis state products generating the irreducible representation $\Delta(S)$ diagonalize S^2 and S_z. In virtue of the mapping chosen, m_S is the highest Cartan weight compatible with the YD, i.e. $m_S = S$.
- If we can show that δ_S (the dimension of $\Delta(S)$) is equal to the multiplicity d_S (the number of states expected in the Clebsch–Gordan expansion to span the space $(S, m_S = S)$), then the product states constructed using the S_N YT cover the entire space belonging to $(S, m_S = S)$, and the theorem is proved (the permutation operator commutes with the ladder

operators, and the spaces of lower Cartan weights generated starting from the eigenspace to the highest Cartan weight also transform according to $\Delta(S)$). Referring to the literature on the irreducible representations of S_N, we find

$$\delta_S = N! \cdot \frac{l_1 + 1 - l_2}{(l_1 + 1)! \cdot l_2!} = N! \cdot \frac{2S + 1}{\left(\frac{N+2S+2}{2}\right)! \cdot \left(\frac{N-2S}{2}\right)!}$$

The expression on the right-hand side corresponds to d_S.   ◇.

Comments.

1. By mapping the monomials for S_N to a set of suitable product spin states, we have established the link between the two-line irreducible representations of S_N and the representations D^S given by the same YD. This link is often referred to as **Schur–Weyl duality**. $\Delta(S)$ is often referred to in the German literature as the *spin rasse*.
2. The Schur–Weyl duality also holds between the irreducible representations of S_N and the representations appearing in the decomposition of the N product $\square^{\otimes N}$ of the natural representation $\square$ of $GL(\mathscr{C}, n)$.

9.5 Multiplets in Atomic Physics

A common problem in atomic physics arises during the modification of the single-electron atomic energy levels brought about by electron–electron interaction. The single-electron energy levels are computed using the configuration Hamiltonian. This is a single-electron Hamiltonian that accounts for the presence of the other electrons using an effective, mean field potential energy. The energy levels resulting are called "shells", as they are filled by electrons within the so-called shell model of atomic physics. The filling proceeds by virtue of a simplified version of the Pauli principle, which predicts that no more than two electrons fill a state with given orbital angular momentum quantum number l and magnetic quantum number m_l. However, the atomic spectra show a more complex pattern, indicating a further splitting of single-electron energy levels into multiplets (known as "Hund's rules"), provided by the so-called multiplet Hamiltonian. This Hamiltonian takes the Coulomb repulsion between the electrons in a shell explicitly into account and is, accordingly, a many-body Hamiltonian. Its diagonalization is performed, typically, within the space provided by the

$(2 \cdot (2l + 1))^N$ products of single-particle wave functions

$$f_{nl}(r) \cdot \{Y_l^{m_l}(\vartheta, \varphi) \cdot Y^{m_s}\}$$

However, not all states must be taken into account because of a stringent requirement about the symmetry of possible eigenfunctions of the many-electron system with respect to particle permutation. This requirement is the content of the most general version of the Pauli principle.

Pauli principle: The eigenfunctions of a N-electron system transform according to the irreducible representation $[1^N]$ of the permutation group S_N, i.e.

$$\Pi(\sigma)\psi(\mathbf{r}_1, m_1, \ldots, \mathbf{r}_N, m_n) = sgn(\sigma) \cdot \psi(\mathbf{r}_1, m_1, \ldots, \mathbf{r}_N, m_n)$$

Comments.

1. In general, the eigenfunctions of Fermions transform according to $[1^N]$, and those of Bosons according to $[N]$.
2. A function transforming according to $[1^N]$ is said to be totally antisymmetric.
3. In virtue of this symmetry requirement, the multiplet Hamiltonian must be only diagonalized within the subspace of the product states that consist of the totally antisymmetric linear combinations. The number of these symmetry-adapted states is, in general, reduced with respect to the $(2 \cdot (2l+1))^N$ product states. Finding them provides therefore an essential reduction of the size of the matrices that remain to be diagonalized.

There are two algorithms that can be used to find these antisymmetric linear combinations. These two algorithms are, for certain aspects, complementary. The first one uses the Schur–Weyl duality to organize the states into multiplets. It is particularly useful for finding the quantum numbers S and L that labels the energy levels resulting from the diagonalization of the multiplet Hamiltonian. The second one is based on the technique of Slater determinants, which finds directly the desired antisymmetric linear combinations. It can be practically implemented for numerical efficiency.

9.5.1   *Schur–Weyl duality: Multiplet analysis*

In this method, one first determines all possible values of S and the corresponding *spin rasse* $\Delta(S)$. This determines the permutation symmetry of the orbital wave functions as well, as they must transform according to an irreducible representation $\Delta(L)$ (L indicates that the irreducible representation refers to the orbital component) such that

$$[1^N] \in \Delta(L) \otimes \Delta(S)$$

Claim.

$$\chi_{\Delta(L)}(\sigma) = sgn(\sigma) \cdot \overline{\chi_{\Delta(S)}(\sigma)}$$

Proof. The representation obtained by multiplying the complex conjugate of the characters of $\Delta(S)$ with the sign of the permutation is irreducible, i.e.

$$\mathscr{M}_\sigma \mid sgn(\sigma) \cdot \chi_{\Delta(S)}(\sigma) \mid^2 = 1$$

However, the average on the left-hand side is exactly the number of times the representation $[1^N]$ is contained in the decomposition of

$$sgn(\sigma) \cdot \overline{\Delta}(S) \otimes \Delta(S)$$

This means that $sgn \cdot \overline{\Delta}(S)$ is the desired representation $\Delta(L)$.   ◇

Comment. In virtue of the Pauli principle, $\Delta(S)$ determines $\Delta(L)$ univocally. For a given S-term scheme, the task is finding those linear combinations of orbital wave functions that transform according to $\Delta(L)$.

Example I: The energy levels of a two-electron system

The $(ns)^2$ configuration: The first example is $(ns)^2$ configuration. This configuration entails two possible cases, $S = 1$ and $S = 0$:

- **S = 0:** $\Delta(S = 0)$ consists of the antisymmetric spin singlet

$$\sqrt{\frac{1}{2}} \left[Y^+(m_1) \cdot Y^-(m_2) - Y^-(m_1) \cdot Y^+(m_2) \right] \Leftrightarrow \sqrt{\frac{1}{2}} (\uparrow\downarrow - \downarrow\uparrow)$$

Accordingly, the orbital wave function must transform according to the identical representation, i.e. one must search for symmetric linear combinations of the product states

$$f_{n,0}(r_1) \cdot f_{n,0}(r_2) \cdot Y_0^0(\Omega_1) \cdot Y_0^0(\Omega_2)$$

This product state is symmetric and transforms according to $D^0 \otimes D^0 = D^0$. We have found the desired orbital wave function for the $(ns)^2$ configuration, and the total wave function

$$f_{n,0}(r_1) \cdot f_{n,0}(r_2) \cdot Y_0^0(\Omega_1) \cdot Y_0^0(\Omega_2) \cdot \sqrt{\frac{1}{2}} \left[Y^+(m_1) \cdot Y^-(m_2) - Y^-(m_1) \cdot Y^+(m_2) \right]$$

is antisymmetric with respect to coordinate permutation. It belongs to the multiplet ^1S. For example, the ground state of a He atom is the configuration $(1s)^2$ and belong to the term scheme for $S = 0$.

• **S = 1**: We turn now to the term scheme for $S = 1$. There are three spin states:

$$\uparrow\uparrow; \sqrt{\frac{1}{2}}\,(\uparrow\downarrow + \downarrow\uparrow)\,;\downarrow\downarrow$$

and they transform according to the identical representation of S_2. Accordingly, one must build an antisymmetric linear combination of the orbital product functions. There is only one product function available in this configuration:

$$f_{n,0}(r_1) \cdot f_{n,0}(r_2) \cdot Y_0^0(\Omega_1) \cdot Y_0^0(\Omega_2)$$

and this function is symmetric. The multiplet ^3S in this configuration, accordingly, violates the Pauli principle and cannot exist.

Comments.

1. This result can be generalized by stating that any attempt of anti-symmetrizing the product wave functions of two electrons having individually exactly the same orbital wave function produces a vanishing wave function, i.e. given two electrons that occupy the same orbital state, they can only form a spin singlet. The original version of the Pauli principle stated that "no more than two electrons can occupy the same orbital state". The group theoretical analysis allows us to refine this statement: "two electrons with the same orbital wave functions must have "opposite spin" (more precisely, they must constitute a singlet).

2. The ^1S multiplet is rendered graphically by filling two boxes symbolizing the ns orbitals with two opposite spins:

   (this symbology should not be confused with a YT).

3. The statement using symmetry-adapted wave functions is a refined version of the original Pauli principle. For the specific configuration under investigation, the application of the Pauli principle has reduced the original degeneracy of the configuration by eliminating one multiplet. In virtue of the Pauli principle, the $1s$ level appears in the $S = 0$-term scheme of the He atom as the ground state, while it is absent in the $S = 1$ term scheme.

The $(np)^2$ configuration: The Pauli principle has eliminated all but one product function in the $(s)^2$ configuration, leaving the spin singlet. We expect similar phenomena, such as reduction of degeneracy and/or elimination of multiplets, to occur in all configurations, and they must be, of course, analyzed case by case in response to some experimental need.

We conduct for the sake of illustrating these phenomena the analysis of the configuration of two electrons in the p-orbital. The Weyl theorem establishes that the multiplets will have either spin quantum number 0 (spin singlets, antisymmetric) or 1 (spin triplets, symmetric). Considering the orbital part of the wave function, we can build three possible total orbital quantum numbers from two $L = 1$ particles:

$$D^1 \otimes D^1 = D^2 \oplus D^1 \oplus D^0$$

We find the corresponding wave function in the table of Clebsch–Gordan coefficients (M. Tanabashi *et al.*, "Particle data group", *Phys. Rev. D* **98**, 030001 (2018), p. 564). There, we note an important property: The $L = 2$ wave functions are symmetric with respect to coordinate exchange, so that the $L = 2$ orbital wave function can only couple with the singlet wave function, producing five antisymmetric wave functions that belong to the quantum numbers $S = 0$ and $L = 2$ and carry the multiplet symbol ^1D. The $L = 1$ wave functions, instead, are antisymmetric with respect to coordinate exchange and can only couple with the triplet spin states giving rise to nine antisymmetric wave functions with the multiplet symbol ^3P. The $L = 0$ wave function is also symmetric and produces the antisymmetric wave function with the multiplet symbol ^1S. The total number of antisymmetric states produced using the Weyl theorem is 15. Without taking the Pauli principle into account, this configuration contains 36 product states.

We conclude that the configurational energy level $2 \cdot E_{np}$ hosts therefore three multiplets: ^1D, ^3P, ^1S. We have a new situation with respect to the $(ns)^2$ configuration: Triplet states exist and provide possible non-vanishing, totally antisymmetric states.

A question now arises: Among all the remaining multiplets, is there an interaction that — at least partially — removes the degeneracy and produces a "multiplet splitting"? This is a key question toward determining which multiplet is the actual ground state of the $(np)^2$ configuration and which ones are excited states. Note that it makes a difference whether, for example the ground state is a triplet or a singlet: In the first case, the ensemble responds *a la* Zeeman to a magnetic field, while in the second, the magnetic field just shifts the energy level. The interaction that removes the degeneracy among the multiplets of the $(np)^2$ configuration is the **Coulomb repulsion between the electrons**. It provides an ordering of the multiplets that goes under the name of "first and second Hund's rules". Accordingly, the multiplet ^3P has a lower energy than the singlets (first Hund's rule). Among the remaining multiplets, ^1D has a lower energy than ^1S (second Hund's rule). These rules are of empirical nature: The Schur–Weyl analysis only provides the possible multiplets but does not order them.

9.5.2 *Slater determinants*

There is a technique that allows us to construct antisymmetric wave functions starting from single-particle wave functions. This is an alternative technique to the one based on the Schur–Weyl duality and was invented by J. C. Slater.[4]

As an example, we consider the electronic configuration $(ns)^2$. The single-particle states arising from the configuration Hamiltonian are

$$\psi_1 \doteq f_n Y_0^0 Y^+ \quad \psi_2 \doteq f_n Y_0^0 Y^-$$

One can construct antisymmetrized product functions starting from these states by means of the so-called Slater determinants:

$$\frac{1}{\sqrt{2}} \cdot \det \begin{pmatrix} \psi_1(\mathbf{r}_1, m_1) & \psi_1(\mathbf{r}_2, m_2) \\ \psi_1(\mathbf{r}_1, m_1) & \psi_1(\mathbf{r}_1, m_1) \end{pmatrix} = 0$$

$$\frac{1}{\sqrt{2}} \cdot \det \begin{pmatrix} \psi_2(\mathbf{r}_1, m_1) & \psi_2(\mathbf{r}_2, m_2) \\ \psi_2(\mathbf{r}_1, m_1) & \psi_2(\mathbf{r}_1, m_1) \end{pmatrix} = 0$$

$$\frac{1}{\sqrt{2}} \cdot \det \begin{pmatrix} \psi_1(\mathbf{r}_1, m_1) & \psi_1(\mathbf{r}_2, m_2) \\ \psi_2(\mathbf{r}_1, m_1) & \psi_2(\mathbf{r}_2, m_2) \end{pmatrix}$$

$$= -\frac{1}{\sqrt{2}} \cdot \det \begin{pmatrix} \psi_2(\mathbf{r}_1, m_1) & \psi_2(\mathbf{r}_2, m_2) \\ \psi_1(\mathbf{r}_1, m_1) & \psi_1(\mathbf{r}_1, m_1) \end{pmatrix}$$

$$= \frac{1}{\sqrt{2}} \cdot [\psi_1(\mathbf{r}_1, m_1)\psi_2(\mathbf{r}_2, m_2) - \psi_2(\mathbf{r}_1, m_1) \cdot \psi_1(\mathbf{r}_2, m_2)]$$

---

[4]J. C. Slater, "The theory of complex spectra", *Phys. Rev.* **34**, 1293 (1929), https://journals.aps.org/pr/abstract/10.1103/PhysRev.34.1293.

From this example, one deduces the rule that Slater determinants involving identical states vanish. This is in line with the original version of the Pauli principle that forbids two electrons having identical quantum numbers. The remaining two non-vanishing determinants produce the same linear, independent antisymmetric product function that describes the spin singlet and represents a configuration where (informally) "two electrons with opposite spin" can occupy identical orbital states.

This tool can be generalized to produce antisymmetric states involving N particles starting from $m \geq N$ single-particle states $\{\psi_i\}$, $i = 1, \ldots, m$ (the pair of variables $(\mathbf{r}, m)$ is rendered symbolically as ξ):

$$\psi^-_{\alpha_1, \alpha_2, \ldots, \alpha_m} = \frac{1}{\sqrt{N!}} \cdot \det \begin{pmatrix} \psi_{\alpha_1}(\xi_1) & \psi_{\alpha_1}(\xi_2) & \cdots & \psi_{\alpha_1}(\xi_N) \\ \psi_{\alpha_2}(\xi_1) & \psi_{\alpha_2}(\xi_2) & \cdots & \psi_{\alpha_2}(\xi_N) \\ \vdots & \vdots & & \vdots \\ \psi_{\alpha_N}(\xi_1) & \psi_{\alpha_N}(\xi_2) & \cdots & \psi_{\alpha_N}(\xi_N) \end{pmatrix}$$

Taking into account that $\alpha_i \neq \alpha_j$, there are

$$\binom{m}{N}$$

such determinants. For instance, the $(ns)^2$ configuration provides two single-particle states ($m = 2$) for two particles ($N = 2$), giving one Slater determinant (see the explicit calculation above). The $(np)^2$ configuration provides $m = 6$ single-particle states, for $N = 2$ electrons. This amounts to

$$\binom{6}{2} = 15$$

non-vanishing Slater determinants, the same number of antisymmetric total wave functions provided by Weyl's method of organizing the various states.

Example II: Three-electron systems

Li has three electrons; therefore, it has two separate term schemes: one corresponding to the totally symmetric spin $\frac{3}{2}$-state, and the other corresponding to the state $\frac{1}{2}$, transforming according to the irreducible representation $[2, 1]$ of S_3. The term scheme with a $\frac{3}{2}$ total spin requires totally antisymmetric orbital wave functions. A state generated from the configuration $(1s)^2 2s$ is therefore prohibited, as it is not possible to build a non-vanishing, totally antisymmetric orbital wave function from it.

This configuration, which is the one with the lowest energy, is therefore excluded from the $\frac{3}{2}$-term scheme. The ground state configuration will therefore lie in the spin-$\frac{1}{2}$ sector.

There are two spin basis states in the Cartan weight $+\frac{1}{2}$:

$$Y^{\frac{1}{2}}_{\frac{1}{2},1} = \sqrt{\frac{1}{6}} \left(\uparrow\uparrow\downarrow + \uparrow\downarrow\uparrow - 2\downarrow\uparrow\uparrow \right)$$

$$Y^{\frac{1}{2}}_{\frac{1}{2},2} = \sqrt{\frac{1}{2}} \left(\uparrow\uparrow\downarrow - \uparrow\downarrow\uparrow \right)$$

As they transform according to $[2,1]$ of S_3, the only possible irreducible representation Δ_L which contains $[1^3]$ in the decomposition of $\Delta_L \otimes [2,1]$ is $[2,1]$. One of the two orbital basis functions with $[2,1]$ symmetry is constructed by applying the projector operator to the product function $\psi_{1s}(1)\psi_{1s}(2)\psi_{2s}(3)$:

$$P_{[2,1]}(\psi_{1s}(1)\psi_{1s}(2)\psi_{2s}(3)) \propto 2\psi_{1s}(1)\psi_{1s}(2)\psi_{2s}(3)$$
$$- \psi_{1s}(2)\psi_{1s}(3)\psi_{2s}(1)$$
$$- \psi_{1s}(3)\psi_{1s}(1)\psi_{2s}(2)$$
$$\doteq [2,1]_1$$

$[2,1]_2$ is obtained by applying any of the permutation to $[2,1]_1$. There will be, accordingly, four total wave functions to the Cartan weight $\frac{1}{2}$ transforming according to $[2,1] \otimes [2,1]$:

$$[2,1]_1 \otimes Y^{\frac{1}{2}}_{\frac{1}{2},1} \quad [2,1]_1 \otimes Y^{\frac{1}{2}}_{\frac{1}{2},2} \quad [2,1]_2 \otimes Y^{\frac{1}{2}}_{\frac{1}{2},1} \quad [2,1]_2 \otimes Y^{\frac{1}{2}}_{\frac{1}{2},2}$$

The next step to finding the total wave function carrying the ground state multiplet ^2S is to find the one linear combination of these four wave functions that is totally antisymmetric with respect to particle permutation. This will be the one wave function of the three-particle ground state with the total spin quantum number $\frac{1}{2}$ and Cartan weight (S_z-quantum number) $+\frac{1}{2}$, which is physically relevant because it obeys the Pauli principle. There will be further three linear combinations; however, two of them transform according to $[2,1]$ one of them according to $[3]$. They are therefore discarded as wave functions for the thee-electron systems. A similar procedure will deliver the totally antisymmetric state with the total spin quantum number $\frac{1}{2}$ and Cartan weight $-\frac{1}{2}$. Together, these two states represents the two basis states of the ground state multiplet ^2S.

Comment. The method of Slater determinants shortens the path to finding these *two* totally antisymmetric states. Starting from the single-particle wave functions

$$\psi_{1s} \uparrow; \psi_{1s} \downarrow; \psi_{2s} \uparrow$$

we find, for example the total antisymmetric wave function to $(S, S_z) = (\frac{1}{2}, +\frac{1}{2})$ by building the Slater determinant

$$\frac{1}{\sqrt{6}} \cdot \begin{vmatrix} \psi_{1s} \uparrow (1) & \psi_{1s} \uparrow (2) & \psi_{1s} \uparrow (3) \\ \psi_{1s} \downarrow (1) & \psi_{1s} \downarrow (2) & \psi_{1s} \downarrow (3) \\ \psi_{2s} \uparrow (1) & \psi_{2s} \uparrow (2) & \psi_{2s} \uparrow (3) \end{vmatrix}$$

where the particle coordinate i denotes $(\mathbf{r}_i, m_i)$. We find the total antisymmetric wave function to $(S, S_z) = (\frac{1}{2}, -\frac{1}{2})$ by building the Slater determinant

$$\frac{1}{\sqrt{6}} \cdot \begin{vmatrix} \psi_{1s} \uparrow (1) & \psi_{1s} \uparrow (2) & \psi_{1s} \uparrow (3) \\ \psi_{1s} \downarrow (1) & \psi_{1s} \downarrow (2) & \psi_{1s} \downarrow (3) \\ \psi_{2s} \downarrow (1) & \psi_{2s} \downarrow (2) & \psi_{2s} \downarrow (3) \end{vmatrix}$$

In atomic physics, these results are often summarized with the symbols

and

Lecture 10

GROUP THEORY AND PHASE TRANSITIONS

In this lecture, we use symmetry arguments to describe the general properties of the order parameter in phase transitions. A deep work on this subject can be found in J. C. Toledano and P. Toledano, *The Landau Theory of Phase Transitions*, World Scientific Lecture Notes in Physics: Volume 3 (1987), https://doi.org/10.1142/0215.

10.1 Introduction

Phase transitions are changes in the state of a system that occur at specific lines and points in, for instance, the pressure–temperature parameter space of a macroscopic system. One example is the transition that produces the crystallization of a liquid into a solid phase. Another example is the collective alignment of an ensemble of spins along one well-defined direction, occurring below a certain characteristic temperature called the Curie temperature.

As an example of a phase transition covered by the symmetry arguments, we think of the rearrangement that atoms at the 110-surface of fcc metals, such as Au, undergo when the temperature is changed.[1] Carefully prepared, the low-temperature atomic order of this surface does not correspond to the ideal geometry one would expect when an fcc crystal is cut to reveal the 110-plane (the so-called primitive one-by-one, or $p(1 \times 1)$, ordered surface). Instead, numerous experimental studies, based on surface technologies such as STM or low-energy electron diffraction, have shown

---

[1] J. C. Campuzano, M. S. Foster, G. Jennings, R. F. Willis, and W. Unertl, Au(110) (1×2)-to-(1×1) phase transition: A physical realization of the two-dimensional Ising model, *Phys. Rev. Lett.* **54**, 2684 (1985), https://journals.aps.org/prl/abstract/10.1103/PhysRevLett.54.2684.

that a "missing row model" of the type shown in the following figure better describes the atomic arrangement at low temperatures (which is said to be a "reconstructed surface").

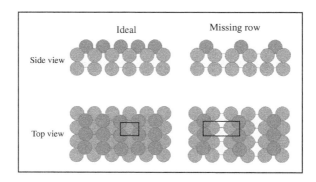

Side and top views of an ideal 110-surface (left) and one with a missing row (right). The unit cells are indicated by black rectangles.

A characteristic of the reconstructed surface is that the surface unit cell is doubled in one direction (the horizontal one in the bottom figures; let us call it the "x"-direction, with the vertical being, accordingly, the y-direction in our convention) with respect to the $p(1 \times 1)$-unit cell.

Following Landau, we describe the state of matter using an atomic probability distribution $\rho(\mathbf{r})$ that peaks at the occupied lattice sites (the shape of this probability density is, for this description, irrelevant). For the ideal surface, for example, one observes peaks of $\rho(x, y)$ at all lattice sites (the following figure, top left for $\rho(x, y = n \cdot a)$).

$\rho(x, y = n \cdot a)$ for the reconstructed surface, instead, peaks at every second lattice site (top right). One possible way of observing the situation of an ideal versus reconstructed surface is to perform a diffraction experiment that measures the intensity I of the diffracted particles (electrons, X-rays, or neutrons). I is proportional to the square of the Fourier transform of $\rho(\mathbf{r})$ (let us call $\eta(G)$ the Fourier transform of $\rho(x, y = n \cdot a)$). In a diffraction experiment on an ideal surface, $\eta(G)$ peaks at values of G corresponding to $\frac{2\pi}{a}$ (bottom left). The diffraction pattern of the reconstructed surface, instead, has peaks at multiples of half $\frac{2\pi}{a}$, i.e. at half integers of the reciprocal lattice vector along the x-direction (bottom right).

The $p(2 \times 1)$-surface reconstruction realizes a particular distribution of the atoms in the lattice which is energetically most favorable. When

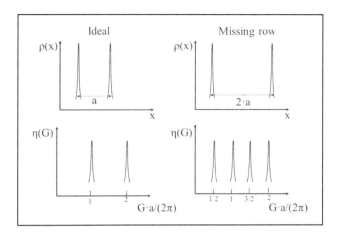

Top left: Sketch of $\rho(x, y = n \cdot a)$ (a: Lattice constant of the ideal surface along the x-direction) for the ideal surface, with peaks repeating with a periodicity of a. Top right: Sketch of $\rho(x, y = n \cdot a)$ for the reconstructed surface, with peaks repeating with a periodicity of $2 \cdot a$. Bottom left: Fourier transform of $\rho(x, y = n \cdot a)$ for an ideal surface. Bottom right: Fourier transform of $\rho(x, y = n \cdot a)$ for a reconstructed surface.

the temperature is increased, entropy promotes disordering of the atomic distribution, and the state of the system corresponds to minimizing its free energy. This means that, at higher temperatures, the atomic distribution can deviate from the low-temperature order. In fact, one observes that, when the temperature is raised, the intensity of the half-order spots decreases (the following figure, left) and almost vanishes at sufficiently high temperatures. The integer-order spots (originating mainly from the subsurface, non-reconstructed layers), instead, are less affected by the temperature.

When one plots the intensity as a function of temperature (right) at the peak of the half-order spots, one observes a very peculiar behavior: at a well-defined temperature of about $650\,\mathrm{K}$, the intensity drops and (almost) vanishes. At higher temperatures, only the integer spots are observed. It has been suggested that, in the high-temperature phase, the atoms sitting in the top rows move to randomly occupy all lattice sites available at the surface, i.e. build a so-called "lattice gas". In this state, the top layer is described, on average, by a uniform distribution $\rho(x, y)$ and contributes a uniform background to the diffraction pattern, which is dominated by the diffraction of the subsurface layers. In terms of symmetry, one can describe this disordering process as a change in symmetry occurring at

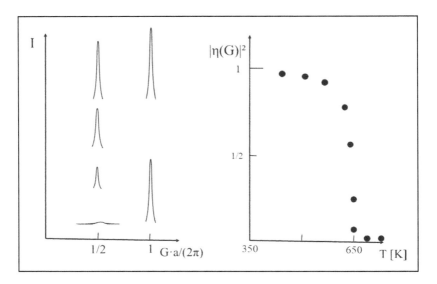

Left: Sketch of $I(G)$ at different temperatures. The temperature decreases from above. Right: Plot of the maximum intensity of the half-order spots as a function of the temperature.

the transition point. The low-temperature phase is invariant with respect to translations by multiples of $2a$ along x. The high-temperature phase, instead, is invariant with respect to all translations along x, i.e. it does not have any particular site at which it peaks.

In the following, we illustrate the symmetry arguments provided by Landau and Lifshitz in order to describe this important class of phenomena, called "second-order phase transitions" (L. Landau, *Phys. Z. Sowjet.* **11**, 26 (1937) and *ibidem*, 545, see also "On the theory of phase transitions", in *Collected Papers of L. D. Landau*, Editor: D. Ter Haar, Pergamon, 1965, pp. 193–216, https://www.sciencedirect.com/science/article/pii/B9780080 105864500341.

10.2 The Landau–Lifshitz Symmetry Rules

First Landau–Lifschitz rule (LL1): The first LL condition establishes the connection between the symmetry groups on both sides of the transition line. Let G_0 be the symmetry group of the high-temperature phase (G_0 can be one of the 230 space groups or even of some continuous groups, such as the full group of translations in a liquid or the full group of rotations O_3 in a Heisenberg exchange ferromagnet). Let $\rho_0(\mathbf{r})$ be the density function describing the high-temperature state. $\rho_0(\mathbf{r})$ is invariant with respect to all symmetry elements of G_0, i.e. transforms according to the "1" representation. For a liquid state, for instance, $\rho_0(\mathbf{r})$ is a constant, invariant with respect to all translations. When crossing the transition line, the density function becomes

$$\rho(\mathbf{r}) = \rho_0(\mathbf{r}) + \delta\rho(\mathbf{r})$$

$\delta\rho(\mathbf{r})$ is zero exactly at the transition point and changes continuously. The symmetry of $\rho(\mathbf{r}) + \delta\rho(\mathbf{r})$ is described by some group G. The continuous change in density imposes that the symmetry elements of $\delta\rho(\mathbf{r})$ are also symmetry elements of G_0. This is the content of the first LL symmetry rule: The symmetry group of the low-temperature phase is some subgroup of the symmetry group of the high-temperature phase, i.e.

$$G \in G_0$$

Second Landau–Lifschitz rule (LL2): The second rule requires the introduction of the Landau functional.

I. The Landau functional: The Landau theory of phase transitions foresees the existence of a functional $\Phi([\rho], p, T)$ as some integral over the density function $\rho(\mathbf{r})$. The functional contains the essential physical constraints in a specific system, such as the various interactions. It is not, by itself, a thermodynamic potential. When P and T are given, Landau suggests that the form of the function ρ describing the state of equilibrium is determined from the condition that $\Phi[\rho]$ should have a minimum. When the equilibrium density function is determined, inserting it into the Landau functional produces the thermodynamic potential in the variables T and P.

II. Order parameters (OPs): One possible way of writing the to-be-minimized Landau functional is to expand the density function $\rho(\mathbf{r})$ into a Fourier series of irreducible invariant functions, transforming according to the irreducible representations (irr. reps.) of the group G_0:

$$\rho(x, y, z) = \sum_{s,i} \eta_{i,s} \varphi_{i,s}(x, y, z)$$

The index s runs over all irr. reps. Γ_s of G_0, and the index i labels the various partner functions transforming according to the d_s-dimensional representation Γ_s. The coefficients of the expansion and the basis functions are supposed to be real quantities. This implies that the matrix elements of the irr. reps. are chosen to be real quantities as well. This expansion is inserted into the functional, and the integral over the spatial coordinates, contained in the known basis functions, is performed. This leaves behind a functional containing the set of coefficients $\{\eta_{i,s}\}$, which must be chosen to minimize $\Phi(\{\eta_{i,s}\}, p, T)$. Above the phase transition, the equilibrium values for all coefficients $\{\eta_{i,s}\}$ belonging to irr. reps. other than the trivial will vanish exactly. In the vicinity of the phase transition, some of them might assume arbitrarily small non-zero values in the low-symmetric phase.

Definition. The set of coefficients $\{\eta_{i,s}\}$ builds the set of OPs.

III. The Taylor expansion: Because of this smallness, the Taylor expansion of the Landau functional in the small parameters $\{\eta_{i,s}\}$ could provide a useful *ansatz* over which to perform the minimization procedure. For writing the Taylor expansion, one must take into account the fact that the functional must be a scalar with respect to the operation of G_0 (as it is the Hamiltonian of the system), i.e. the functional is invariant with respect to all operations of G_0. As the Taylor expansion consists of successive powers of the coefficients $\{\eta_{i,s}\}$, the expansion must contain **invariant polynomials** of these coefficients, i.e. polynomials in the coefficients $\{\eta_{i,n}\}$ that transform according to the "1" representation.

To determine these polynomials, we need to find the transformation law of the Fourier coefficients $\{\eta_{i,s}\}$ under the symmetry operations of G_0. For the sake of illustration, we consider a transition where G_0 is related to some isometries and write $(g \equiv (R \mid \mathbf{t}),\ \eta_{i,s} = \int d^3\mathbf{x}\rho(\mathbf{x}) \cdot \varphi_{i,s}(\mathbf{x}))$

$$O_g\eta_{i,s} = \int d^3\mathbf{x}\,\underbrace{(O_g\rho(\mathbf{x}))}_{\rho(g^{-1}\mathbf{x})}\cdot\varphi_{i,s}(\mathbf{x}) = \int d^3\mathbf{y}\rho(\mathbf{y})\cdot\underbrace{\varphi_{i,s}(g\mathbf{y})}_{\sum_j \overline{\Gamma^s_{i,j}}\varphi_{j,s}(\mathbf{y})}$$

$$= \sum_j \overline{\Gamma^s_{i,j}}\eta_{j,s}$$

(the use of complex conjugates has, to be exact, only a formal significance, as we have assumed that the matrix elements of the irr. reps. are real numbers).

The expansion is written as

$$\Phi(\{\eta_{i,s}\}, p, T) = \Phi_0(T, p)$$

$$+ \sum_{l,s,i} a_{l,s}(T, P) \cdot P_l^1(\{\eta_{i,s}\})$$

$$+ \sum_{l,s,i} A_{l,s}(T, P) \cdot P_l^2(\{\eta_{i,s}\})$$

$$+ \sum_{l,s,i} B_{l,s}(T,P) \cdot P_l^3(\{\eta_{i,s}\})$$

$$+ \sum_{l,s,i} D_{l,s}(T,P) \cdot P_l^4(\{\eta_{i,s}\})$$

$$+ \cdots$$

The sum runs over the indices s and i. One needs a further index l that takes into account possibly the existence of more polynomials with the same degree. P_l^n are the invariant polynomials of power n in the coefficients $\{\eta_{i,s}\}$. In the spirit of the Taylor expansion, only the lowest powers are indicated. The expansion coefficients $a_{l,s}(T,P)$, $A_{l,s}$, $B_{l,s}$ and $D_{l,s}$ are suitable functions of the thermodynamic variables p, T. In solving problems, these functions remain parameters to adapt to the specific situation.

Zeroth-order term: The zeroth-order term in the expansion of $\Phi(\{\eta_{i,s}\}, p, T)$ is a function of T, p only. $\Phi_0(T, p)$ provides the non-singular part of the thermodynamic potential $\Phi(p, T)$ that varies smoothly at the phase transition. It is associated with the zeroth power of the coefficients $\{\eta_{i,s}\}$, which is a constant and therefore certainly invariant with respect to the irr. rep. "1".

First-order terms: The only first-order invariant is the coefficient $\eta_{1,1}$ transforming according to the "1" representation (the remaining coefficients, to the power of one, are orthogonal to the space spanned by the "1" representation, as they transform according to a representation different from the "1" one):

$$P_l^1(\{\eta_{i,s}\}) = \eta_{1,1}$$

The appearance of the coefficient $\eta_{1,1}$ at the phase transition does not change the symmetry. The change in the density function, which is not accompanied by a symmetry change is called a "relaxation". This relaxation is taken into account by suitably smoothly modifying $\Phi_0(p, T)$ at the phase transition.[2]

---

[2]Note that the application of an external "field" might produce the appearance of order-one powers of the OPs associated with irr. rep. other than the "1" representation. We refer to more specialized literature for the treatment of this situation.

Second-order terms: We show later that only one second-degree polynomial can be built for a given irr. rep.,

$$P_l^2(\{\eta_{i,s}\}) = \sum_i \eta_{i,s}^2$$

At the transition point, the coefficients $\eta_{i,s}$ are all vanishing. For the minimum of Φ in the high-symmetric phase to be $\eta_{i,s} = 0$ and in the low-symmetric phase to be some non-zero values of the coefficients $\eta_{i,s}$, all A_s must be non-negatives in the high-symmetric phase and some of them should become negative in the low-symmetric phase. Note that it is very improbable that more than one coefficient vanishes along a line in the P–T plane. Such a simultaneous vanishing of two coefficients can only occur in an isolated point of the plane, which represents the intersection of two lines of continuous phase transitions. The change in sign of one of the coefficients means that at one side of the transition point, all coefficients are positive, and the value minimizing Φ is $\eta_{s,i} = 0$. On this side, the symmetry group G_0 is realized. On the other side, instead, one coefficient, A_{s_0}, assumes a finite, small, negative value in the vicinity of the transition point. There will be a restricted number of elements of G_0 that, when operated on the order parameter vector $(\eta_{1,s_0}, \eta_{2,s_0}, \ldots)$, transform it into itself. These elements realize the subgroup G_0 in the lower-symmetric phase. Typically, therefore, the situation is one where all non-changing sign coefficients are set to 0 and only the irr. rep. Γ^{s_0}, realized on the low-symmetry side of the transition point, is kept in the expansion. This is called **active** representation. The phase transition line is therefore determined by the equation

$$A_{s_0}(T_c, P_c) = 0$$

The second LL rule is expressed by the following sentence:

"LL2: A continuous (second-order) phase transition, where the density function changes continuously, leads to a state of the system where one OP vector

$$\left(\eta_{1,s_0}, \eta_{2,s_0}, \ldots, \eta_{d_{s_0},s_0}\right)$$

is transformed into itself by the subgroup G".

Hereinafter, we drop the index s_0, as only one irr. rep., the active one, is of concern. The vector $(\eta_1, \ldots, \eta_d)$, carrying the active irr. rep. Γ of the group G_0, appears below the transition point and is called the OP of the system, and the corresponding space is the OP space. Depending on the dimension of Γ, the OP can have more than one component. The appearance of an OP below the transition point establishes a new symmetry group G.

In the literature, one often introduces the quantities

$$\eta_i \doteq \eta \cdot \gamma_i$$

with $\sum_i \gamma_i^2 = 1$ defining a unit vector with "amplitude" η within a d-dimensional space. The expansion of the thermodynamic potential is then written as

$$\Phi(T, P, \{\eta, \gamma_1, \ldots, \gamma_d\}) = \Phi_0(p, T)$$
$$+ \eta^2 \cdot A(T, P)$$
$$+ \eta^3 \cdot \sum_l B_l(T, P) \cdot P_l^3(\{\gamma_i\})$$
$$+ \eta^4 \cdot \sum_l D_l(T, P) \cdot P_l^4(\{\gamma_i\})$$
$$+ \cdots$$

Third-order terms and LL3: The third LL rule is concerned with the role of the third-order terms. A model including a third-order term is considered in detail in the original papers by Landau (L. Landau, Zur Theorie der Phasenumwandlungen I, *Phys. Z. Sowjet.* **11**, 26 (1937); L. Landau, Zur Theorie der Phasenumwandlungen II, *Phys. Z. Sowjet.* **11**, 545 (1937)). A simple argument for estimating the role of the third-order term uses a one-component OP and

$$\Phi(T, P, \eta) = \Phi_0(p, T) + A(T, P) \cdot \eta^2 + B(T, P)\eta^3 + \cdots$$

The graph of the Landau potential as a function of η in the high-symmetric phase $(A > 0, B > 0)$ reveals that $\eta = 0$ is only a **local** minimum. The finite values of η decreases the Landau potential indefinitely toward negative values. This is an unphysical situation produced by the third-order term. More precise considerations, applied in the presence of a third-order term, lead to a minimum at the transition temperature being at a finite value for η and thus to a discontinuous (first-order) phase transition, i.e. a discontinuous change in the density function. In this case, the groups on the two sides of the phase transition need not be related. Thus, the Landau condition (LL3) for a phase transition being continuous and relating a group to one of its subgroups is as follows:

"LL3: A continuous phase transition is characterized by the absence of third-order terms in the expansion of the Landau functional".[3]

Fourth-order terms: In contrast to the third-order terms, a fourth order term can be positive in any direction and at any values of p, T and therefore can sustain a second-order phase transition. The fourth-degree terms are essential for determining the actual direction of the OP in the OP space. As the second-order term is independent of γ_i, these quantities must be determined by the requirement that the fourth-order terms be minimized with respect to γ_i. The amplitude η of the OP is then determined by minimizing globally Φ.

10.3 Invariant Polynomials

Invariant polynomials play a central role in the Landau theory of phase transitions. They are used to build the Landau functional. Polynomials of the coefficients $\gamma_1, \ldots, \gamma_d$ transforming according to the active representation Γ are products of these coefficients. They are thus to be found as symmetry-adapted vectors transforming according to the "1" representation in the reduction of

$$\underbrace{\Gamma \otimes \Gamma \otimes \cdots \otimes \Gamma}_{\text{n-times}} = \Gamma^{\otimes n} \quad \text{written} \quad \Gamma^n \quad \text{for simplicity.}$$

This product can be rendered with the YD method:

$$\Gamma^1 \rightarrow \underbrace{\boxed{}}_{YD} \rightarrow \underbrace{\boxed{1}, \boxed{2}, \ldots, \boxed{d}}_{YT}$$

$$\Gamma^2 = \boxed{1} \otimes \boxed{1} \rightarrow \underbrace{\boxed{}}_{YD} \rightarrow \underbrace{\boxed{1\,1}, \boxed{1\,2}, \ldots, \boxed{d\,d}}_{YT}$$

$$\rightarrow \underbrace{\boxed{\begin{array}{c}\\\end{array}}}_{YD} \rightarrow \underbrace{\boxed{\begin{array}{c}1\\2\end{array}}, \boxed{\begin{array}{c}1\\3\end{array}} \cdots}_{YT}$$

---

[3]The absence of third-order invariants can be due to the impossibility of finding third-degree invariant polynomials for the group G_0. This absence, due to symmetry reasons, will hold at any temperature and pressure. It might also be possible, at some isolated point of the P–T-plane, that the coefficient of the third-order term vanishes "accidentally". In this case, the Curie point is the intersection of the lines $A(T, P) = 0$ and $B(T, P) = 0$ and is therefore an isolated point in the P–T-plane.

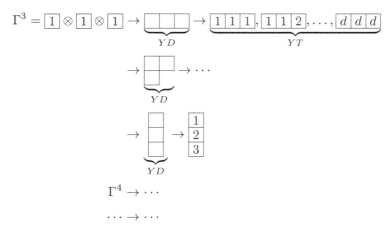

The Young tableau $\boxed{1\,1}$, for instance, provides the monomial $\gamma_1\gamma_1'$, which is one of the basis monomials $\{\gamma_i\gamma_j'\}$ in the product space of $\boxed{}$. When the prime variable is set to have the same value as the unprimed, it provides one of the monomials that, potentially, can enter an invariant second-order polynomial of the coefficients. The Young tableau $\begin{smallmatrix}\boxed{1}\\[-2pt]\boxed{2}\end{smallmatrix}$ provides the antisymmetric monomial $\gamma_1\gamma_2' - \gamma'1\gamma_2$. When the primed and unprimed variables are set equal, the monomial vanishes. We conclude that YD with vertical boxes can be discarded as a supply for invariant polynomials (this is equivalent to saying that the YD representations with vertical boxes do not contain the representation "1" in their decomposition into irr. reps. of G_0).

We are therefore left with the problem of:

a. finding the number of times, n, the representation "1" appears in the decomposition of the representations $\boxed{}$, $\boxed{}$, $\boxed{}$, $\ldots$

b. projecting out the invariant components from the corresponding Young tableaus.

We now solve these tasks for $\boxed{}$ and $\boxed{}$ as examples.

Γ^1. In view of the orthonormality properties of the characters, we have

$$n_{''1''} = \mathcal{M}_g 1 \cdot \chi_\Gamma(g) = 1 \quad \text{for } \Gamma ='' 1''$$

$$n_{''1''} = \mathcal{M}_g 1 \cdot \chi_\Gamma(g) = 0 \quad \text{for } \Gamma \neq'' 1''$$

Accordingly, the invariant monomial is just $\gamma_{1,1}$.

Γ^2. The character of the representation Γ^2 is $\chi_\Gamma^2(g)$. We compute

$$n_{''1''} = \mathcal{M}_g 1 \cdot \chi_\Gamma^2(g) = \mathcal{M}_g(\overline{\chi_\Gamma(g)}) \cdot (\chi_\Gamma(g)) = 1$$

because Γ is an irr. rep. This result means that, for any irreducible active representation, there is only one invariant polynomial of second degree.

We project out from γ_1^2 the following polynomial:

$$P_{''1''}\gamma_1^2 = \mathcal{M}_g \sum_{i,j} \Gamma_{1i}(g)\gamma_i\Gamma_{1j}(g)\gamma_j$$

$$= \sum_{i,j} \underbrace{\mathcal{M}_g\Gamma_{1i}(g)\Gamma_{j1}(g)}_{\delta_{ij}} \gamma_i\gamma_j$$

$$= \sum_{i,j} \delta_{ij}\gamma_i\gamma_j = \sum_i \gamma_i^2$$

The only second-degree invariant polynomial is, for any representation,

$$\sum_{i=1}^{d} \gamma_i^2$$

The method of invariant polynomials is explained in great detail in J. C. Toledano and P. Toledano, *The Landau Theory of Phase Transitions*, World Scientific Lecture Notes in Physics: Volume 3 (1987), https://doi.org/10.1142/0215.

10.4   Magnetic Phase Transitions

The most famous case of the application of the Landau theory of phase transitions is the classical Ising magnet on a lattice (specified by an index i) described by the Hamilton function

$$H = -J \sum_{i,j} \sigma_i \cdot \sigma_j$$

where the variable σ takes the value 1 or -1. The symmetry group of this system consists of two operations: the identity operation E and the inversion I, which simultaneously transforms all σ_i into $-\sigma_i$. This group — often called Z_2 in the relevant literature — has the following character table.

$G_0 = Z_2$	E	I
g	1	1
u	1	$\bar{1}$

The only active irr. rep. is the "u" one. Its basis state is the one-dimensional OP η, which changes sign upon inversion. In the lower-symmetric phase, the only subgroup that can be realized is the subgroup $\{E\}$, as E is the only element leaving η invariant.

The second-order monomial is η^2. The third-order monomial is absent: η^3 changes sign upon inversion, i.e. it is not invariant with respect to G_0. A fourth-order term η^4 can appear in the Landau functional $\Phi[\eta], T$, which, in the lowest order, is written as

$$\Phi[T, \eta] = \Phi_0(T) + A(T, P)\eta^2 + D(T, P)\eta^4$$

At one side of the Curie point (which we take to be the high-temperature side), $A > 0$, provided $D > 0$, the functional has a minimum at $\eta = 0$, i.e. the system has the full Z_2 symmetry. At the Curie point, $A = 0$, and below it, $A < 0$. From the minimization of Φ, i.e. from the equation $\partial\Phi/\partial\eta = 0$, we find

$$A\eta + 2D\eta^3 = 0$$

with two solutions: $\eta = 0$ and $\eta = \pm\sqrt{-\frac{A}{2D}}$. The solution $\eta = 0$ represents, below the Curie point, a maximum of Φ. The other two solutions are the minima of the Landau functional, and both realize a symmetry-broken state below the phase transition. If the system assumes one of the minima, its symmetry group will consist of only the identity operation. The remaining operation of the full symmetry group transforms one minimum into the other. Thus, the existence of a larger symmetry group above the phase transition manifests itself below the phase transition through the existence of different states, which are related by operations of the full symmetry group Z_2.

The application of the Landau theory to a Heisenberg exchange ferro-magnet follows the same arguments. The Heisenberg classical ferromagnet is defined by the Hamiltonian

$$H = -J\sum_{i,j} \mathbf{n}_i \cdot \mathbf{n}_j$$

where the variables $\mathbf{n}$ are unit (pseudo) vectors in a three-dimensional Euclidean space. The symmetry group of this Hamiltonian is the full rotation group O_3, which includes improper rotations. We search for a phase transition where the IR D_1^+ is realized below the Curie point. Third-order degree polynomials, which contain odd powers of the basis functions of D_1^+, cannot be excluded from the O_3 symmetry alone. However, the time reversal symmetry represents a further symmetry element of the Hamiltonian, and the odd-order polynomials of the OP $\eta \cdot (n_x, n_y, n_z)$ are not invariant with respect to time reversal symmetry and cannot appear in Φ. We have at least one invariant of the fourth degree $((\eta^2)^2)$, which is independent of the direction of the OP. In fact, as the group O_3 contains rotations by arbitrary angles, we come to the conclusion that this is the only fourth-degree invariant polynomial. Thus, below the phase transition, we will have a continuous degeneracy of the system corresponding to the surface of a sphere of radius 1, as any direction minimizes the fourth-order invariant. This continuous degeneracy again reflects the continuous symmetry of the system above the transition point. The expansion of Φ contains only the amplitude of the OP, and its minimization proceeds along the same lines encountered for the Ising model. Once the system has assumed one of the many degenerate states corresponding to some direction of the OP, the symmetry of the system is lowered from O_3 to O_2 because the state of the system below the transition point will only be invariant with respect to rotations around the direction of the OP.

10.5   Landau's Model of the Liquid–Solid Phase Transition

(This is a tutorial rendering of the paper by S. Alexander and J. McTague, "Should all crystals be bcc? Landau theory of solidification and crystal nucleation", *Phys. Rev. Lett.* **41**, 702 (1978), https://journals.aps.org/prl/abstract/10.1103/PhysRevLett.41.702.) The symmetry group of the liquid is primarily the full group T of continuous translations in space and the full group of rotations. The irr. reps. of this direct product group are characterized by $\mathbf{k}$-vectors with endpoints lying on the surface of a sphere with a variable radius. The $\mathbf{k} = 0$ representation contains functions remaining invariant with respect to translations and rotations, i.e. it is the "1" representation. In the solid phase, we expect each irr. rep. to "condensate" into the star of $\mathbf{k}$ within the first Brillouin zone. Since the liquid is isotropic, $\rho_0 = $ const. At the point of transition, ρ becomes $\rho_0 + \delta\rho$, where $\delta\rho$ will, in general, assume the form of a Fourier integral

$$\delta\rho(\mathbf{r}) = \int \eta(\mathbf{k})e^{-i\mathbf{k}\cdot\mathbf{r}}d\mathbf{k}$$

(to ensure the reality of $\delta\rho$, we impose $\eta(-\mathbf{k}) = \eta^*(\mathbf{k})$). The integral should be taken over successive spherical surfaces, with the variable $\mathbf{k}$ extending over all $\mathbf{k}$-vectors in the three-dimensional space. These spherical surfaces represent the geometrical loci of the irr. reps. of the group $T \times O_3$. The transformation property of the OP $\eta(\mathbf{k})$ under translation is given by

$$O_{\mathbf{t}}\eta(\mathbf{k}) = \int d^3\mathbf{x} \underbrace{O_{\mathbf{t}}\rho(\mathbf{x})}_{\rho(\mathbf{x}-\mathbf{t})} e^{-i\mathbf{k}\cdot\mathbf{x}} = \int d^3\mathbf{y}\rho(\mathbf{y})e^{-i\mathbf{k}(\mathbf{y}+\mathbf{t})} = e^{-i\mathbf{k}\cdot\mathbf{t}}\eta(\mathbf{k})$$

Accordingly, the monomials of the type

$$\eta(\mathbf{k}_1) \cdot \eta(\mathbf{k}_2) \cdot \eta(\mathbf{k}_3) \cdots$$

are only invariant if

$$\mathbf{k}_1 + \mathbf{k}_2 + \cdots = 0$$

Accordingly, terms linear in $\eta_{\mathbf{k}}$ cannot appear in the expansion of Φ.

At the phase transition point, we expect that one irr. rep. belonging to some $\mathbf{G}$ is realized. The second-order term reads

$$A(T,P,G)\int_{|\mathbf{k}|=G} d^3\mathbf{k} \cdot \eta(\mathbf{k}) \cdot \eta(-\mathbf{k}) = A(T,P,G) \cdot \underbrace{\int_{|\mathbf{k}|=G} d^3\mathbf{k} \cdot |\eta(\mathbf{k})|^2}_{\doteq \eta^2(\mathbf{G})}$$

The second-order term contains the square of the OP $\eta(G)$.

The third-order terms are not required by symmetry to be absent but have to be considered under the requirement that $|\mathbf{k}_i| = G$ and

$$\mathbf{k}_1 + \mathbf{k}_2 + \mathbf{k}_3 = 0$$

i.e. the three vectors build an equilateral triangle. In all third-order terms, these triangles have equal size and differ only by their orientation in space. Again, because of the rotational symmetry, the coefficients of the third-order terms depend only on the size, not on the orientation of the triangles. Therefore, all coefficients in the third-order terms are equal: Their common value will be denoted by $B(T,P,G)$. Thus, the third-order contribution is written as

$$B(T,P,G) \cdot \int\int\int_{|\mathbf{k}_i|=G} d^3\mathbf{k}_1 d^3\mathbf{k}_2 d^3\mathbf{k}_3 \delta(\mathbf{k}_1+\mathbf{k}_2+\mathbf{k}_3) \cdot \eta(\mathbf{k}_1) \cdot \eta(\mathbf{k}_2) \cdot \eta(\mathbf{k}_3)$$

We now proceed to give an example of how a calculation based on a Landau functional could proceed. We limit ourselves to a simple model of this phase transition where only three vectors contribute to the third-order term. We also neglect higher-order terms for simplicity. The simplest choice is that the crystallization process selects

$$\mathbf{k}_1 = G \cdot (1,0,0) \quad \mathbf{k}_2 = G \cdot \left(-\frac{1}{2}, \frac{\sqrt{3}}{2}, 0\right) \quad \mathbf{k}_3 = G \cdot \left(-\frac{1}{2}, -\frac{\sqrt{3}}{2}, 0\right)$$

(Note that any other set arising from a general rotation also provides a possible crystal structure with the same thermodynamic potential. This corresponds to the fact that the crystal structure generated by the solidification process can be oriented in any spatial direction.) We parametrize the amplitude of each component by the parameter $\eta(G)$ and write

$$\eta^2(\mathbf{k}_i) = \frac{\eta^2(G)}{3} \quad \text{i.e.} \quad \eta(\mathbf{k}_i) = \pm \left(\frac{\eta^2(G)}{3} \right)^{\frac{1}{2}}$$

As we have imposed that $\eta(-\mathbf{k}) = \overline{\eta(\mathbf{k})}$, the negatives of the three vectors chosen also contribute to the third-order term. Depending on the sign of $\eta(\mathbf{k}_i)$, the third-order contribution reads

$$\pm 2 \cdot B(T, P, G) \cdot \frac{\eta^3(G)}{3^{\frac{3}{2}}}$$

The total Landau functional to be minimized with respect to $\eta(G)$ in order to find the equilibrium value of $\eta(G)$ is written as

$$A \cdot \eta^2(G) + B \cdot 2 \cdot \frac{\eta^3(G)}{3^{\frac{3}{2}}} \quad \text{with} \quad \eta(\mathbf{k}_i) = + \left(\frac{\eta^2(G)}{3} \right)^{\frac{1}{2}}$$

and

$$A \cdot \eta^2(G) - B \cdot 2 \cdot \frac{\eta^3(G)}{3^{\frac{3}{2}}} \quad \text{with} \quad \eta(\mathbf{k}_i) = - \left(\frac{\eta^2(G)}{3} \right)^{\frac{1}{2}}$$

Setting the derivatives to zero gives

$$\eta(G) = \mp \frac{A}{B} \cdot \frac{1}{\sqrt{3}} \quad \Phi_{\mp}(A, B, G) = \frac{1}{27} \cdot \frac{A^3}{B^2}$$

Both solutions lead to the same thermodynamic potential. The two crystal structures arising from crystallization and corresponding to

$$\eta(\mathbf{k}_i) = \pm \frac{1}{3} \cdot \frac{|A|}{|B|}$$

are degenerate.

We compute the density function $\rho(\mathbf{r})$ for the two structures corresponding to the two signs of the OP as follows:

$$\rho(G, +\eta(G), x, y, z)$$

$$= \frac{1}{3} \cdot \frac{|A|}{|B|} \cdot \left(\cos(G \cdot x) + 2 \cos \left(\frac{1}{2} \cdot G \cdot x \right) \cdot \cos \left(\frac{\sqrt{3}}{2} \cdot G \cdot y \right) \right)$$

This density function represents a rod-like structure along the z-direction with a two-dimensional periodicity in the xy-plane. The bright sites (left-hand side of the following figure) locate a triangular lattice.

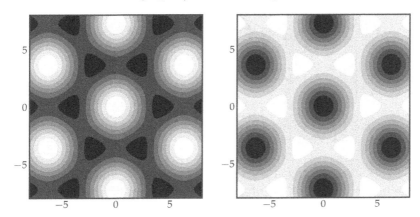

Contour plot of the density function describing a triangular lattice (left) and a honeycomb lattice (right). The bright regions are locations with high density. Plots provided by G. Pescia.

$$\rho(G, -\eta(G), x, y, z)$$

$$= -\frac{1}{3} \cdot \frac{|A|}{|B|} \cdot \left(\cos(G \cdot x) + 2 \cos\left(\frac{1}{2} \cdot G \cdot x\right) \cdot \cos\left(\frac{\sqrt{3}}{2} \cdot G \cdot y\right) \right)$$

This density function represents a rod-like structure along the z-direction with a two-dimensional periodicity in the xy-plane. The bright sites (right-hand side of the figure) locate a honeycomb lattice.

Alexander and McTague consider the case of more vectors $\mathbf{k}_i$ being realized and find that such vectors can be essentially seen as terminating at the edges of a three-dimensional polyhedron with faces consisting of identical triangles, as shown in the following figure.

The octahedron on the left generates a fcc Brillouin zone, which corresponds to a bcc lattice. The one on the right is a regular icosahedron and generates icosahedral quasi-crystals. One can show that the bcc lattice has the lowest value of the thermodynamic potential and is therefore the most probable equilibrium crystal structure that can appear when a liquid is solidified (the title of the paper is "Should All Crystals Be bcc? Landau Theory of Solidification and Crystal Nucleation"). Note that a detailed treatment of this problem (L. Landau, Zur Theorie der Phasenumwandlungen II, *Phys. Z. Sowjet.* **11**, 545 (1937)) and S. Alexander and J. McTague,

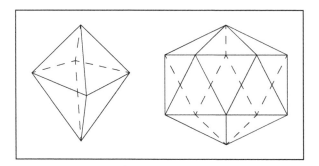

The orientation of the six vectors solving $\mathbf{k}_1 + \cdots + \mathbf{k}_6 = 0$ (left) and of the 15 vectors solving $\mathbf{k}_1 + \cdots + \mathbf{k}_{15} = 0$ (right).

Phys. Rev. Lett. **41**, 702 (1978)) requires, besides determining the extrema of the Landau functional, verifying that these represent true minima of the functional. In addition, to distinguish equilibrium states from metastable ones, the actual values of the Landau functional computed at the various solutions must be compared. All this requires a more comprehensive approach that includes terms of order higher than the ones considered here and is beyond the scope of this lecture.

10.6　The Method of Invariant Polynomials Applied to the C_{4v} Symmetry Group

We study the case of a physical system with point group symmetry C_{4v} in the high-temperature phase. We need, of course, the character table. Some basis functions are also welcome (and we will use them). The transformation of the coordinates (x, y) under the symmetry elements is also given.

G_Δ	E	$C_{2z}\ (=\bar{x}\bar{y})$	$C_{4z}^{-1}(=\bar{y}x)$ $C_4(=y\bar{x})$	$IC_{2x}(=\bar{x}y)$ $IC_{2y}(=x\bar{y})$	$IC_{2xy}(=\bar{y}\bar{x})$ $IC_{2y\bar{y}}\ (=yx)$	Basis functions
Δ_1	1	1	1	1	1	
Δ_1'	1	1	1	$\bar{1}$	$\bar{1}$	
Δ_2	1	1	$\bar{1}$	1	$\bar{1}$	
$\Delta_{2'}$	1	1	$\bar{1}$	$\bar{1}$	1	
Δ_5	2	$\bar{2}$	0	0	0	x, y

We select, as an active irr. rep., the Δ_5 representation. The OP will be a two-dimensional one: (η_1, η_2), transforming as the coordinates (x, y). Let us now construct the Landau functional that describes phase transitions from the phase with C_{4v} symmetry into a phase carrying the OP (η_1, η_2).

• Invariant polynomials of order 2: We already know that the invariant polynomial of order 2 will be $\eta_1^2 + \eta_2^2$, but we want to make sure that we know how to proceed in a concrete situation. The following three monomials of order 2 can be built with a two-dimensional OP:

$$\eta_1^2 \quad \eta_1 \cdot \eta_2 \quad \eta_2^2$$

These monomials carry a three-dimensional representation — let us call it the representation Γ^2 — of C_{4v}. The trace can be obtained by looking at the transformation properties of the monomials, given that η_1 and η_2 transform as (x, y).

$$C_{2z}\eta_1^2 = (-\eta_1)^2 \quad C_{2z}\eta_1\eta_2 = -\eta_1 \cdot (-\eta_2) \quad C_{2z}\eta_2^2 = (-\eta_2)^2$$

The trace of the matrix representing C_{2z} in the representation Γ^2 is therefore 3. Similarly, we can compute the traces for all other elements and arrive at the following.

G_Δ	E	$C_{2z}\ (=\bar{x}\bar{y})$	$C_{4z}^{-1}(=\bar{y}x)$ $C_4(=y\bar{x})$	$IC_{2x}(=\bar{x}y)$ $IC_{2y}(=x\bar{y})$	$IC_{2xy}(=\bar{y}\bar{x})$ $IC_{2y\bar{y}}\ (=yx)$	Basis functions
Γ^2	3	3	$\bar{1}$	1	1	$\eta_1^2, \eta_1 \cdot \eta_2, \eta_2^2$

A look at the character table reveals that the decomposition of Γ^2 into irr. reps. contains Δ_1 only once. We find the symmetry-adapted basis function by using the projector method:

$$P^{\Delta_1}\eta_1^2 = \eta_1^2 + \eta_1^2 + \eta_2^2 + \eta_2^2 + \eta_1^2 + \eta_1^2 + \eta_2^2 + \eta_2^2 \propto \eta_1^2 + \eta_2^2$$

This is the result we expected from our general arguments.

• Invariant polynomials of order 3: The following four monomials of order 3 can be built with a two-dimensional OP:

$$\eta_1^3 \quad \eta_1^2 \cdot \eta_2 \quad \eta_1 \cdot \eta_2^2 \quad \eta_2^3$$

These monomials carry a four-dimensional representation — let us call it the representation Γ^3 — of C_{4v}. The trace can be obtained by looking at the transformation properties of the monomials, given that η_1 and η_2 transform as (x, y).

$$C_{2z}\eta_1^3 = (-1) \cdot \eta_1^3 \quad C_{2z}\eta_1^2\eta_2 = (-\eta_1)^2 \cdot (-\eta_2) = -1 \cdot \eta_1^2 \cdot \eta_2 \quad \text{etc.}$$

The trace of the matrix representing C_{2z} in the representation Γ^3 is therefore $\bar{4}$. Similarly, we can compute the traces for all other elements and arrive at the following table.

G_Δ	E	C_{2z} $(=\bar{x}\bar{y})$ $C_4(=y\bar{x})$	$C_{4z}^{-1}(=\bar{y}x)$ $IC_{2y}(=x\bar{y})$	$IC_{2x}(=\bar{x}y)$	$IC_{2xy}(=\bar{y}\bar{x})$ $IC_{2y\bar{y}}\ (=yx)$	Basis functions
Γ^3	4	$\bar{4}$	0	0	0	$\eta_1^3, \eta_1^2\cdot\eta_2, \eta_1\cdot\eta_2^2, \eta_2^3$

A look at the character table reveals that the decomposition of Γ^3 into irr. reps. DOES NOT contain Δ_1. This means that third order polynomials are forbidden by symmetry in the Landau functional.

• Invariant polynomials of order 4: There are five monomials of order 4 that can be built with a two-dimensional OP:

$$\eta_1^4 \quad \eta_1^3\cdot\eta_2^1 \quad \eta_1^2\cdot\eta_2^2 \quad \eta_1\cdot\eta_2^3 \quad \eta_2^4$$

These monomials carry a five-dimensional representation — let us call it the representation Γ^4 — of C_{4v}. The trace can be obtained by looking at the transformation properties of the monomials, given that η_1 and η_2 transform as (x,y), in a way similar to the one we have used before. Finally, we get the characters as follows.

G_Δ	E	C_{2z} $(=\bar{x}\bar{y})$ $C_4(=y\bar{x})$	$C_{4z}^{-1}(=\bar{y}x)$ $IC_{2y}(=x\bar{y})$	$IC_{2x}(=\bar{x}y)$	$IC_{2xy}(=\bar{y}\bar{x})$ $IC_{2y\bar{y}}\ (=yx)$	Basis functions
Γ^4	5	5	1	1	1	$\eta_1^4, \eta_1^3\cdot\eta_2^1$ $\eta_1^2\cdot\eta_2^2, \eta_1\cdot\eta_2^3, \eta_2^4$

A look at the character table reveals that the decomposition of Γ^4 into irr. reps. contains Δ_1 TWICE. This means that there are two fourth-order polynomials in the Landau functional. We find them using the projector method:

$$P^{\Delta_1}\eta_1^4 \propto \eta_1^4 + \eta_2^4 \quad P^{\Delta_1}\eta_1^2\cdot\eta_2^2 \propto \eta_1^2\cdot\eta_2^2$$

The Landau functional describing the phase transition is given by

$$\Phi(\eta_1,\eta_2,T,p) = \Phi_0(T,P) + A(T,p)\cdot(\eta_1^2+\eta_2^2) + D_1(T,P)\cdot(\eta_1^4+\eta_2^4)$$
$$+ D_2(T,p)\cdot\eta_1^2\cdot\eta_2^2$$

We find the extremal values of this function by finding the zero of the coupled algebraic system of equations:

$$\frac{\partial\Phi}{\partial\eta_1} = 2A\eta_1 + 4D_1\eta_1^3 + 2D_2\eta_1\cdot\eta_2^2 = 0$$

$$\frac{\partial\Phi}{\partial\eta_2} = 2A\eta_2 + 4D_1\eta_2^3 + 2D_2\eta_1^2\cdot\eta_2 = 0$$

i.e.

$$\eta_1 \left(2A + 4D_1\eta_1^2 + 2D_2\eta_2^2\right) = 0 \quad \eta_2 \left(2A + 4D_1\eta_2^2 + 2D_2\eta_1^2\right) = 0$$

One solution is $\eta_1 = \eta_2 = 0$. This solution is a minimum if $A > 0$ because in that case, the determinant of the Hesse matrix

$$\begin{pmatrix} \frac{\partial^2 \Phi}{\partial \eta_1^2} & \frac{\partial^2 \Phi}{\partial \eta_1 \partial \eta_2} \\ \frac{\partial^2 \Phi}{\partial \eta_2 \partial \eta_1} & \frac{\partial^2 \Phi}{\partial \eta_2^2} \end{pmatrix}$$

is positive and $\frac{\partial^2 \Phi}{\partial \eta_1^2} > 0$. For $A < 0$, the solution is a maximum of the function, as the determinant is still positive but $\frac{\partial^2 \Phi}{\partial \eta_1^2} < 0$. We set $A = 0$ at the transition point $T = T_c$ and choose $A > 0$ on one side of the transition and $A < 0$ on the other side. This can be realized by, for example, taking $A \propto (T - T_C)$ (in which case the high-symmetric phase with the solution $\eta = 0$ is on the high-temperature side of the transition). We can also have $A \propto (T_C - T)$, in which case the phase transition is the so-called inverse phase transition because the high-symmetric phase is on the low-temperature side.[4] In the following, we assume $A \propto (T - T_C)$. In this case, the solutions with a finite value of η in the low-temperature phase minimize the Landau potential and correspond to equilibrium states of the system below the phase transition. Solving the coupled equations, we find four such solutions:

$$\sqrt{-\frac{2A}{4D_1 + 2D_2}}(1,1) \quad \sqrt{-\frac{2A}{4D_1 + 2D_2}}(1,-1)$$

$$\sqrt{-\frac{2A}{4D_1 + 2D_2}}(-1,1) \quad \sqrt{-\frac{2A}{4D_1 + 2D_2}}(-1,-1)$$

The symmetry elements of each of these four vectors build the subgroup of C_{4v} realized in the symmetry-broken phase: $\{E, \sigma\}$, with σ being a reflection in the lines parallel to $(1,1)$ and $(-1,1)$. We assume, for the sake of illustration, that the "liquid" phase with C_{4v} symmetry is represented by a density function of the type $\frac{|x|+|y|}{x^2+y^2}$ and that the basis functions of the

---

[4]N. Saratz, A. Lichtenberger, O. Portmann, U. Ramsperger, A. Vindigni, and D. Pescia, *Phys. Rev. Lett.* **104**, 077203 (2010), https://journals.aps.org/prl/abstract/10.1103/PhysRevLett.104.077203.

rep. Δ_5 are $\frac{x}{x^2+y^2}$ and $\frac{y}{x^2+y^2}$. Examples of the density functions realized in the liquid and symmetry-broken states at some low temperatures are given in the following figure.

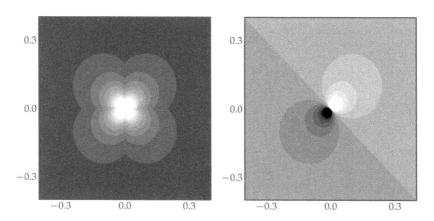

Left: Contour plot of the density function distribution with C_{4v} symmetry, as it can be realized in the "liquid" state of the system. Right: A density function distribution with a finite-order parameter realizing the (1,1) solution. The symmetry is reduced to a subgroup of C_{4v}. Plots provided by G. Pescia.

PART II

Exercises

11

EXERCISES TO LECTURE 1

Question 1: The point group C_{3v}

The C_{3v} group is the so-called pyramidal group and describes the symmetry of a pyramid with an equilateral triangle as its base.

A. Find all symmetry operations of (C_{3v}). Name them using the Schönflies notation. Describe the C_{3v} group as a group of permutations of the numbers $(1, 2, 3)$, i.e. as the S_3 group. Construct two-dimensional matrices describing the transformation of suitably chosen basis vectors under the symmetry elements of C_{3v}.

B. Construct the composition table and find the conjugacy classes.

Question 2: Cyclic groups

In a *cyclic* group, the group elements are generated by the power of one single element, called the *generator* of the group.

A. Show that a cyclic group is Abelian.

B. Consider the set of complex numbers $\{1, i, -1, -i\}$. Endow it with a multiplication rule so that it becomes a group, and construct the corresponding composition table.

C. Show that it is a cyclic group.

D. Consider now the set $C_4 = \{E, C_4, C_4^2, C_4^3\}$, and show that it has the same composition table as $\{1, i, -1, -i\}$.

E. Find the generator.

F. Draw a figure with the point group symmetry C_4.

Question 3: The point groups C_{2v} and C_4

A. Find the symmetry operations for both C_{2v} and C_4 groups, name them using the Schönflies notation, construct the corresponding composition tables, describe them as subgroups of permutations of four objects (1, 2, 3, 4) (the S_4 group), and construct two-dimensional matrices describing the transformation of suitably chosen basis vectors under the symmetry elements of C_{2v} and C_4.

B. Find the conjugacy classes.

Question 4: Subgroups

Find all subgroups of C_{4v} (the symmetry group of a square-based pyramid). *Hint: A convenient way of determining subgroups of the symmetry group of a given physical system is by distorting the system in various ways and searching for the symmetry elements of the distorted system. As an example, distort the square base of the pyramid so that it becomes a rectangle. Which subgroup is then produced?*

Question 5: Prove the following propositions

1. Show that two perpendicular twofold rotation axes imply the presence of a third twofold axis perpendicular to them. *Hint: Take the two perpendicular rotation axes along x and y, construct the rotation matrices, and show that their multiplication produces a rotation about z.*

2. By explicit construction of composition tables, show that there are exactly two different composition tables for groups of order 4. Give an example for both types of groups.

3. Show that the so-called *left translation* defined by the map $f(s) = a \circ s$, $\forall s$ in G with $a \in G$ being fixed, is a bijective map, i.e. it is injective and surjective.

Question 6: Embedding in O_h

Consider the table giving the transformation of xyz under the 48 operations of the group O_h.

Find, among these operations, those building the subgroups C_{2v}, C_{3v}, and C_{4v}.

Hint: To solve this problem, inscribe a suitable rectangle (triangle and square, respectively) into the cube and collect those operations that transform the rectangle (triangle and square) into itself by simultaneously leaving the normal to the rectangle (triangle and square) invariant.

Solution 1: The point group C_{3v}

A. The symmetry operations, the group of permutation of three objects $(1, 2, 3)$, and the two-dimensional matrices describing the transformation of C_{3v} are shown in the following figure.

E	C_3	$C_3^2 \equiv C_3^{-1}$	σ_1	σ_2	σ_3
$\begin{pmatrix} 1 & 2 & 3 \\ 1 & 2 & 3 \end{pmatrix}$	$\begin{pmatrix} 1 & 2 & 3 \\ 3 & 1 & 2 \end{pmatrix}$	$\begin{pmatrix} 1 & 2 & 3 \\ 2 & 3 & 1 \end{pmatrix}$	$\begin{pmatrix} 1 & 2 & 3 \\ 1 & 3 & 2 \end{pmatrix}$	$\begin{pmatrix} 1 & 2 & 3 \\ 3 & 2 & 1 \end{pmatrix}$	$\begin{pmatrix} 1 & 2 & 3 \\ 2 & 1 & 3 \end{pmatrix}$
$\begin{pmatrix} 1 & 0 \\ 0 & 1 \end{pmatrix}$	$\begin{pmatrix} \frac{-1}{2} & \frac{-\sqrt{3}}{2} \\ \frac{\sqrt{3}}{2} & \frac{-1}{2} \end{pmatrix}$	$\begin{pmatrix} \frac{-1}{2} & \frac{\sqrt{3}}{2} \\ \frac{-\sqrt{3}}{2} & \frac{-1}{2} \end{pmatrix}$	$\begin{pmatrix} \frac{1}{2} & \frac{\sqrt{3}}{2} \\ \frac{\sqrt{3}}{2} & \frac{-1}{2} \end{pmatrix}$	$\begin{pmatrix} \frac{1}{2} & \frac{-\sqrt{3}}{2} \\ \frac{-\sqrt{3}}{2} & \frac{-1}{2} \end{pmatrix}$	$\begin{pmatrix} -1 & 0 \\ 0 & 1 \end{pmatrix}$

From top to bottom: The elements of C_{3v} as symmetry operations of an equilateral triangle, their labeling, their representation as permutation of numbers, and their representations as 2×2 matrices.

B. On the basis of A, one can compute the following composition table.

C_{3v}	E	C_3	C_3^{-1}	σ_1	σ_2	σ_3
E	E	C_3	C_3^{-1}	σ_1	σ_2	σ_3
C_3	C_3	C_3^{-1}	E	σ_3	σ_1	σ_2
C_3^{-1}	C_3^{-1}	E	C_3	σ_2	σ_3	σ_1
σ_1	σ_1	σ_2	σ_3	E	C_3	C_3^{-1}
σ_2	σ_2	σ_3	σ_1	C_3^{-1}	E	C_3
σ_3	σ_3	σ_1	σ_2	C_3	C_3^{-1}	E

To find the conjugacy classes, we point out that C_3 and C_3^{-1} belong to the same conjugacy class, as applying a reflection σ changes the handedness of the coordinate system. σ_1, σ_2, and σ_3 belong to the same conjugacy class as a rotation C_3 brings the different reflection planes into each other. Thus, the three conjugacy classes of C_{3v} are

$$\{E\} \quad \{C_3, C_3^{-1}\} \quad \{\sigma_1, \sigma_2, \sigma_3\}$$

Solution 2: Cyclic groups

A. Consider two arbitrary elements $a, b \in G$:

$$a \cdot b = g^{n_a} g^{n_b} = \underbrace{(g \cdots g)}_{n_a} \cdot \underbrace{(g \cdots g)}_{n_b} = \underbrace{(g \cdots g)}_{n_b} \cdot \underbrace{(g \cdots g)}_{n_a}$$

$$= g^{n_b} g^{n_a} = b \cdot a$$

B. The composition table of $\{\{1, i, -1, -i\}, \cdot\}$ reads

$\cdot$	1	i	$\bar{1}$	$\bar{i}$
1	1	i	$\bar{1}$	$\bar{i}$
i	i	$\bar{1}$	$\bar{i}$	1
$\bar{1}$	$\bar{1}$	$\bar{i}$	1	i
$\bar{i}$	$\bar{i}$	1	i	$\bar{1}$

C. Through the successive powers of i, one can generate all the other elements of $\{1, i, -1, -i\}$.

D. By replacing the element $i \rightarrow C_4$, one can show that $\{1, i, -1, -i\}$ and C_4 are isomorphic:

$\{1, i, \bar{1}, \bar{i}\}$	isomorphic	C_4	E	C_4	C_4^2	C_4^3
1	$\longleftrightarrow$	1	E	C_4	C_4^2	C_4^3
i	$\longleftrightarrow$	C_4	C_4	C_4^2	C_4^3	E
$\bar{1}$	$\longleftrightarrow$	C_4^2	C_4^2	C_4^3	E	C_4
$\bar{i}$	$\longleftrightarrow$	C_4^3	C_4^3	E	C_4	C_4^2

E. Since C_4 and $\{1, i, -1, -i\}$ are isomorphic, the generator i found for $\{1, i, -1, -i\}$ is replaced by C_4.

F. The possible objects that are members of the C_4 group belong to the class of the so called "Chiral objects": Their symmetry group contains neither the inversion nor a mirror plane as the symmetry element. A drawing with C_4 symmetry is shown next. It is a polyomino with 12 squares. Polyominoes are planar figures obtained by joining one or more equal squares. They are

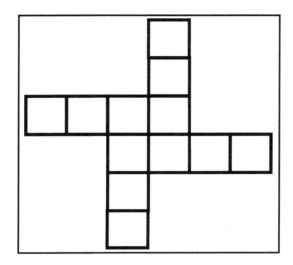

A drawing with C_4-symmetry group.

popular as mathematical and puzzle games. For more information see e.g. https://en.wikipedia.org/wiki/Polyomino. A further (in)famous drawing with C_4 symmetry is the swastika, that was used as a symbol for the NSDAP in Hitler's Germany. For an history of the swastika as symbol see e.g. https://en.wikipedia.org/wiki/Swastika. An object with C_4 symmetry is e.g. the central propeller of the RMS Olympic, a photograph of which can be found e.g. in https://en.wikipedia.org/wiki/RMS_Olympic.

Solution 3: The point groups C_{2v} and C_4

A. The symmetry operations (a) and the composition tables (b) read

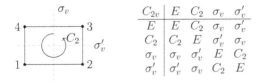

Left: The operations of C_{2v}. Right: The multiplication table.

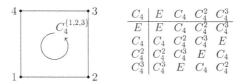

Left: The operations of C_4. Right: The multiplication table.

The permutation of four objects $(1, 2, 3, 4)$ (c) and the two-dimensional matrices describing the symmetry transformations (d) are listed as follows:

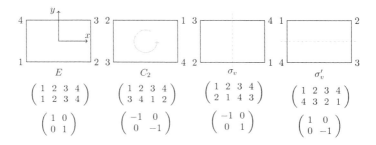

From top to bottom: The elements of C_{2v} as symmetry operations of an equilateral triangle, their labeling, their representation as permutation of numbers, and their representations as 2×2 matrices.

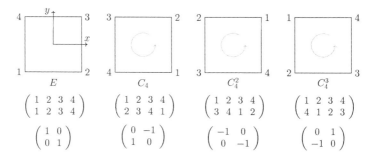

From top to bottom: The elements of C_4 as symmetry operations of an equilateral triangle, their labeling, their representation as permutation of numbers, and their representations as 2×2 matrices.

B. The groups are Abelian: Each element builds a class on its own.

Solution 4: Subgroups

Consider a square-based pyramid with nodes A, B, C, D, and T.

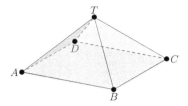

Pyramid with a square base.

We distort the square base with the aim of lowering the symmetry. The following figures illustrate all possible distortions, resulting in four different subgroups of C_{4v} (six, including the trivial subgroups of C_{4v} and the identity operation).

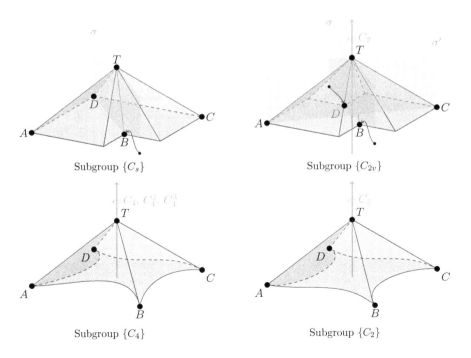

Subgroup $\{C_s\}$ Subgroup $\{C_{2v}\}$

Subgroup $\{C_4\}$ Subgroup $\{C_2\}$

The four possible non-trivial subgroups of C_{4v}.

Solution 5: Prove the following propositions

1. Consider the three-dimensional Cartesian coordinate system labeled with the axes x, y and z. Without loss of generality, the x- and y-axes will both be twofold rotation axes C_2. The corresponding rotation matrices are written as

$$C_{2x} = \begin{pmatrix} 1 & 0 & 0 \\ 0 & -1 & 0 \\ 0 & 0 & -1 \end{pmatrix} \quad C_{2y} = \begin{pmatrix} -1 & 0 & 0 \\ 0 & 1 & 0 \\ 0 & 0 & -1 \end{pmatrix}$$

Their multiplication must also be a symmetry element:

$$C_{2x}C_{2y} = \begin{pmatrix} 1 & 0 & 0 \\ 0 & -1 & 0 \\ 0 & 0 & -1 \end{pmatrix} \cdot \begin{pmatrix} -1 & 0 & 0 \\ 0 & 1 & 0 \\ 0 & 0 & -1 \end{pmatrix} = \begin{pmatrix} -1 & 0 & 0 \\ 0 & -1 & 0 \\ 0 & 0 & 1 \end{pmatrix}$$

This last matrix describes a twofold rotation about the z-axis.

2. We build the possible composition tables starting from

	e	a	b	c
e	e	a	b	c
a	a			
b	b			
c	c			

Case 1. Assume $a \circ a = b$. Then, one can fill the table only in such a way that the following table results (the filling *must* obey the rearrangement theorem, according to which the elements appear only once along the columns or along the lines of the composition table):

	e	a	b	c
e	e	a	b	c
a	a	b	c	e
b	b	c	e	a
c	c	e	a	b

This is a so-called cyclic group: One obtains the lines of the composition table by cycling the elements from the previous line — remember that a cycle of four elements (i_1, i_2, i_3, i_4) is the permutation:

$$\begin{pmatrix} i_1 & i_2 & i_3 & i_4 \\ i_2 & i_3 & i_4 & i_1 \end{pmatrix}$$

As a consequence, the elements of the groups are produced by the powers of one element. In this case, $a^0 = e, a^1 = a, a^2 = b, a^3 = c$. An example of a cyclic group with four elements is C_4. One can verify that the composition table of C_4 is indeed isomorphic to the one we have just constructed.

Case 2. Assume now $a \circ a = e$. We can continue filling the composition table and find one possible table in this second case (case 2a):

	e	a	b	c
e	e	a	b	c
a	a	e	c	b
b	b	c	a	e
c	c	b	e	a

Here, a subtle situation arises. This table is not really something new; it is again the table of a cyclic group, with the only difference from case 1 being that the generator is the element b rather than a. The group produced in case 2a is isomorphic to the group produced in case 1: Both are cyclic.

Within case 2, we find a second table (case 2b):

	e	a	b	c
e	e	a	b	c
a	a	e	c	b
b	b	c	e	a
c	c	b	a	e

This is the Cayley table of the so-called Klein 4-group. A physical realization of this group is the point group C_{2v}.

3. We first prove that the map $f_a(s) = a \circ s$ is injective.

Suppose that $f_a(s) = f_a(s')$ with $s \neq s'$. We have

$$s = e \circ s = a^{-1} \circ a \circ s = a^{-1} \circ f_a(s) = a^{-1} \circ f_a(s') = a^{-1} \circ a \circ s' = e \circ s' = s'$$

contrary to the assumption.

We show now that the map is surjective:

Consider an arbitrary $s' \in G$. We must show that one can find an s such that $f_a(s) = s'$. Take $s = a^{-1} \circ s'$. Then,

$$f_a(s) = a \circ s = a \circ a^{-1} \circ s' = e \circ s' = s'$$

Solution 6: Embedding in O_h

Type	Operation	Coordinate	Type	Operation	Coordinate
C_1	C_1	xyz	I	I	$\bar{x}\bar{y}\bar{z}$
C_4^2	C_{2z}	$\bar{x}\bar{y}z$	IC_4^2	IC_{2z}	$xy\bar{z}$
	C_{2x}	$x\bar{y}\bar{z}$		IC_{2x}	$\bar{x}yz$
	C_{2y}	$\bar{x}y\bar{z}$		IC_{2y}	$x\bar{y}z$
C_4	C_{4z}^{-1}	$\bar{y}xz$	IC_4	IC_{4z}^{-1}	$y\bar{x}\bar{z}$
	C_{4z}	$y\bar{x}z$		IC_{4z}	$\bar{y}x\bar{z}$
	C_{4x}^{-1}	$x\bar{z}y$		IC_{4x}^{-1}	$\bar{x}z\bar{y}$
	C_{4x}	$xz\bar{y}$		IC_{4x}	$\bar{x}\bar{z}y$
	C_{4y}^{-1}	$zy\bar{x}$		IC_{4y}^{-1}	$\bar{z}\bar{y}x$
	C_{4y}	$\bar{z}yx$		IC_{4y}	$z\bar{y}\bar{x}$
C_2	C_{2xy}	$yx\bar{z}$	IC_2	IC_{2xy}	$\bar{y}\bar{x}z$
	C_{2xz}	$z\bar{y}x$		IC_{2xz}	$\bar{z}y\bar{x}$
	C_{2yz}	$\bar{x}zy$		IC_{2yz}	$x\bar{z}\bar{y}$
	$C_{2x\bar{y}}$	$\bar{y}x\bar{z}$		$IC_{2x\bar{y}}$	yxz
	$C_{2\bar{x}z}$	$\bar{z}\bar{y}\bar{x}$		$IC_{2\bar{x}z}$	zyx
	$C_{2y\bar{z}}$	$\bar{x}\bar{z}\bar{y}$		$IC_{2y\bar{z}}$	xzy
C_3	C_{3xyz}^{-1}	zxy	IC_3	IC_{3xyz}^{-1}	$\bar{z}\bar{x}\bar{y}$
	C_{3xyz}	yzx		IC_{3xyz}	$\bar{z}\bar{x}\bar{y}$
	$C_{3x\bar{y}z}^{-1}$	$z\bar{x}\bar{y}$		$IC_{3x\bar{y}z}^{-1}$	$\bar{z}xy$
	$C_{3x\bar{y}z}$	$\bar{y}\bar{z}x$		$IC_{3x\bar{y}z}$	$yz\bar{x}$
	$C_{3x\bar{y}\bar{z}}^{-1}$	$\bar{z}\bar{x}y$		$IC_{3x\bar{y}\bar{z}}^{-1}$	$zx\bar{y}$
	$C_{3x\bar{y}\bar{z}}$	$\bar{y}z\bar{x}$		$IC_{3x\bar{y}\bar{z}}$	$y\bar{z}x$
	$C_{3xy\bar{z}}^{-1}$	$\bar{z}x\bar{y}$		$IC_{3xy\bar{z}}^{-1}$	$z\bar{x}y$
	$C_{3xy\bar{z}}$	$y\bar{z}\bar{x}$		$IC_{3xy\bar{z}}$	$\bar{y}zx$

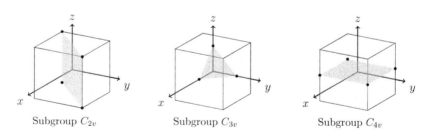

Subgroup C_{2v} Subgroup C_{3v} Subgroup C_{4v}

Embedding triangle, rectangle, and square into a cube.

The figure provides the rectangle (triangle and square), which can be inscribed into the cube to find those elements that transform the rectangle (triangle and square) into itself from the O_h operation table.

Subgroup C_{2v}: The normal to the rectangle is along the bisectrix of the xy-plane. Referring to the table of operation, we are looking for those symmetry transformations that leave xy invariant or transform them into yx:

$$C_1 \quad C_{2xy} \quad IC_{2z} \quad IC_{2x\bar{y}}$$

Subgroup C_{3v}: The normal to the triangle is along xyz. We need to select those operations that leave xyz invariant or transform them into a permutation. These are

$$C_1, C_{3xyz}^{-1}, C_{3xyz}, IC_{2x\bar{y}}, IC_{2\bar{x}z}, IC_{2y\bar{z}}$$

Subgroup C_{4v}: The normal to the square is along z, so we select all those operations that transform z into itself:

$$C_1, C_{4z}, C_{2z}, C_{4z}^{-1}, IC_{2y}, IC_{2x}, IC_{2x\bar{y}}, IC_{2xy}$$

12

EXERCISES TO LECTURE 2

Question 1: Coefficients of decomposition

Consider a group G with an average and a finite-dimensional representation $T(g)$. Show that for the coefficients n_r in the decomposition of T into irreducible irreducible representations T_r ($T = \oplus n_r T_r$), the following relations hold true:

$$\mathscr{M}_G \mid \chi_T(g) \mid^2 = \sum_r n_r^2$$

Hint: Use the orthogonality of the characters of irreducible representations.

Question 2: Two propositions on characters

Given is a group G with an average and an n-dimensional representation T. Let the representation of the elements of the conjugate class $\{g\} \neq \{e\}$ have the character n (with e being the identity element).

1. Show that the matrix representation of the elements of this class is necessarily the identity matrix $\mathbb{1}$.
2. Use (a) to show that an n-dimensional representation T is one-to-one if and only if only e has the character n.

Question 3: Two-element group

A. Construct the multiplication table of a two-element group.
B. Construct its character table.
C. Use the multiplication table and the character table to construct all possible two-dimensional representations of the two-element group. *Hint: Use the fact that a two-dimensional representation can be decomposed into the sum of the two one-dimensional irreducible representations that you*

have just found. Take into account that apparently different representations can be equivalent, i.e. they can be reduced to the same representation by a similarity transformation (this is the most cumbersome part of the exercise).

Question 4: Character table of cyclic groups

A. Construct the multiplication table of the group C_3 (order the elements such that the identity appears along the diagonal).

B. The bulk of the table contains nine entries. Consider these entries as variables, and compute formally the determinant (or "circulant", as the multiplication table is "cyclic") of the multiplication table, seen as a 3×3 matrix.

C. Construct the character table of the group C_3.

D. Frobenius[1] (1849–1017) (following a suggestion by Dedekind) developed the theory of characters (and ultimately the representation theory of groups) because he noticed that the determinant of the multiplication table of any finite group can be factorized into homogeneous polynomials of their entries that contain the characters of the irreducible representations as coefficients. Show that this is the case for the circulant of the C_3 multiplication table.

E. Find a two-dimensional representation of the group, and prove its irreducibility or else reduce it.

F. Generalize the character table to any cyclic group of order n, i.e. C_n.

Question 5: Tower representation of C_{2v} and C_{3v}

A. Construct the character tables of C_{2v} and C_{3v}.

B. Write the matrices for the "tower" (permutation) representations of C_{2v} and C_{3v}. *Hint: For this purpose, find a suitable "object" that can be used for the "tower" representation.*

C. Find their characters.

D. Find their irreducible components.

Question 6: The group $C_{\infty v}$

The group $C_{\infty v}$ is one of the so-called Curie groups and consists of all proper rotations about a fixed axis and an infinite number of reflection planes of the type σ_v. It is particularly important in chemistry, as it is the symmetry group of many linear molecules. In the language of planar isometries, it is

---

[1]Ferdinand Georg Frobenius (1849–1917), mathematician, see e.g. https://en.wikipedia.org/wiki/Ferdinand_Georg_Frobenius.

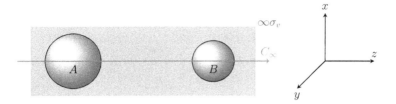

Graphical illustration of the group $C_{\infty v}$.

the group O_2, which also provides a natural, two-dimensional representation of the elements of $C_{\infty v}$.

A. From the lecture, find the appropriate 2×2 matrices that provide the natural representation of O_2.

B. How many parameters are necessary to describe the elements of O_2? What is their range of values? Conclude that the group is compact.

C. The group is compact and has therefore an average. How is the functional written?

D. This question is not relevant for the remainder of the exercise, but it is interesting. How is the composition law written in terms of the parameters describing the group? *Hint: The functions describing the composition law are not continuous; therefore, the group is not a topological group. Yet, it has an average.*

E. Find the conjugacy classes of O_2. Pay attention to the rules for finding the conjugate elements described in the lecture.

F. Construct as many as possible representations of O_2, and test them for irreducibility. *Hint: Is it possible for the two-dimensional natural representation to be irreducible even if SO_2 has only one-dimensional irreducible representations?*

G. Consider now the monomials made of the x, y, z coordinates: $1, x, y, z, x^2, y^2, z^2, xy, xz, yz$. Construct symmetry-adapted polynomials for the suitable irreducible representations. Hint: Some polynomials can be assigned just by looking at the figure and finding out how the monomials would transform upon application of the elements of O_2. For instance, z is certainly invariant with respect to the operations of O_2 and constitutes a basis to a certain irreducible representation (which one?). x, instead, is transformed into some linear combination of x and y, so that it must belong to a two-dimensional invariant subspace with basis functions x and y. To which irreducible representation do they belong? At least in one case (to generate $x^2 - y^2$ from, say, x^2), you must use the projector technique.

Solution 1: Coefficients of decomposition

$$\mathcal{M}_G \left| \chi_T(g) \right|^2 \equiv \mathcal{M}_G \chi_T^\dagger(g) \chi_T(g) = \mathcal{M}_G \sum_k T_{kk}(g)^\dagger T_{kk}(g)$$

$$= \mathcal{M}_G \sum_{r,r'} n_r \cdot n_{r'} \chi_r^\dagger(g) \chi_{r'}(g)$$

$$= \sum_{r,r'} n_r \cdot n_{r'} \cdot \underbrace{\mathcal{M}_G \chi_r^\dagger(g) \chi_{r'}(g)}_{\text{orthogonality of characters} \rightarrow = \delta_{r,r'}}$$

$$= \sum_r n_r^2$$

Solution 2: Two propositions on characters

1. As the group has an average, $T(g)$ is a unitary matrix. It can be diagonalized with n diagonal matrix elements $e^{i\theta_1}, \ldots, e^{i\theta_n}$. The character

$$\sum_j e^{i\theta_j}$$

can only be equal to n if $\theta_j = 0 \; \forall j$. This produces

$$T(g) = \mathbb{1}$$

For all other elements b of this class, we have

$$T(c)\mathbb{1}T(c)^{-1} = \mathbb{1} \quad \diamond$$

2. Assume the representation is one-to-one. Then, only $T(e) = \mathbb{1}$, and not other elements, i.e. only $T(e)$ has the character n. Assume now that $\chi(g \neq e) = n$. Then, the elements of the class $\{g\}$ all belong to the kernel of the representation, which cannot be faithful. $\diamond$

Solution 3: Two-element group

A. The multiplication table for the two elements e, a has four entries, one of which, at first glance, is unknown:

$\circ$	e	a
e	e	a
a	a	x

The rearrangement theorem imposes $x = e$.

B. There are two classes, $\{e\}$ and $\{a\}$. The trivial irreducible representation has characters 1 for each class. The second irreducible representation has characters 1 and $\bar{1}$.

G	e	a
Γ_g	1	1
Γ_u	1	$\bar{1}$

C. In the following table, we provide the possible characters for a two-dimensional representation:

G	e	a
Γ_g	1	1
Γ_u	1	$\bar{1}$
$T_1 = 2\Gamma_g$	2	2
$T_2 = 2\Gamma_u$	2	$\bar{2}$
$T_3 = \Gamma_g + \Gamma_u$	2	0

Matrix representations:

- T_1. $T_1(e) = \mathbb{1}$. *Ansatz:*

$$T_1(a) = \begin{pmatrix} 1 & x \\ y & 1 \end{pmatrix}$$

From

$$T_1(a)T_1(a) = \mathbb{1}$$

we obtain $x = y = 0$ and $T_1(a) = \mathbb{1}$.
- *Ansatz:*

$$T_2(a) = \begin{pmatrix} \bar{1} & x \\ y & \bar{1} \end{pmatrix}$$

As $T_2(a)T_2(a) = \mathbb{1}$, we obtain $T_2(e) = \mathbb{1}$,

$$T_2(a) = \begin{pmatrix} \bar{1} & 0 \\ 0 & \bar{1} \end{pmatrix}$$

- For $T_3(a)$, we make the *ansatz*

$$T_3(a) = \begin{pmatrix} x & y \\ y' & \bar{x} \end{pmatrix}$$

As $T_3(a)T_3(a) = T_3(e)$, we obtain either $y = y' = 0$, $x = 1$ and

$$T_{3,I}(a) = \begin{pmatrix} 1 & 0 \\ 0 & \bar{1} \end{pmatrix}$$

or, as a second solution,

$$T_{3,II}(a) = \begin{pmatrix} x & \frac{1-x^2}{y'} \\ y' & \bar{x} \end{pmatrix}.$$

We show that the matrices $T_{3,II}(a)$ and $T_{3,I}(a)$ are related by a similarity transformation, i.e. they are equivalent:

$$\tilde{T}_{3II}(a) = S^{-1}T_{3I}(a)S = \begin{pmatrix} a & b \\ c & d \end{pmatrix} \begin{pmatrix} 1 & 0 \\ 0 & -1 \end{pmatrix} \begin{pmatrix} a & b \\ c & d \end{pmatrix}^{-1}$$

$$= \frac{1}{ad - bc} \begin{pmatrix} ad + bc & -2ab \\ 2cd & -(ad+bc) \end{pmatrix} \equiv \begin{pmatrix} x & \frac{1-x^2}{y'} \\ y' & -x \end{pmatrix},$$

where $x = (ad + bc)/(ad - bc)$ and $y' = 2cd/(ad - bc)$.

Solution 4: Character table of cyclic groups

A.

$\circ$	e	C_3	C_3^2
e	e	C_3	C_3^2
C_3^2	C_3^2	e	C_3
C_3	C_3	C_3^2	e

You will note that the next line in the multiplication table can be obtained from the previous line by cycling. Such a multiplication table is common to cyclic groups.

B. The determinants of such "cyclic" matrices are called **circulants**. Let us assign to the elements e, C_3, C_3^2 the variables X_0, X_1, X_3. The circulant is written as

$$X_0^3 + X_1^3 + X_2^3 - 3X_0X_1X_2$$

C.

C_3	E	C_3	C_3^2
Γ_0	1	1	1
Γ_1	1	a	b
Γ_2	1	c	d

a, b, c, d are to be determined. From

$$b = a^2 \quad d = c^2$$

(multiplication table) and

$$1 + a + a^2 = 0 \quad 1 + c + c^2 = 0$$

(orthogonality of characters), we obtain

$$a \doteq \omega = +e^{+2\pi i/3} \quad c = e^{+4\pi i/3}$$

so that the character table is constructed as

C_3	E	C_3	C_3^2
Γ_0	1	1	1
Γ_1	1	ω	ω^2
Γ_2	1	ω^2	ω

D. One verifies that

$$X_0^3 + X_1^3 + X_2^3 - 3X_0 X_1 X_2 = (1 \cdot X_0 + 1 \cdot X_1 + 1 \cdot X_2)$$
$$\cdot (1 \cdot X_0 + \omega \cdot X_1 + \omega^2 \cdot X_2)$$
$$\cdot (1 \cdot X_0 + \omega^2 \cdot X_1 + \omega \cdot X_2)$$

E. We transform a pair of orthogonal vectors along the directions x and y:

$$T_{\Gamma'}(E) = \begin{pmatrix} 1 & 0 \\ 0 & 1 \end{pmatrix} \quad T_{\Gamma'}(C_3) = \begin{pmatrix} \frac{-1}{2} & \frac{-\sqrt{3}}{2} \\ \frac{\sqrt{3}}{2} & \frac{-1}{2} \end{pmatrix} \quad T_{\Gamma'}(C_3^2) = \begin{pmatrix} \frac{-1}{2} & \frac{\sqrt{3}}{2} \\ \frac{-\sqrt{3}}{2} & \frac{-1}{2} \end{pmatrix}$$

from which we can obtain the characters of the two-dimensional representation:

C_3	E	C_3	C_3^2
Γ_0	1	1	1
Γ_1	1	$e^{2\pi i/3}$	$(e^{2\pi i/3})^2$
Γ_2	1	$(e^{2\pi i/3})^2$	$e^{2\pi i/3}$
Γ'	2	-1	-1

One can recognize that the only possibility of obtaining the characters of the last line is by summing the characters of Γ_1 and Γ_2. Accordingly, $\Gamma' = \Gamma_1 \oplus \Gamma_2$.

F. We can generalize the character table of any cyclic group of order n as follows:

C_n	$C_n^0 = E$	C_n^1	$\ldots$	C_n^k	$\ldots$	C_n^{n-1}
Γ_0	1	1		$\ldots$		1
Γ_1	1	$e^{\frac{2\pi i}{n}}$	$\ldots$	$e^{\frac{2\pi i}{n}k}$	$\ldots$	$e^{\frac{2\pi i}{n}(n-1)}$
$\vdots$	$\vdots$	$\vdots$		$\vdots$		$\vdots$
Γ_j	1	$\left(e^{\frac{2\pi i}{n}}\right)^j$		$\left(e^{\frac{2\pi i}{n}k}\right)^j$		$\left(e^{\frac{2\pi i}{n}(n-1)}\right)^j$
$\vdots$	$\vdots$	$\vdots$		$\vdots$		$\vdots$
Γ_{n-1}	1	$\left(e^{\frac{2\pi i}{n}}\right)^{(n-1)}$	$\ldots$	$\left(e^{\frac{2\pi i}{n}k}\right)^{(n-1)}$	$\ldots$	$\left(e^{\frac{2\pi i}{n}(n-1)}\right)^{(n-1)}$

Solution 5: Tower representation of C_{2v} and C_{3v}

A. The conjugacy classes are

$$C_{2v} : \{E\}, \{C_2\}, \{\sigma_v\}, \{\sigma_v'\};$$
$$C_{3v} : \{E\}, \{C_3, C_3^{-1}\}, \{3\sigma\}$$

and using the rules, we find the following character tables in the BSW and M notations[2]:

C_{2v} BSW	M	E	C_2	σ_v	σ_v'
Σ_1	A_1	1	1	1	1
Σ_1'	A_2	1	1	-1	-1
Σ_2	B_1	1	-1	1	-1
Σ_2'	B_2	1	-1	-1	1

C_{3v} BSW	M	E	$C_3^{\pm}$	3σ
Λ_1	A_1	1	1	1
Λ_2	A_2	1	1	-1
Λ_3	E	2	-1	0

B.,C. The matrices corresponding to the tower representation and the resulting characters are as follows:

We summarize what we obtained above:

C_{2v}	E	C_2	σ_v	σ_v'
Σ_1	1	1	1	1
Σ_1'	1	1	-1	-1
Σ_2	1	-1	1	-1
Σ_2'	1	-1	-1	1
Σ_T	4	0	0	0

C_{3v}	E	$C_3^{\pm}$	3σ
Λ_1	1	1	1
Λ_2	1	1	-1
Λ_3	2	-1	0
Λ_T	3	0	1

---

[2]There are several ways of labeling irreducible representations. The one typically used in solid-state physics is the Γ *notation*, introduced by the German physicist Hans Bethe (1929), followed by Bouckaert, Smoluchowski, and Wigner (http://journals.aps.org/pr/abstract/10.1103/PhysRev.50.58) (BSW). The other notation was introduced by the American physicist and chemist Robert S. Mulliken (https://journals.aps.org/pr/pdf/10.1103/PhysRev.43.279) (M) and is usually found in molecular and chemical physics and may be called the *chemical notation*.

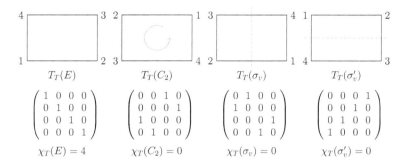

Top line: The elements of C_{2v} are represented as symmetry operations of a rectangle. Middle line: The tower matrix representations and their labels. Bottom line: The characters of the tower representations.

$$T_T(E)$$

$$\begin{pmatrix} 1 & 0 & 0 \\ 0 & 1 & 0 \\ 0 & 0 & 1 \end{pmatrix}$$

$$\chi_T(E) = 3$$

$$T_T(C_3)$$

$$\begin{pmatrix} 0 & 0 & 1 \\ 1 & 0 & 0 \\ 0 & 1 & 0 \end{pmatrix}$$

$$\chi_T(C_3) = 0$$

$$T_T(C_3^{-1})$$

$$\begin{pmatrix} 0 & 1 & 0 \\ 0 & 0 & 1 \\ 1 & 0 & 0 \end{pmatrix}$$

$$\chi_T(C_3^{-1}) = 0$$

$$T_T(\sigma_1)$$

$$\begin{pmatrix} 1 & 0 & 0 \\ 0 & 0 & 1 \\ 0 & 1 & 0 \end{pmatrix}$$

$$\chi_T(\sigma_1) = 1$$

$$T_T(\sigma_2)$$

$$\begin{pmatrix} 0 & 0 & 1 \\ 0 & 1 & 0 \\ 1 & 0 & 0 \end{pmatrix}$$

$$\chi_T(\sigma_2) = 1$$

$$T_T(\sigma_3)$$

$$\begin{pmatrix} 0 & 1 & 0 \\ 1 & 0 & 0 \\ 0 & 0 & 1 \end{pmatrix}$$

$$\chi_T(\sigma_3) = 0$$

Top line: The elements of C_{3v} are represented as symmetry operations of an equilateral triangle. Middle line: The tower matrix representations and their labels. Bottom line: The characters of the tower representations.

D. Using the decomposition theorem, we conclude that

$$\text{for } C_{2v}: \qquad \Sigma_T = \Sigma_1 \oplus \Sigma_1' \oplus \Sigma_2 \oplus \Sigma_2',$$
$$\text{for } C_{3v}: \qquad \Lambda_T = \Lambda_1 \oplus \Lambda_3.$$

Solution 6: The group $C_{\infty v}$

A. The natural representation of O_2 is, for example, the set of all matrices

$$\begin{pmatrix} \cos\varphi & -\sin\varphi \\ \sin\varphi & \cos\varphi \end{pmatrix} \cup \begin{pmatrix} 1 & 0 \\ 0 & -1 \end{pmatrix} \cdot \begin{pmatrix} \cos\varphi & -\sin\varphi \\ \sin\varphi & \cos\varphi \end{pmatrix}$$

B. Two parameters are needed to describe the elements of the group: $\varphi \in [0, 2\pi]$ and the discrete variable $m = \pm 1$, identifying the determinants of the matrices. The proper rotations are labeled by the variables $(\varphi, m = 1)$ and the improper rotations by the variables $(\varphi, m = -1)$. The parameter space is a compact set but disconnected: Graphically, it can be represented by two disjoint segments in the (m, φ)-plane.

C.

$$\frac{1}{4\pi} \sum_{m=-1}^{1} \int_0^{2\pi} f(\varphi, m) d\varphi = \frac{1}{4\pi} \int_0^{2\pi} [f(\varphi, m = 1) + f(\varphi, m = -1)] d\varphi$$

D.

$$\varphi_c = (\varphi_a + \varphi_b) \quad m_c = m_a \cdot m_b$$

E. All the reflection planes $\{\infty\sigma_v\}$ can be brought into each other by the rotations and thus belong to the same conjugacy class. Furthermore, in comparison to the case of the finite cyclic group C_n, the elements with rotation angles φ and $-\varphi$ belong to the same conjugacy class, as there is always a reflection plane which changes the handedness of the coordinate system. The conjugacy classes are, schematically,

$$\{E\}, \{C_{\pm\varphi_1}\}, \{C_{\pm\varphi_2}\}, \ldots, \{\infty\sigma_v\}$$

F. The natural representation is a two-dimensional representation of O_2, and it is irreducible: $\frac{1}{4\pi} \int_0^{2\pi} 4\cos^2 \varphi d\varphi = 1$ (the character of improper rotations is 0). Further irreducible representations are the trivial representation and the determinantal representation, where the proper rotations are represented by 1 and the improper rotations by -1. The functions $e^{\pm ima}$, which belong to the irreducible representations $\Gamma^{\pm m}$ of SO_2, are transformed into each other by the reflections. Thus, new invariant, two-dimensional subspaces are formed, which carry two-dimensional representations of O_2 with an index of m. The character table of $C_{\infty v}$ is given as follows:

$C_{\infty v}$	E	$C_{\pm\varphi}$	$\ldots$	$\infty\sigma_v$
$A_1(\Sigma^+)$	1	1	$\ldots$	1
$A_2(\Sigma^-)$	1	1	$\ldots$	-1
$E_1(\Pi)$	2	$2\cos(\varphi)$	$\ldots$	0
$E_2(\Delta)$	2	$2\cos(2\varphi)$	$\ldots$	0
$\vdots$	$\vdots$	$\vdots$	$\vdots$	$\vdots$
E_m	2	$2\cos(m\varphi)$	$\ldots$	0
$\vdots$	$\vdots$	$\vdots$	$\vdots$	$\vdots$

Now, the question is: Are these all irreducible representations of O_2? As $C_{\pm\varphi}$ must belong to the same class, the character must be even in φ.

The set $\cos n\varphi$, with $n = 0, 1, 2, \ldots$ is complete in the space of even functions in the interval $[0, 2\pi]$, so that the irreducible representations found are all irreducible representations of O_2.

G. One can assign, by inspection, symmetry-adapted polynomials to

$$A_1 \quad \leftarrow (z; x^2 + y^2 + z^2) \quad E_1 \leftarrow (x, y); (xz, yz)$$

We apply the projector technique to find

$$P_{E_2(\Delta)}x^2 \approx \int_0^{2\pi} d\varphi \cos 2\varphi \cdot \underbrace{O_\varphi x^2}_{(x \cos \varphi - y \sin \varphi)^2} \approx x^2 - y^2$$

A further polynomial in the space E_2 is obtained by, for example, rotating $x^2 - y^2$ by $\frac{\pi}{4}$ and written as xy.

13

EXERCISES TO LECTURE 3

The scope of this set of exercises is to get acquainted with the method of infinitesimal generators.

Question 1: SO₂ as exponential map

Show by explicitly summing the Taylor series that

$$e^{-i\sigma_y \varphi} = \begin{pmatrix} \cos(\varphi) & -\sin(\varphi) \\ \sin(\varphi) & \cos(\varphi) \end{pmatrix} \qquad \sigma_y = \begin{pmatrix} 0 & -i \\ i & 0 \end{pmatrix}$$

Question 2: Infinitesimal generators and exponential maps of Lie groups and non-Lie groups

A. Show that the set of matrices

$$A(t) = \begin{pmatrix} 1 & t \\ 0 & 1 \end{pmatrix} \qquad t \in \mathbb{R}$$

forms a group. Find its generator and integrate it to obtain the group elements.

B. Consider the continuous matrices

$$\begin{pmatrix} t & 0 \\ 0 & 1 \end{pmatrix}$$

Determine for which parameters t the set forms a group. Determine further the infinitesimal generator, and integrate it using the exponential map to generate the group.

Question 3: Infinitesimal generator of the scaling transformation

The scaling transformation $x' = t \cdot x$, with $t \in \mathbb{R} \backslash \{0\}$, defines a group with the composition law $t_3 = t_1 \cdot t_2$.

Taking into account that within the function space, a representation of this group is defined by $O_t f(x) = f(\frac{x}{t})$, find the infinitesimal generator of this representation and produce the exponential map.

Question 4: Differential equation for one-parameter group

To find the one-parameter group $A(t)$ generated by the infinitesimal generator I (with I being an $n \times n$ matrix), one can use the exponential map

$$A(t) := e^{-i \cdot I \cdot t}$$

An alternative and equivalent way of constructing the group is to solve the system of differential equations

$$\frac{d}{dt} A(t) = -i \cdot I \cdot A(t)$$

with the initial condition $A(0) = E$.

Construct, using both methods, the one-parameter continuous group generated by

$$I = \begin{pmatrix} i & i \\ 0 & i \end{pmatrix}$$

Question 5: Adjoint representation of a Lie algebra

We define a map in the set of elements $\{x\}$ of a Lie algebra by the relation

$$F(a)(x) \doteq [a, x]$$

with $a \in \{x\}$ being a given fixed element of the Lie algebra.

Show that this map is a representation of the elements $\{a\}$ of the Lie algebra, the so-called "adjoint representation".

Solution 1: SO_2 as exponential map

First, we calculate the square of the generator σ_y:

$$\sigma_y = \begin{pmatrix} 0 & -i \\ i & 0 \end{pmatrix} \Rightarrow \sigma_y^2 = \begin{pmatrix} 0 & -i \\ i & 0 \end{pmatrix}\begin{pmatrix} 0 & -i \\ i & 0 \end{pmatrix} = \begin{pmatrix} 1 & 0 \\ 0 & 1 \end{pmatrix}$$

$$\Rightarrow \begin{cases} \sigma_y^n = E & \text{if } n \text{ even} \\ \sigma_y^n = \sigma_y & \text{if } n \text{ odd} \end{cases}$$

Knowing the property of σ_y^n, we can sum up the terms appearing in the Taylor expansion of $e^{-i\sigma_y\varphi}$:

$$e^{-i\sigma_y\varphi} = \sum_{n=0}^{\infty} \frac{(-i\varphi)^n}{n!} \sigma_y^n$$

$$= \underbrace{\sum_{l=0}^{\infty} \frac{(-i\varphi)^{2l}}{(2l)!}}_{\cos\varphi} \cdot \sigma_y^2 + \underbrace{\sum_{l=0}^{\infty} \frac{(-i\varphi)^{(2l+1)}}{(2l+1)!}}_{(-i)\sin\varphi} \cdot \sigma_y$$

$$= \cos\varphi \begin{pmatrix} 1 & 0 \\ 0 & 1 \end{pmatrix} - i\sin\varphi \begin{pmatrix} 0 & -i \\ i & 0 \end{pmatrix}$$

$$= \begin{pmatrix} \cos(\varphi) & -\sin(\varphi) \\ \sin(\varphi) & \cos(\varphi) \end{pmatrix} \quad \square$$

Solution 2: Infinitesimal generators and exponential maps of Lie groups and non-Lie groups

A.

• Composition law:

$$\begin{pmatrix} 1 & t_1 \\ 0 & 1 \end{pmatrix}\begin{pmatrix} 1 & t_2 \\ 0 & 1 \end{pmatrix} = \begin{pmatrix} 1 & t_1 + t_2 \\ 0 & 1 \end{pmatrix} \Leftrightarrow A(t_1) \cdot A(t_2) = A(t_1 + t_2)$$

• Identity element:

$$E = \begin{pmatrix} 1 & 0 \\ 0 & 1 \end{pmatrix}$$

- Inverse element:

$$A(t)^{-1} = A(-t) = \begin{pmatrix} 1 & -t \\ 0 & 1 \end{pmatrix}$$

- The generator of the group is written as:

$$\frac{-1}{i}\frac{d}{dt}\begin{pmatrix} 1 & t \\ 0 & 1 \end{pmatrix}\bigg|_{t=0} = \frac{-1}{i}\begin{pmatrix} 0 & 1 \\ 0 & 0 \end{pmatrix} \doteq I$$

It has the property

$$I^n = \begin{pmatrix} 0 & 0 \\ 0 & 0 \end{pmatrix} \text{ for } n > 1;$$

The use of the exponential map produces the elements of the group:

$$\exp(-i \cdot I \cdot t) = \begin{pmatrix} 1 & 0 \\ 0 & 1 \end{pmatrix} + \begin{pmatrix} 0 & 1 \\ 0 & 0 \end{pmatrix} \cdot t = \begin{pmatrix} 1 & t \\ 0 & 1 \end{pmatrix}$$

B. In order to establish the domain of definition of the parameter t, we test for the group properties:

$$\begin{pmatrix} t_1 & 0 \\ 0 & 1 \end{pmatrix}\begin{pmatrix} t_2 & 0 \\ 0 & 1 \end{pmatrix} = \begin{pmatrix} t_1 \cdot t_2 & 0 \\ 0 & 1 \end{pmatrix}$$

$$E = \begin{pmatrix} 1 & 0 \\ 0 & 1 \end{pmatrix} = A(t = 1)$$

$$A(t)^{-1} = A\left(\frac{1}{t}\right) = \begin{pmatrix} \frac{1}{t} & 0 \\ 0 & 1 \end{pmatrix} \qquad \Rightarrow t \in \mathbb{R}\setminus\{0\}$$

- The identity is identified by $t = 1$.
- The law of composition in the parameter space is written as

$$t_3 = t_1 \cdot t_2$$

The group is a Lie group only in the vicinity of the identity since, at $t = 0$, the composition law is not analytic (and not even continuous).

- The infinitesimal generator is written as

$$\frac{-1}{i}\frac{d}{dt}\begin{pmatrix} t & 0 \\ 0 & 1 \end{pmatrix}\Bigg|_{t=1} = \frac{-1}{i}\begin{pmatrix} 1 & 0 \\ 0 & 0 \end{pmatrix} \doteq I$$

- The exponential map is written as

$$
\begin{aligned}
e^{(-i \cdot I \cdot (x-1))} &= \exp\left(\begin{pmatrix} 1 & 0 \\ 0 & 0 \end{pmatrix} \cdot (x-1)\right) \\
&= \begin{pmatrix} 1 & 0 \\ 0 & 1 \end{pmatrix} + \begin{pmatrix} 1 & 0 \\ 0 & 0 \end{pmatrix} \cdot (x-1) + \begin{pmatrix} 1 & 0 \\ 0 & 0 \end{pmatrix} \cdot \frac{(x-1)^2}{2} + \cdots \\
&= \begin{pmatrix} 1 & 0 \\ 0 & 1 \end{pmatrix} + \begin{pmatrix} e^{(x-1)} - 1 & 0 \\ 0 & 0 \end{pmatrix} \\
&= \begin{pmatrix} e^{(x-1)} & 0 \\ 0 & 1 \end{pmatrix}
\end{aligned}
$$

With $x \in \mathscr{R}$, the exponential map produces only those elements of the group with $t > 0$. One can use the exponential map to enter the range $t < 0$ by choosing $x - 1 = i\pi + \ln t$, $t > 0$. The parameter space for the exponential map consists of two disjoint infinite segments in the complex plane.

Solution 3: Infinitesimal generator of the scaling transformation

The identity element is identified by $t = 1$. The Taylor expansion of $f(\frac{x}{t})$ at $t = 1$ is given by

$$f(x) - (t-1) \cdot x \cdot f'(x) + 1/2 \cdot (t-1)^2 \cdot \left[x^2 \cdot f''(x) + 2 \cdot x \cdot f'(x)\right] \pm \cdots$$

The infinitesimal generator can be obtained as

$$i \cdot x \cdot \frac{d}{dx}$$

The exponential map is written as

$$e^{-x \cdot \frac{d}{dx} \cdot (t-1)} = 1 - (t-1) \cdot x \cdot \frac{d}{dx} + 1/2 \cdot (t-1)^2 \cdot \left[x^2 \frac{d^2}{dx^2} + x\frac{d}{dx}\right] \pm \cdots$$

Note that the exponential map, applied to a function $f(x)$, does not precisely give the Taylor expansion of the original function.

Solution 4: Differential equation for one-parameter group

On one side, one can directly construct the group by using the exponential map

$$A(t) = \exp \begin{pmatrix} 1 & 1 \\ 0 & 1 \end{pmatrix} t = \begin{pmatrix} 1 & 0 \\ 0 & 1 \end{pmatrix} + \begin{pmatrix} 1 & 1 \\ 0 & 1 \end{pmatrix} t + \begin{pmatrix} 1 & 2 \\ 0 & 1 \end{pmatrix} \frac{t^2}{2} + \cdots$$

$$= \begin{pmatrix} e^t & te^t \\ 0 & e^t \end{pmatrix}$$

On the other side,

$$\frac{d}{dt} A(t) = -i \cdot I \cdot A(t) \Leftrightarrow$$

$$\frac{d}{dt} \begin{pmatrix} a_1 & a_2 \\ a_3 & a_4 \end{pmatrix} = \begin{pmatrix} 1 & 1 \\ 0 & 1 \end{pmatrix} \begin{pmatrix} a_1 & a_2 \\ a_3 & a_4 \end{pmatrix}$$

$$= \begin{pmatrix} a_1 + a_3 & a_2 + a_4 \\ a_3 & a_4 \end{pmatrix}$$

which can be integrated elementarily to

$$\begin{cases} a_3 = C_3 e^t \\ a_4 = C_4 e^t \end{cases} \quad \begin{cases} \frac{da_1}{dt} = a_1 + C_3 e^t \\ \frac{da_2}{dt} = a_2 + C_4 e^t \end{cases}$$

i.e.

$$a_1 = C_3 t e^t + C_1 e^t$$

$$a_2 = C_4 t e^t + C_2 e^t.$$

We require that $A(t = 0) = E$, so that $C_1 = 1$, $C_2 = 0$, $C_3 = 0$, and $C_4 = 1$, and finally,

$$A(t) = \begin{pmatrix} e^t & te^t \\ 0 & e^t \end{pmatrix}$$

Solution 5: Adjoint representation of a Lie Algebra

a. We show that

$$F(a + b) = F(a) + F(b)$$

Proof:

$$F(a + b)(x) = [a + b, x] = [a, x] + [b, x] = F(a)(x) + F(b)(x) \qquad \diamond$$

b. We show that

$$F(\lambda a) = \lambda F(a)$$

Proof:

$$F(\lambda a)(x) = [\lambda a, x] = \lambda [a, x] = \lambda F(x) \qquad \diamond$$

c. We show that

$$[F(a), F(b)] = F([a, b])$$

Proof (uses Jacobi identity: see e.g. https://en.wikipedia.org/wiki/Jacobi_identity):

$$
\begin{aligned}
[F(a), F(b)](x) \quad &= \quad (F(a)F(b) - F(b)F(a))(x) = [a, [b, x]] - [b, [a, x]] \\
&= \quad [a, [b, x]] + [b, [x, a]] \\
&\underset{\text{Jacobi identity}}{=} \quad -[x, [a, b]] = [[a, b], x] = F([a, b])(x) \qquad \diamond
\end{aligned}
$$

14

EXERCISES TO LECTURE 4

The scope of this set of exercises is to solve some practical problems on the application of symmetry so that one can learn the "technology" associated with their solution in detail.

Question 1: Molecular eigenmodes

Consider a molecule with two equal atoms residing within the xy-plane. The spring joining the atoms has a spring constant of f. The mass of each atom is m. Note that there are typically two symmetry groups associated with diatomic molecules, either $C_{\infty v}$ or $D_{\infty h}$. To keep the problem relatively simple, we impose on the molecule a C_2 symmetry. Although finding the vibrational modes and frequencies is, in this case, numerically trivial, it is worthwhile to solve it using the methods of group theory in order to become familiar with the important steps.

A. Consider that the motion is restricted to the xy-plane. Write down the four Newton equations governing the deviation of the atomic position from its equilibrium position (with an equilibrium distance of a).

B. Transform, using the eigenmode *ansatz*, the differential equations into an eigenvalue equation of a force matrix (we called it the "F" matrix in the lecture).

C. Find the matrices and the characters of the "vectorial tower representation".

D. Decompose the vectorial tower representation into irreducible representations of C_2.

E. Find the symmetry-adapted vectors, and use them to solve the eigenvalue problem of the force matrix, i.e. find the eigenfrequencies and eigenmodes.

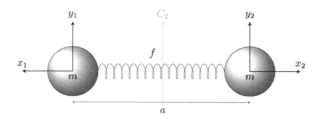

Sketch of the diatomic molecule with vectors describing the planar displacement of the atoms.

Question 2: Eigenvalue problem I

Consider the *function space of 2π periodic functions* $f(\varphi) = f(\varphi + 2\pi)$ and the operator

$$M \doteq -i\frac{d}{d\varphi}$$

acting within this space.

A. Define a representation of the group C_∞ within the function space of periodic functions $f(\varphi)$.

B. Find the characters of the irreducible representation of the symmetry group of the circle C_∞.

C. Project out, starting from a general function $f(\varphi)$, the symmetry-adapted wave function transforming according to the irreducible representation of C_∞.

D. Use these symmetry-adapted functions to solve the eigenvalue problem of M.

Question 3: Eigenstates of ring-like molecule

We consider a ring-like molecule consisting of n objects arranged in a circle and equally spaced. Let their vibrational modes perpendicular to the ring plane be described by the lattice version of the Helmholtz equation

$$-\frac{1}{a^2}\frac{\partial^2}{\partial\varphi^2}u(\varphi) = \frac{\omega^2}{c^2}u(\varphi)$$

with a being the radius of the ring, ω the desired vibrational frequencies, c the velocity of wave propagation, and u the amplitude of vibration.

A. Find the lattice version of the operator $-\frac{\partial^2}{\partial\varphi^2}u(\varphi)$. For this purpose, write down the Taylor expansion of $u(\varphi \pm \frac{2\pi}{n})$, and solve the two resulting equations for $\frac{\partial^2}{\partial\varphi^2}u(\varphi)$.

B. In the lattice (finite element) version, the vibrational amplitude becomes a "weight" attached to each element of the molecule. The Helmholtz equation becomes an eigenvalue equation of a matrix M in an n-dimensional vector space with n basis vectors

$$e_k = (0,0,\ldots, \underbrace{1}_{k\text{th site}}, \ldots, 0) \quad (k = 0,\ldots, n-1)$$

Write the matrix M using the lattice version of the operator you have just derived. Do not forget that the object at site 0 is coupled with the object at site $n-1$ because the ring-like shape mimics periodic boundary conditions!

C. The symmetry group of the molecule is the cyclic group C_n. Accordingly, the matrix M of this problem commutes with all elements of the tower representation Γ_T of this group. Recall that this group has n irreducible representations $\Gamma_0, \ldots, \Gamma_{n-1}$ and keep its character table in mind. Use the projector technique in the representation space of the tower representation to compute n symmetry-adapted vectors. Hint: The symmetry-adapted vector transforming according to the irreducible representation Γ_j is found by applying the projector P_{Γ_j} onto, for example, the vector e_0. Recall that $\Gamma_T(C_n^k)e_0 = e_k$.

D. Use the jth symmetry-adapted vector to compute the jth eigenvalue of the matrix M, $j = 0, 1, \ldots, n-1$.

Question 4: Eigenvalue problem II

A. Solve the following eigenvalue problem using symmetry arguments:

$$b \cdot x_{i-1} + ax_i + bx_{i+1} = \lambda x_i$$

The boundary condition is $x_{i\pm4} = x_i$.

Solution 1: Molecular eigenmodes

A. Referring to the sketch in Question 1, we can write down the four Newton equations:

$$m\ddot{x}_1 = -f(x_1 + x_2),$$
$$m\ddot{x}_2 = -f(x_1 + x_2),$$

$$m\ddot{y}_1 = 0,$$

$$m\ddot{y}_2 = 0.$$

B. Using the eigenmode *ansatz*, we obtain

$$\omega^2 \cdot \begin{pmatrix} x_1^0 \\ x_2^0 \\ y_1^0 \\ y_2^0 \end{pmatrix} = \underbrace{\begin{pmatrix} f/m & f/m & 0 & 0 \\ f/m & f/m & 0 & 0 \\ 0 & 0 & 0 & 0 \\ 0 & 0 & 0 & 0 \end{pmatrix}}_{\doteq F} \cdot \begin{pmatrix} x_1^0 \\ x_2^0 \\ y_1^0 \\ y_2^0 \end{pmatrix}$$

where ω denotes the eigenfrequency and F is the force matrix.
C. In the space of displacements, we generate the vectorial tower representation of C_2. To find the matrices corresponding to the symmetry elements E and C_2, we need to consider their actions upon each displacement vector. The rotation C_2 maps x_1 onto x_2 and vice versa, and y_1 onto y_2 and vice versa. Thus, the matrix representation reads

$$T_{V.T.}(C_2) = \begin{pmatrix} 0 & 1 & 0 & 0 \\ 1 & 0 & 0 & 0 \\ 0 & 0 & 0 & 1 \\ 0 & 0 & 1 & 0 \end{pmatrix},$$

with a character of 0.
D. Recalling the character table of C_2

C_2	E	C_2
Σ_1	1	1
Σ_2	1	-1
$\Sigma_{V.T.}$	4	0

we observe that

$$\Sigma_{V.T.} = 2\Sigma_1 \oplus 2\Sigma_2$$

E. We obtain the symmetry-adapted vectors transforming according to Σ_1 and Σ_2 by the projector technique:

$$\vec{e}_{\Sigma_1} \doteq P^{(\Sigma_1)} \begin{pmatrix} x_1^0 \\ x_2^0 \\ y_1^0 \\ y_2^0 \end{pmatrix} = \frac{1}{2} \left[1 \cdot T_{V.T.}(E) \cdot \begin{pmatrix} x_1^0 \\ x_2^0 \\ y_1^0 \\ y_2^0 \end{pmatrix} + 1 \cdot T_{V.T.}(C_2) \cdot \begin{pmatrix} x_1^0 \\ x_2^0 \\ y_1^0 \\ y_2^0 \end{pmatrix} \right]$$

$$= \frac{1}{2}\begin{pmatrix} x_1^0 + x_2^0 \\ x_2^0 + x_1^0 \\ y_1^0 + y_2^0 \\ y_2^0 + y_1^0 \end{pmatrix} \equiv \frac{1}{2}\begin{pmatrix} a \\ a \\ b \\ b \end{pmatrix},$$

$$\vec{e}_{\Sigma_2} \doteq P^{(\Sigma_1)}\begin{pmatrix} x_1^0 \\ x_2^0 \\ y_1^0 \\ y_2^0 \end{pmatrix} = \frac{1}{2}\left[1 \cdot T_{V.T.}(E) \cdot \begin{pmatrix} x_1^0 \\ x_2^0 \\ y_1^0 \\ y_2^0 \end{pmatrix} - 1 \cdot T_{V.T.}(C_2) \cdot \begin{pmatrix} x_1^0 \\ x_2^0 \\ y_1^0 \\ y_2^0 \end{pmatrix}\right]$$

$$= \frac{1}{2}\begin{pmatrix} x_1^0 - x_2^0 \\ x_2^0 - x_1^0 \\ y_1^0 - y_2^0 \\ y_2^0 - y_1^0 \end{pmatrix} \equiv \frac{1}{2}\begin{pmatrix} c \\ -c \\ d \\ -d \end{pmatrix}.$$

We insert them into the eigenvalue equation $\omega^2 \vec{e} = F\vec{e}$. The Σ_1-sector reads

$$\begin{pmatrix} f/m & f/m & 0 & 0 \\ f/m & f/m & 0 & 0 \\ 0 & 0 & 0 & 0 \\ 0 & 0 & 0 & 0 \end{pmatrix} \cdot \vec{e}_{\Sigma_1} = \frac{f}{m}\begin{pmatrix} a \\ a \\ 0 \\ 0 \end{pmatrix} \stackrel{!}{=} \frac{\omega^2}{2}\begin{pmatrix} a \\ a \\ b \\ b \end{pmatrix}.$$

This equation has two solutions:

$$\omega^2 = 2f/m \quad \text{Eigenvector} \quad \begin{pmatrix} 1 & 1 & 0 & 0 \end{pmatrix}$$
$$\omega = 0 \quad \text{Eigenvector} \quad \begin{pmatrix} 0 & 0 & 1 & 1 \end{pmatrix}$$

Let us now consider the Σ_2-sector:

$$\begin{pmatrix} f/m & f/m & 0 & 0 \\ f/m & f/m & 0 & 0 \\ 0 & 0 & 0 & 0 \\ 0 & 0 & 0 & 0 \end{pmatrix} \cdot \vec{e}_{\Sigma_2} = \frac{f}{m}\begin{pmatrix} 0 \\ 0 \\ 0 \\ 0 \end{pmatrix} \stackrel{!}{=} \frac{\omega^2}{2}\begin{pmatrix} c \\ -c \\ d \\ -d \end{pmatrix}$$

The solution is double degenerate:

$$\omega = 0 \quad \text{Eigenvectors} \quad \begin{pmatrix} 1 & \bar{1} & 0 & 0 \end{pmatrix} \quad \text{and} \quad \begin{pmatrix} 0 & 0 & 1 & \bar{1} \end{pmatrix}$$

The following figure is an illustration of the different modes of our diatomic molecule with C_2 symmetry:

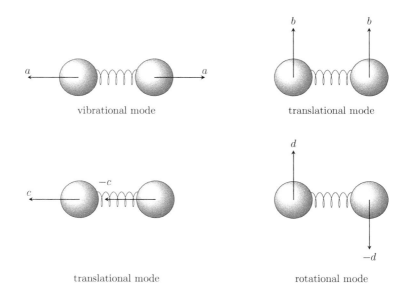

Sketch of the eigenmodes of the diatomic molecule.

Solution 2: Eigenvalue problem I

A. A possible representation of the group C_∞ within the function space of periodic functions $f(\varphi)$ is given by the choice

$$O_\alpha f(\varphi) \doteq f(\varphi - \alpha)$$
$$\Rightarrow \quad O_\beta O_\alpha f(\varphi) = O_\beta f(\varphi - \alpha) = f(\varphi - \alpha - \beta) = O_{\alpha+\beta} f(\varphi) \qquad \checkmark$$

B. The character table of C_∞ reads

C_∞	C_∞^α
Γ_0	1
$\Gamma_{\pm 1}$	$e^{\pm i\alpha}$
$\vdots$	$\vdots$
$\Gamma_{\pm m}$	$e^{\pm im\alpha}$
$\vdots$	$\vdots$

C. Recall that group C_∞ is not finite. In this case, the average for a 2π periodic function $f(\varphi)$ is defined as

$$\mathcal{M}_{\varphi \in C_\infty} f(\varphi) \doteq \frac{1}{2\pi} \int_0^{2\pi} f(\varphi) \mathrm{d}\varphi.$$

Starting from a general function $f(\varphi)$, the symmetry-adapted wave function transforming according to the irreducible representation Γ_m of C_∞ is

$$P^{(\Gamma_m)} f(\varphi) = \frac{1}{2\pi} \int_0^{2\pi} \mathrm{e}^{-im\alpha} f(\varphi - \alpha) \, \mathrm{d}\alpha$$

$$= \mathrm{e}^{-im\varphi} \frac{1}{2\pi} \int_0^{2\pi} \mathrm{e}^{im(\varphi - \alpha)} f(\varphi - \alpha) \, \mathrm{d}\alpha$$

$$\Rightarrow \quad P^{(\Gamma_m)} f(\varphi) \propto \mathrm{e}^{-im\varphi} \cdot A(m).$$

D. From

$$O_\alpha \left[-i \cdot \frac{\mathrm{d}}{\mathrm{d}\varphi} \right] O_\alpha^{-1} f(\varphi) = -iO_\alpha \frac{\mathrm{d}}{\mathrm{d}x} f(x)|_{x = \varphi + \alpha} = -i \frac{\mathrm{d}}{\mathrm{d}x} f(x)|_{x = \varphi}$$

$$= -i \frac{\mathrm{d}}{\mathrm{d}\varphi} f(\varphi),$$

it follows that M commutes with O_α. Therefore, M and the irreducible representation of C_∞ have common eigenspaces $\{ \mathrm{e}^{-im\varphi} \cdot A(m) \}$.

Applying M to these eigenfunctions, we obtain the eigenvalues E_m of M:

$$E_m \cdot \mathrm{e}^{-im\varphi} \cdot A(m) \stackrel{!}{=} -i \frac{\mathrm{d}}{\mathrm{d}\varphi} \mathrm{e}^{-im\varphi} \cdot A(m) = -m \, \mathrm{e}^{-im\varphi} \cdot A(m)$$

$$\Rightarrow \quad E_m = -m \in \mathbb{Z}.$$

Solution 3: Eigenstates of ring-like molecule

A.

$$u\left(\varphi \pm \frac{2\pi}{n}\right) = u(\varphi) + \frac{\mathrm{d}u(\varphi)}{\mathrm{d}\varphi} \cdot \pm \frac{2\pi}{n} + \frac{1}{2} \frac{\mathrm{d}^2 u(\varphi)}{\mathrm{d}\varphi^2} \cdot \left(\frac{2\pi}{n}\right)^2$$

Summing the equations for $u(\varphi + \frac{2\pi}{n})$ and $u(\varphi - \frac{2\pi}{n})$ and solving for the second derivative, we obtain

$$-\frac{1}{a^2} \frac{\partial^2}{\partial \varphi^2} u(\varphi) = \frac{2 \cdot u(\varphi) - u\left(\varphi + \frac{2\pi}{n}\right) - u\left(\varphi - \frac{2\pi}{n}\right)}{a^2 \cdot \left(\frac{2\pi}{n}\right)^2}$$

B. The operator produces a diagonal matrix element $\frac{1}{a^2 \cdot (\frac{2\pi}{n})^2} \cdot 2$ and fills the nearest diagonal elements with $-\frac{1}{a^2 \cdot (\frac{2\pi}{n})^2} \cdot 1$. Accordingly, the operator M reads

$$
M = \frac{1}{a^2 \cdot \left(\frac{2\pi}{n}\right)^2} \cdot
\begin{pmatrix}
2 & \bar{1} & 0 & \cdots & & 0 & \bar{1} \\
\bar{1} & 2 & \bar{1} & 0 & \cdots & & 0 \\
0 & \bar{1} & 2 & \bar{1} & 0 & \cdots & 0 \\
\vdots & \ddots & \ddots & \ddots & \ddots & \ddots & \vdots \\
0 & & 0 & \bar{1} & 2 & \bar{1} & 0 \\
0 & & \cdots & 0 & \bar{1} & 2 & \bar{1} \\
\bar{1} & 0 & & \cdots & 0 & \bar{1} & 2
\end{pmatrix}
$$

C. The character table of the cyclic group of order n is constructed as

C_n	$C_n^0 = E$	C_n^1	$\cdots$	C_n^k	$\cdots$	C_n^{n-1}
Γ_0	1	1		$\cdots$		1
Γ_1	1	$e^{\frac{2\pi i}{n}}$	$\cdots$	$e^{\frac{2\pi i}{n} k}$	$\cdots$	$e^{\frac{2\pi i}{n}(n-1)}$
$\vdots$	$\vdots$	$\vdots$		$\vdots$		$\vdots$
Γ_j	1	$(e^{\frac{2\pi i}{n}})^j$		$(e^{\frac{2\pi i}{n} j})^k$		$(e^{\frac{2\pi i}{n}(j)})^{n-1}$
$\vdots$	$\vdots$	$\vdots$		$\vdots$		$\vdots$
Γ_{n-1}	1	$(e^{\frac{2\pi i}{n}})^{(n-1)}$	$\cdots$	$(e^{\frac{2\pi i}{n}(n-1)})^k$	$\cdots$	$(e^{\frac{2\pi i}{n}(n-1)})^{(n-1)}$

$$
\begin{aligned}
P^{(\Gamma_j)} \vec{e}_0 &= \frac{1}{n}[1 \cdot T(E) + (e^{-\frac{2\pi i}{n}})^j \cdot T(C_n^1) + \cdots + (e^{-\frac{2\pi i}{n} j})^k \cdot T(C_n^k) + \cdots \\
&\quad + (e^{-\frac{2\pi i}{n}(j)})^{n-1} \cdot T(C_n^{n-1})]\,\vec{e}_0 \\
&= \frac{1}{n}[1 \cdot \begin{pmatrix} 1 & 0 & 0 & \cdots & 0 \end{pmatrix} + (e^{-\frac{2\pi i}{n}})^j \cdot \begin{pmatrix} 0 & 1 & 0 & \cdots & 0 \end{pmatrix} + \cdots \\
&\quad + (e^{-\frac{2\pi i}{n} j})^k \cdot \underbrace{\begin{pmatrix} 0 & 0 & \cdots & 0 & 1 & 0 & \cdots & 0 \end{pmatrix}}_{=\vec{e}_k} \\
&\quad + (e^{-\frac{2\pi i}{n}(j)})^{n-1} \cdot \begin{pmatrix} 0 & 0 & 0 & \cdots & 0 & 1 \end{pmatrix}] \\
&= \frac{1}{n}\sum_{k=0}^{n-1}(e^{-\frac{2\pi i}{n} j})^k e_k
\end{aligned}
$$

D. The first equation in

$$
M \frac{1}{n}\sum_{k=0}^{n-1}(e^{-\frac{2\pi i}{n} j})^k e_k = \frac{\omega^2}{c^2}\frac{1}{n}\sum_{k=0}^{n-1}(e^{-\frac{2\pi i}{n} j})^k e_k
$$

provides the desired eigenfrequencies. The equation reads

$$\frac{1}{a^2 \left(\frac{2\pi}{n}\right)^2}(2 \cdot (e^{-\frac{2\pi i}{n}j})^0 + 1 \cdot (e^{-\frac{2\pi i}{n}j})^1 + 1 \cdot (e^{-\frac{2\pi i}{n}j})^{n-1}) = \frac{\omega^2}{c^2} \cdot (e^{-\frac{2\pi i}{n}j})^0$$

The eigenfrequencies we are interested in obey

$$\omega^2 = \frac{2c^2}{a^2 \left(\frac{2\pi}{n}\right)^2} \cdot \left(1 + \cos \frac{2\pi}{n} \cdot j\right)$$

Solution 4: Eigenvalue problem II

A. This is only apparently a trivial problem. The problem can be cast as an eigenvalue equation of a matrix that we call the H matrix:

$$\begin{pmatrix} a & b & 0 & b \\ b & a & b & 0 \\ 0 & b & a & b \\ b & 0 & b & a \end{pmatrix} \begin{pmatrix} x_1 \\ x_2 \\ x_3 \\ x_4 \end{pmatrix} = \lambda \cdot \begin{pmatrix} x_1 \\ x_2 \\ x_3 \\ x_4 \end{pmatrix}$$

To use symmetry arguments, we need to find possible symmetry groups of the matrix. The most obvious symmetry to look for is provided by the periodic boundary conditions. If we define a lattice with the lattice points $\ldots, x_{-1}, x_0, x_1, x_2, x_3, x_4, \ldots$ aligned along a straight line, the operating mode of the matrix can be viewed as an object centered at x_i and having "wings" toward the left and right, see the following figure.

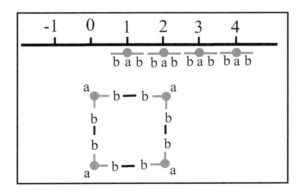

Sketch of the matrix operator on a linear lattice (top) and on a square (bottom).

The "object" has translational symmetry modulo "4". This can be rendered by winding the linear lattice "modulo 4" into a closed loop, i.e. into a lattice with four points occupying the edges of a square. The symmetry group of the object possibly representing our operator is, accordingly, the cyclic group C_4. The next step is to establish a representation of C_4 onto the four-dimensional space where the matrix is defined. The basis vectors consist of a weight "1" on one of the four lattice points and a weight "0" on the remaining three — we called this the tower representation. Once this representation is established, we must check that its matrices commute with the operator: This is the exact mathematical proof that the symmetry group C_4 is indeed the symmetry group of H. This check is sometimes tedious, but our interpretation of the operator H with a suitable object gives us a good feeling about the result of this check.

Once commutativity has been established, the rules of group theory tell us how to proceed further. First, one has to establish the character of the tower representation. As C_4 rotates the lattice points, the tower representation of the elements other than the identity has "0" character. We display this result in the character table of C_4:

C_4	E	C_4	C_4^2	C_4^3	
Γ_0	1	1	1	1	
Γ_1	1	i	-1	$-i$	
Γ_2	1	-1	1	-1	
Γ_3	1	-1	-1	i	
Γ_T	4	0	0	0	$= 2\Gamma_0 \oplus 2\Gamma_2$

This result tells us that, if we compute by the projector method the symmetry-adapted vectors to the irreducible representations Γ_0 and Γ_2 appearing in the decomposition of Γ_T, we will reach a block diagonalization of H into two 2×2 blocks that must be finally "manually" (i.e. with no possible further help from symmetry) diagonalized to find the eigenvalues. However, the block diagonalization has reduced the determinantal equations to be solved into two quadratic equations, in contrast to the original quartic equation implied by the size of the original matrix H.

This result does not, however, entail the full potential of the use of symmetry. In fact, looking at the object representing the operator on the square, one might be tempted to search for further symmetry elements, i.e. to expand the symmetry group. As a matter of fact, the two-dimensional dihedral group D_4 (the point group C_{4v}) hosts further symmetry elements of the object and provides an expansion of the symmetry group. One can construct the tower representation of these further symmetry elements and convince oneself that they also commute with H. Accordingly, we can construct the following table, describing the full symmetry property of the problem:

Character table of the group C_{4v}. Along the bottom line, the characters of the tower representations are given, together with its decomposition into irreducible representations. The symmetry-adapted basis vectors (as extensively discussed in Lecture 4) are also indicated.

C_{4v}	e	$2C_4$	C_2	$2m$	$2m'$	
Δ_1	1	1	1	1	1	$(1,1,1,1)$
$\Delta_{1'}$	1	1	1	$\bar{1}$	$\bar{1}$	
Δ_2	1	$\bar{1}$	1	1	$\bar{1}$	
$\Delta_{2'}$	1	$\bar{1}$	1	$\bar{1}$	1	$(1,\bar{1},1,\bar{1})$
Δ_5	2	0	$\bar{2}$	0	0	$\left\{ \begin{matrix} (1,0,\bar{1},0) \\ (0,1,0,\bar{1}) \end{matrix} \right\}$
Δ_T	4	0	0	0	2	$\Delta_1 \oplus \Delta_{2'} \oplus \Delta_5$

The result of being able to find further symmetry elements is that the symmetry-adapted vectors transforming according to Δ_1, Δ_2', and Δ_5 will produce a block matrix H consisting of three blocks: two one-dimensional blocks and one two-dimensional block. This result represents a further simplification of the problem, which reduces to the solution of only one quadratic equation with respect to the original quartic one and the two quadratic equations resulting from the C_4 symmetry. There is one more important simplification: The two-dimensional block carries an irreducible representation, so that when symmetry-adapted vectors transforming according to Δ_5 are used, it will be directly diagonalized without the need for solving any quadratic equation.

Let us now use the symmetry-adapted vectors to find the eigenvalues:

$$\begin{pmatrix} a & b & 0 & b \\ b & a & b & 0 \\ 0 & b & a & b \\ b & 0 & b & a \end{pmatrix} \begin{pmatrix} 1 \\ 1 \\ 1 \\ 1 \end{pmatrix} = (a+b) \cdot \begin{pmatrix} 1 \\ 1 \\ 1 \\ 1 \end{pmatrix}$$

$$\begin{pmatrix} a & b & 0 & b \\ b & a & b & 0 \\ 0 & b & a & b \\ b & 0 & b & a \end{pmatrix} \begin{pmatrix} 1 \\ \bar{1} \\ 1 \\ \bar{1} \end{pmatrix} = (a-b) \cdot \begin{pmatrix} 1 \\ \bar{1} \\ 1 \\ \bar{1} \end{pmatrix}$$

$$\begin{pmatrix} a & b & 0 & b \\ b & a & b & 0 \\ 0 & b & a & b \\ b & 0 & b & a \end{pmatrix} \begin{pmatrix} 1 \\ 0 \\ \bar{1} \\ 0 \end{pmatrix} = a \cdot \begin{pmatrix} 1 \\ 0 \\ \bar{1} \\ 0 \end{pmatrix}$$

15

EXERCISES TO LECTURE 5

The scope of this set of exercises is to learn how to assign the symmetry elements of wallpaper patterns.

Question 1: Wallpaper patterns

Describe the symmetry elements of the following wallpaper patterns, and assign them to one of the 17 two-dimensional space groups.

Hints: We need a strategy for determining the space group of the patterns. The strategy proceeds as follows:

1. The first step entails the search for the type of lattice and the drawing of the unit cell. The unit cell contains that section of the pattern that is repeated upon translations by integer multiples of the basis vectors. This step leads us toward one of the crystal systems (oblique, rectangular, square, or triangular). The unit cell will be marked in green in the solutions.
2. The next step entails the search for rotational and reflectional symmetry elements. In the solutions, the locations of the fourfold rotational elements are marked by full red squares (triangles and hexagons for three- and sixfold rotational elements), and those of the twofold rotational elements are indicated by diamonds. Pure reflection lines are given by red straight lines. These elements determine the crystal class, i.e. the point group of the pattern.
3. Once the crystal class is determined, the number of potential space groups is limited to a small set.
4. The final selection leading to the space group depends on the existence of glide reflections (marked in blue in the solutions).

Pattern 1: The wallpaper pattern is taken from Owen Jones, "Grammatik der Ornamente", Day Denicke, London Leipzig, 1856, p. 165, https://doi.org/10.11588/diglit.17930.

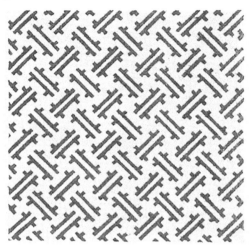

Solution:

The unit cell is a square (square lattice). The pattern has four- and twofold rotational elements together with reflection lines (in red), so that the crystal class is D_4. One can also identify glide reflection lines (blue). The space group is necessarily $p4gm$.

Pattern 2: The wallpaper pattern is taken from Owen Jones, "Grammatik der Ornamente", Day Denicke, London Leipzig, 1856, p. 44, Universitätsbibliothek Heidelberg, https://doi.org/10.11588/diglit.17930.

Solution:

The unit cell is rectangular. The pattern has twofold rotational elements together with reflection lines (in red), so that the crystal class is D_2. There are no further symmetry elements. The space group is $p2mm$.

Pattern 3: The wallpaper pattern is taken from Owen Jones, "Grammatik der Ornamente", Day Denicke, London Leipzig, 1856, p. 44, https://doi.org/10.11588/diglit.17930.

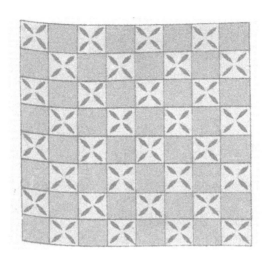

Solution:

The lattice is oblique. One recognizes only twofold rotational centers. The crystal class is C_2. The space group is necessarily $p2$.

Pattern 4: The wallpaper pattern is taken from Owen Jones, "Grammatik der Ornamente", Day Denicke, London Leipzig, 1856, p. 45, https://doi.org/10.11588/diglit.17930.

Solution:

The lattice is rectangular. One recognizes twofold rotational centers and horizontal reflection lines (in red). The crystal class is D_2. The vertical reflection lines are glides. The space group is $p2mg$.

Pattern 5: The wallpaper pattern is taken from Owen Jones, "Grammatik der Ornamente", Day Denicke, London Leipzig, 1856, p. 45, https://doi.org/10.11588/diglit.17930.

Solution:

The pattern belongs to the square system. It has four- and twofold rotational centers and reflection lines. The crystal class is D_4. There is a net of glide reflection lines. The space group is $p4mg$.

Pattern 6: The wallpaper pattern is taken from Owen Jones, "Grammatik der Ornamente", Day Denicke, London Leipzig, 1856, p. 45, https://doi.org/10.11588/diglit.17930.

Solution:

The pattern belongs to the oblique crystal system. There are twofold rotational centers. The crystal class is necessarily C_2. The space group is $p2$.

Pattern 7: The wallpaper pattern is taken from Owen Jones, "Grammatik der Ornamente", Day Denicke, London Leipzig, 1856, p. 45, https://doi.org/10.11588/diglit.17930.

Solution:

The pattern belongs to the square crystal system. It has four- and twofold rotational centers. The crystal class is C_4. The space group is $p4$.

Pattern 8: The wallpaper pattern is taken from Owen Jones, "Grammatik der Ornamente", Day Denicke, London Leipzig, 1856, p. 46, https://doi. org/10.11588/diglit.17930.

Solution:

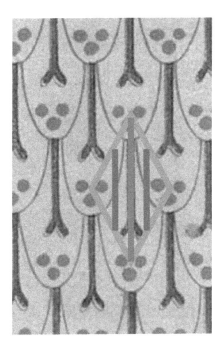

The lattice is centered rectangular. We recognize vertical reflection lines (red) and glide reflection lines (blue) but not rotational elements. The crystal class is D_1. The space group is *cm*.

Pattern 9: The wallpaper pattern is taken from Owen Jones, "Grammatik der Ornamente", Day Denicke, London Leipzig, 1856, p. 165, https://doi.org/10.11588/diglit.17930.

Solution:

The crystal system is square. There are four- and twofold rotational elements, together with reflection lines. The crystal class is D_4. The glide lines are pure reflections combined with primitive translations, so that the space group is $p4mm$.

Pattern 10: The wallpaper pattern is taken from a photography of a garden furniture.

Solution:

The crystal system is rectangular. The unit cell is that of the centered lattice. There are twofold rotational elements, together with reflection lines. The crystal class is D_2. The glides originate from the combination of pure reflections with primitive translations. The space group is $c2mm$.

PART III
Projects

16

PROJECT FOR LECTURE 2: FUNCTIONAL ANALYTIC AND NUMERICAL ASPECTS OF EIGENVALUE PROBLEMS

16.1 Introduction

Many problems in the natural and social sciences can be cast into the form of an eigenvalue equation of the type

$$Lf = \lambda \cdot f$$

The operator L is, for example, a linear differential or integral operator, and the eigenvalue equation is typically an Euler–Lagrange equation resulting from the minimization of an appropriate Lagrangian function (classical or quantum mechanical). It acts on a set of functions f that describe possible states of a system (e.g. the vibrational modes of a string or the states of an electron subject to the Coulomb interaction of a proton). f is defined on a set of variables q that describe the degrees of freedom of the system. q can assume continuous values, such as the coordinate $x \in [0, L]$ of the vibrating string of length L, or discrete values, such as the z-component of spin. The desired set of λ values represents measurable real values characterizing particularly important states of the system, called eigenstates. λ are, for example, the eigenfrequencies of a vibrating string or the energy eigenvalues of a quantum mechanical system. The eigenvalue equation is often accompanied by some boundary conditions — imposed by some particular requirements of the problem — that restrict the set of functions over which the solution of the eigenvalue equation must be sought. For instance, one may seek the eigenfrequencies of a vibrating string, imposing the "boundary conditions" that the amplitude of vibration at certain locations is vanishing at all time. The eigenfrequencies depend on the boundary conditions: The same operator might assume completely different eigenvalues and eigenfunctions if the boundary conditions are changed. By virtue of the boundary conditions, the eigenvalue equation becomes what is known as an eigenvalue problem. The existence of boundary conditions defines, in technical terms, the domain of definition specific to the operator in the given physical situation.

The eigenvalue problem we have in mind was formally discussed originally by J. von Neumann, *Math. Ann.* **102**, 49–131 (1929). It has numerous exact properties that were developed over many decades by a branch of mathematics called "functional analysis". In the first part of this project, we present important results on these functional analytic properties. In the second part, we introduce two numerical methods for solving such eigenvalue problems in practice: the finite difference method and the finite element method. We also discuss the use of symmetry arguments and group theory to reduce the computational complexity. Finally, we apply these numerical methods to concrete problems, as a way of illustration.

16.2   Functional Analytic Aspects of the Eigenvalue Problem

One of the important subjects of interest in modern mathematics is to ascertain, through general and exact methods, the conditions under which important equations have a solution. In this project, the questions we refer to are: for which class of operators does a solution of the eigenvalue problem exist, and whether the eigenvalues constitute a discrete or continuous set.

16.2.1   *Hilbert spaces*

Computations require establishing an algebra within the domain of definition of the operator L:

A. The operator L is, typically, a linear operator. This means that if $f_1(q)$ and $f_2(q)$ are solutions of the eigenvalue equation, so is the linear combination of both. This property is called the **superposition principle**:

$$f_1(q), f_2(q) \in \{f\} \quad \Rightarrow \quad \alpha_1 \cdot f_1(q) + \alpha_2 \cdot f_2(q) \in \{f\}$$

(α_i are complex numbers). The set $\{f\}$ becomes a **vector space**.

B. The elements of $\{f\}$ are "multiplied" using the Hermitic (Hilbert) scalar product. The scalar product is an operation that associates to two elements f and g of the set $\{f\}$ a **finite** number ("scalar quantity")

$$(f, g)$$

such that it satisfies the following:

a.

$$(f, g) = \overline{(f, g)}$$

b. Linearity:

$$(\lambda_1 f_1 + \lambda_2 f_2, g) = \bar{\lambda}_1 (f_1, g) + \bar{\lambda}_2 (f_2, g)$$
$$(f, \lambda_1 g_1 + \lambda_2 g_2) = \lambda_1 (f, g_1) + \lambda_2 (f, g_2)$$

c. Positivity:

$$(f, f) \geq 0 \quad (f, f) = 0 \Longleftrightarrow f = 0$$

Comments.

1. The set $\{f\}$ becomes a vector space with a scalar product.

2. The scalar product can be used to measure the "length" (norm) $\| f \|$ of an element f and the distance $d(f, g)$ between two vectors:

$$\| f \| \doteq \sqrt{(f, f)} \quad d(f, g) \| f - g \|$$

3. Two vectors, f and g, are said to be orthogonal if

$$(f, g) = 0$$

and a set of n vectors, $\{f_1, \ldots, f_n\}$, with the property

$$(f_i, f_j) = \delta_{ij}$$

is called an orthonormal set (ONS).

C. A practical property is the possibility of expanding f as a linear combination of so-called **basis vectors** $\{f_i\}$:

$$f = \sum_{i=1}^{d} c_i \cdot f_i$$

c_i are the expansion coefficients, or, in other words, the "coordinates" of f with respect to $\{f_i\}$. The basis vectors are usually constructed (e.g. using the Gram–Schmidt orthonormalization algorithm, see e.g. https://encyclopediaofmath.org/index.php?title=Orthogonalization) to be orthonormalized, so that

$$c_i = (f_i, f)$$

d is the dimension of the vector space and can be, as in the case of a function space, infinite. If this expansion exists for any f, the set $\{f\}$ becomes a **complete** vector space with a Hermitic scalar product (called the Hilbert space).

Comments.

1. When d is infinite, one needs to define a convergence criterion for the infinite sum. For $d = \infty$, the equality in the expansion means that

$$\lim_{n \to \infty} \| f - \sum_{i=1}^{n} (f_i, f) f_i \|^2 = 0$$

(norm-wise convergence) or, equivalently,

$$\| f \|^2 = \sum_{i=1}^{\infty} | (f_i, f) |^2$$

This last expression is known as the completion relation (or the Parseval identity).

2. The equality

$$f = \sum_{i=1}^{\infty} c_i \cdot f_i$$

does not necessarily imply that the left- and right-hand sides of this equation are equal at any point q (point-wise convergence). In fact, it is not difficult to imagine that one could alter the graph of a continuous function $f(q)$ point-wise to produce an infinite number of singular functions that differ from the function f "point-wise" (more precisely, within what is known as the "Lebesgue null set"[1]) but

---

[1] The famous Dirichlet function provides an example of such a Lebesgue null set and also introduces the need to use Lebesgue integration when defining functions in a Hilbert space. This function is almost everywhere zero in the interval $[0, 1]$ except at the position of the rational values in the same interval, where it has the value of "1". This function is not Riemann-integrable, but it can be integrated — like all functions which have a finite or countably infinite set of point discontinuities — through the Lebesgue method, in this case to give a vanishing integral, as the function is almost everywhere vanishing.

have the same expansion coefficients as the function f. Accordingly, the expansion coefficients generate an entire class of "equivalent" functions.[2]

3. According to Moretti (Valter Moretti, Do continuous wavefunction form a Hilbert space? https://physics.stackexchange.com/q/719681), the classes contain "true monsters". For all practical purposes, however, (except for advanced "proof mathematics"), one deals with representatives of the classes which are typically continuous and at least once differentiable, so that they can be, for instance, solutions of the Schrödinger equation.

4. The existence of an expansion for any f shows that the system of basis functions used is complete, i.e. no basis function was left out.

5. It also establishes an isometry between any Hilbert space and the Hilbert space $(c_1, c_2, \ldots)$ with $\sum_i |c_i|^2 < \infty$.

6. The infinite linear combinations of continuous CONS functions might produce discontinuous functions (recall the construction of, for instance, a square signal using a suitable Fourier series). Accordingly, classes with a discontinuous function as a representative can also belong to the Hilbert space of a quantum mechanical system. They are, however, exotic instances in practical physical problems related to atoms, molecules, and solids (see again the comments by V. Moretti).

7. The finding of a CONS is important for performing both exact and numerical computations. Completeness means that no basis state has been neglected, so that the knowledge of the basis vectors in a vector space is enough to describe, by means of a linear superposition, any other element of the vector space. Accordingly, the solving of the Schrödinger equation and, in general, of scientific problems "reduces" to finding a suitable CONS. Using the words of Feynman, physical research is about finding the "base states of the world".

---

[2]Two functions $f(q)$ and $g(q)$ belonging to the same class are considered equivalent, $f \sim g$, if $\{q : f(q) - g(q) \neq 0\}$ is a Lebesgue null set. In other words, $f \sim g$ if $f(q) = g(q)$ for almost all q. This relation is an equivalence relation and therefore allows us to build equivalence classes $[f]$. The set of all equivalence classes $\{[f]\}$ is a vector space with the Hermitic metric:

$$[f] + [g] \doteq [f + g]; \quad \alpha[f] \doteq [\alpha \cdot f]; \quad ([f], [g]) \doteq (f, g)$$

i.e. the operations between the classes can be performed using their representatives.

Examples of CONS:

1. n orthogonal vectors in a vector space with finite dimensions n.
2. The set of infinite-dimensional vectors

$$f_1 = (1,0,0,\ldots), f_2 = (0,1,\ldots),\ldots, f_n = (0,0,\ldots,1,0,0,\ldots),\ldots$$

in the Hilbert space ℓ_2 of all sequences $(a_1, a_2, \ldots)$ for which $\sum_{i=0}^{\infty} | a_i |^2 < \infty$.

3. The set

$$\left\{ \sqrt{\frac{2}{L}} \cos \frac{(2n+1)\pi}{L} \cdot q, \sqrt{\frac{2}{L}} \sin \frac{(2n)\pi}{L} \cdot q \right\} \quad ; \quad n = 1,2,3,\ldots$$

constitutes a CONS in the vector space of functions that vanishes at $-\frac{L}{2}, \frac{L}{2}$.

4. A famous theorem by Fourier establishes that the plane waves

$$u_k(x) = \frac{1}{\sqrt{L}} \cdot e^{i \cdot k \cdot x}$$

with $k = \frac{2\pi}{L} \cdot n$ and $n \in \mathscr{Z}$ constitute a CONS in the function space of L-periodic functions. The orthonormality relation is written as

$$(u_q, u_k) = \frac{1}{L} \int_{-\frac{L}{2}}^{\frac{L}{2}} dx e^{-i \cdot q \cdot x} e^{i \cdot k \cdot x} = \delta_{qk}$$

5. The eigenfunctions

$$\left\{ \sqrt{\frac{1}{\sqrt{\pi}} \cdot \frac{1}{2^n \cdot n!}} \cdot H_n(x) \cdot e^{-\frac{x^2}{2}} \right\}$$

of a one-dimensional oscillator constitute a CONS in the Hilbert space of square-integrable functions in the variable $x \in \mathscr{R}$.

6: The Fourier Transform: Mathematics endows the set of square-integrable functions with basis functions labeled by an index that takes continuum values. One example of this set is the technique known as the Fourier transform. This is a kind of "improper" CONS that requires a special treatment. The starting point is the Fourier expansion of L-periodic functions, which uses the basis functions $u_k(x) = \frac{1}{\sqrt{L}} \cdot e^{i \cdot k \cdot x}$. For any periodic function, the Fourier series is written as

$$f(x) = \sum_k \tilde{f}(k) \cdot u_k(x)$$

with

$$\tilde{f}(k) = (u_k, f) = \int_{-\frac{L}{2}}^{\frac{L}{2}} dx \, \overline{u_k(x)} \cdot \psi(x)$$

We would like to find out how the Fourier series develops in the limit $L \to \infty$, so that it covers the space of physical functions which are defined over the entire configuration space $\mathscr{R}$ while also not imposing periodic boundary conditions. For this purpose, we rewrite

$$f(x) = \sum_k \tilde{f}(k) u_k(x)$$

as

$$f(x) = \frac{1}{\sqrt{2\pi}} \sum_k \frac{\sqrt{L}}{\sqrt{2\pi}} \cdot \tilde{f}(k) \cdot \frac{2\pi}{L} \cdot e^{ikx} = \frac{1}{\sqrt{2\pi}} \sum_k \tilde{f}'(k) \Delta_k e^{ikx}$$

with

$$\tilde{f}'(k) = \frac{1}{\sqrt{2\pi}} \int_{-\frac{L}{2}}^{\frac{L}{2}} dx \cdot \psi(x) \cdot e^{-ikx}$$

and $\Delta_k = \frac{2\pi}{L}$. Now, let $L \to \infty$. Then, $\Delta_k \to 0$ and the Riemann sum

$$\frac{1}{\sqrt{2\pi}} \sum_k \tilde{f}'(k) \Delta_k e^{ikx}$$

formally converges to

$$f(x) = \frac{1}{\sqrt{2\pi}} \int_{-\infty}^{\infty} dk \cdot \tilde{f}(k) \cdot e^{ikx}$$

with

$$\tilde{f}(k) = \frac{1}{\sqrt{2\pi}} \int_{-\infty}^{\infty} dx \cdot \psi(x) \cdot e^{-ikx}$$

This equation tells us, for example, that any square-integrable function $f(x)$ can be expressed as a **Fourier integral**. The square-integrable function $\tilde{f}(k)$ is its Fourier transform. The Parseval identity is written as

$$\int_{-\infty}^{\infty} dx \mid f(x) \mid^2 = \int_{-\infty}^{\infty} dk \mid \tilde{f}(k) \mid^2$$

and is called the Plancherel relation. The Fourier transform is an example of an expansion of a function into a set of basis functions labeled by a continuous index, namely k. In fact, one can interpret the expression

$$f(x) = \frac{1}{\sqrt{2\pi}} \int_{-\infty}^{\infty} dk \cdot \tilde{f}(k) \cdot e^{ikx}$$

formally as the superposition of basis states u_k with

$$u_k(x) = \frac{1}{\sqrt{2\pi}} \cdot e^{ik \cdot x} \quad ; \quad k \in \mathscr{R}$$

The orthonormality relation between these basis states is written as

$$(u_{k'}, u_k) = \frac{1}{2\pi} \int_{\mathscr{R}} dx \, e^{-ik'x} \cdot e^{ikx} = \delta(k' - k)$$

$\delta(k' - k)$ is the so-called Dirac Delta function: It is not a function of k, k' in the usual sense, but a "generalized function". Such generalized functions (also called "distribution") were first introduced by Dirac, and later, a very rigorous mathematical framework was worked out by L. Schwartz and I. M. Gel'fand.

16.2.2 *Spectral theorem for finite-dimensional Hilbert spaces*

Spectral theorems specify exactly for which class of operators the eigenvalue equation has a complete set of eigenfunctions and eigenvalues. They play a key role, for instance, in the evolution of the formal aspects of quantum mechanics and provide a formal framework for current research in quantum technologies.

Definition. An $n \times n$ matrix L is called Hermitic (or symmetric) if

$$(Lx, x) = (x, Lx)$$

for any $x \in V$.

Comment. For the matrix elements of a Hermitic matrix, we have the relation

$$\overline{L_{ji}} = L_{ij}$$

Theorem. *An $n \times n$ Hermitic (or symmetric) matrix has n real eigenvalues and n corresponding mutually orthogonal eigenvectors, i.e. it can be* *converted into the diagonal form, with all diagonal matrix elements (the eigenvalues of the matrix) being real.*

Proof.

(a) We prove first that all eigenvalues, if they exist, are real. Inserting $Lx = \lambda x$ in $(Lx, x) - (x, Lx) = 0$, we obtain

$$(\overline{\lambda} - \lambda) \cdot (x, x) = 0 \quad \text{or} \quad \overline{\lambda} = \lambda$$

(b) The proof that n eigenvalues exist is typically performed by induction. First, we show that the theorem holds for $n = 1$. In $n = 1$, L is the multiplication by a scalar: The scalar itself is the eigenvalue, and the number "1" is the eigenvector. We then assume that the theorem holds for dimension $n - 1$. Because of the fundamental theorem of algebra, which asserts that the equation

$$\det(L - \lambda \mathbb{1}) = 0$$

has at least one non-zero solution, we know that at least one non-zero eigenvalue λ_n exists, and we call the corresponding eigenfunction f_n. We now perform the Gram–Schmidt orthogonalization procedure to construct a set of vectors, $f_1, \ldots, f_{n-1}$, orthogonal to f_n. This means that the matrix elements

$$L_{nj} \doteq (f_n, L f_j) = (L f_n, f_j) = \lambda_n (f_n, f_j) = 0 = \overline{L_{jn}}$$

are vanishing and in the orthonormal basis $f_1, f_2, \ldots, f_n$, the matrix representation of L reads

$$\begin{pmatrix} [A_{n-1}] & [0] \\ [0] & \lambda_n \end{pmatrix}$$

where $[\,]$ indicates an $(n-1) \times (n-1)$ block matrix. As we assumed that the $[A_{n-1}]$ matrix can be diagonalized, we have proven the spectral theorem for finite Hermitic matrices.

Comments.

1. The space belonging to one single eigenvalue might be one-dimensional or multi-dimensional. The number n_λ of eigenvectors to the eigenvalue λ is called the degeneracy of the eigenvalue. Given n_λ linearly independent vectors within the eigenspace of λ, the Gram–Schmidt orthonormalization algorithm allows us to construct an orthonormal set within each

finite eigenspace itself. The entire set of orthonormalized eigenvectors constitutes a CONS.

2. In virtue of this theorem, in the basis set $\{f_i\}$ in which L is diagonal, we have, for any x

$$x = \sum_{k=1}^{m}\sum_{n=1}^{n_k}\underbrace{(f_{k,n},x)}_{a_{kn}}f_{k,n}$$

with n_k being the degeneracy of the kth eigenvalue, and the completeness relation reads

$$\| x \| = \sqrt{\sum_{k,n}| a_{kn} |^2}$$

This relation expresses the fact that the set $\{a_{kn}\}$ is enough to express the length of any vector, i.e. no coordinate (no basis vector) was left out.

16.2.2.1　*The transition to infinite-dimensional Hilbert spaces*

When trying to generalize this theorem to infinite-dimensional Hilbert spaces, one encounters difficulties. For example, the generalization of Hermitic matrices to infinite-dimensional spaces is the so-called **symmetric** operators — an operator L is called symmetric if

$$(Lf,g) = (f,Lg)$$

with f and g in the domain of definition of L. The problem is that the eigenvalue equation of symmetric operators might not necessarily have solutions! Here are some points to remember:

1. The set of eigenvalues can be empty: Consider, for example, the differential operator

$$p \doteq -i\frac{d}{dx}$$

and define it within the function space of differentiable, square-integrable functions $f(x)$ in the interval $[0,L]$. The operator is not symmetric:

$$(f,pg) = \int_0^L dx\overline{f}(x)\cdot\left[-i\frac{d}{dx}g(x)\right] \overset{\text{P.I.}}{=}$$
$$-i\,\overline{f}(x)\cdot g(x)\,|_0^L + \int_0^L dx\overline{\left(-i\frac{d}{dx}f(x)\right)}\cdot g(x)$$

$$= -i\overline{f}(x) \cdot g(x) \mid_0^L + (pf, g)$$
$$\neq (pf, g)$$

as the constant $-i\overline{f(x)} \cdot g(x) \mid_0^L$ arising from the integration by parts is, in general, not vanishing. However, the operator can be rendered symmetric by imposing some boundary conditions. A possible set of boundary conditions involves allowing only functions that vanish at $x = 0$ and $x = L$, so that the constant $-i\overline{f(x)} \cdot g(x) \mid_0^L$ arising from the integration by parts is made to be vanishing. However, with this boundary condition, the eigenvalue equation

$$-i\frac{d}{dx}f(x) = \lambda f(x)$$

has no solution, as the only function solving the differential equation and fulfilling the boundary condition is the function $f(x) \equiv 0$, which cannot be considered an eigenfunction. Thus, the operator, with the boundary condition introduced, has neither eigenfunctions nor eigenvalues.

2. The set of eigenvalues is countably infinite: Consider the operator again. With periodic boundary conditions, the operator is symmetric and has an infinite set of eigenvalues and eigenfunctions:

$$\lambda_n = n \in \mathscr{L} \quad f_n = \frac{1}{\sqrt{L}}e^{i\lambda_n \cdot x}$$

$\{f_n\}$ is, by virtue of the Fourier theorem, a complete set in the Hilbert space of L-periodic functions. For this operator, with these specific boundary conditions, the finite-dimensional expansion theorem holds in an infinite-dimensional Hilbert space as well.

3. Symmetric operators with a continuum set of eigenvalues: There are situations where the problem $Lf = \lambda f$ cannot be solved *exactly* by square-integrable functions, but only in the approximate sense of Weyl, i.e. there is a (Weyl) sequence $(f_n)_0^\infty$ of square-integrable functions such that

$$\| Lf_n - \lambda f_n \| < \epsilon \quad n \quad \text{sufficiently large}$$

The values of λ obtained in this way typically occupy some continuous interval.

As an example, we study the position operator q_x defined as

$$q_x f(x) \doteq x \cdot f(x)$$

$x \in [-\infty, \infty]$. It is a symmetric operator. The eigenvalue equation is written as

$$x \cdot f(x) = \lambda f(x)$$

We search for a discrete set of eigenvalues by attempting to solve the equation

$$(x - \lambda) \cdot f(x) = 0 \quad \forall x \quad ; \quad f(x) = \text{square integrable}$$

This equation admits, for any given λ, only solutions that vanish throughout the x-axis except at $x = \lambda$, i.e. the square-integrable function solving this equation is only the "zero" function. We conclude that the position operator neither has square-integrable eigenfunctions nor discrete eigenvalues.

We search now for a solution in a continuous range of the real axis. We consider the sequence of square-integrable functions with norm 1:

$$f_n(x) = n^{\frac{1}{2}} \cdot \pi^{-\frac{1}{4}} e^{-\frac{n^2(\cdot x - \lambda)^2}{2}}$$

We then compute that

$$\| (x - \lambda) f_n \|^2 = n \cdot \pi^{-\frac{1}{2}} \int_{-\infty}^{\infty} dx (x - \lambda)^2 \cdot e^{-n^2 \cdot (x-\lambda)^2} = \frac{1}{2n^2}$$

As

$$\lim_{n \to \infty} \| (x - \lambda) f_n \|^2 = 0 \quad \forall \lambda$$

the eigenvalue equation has an approximate solution for all real values of λ, i.e. the continuous spectrum covers the entire set of real values.

4. Generalized eigenvalue problem: This example shows that finding the continuous spectrum by explicit construction of Weyl sequences might be a difficult process. A more useful *ansatz* consists of rewriting the eigenvalue equation as a test integral over a suitable set of test functions $\{\varphi\}$ that fulfill the boundary conditions:

$$Lf - \lambda f = 0 \quad \Rightarrow \quad (\varphi, Lf - \lambda f) = 0$$

The equation on the right-hand side is equivalent to the original eigenvalue problem. However, it has the property that one can use the method of distributions to seek those solutions that belong to the continuous spectrum, avoiding the cumbersome construction of sequences.

Example: Generalized eigenvalue equation for the position operator in one dimension: We rewrite the equation

$$(x - \lambda) \cdot f_\lambda(x) = 0$$

as

$$\int dx (x - \lambda) \cdot f_\lambda(x) \cdot \varphi(x) = 0$$

where $\varphi(x)$ is a suitable "test function". This equation is solved by searching for the Fourier transform of the desired solution:

$$f_\lambda(x) = \frac{1}{\sqrt{2\pi}} \int dk f_\lambda(k) \cdot e^{i \cdot k \cdot x}$$

Inserting this expression in the generalized eigenvalue equation, we obtain

$$\int dx (x - \lambda) \cdot f_\lambda(x) \cdot \varphi(x) = \int dx (x - \lambda) \cdot \frac{1}{\sqrt{2\pi}} \int dk f_\lambda(k) \cdot e^{i \cdot k \cdot x} \cdot \varphi(x)$$

$$= \frac{1}{\sqrt{2\pi}} \int dk \cdot f_\lambda(k) \int dx (x - \lambda) \cdot e^{i \cdot k \cdot x} \cdot \varphi(x)$$

The right-hand side can be rewritten as

$$\frac{1}{\sqrt{2\pi}} \int dk \cdot f_\lambda(k) \frac{e^{i \cdot k \cdot \lambda}}{i} \cdot \int dx \cdot i \cdot (x - \lambda) e^{ik \cdot (x-\lambda)} \varphi(x)$$

$$= \frac{1}{\sqrt{2\pi}} \int dk \cdot f_\lambda(k) \frac{e^{i \cdot k \cdot \lambda}}{i} \frac{d}{dk} \underbrace{\int dx e^{ik \cdot (x-\lambda)} \varphi(x)}_{\sqrt{2\pi} \tilde{\varphi}_\lambda(k)}$$

The remaining integral

$$\int dk \cdot f_\lambda(k) \cdot e^{i \cdot k \cdot \lambda} \frac{d}{dk} \tilde{\varphi}_\lambda(k)$$

is evaluated through partial integration to obtain

$$-\int dk \cdot \frac{d}{dk} (f_\lambda(k) \cdot e^{i \cdot k \cdot \lambda}) \tilde{\varphi}_\lambda(k)$$

As this integral must vanish for any test function, we find that the original generalized eigenvalue equation is solved by

$$\frac{d}{dk}(f_\lambda(k) \cdot e^{i \cdot k \cdot \lambda}) = 0 \Leftrightarrow f_\lambda(k) = c \cdot e^{-i \cdot k \cdot \lambda}$$

This is the Fourier transform of the originally desired solution $f_\lambda(x)$, which is then written as

$$f_\lambda(x) \propto \int dk \cdot e^{i \cdot k(x-\lambda)} \propto \delta(x - \lambda)$$

i.e. the eigenfunctions of the continuous spectrum are the generalized Dirac Delta functions.

5. Proposition: *The eigenvalues of a symmetric operator L are real.*

Proof. Choose a test function φ that obeys the boundary conditions, so that

$$(\varphi, Lf) - (L\varphi, f) = 0$$

The eigenvalue equation is rewritten as

$$\lambda(\varphi, f) - \underbrace{(\varphi, Lf)}_{\overline{\lambda} \cdot (f, \varphi)} = (\lambda - \overline{\lambda}) \cdot (\varphi, f) = 0 \quad \Rightarrow \lambda = \overline{\lambda}$$

6. Self-adjoint operators.

The anomalies observed in an infinite-dimensional Hilbert space when dealing with symmetric operators show that being a symmetric operator is a necessary condition for representing the operator of a physical problem,[3] as symmetric operators have the attractive property of having real eigenvalues. However, it is not a sufficient condition for describing a well-defined eigenvalue problem. The spectral theorem by Hilbert and von Neumann finds exactly the subset of the symmetric operators that has a complete spectral decomposition in an infinite-dimensional Hilbert space and corresponds therefore to observables, which are known as **self-adjoint operators**. For self-adjoint operators, mathematics offers a bouquet of theorems that hold general significance in finding solutions to eigenvalue problems.

---

[3]In the language of quantum mechanics, an "observable".

Definition. The operator that solves the equation

$$(L^\dagger g, f) = (g, Lf)$$

is called the adjoint operator to L.

Definition. If $L^\dagger = L$ *and* the domain of definition of L are identical to the domain of definition of $L^\dagger$, then the operator is said to be **self-adjoint**.

Comments.

1. It is possible that, formally, $L^\dagger = L$, but the domain of definition of $L^\dagger$ is larger than that of L. Then, the operator is symmetric but not self-adjoint.

Examples.

1. The momentum operator: For the operator $p = -i\frac{d}{dx}$ in the interval $[a, b]$, we compute $p^\dagger$ using integration by parts:

$$(g, pf) = -i\left[f(b)\overline{g(b)} - f(a)\overline{g(a)}\right] + \left(-i\frac{dg}{dx}, f\right)$$
$$= (p^\dagger g, f)$$

We can distinguish the following cases:

- No boundary conditions at a and b: Because of the boundary term, $p^\dagger \neq p$.
- Dirichlet boundary conditions: If we require $f(0) = f(L) = 0$, the boundary term evaluates to 0 and $p^\dagger = -i\frac{d}{dx} = p$, i.e. p is certainly symmetric. However, the boundary terms vanish even if $g(a) \neq g(b) \neq 0$, i.e. $D(g)$ is a larger set than $D(f)$. With these boundary conditions, p is not self-adjoint.
- Generalized periodic boundary conditions: We now require

$$f(b) = f(a) \cdot e^{i\theta}$$

with some real θ (the case $\theta = 0$ is called the Born–von Karman boundary conditions). With these boundary conditions, the boundary terms become

$$f(a) \cdot e^{i\theta}\overline{g(b)} - f(a)\overline{g(a)}$$

In order for this boundary term to vanish, we must have

$$g(b) = e^{i\theta}g(a)$$

i.e. the set of functions in the domain of $p^\dagger$ must obey the same boundary conditions as the set of functions in the domain of p. With these boundary conditions, p is self-adjoint.

2. The Hamilton operator for atoms and molecules is self-adjoint (T. Kato, *Trans. Am. Math. Soc.* **70**, 195 (1951)).

16.2.3   *Hilbert–von Neuman spectral theorem (for laymen)*

(See, for example, E. Onofri, *Lezioni sulla Teoria degli Operatori Lineari*, 1984, p. 122, https://books.google.ch/books?id=c1Tx9u9bJVQC)

I. For a self-adjoint operator L, there is a set of **real** numbers λ and a set of (Weyl) sequences $f_n(\lambda) \in D(L)$ such that
   a. $\| f_n \| = 1$
   b.

$$\lim_{n \to \infty} \|Lf_n(\lambda) - \lambda \cdot \mathbb{1} f_n(\lambda)\| = 0$$

II. The set $\{\lambda\}$ constitutes the spectrum $\sigma(L)$ of L. $\sigma(L)$ has two components: a discrete one and a continuous one. The discrete and continuous spectra belong to a disjoint interval of the real axis.

III. In the discrete sector, the Weyl sequence converges to the elements $f_i(\lambda) \in D(L)$, with $i = 1, 2, \ldots, d(\lambda)$. The eigenstates $\{f_i\}$ constitute the eigenspace to λ, with a dimension of $d(\lambda)$, which is called the degeneracy of the eigenvalue λ.[4] In the continuous sector, the Weyl sequence converges to an element $f(\lambda)$ that solves formally the eigenvalue equation but is outside $D(L)$.[5] These functions are "improper" eigenfunctions.

IV. The set of eigenfunctions in the discrete *and* continuous sectors constitutes a complete orthonormal set: Any function in the domain of definition of the operator can be expanded as a linear combination of the eigenfunctions of L. The linear combination of improper eigenfunctions is an integral, *a la* the Fourier transform.

---

[4] The bound states of the hydrogen atom belong, for instance, to this sector.
[5] The scattering states of a quantum mechanical system belong to this sector.

16.2.4   *Commuting self-adjoint matrices*

At the end of this section, we mention an important theorem regarding **commuting** self-adjoint operators, which can be demonstrated in a simple way when the Hilbert space is finite-dimensional (in a finite-dimensional space, self-adjoint matrices are diagonalizable with certainty and the eigenvalues constitute a discrete set).

Theorem. *A set of self-adjoint matrices, $A_1, \ldots, A_r$, is simultaneously diagonalizable.*

Proof. The proof is by induction. As A_1 is self-adjoint, it can be diagonalized. We assume that the theorem holds for $A_1, A_2, \ldots, A_{r-1}$, i.e. for the matrices $A_1, \ldots, A_{r-1}$, we construct a set of mutually orthogonal spaces $V_1, \ldots, V_n$, so that for the vector space V, we have

$$V = \oplus_n V_n$$

Within the spaces V_n (which can be more than one-dimensional, depending on the degeneracy of the corresponding eigenvalues), a CONS exists, with vectors v_i^p. The index p labels the orthogonal spaces, and the index i the basis vectors within each subspace. With respect to this CONS, $A_1, \ldots, A_{r-1}$ are all diagonal. We now show that A_r can be diagonalized. Consider, for example, the vector $A_r v_i^p$. From

$$A_1(A_r v_i^p) = A_r A_1 v_i^p = A_r \lambda_p v_i^p = \lambda_p A_r v_i^p$$

we obtain that

$$A_r v_i^p \in V_p$$

i.e. V_p is closed with respect to A_r. As A_r is self-adjoint, the matrix of A_r constructed with the vectors v_i^p is certainly diagonalizable, i.e. there are linear combinations of the set $\{v_i^p\}$ that render the matrix of A_r diagonal. These linear combinations also maintain the diagonal form of the remaining operators $A_1, \ldots, A_{r-1}$ and can be used as a new set to diagonalize the entire set $A_1, \ldots, A_r$. $\diamond$

Comment. Once a matrix has been diagonalized, the commuting matrix does not need to be also diagonal. This theorem says that it is possible to find a suitable set of common eigenvectors.

16.3   Numerical Approaches to Eigenvalue Problems

Many problems in science can be reduced to one-dimensional eigenvalue problems, known as the general Sturm–Liouville eigenvalue problem. These consist of a homogeneous linear differential equation,

$$\frac{d}{dx}\left[p(x)\frac{d\psi}{dx}\right] - q(x)\cdot\psi + \lambda\cdot r(x)\cdot\psi = 0$$

with boundary conditions

$$\alpha_1\psi(a) + \alpha_2\psi'(a) = 0$$

$$\beta_1\psi(b) + \beta_2\psi'(b) = 0$$

where $p(x), q(x)$, and $r(x)$ are smooth within the fundamental interval $[a, b]$ (with $r(x), p(x)$ being strictly positive). λ is the desired eigenvalue. The coefficients α_1, α_2 and, respectively, β_1, β_2 cannot vanish simultaneously. The special boundary conditions $\psi(a) = \psi(b) = 0$ are called Dirichlet boundary conditions. $\psi'(a) = \psi'(b) = 0$ are the von Neumann boundary conditions. To be specific, in the following we discuss numerical approaches to the Sturm–Liouville eigenvalue problems. The next theorem provides a general framework.

Theorem (Sturm-Liouville operator. Encyclopedia of Mathematics. http://encyclopediaofmath.org/index.php?title=Sturm-Liouville_operator&oldid=44406).

a. *The Sturm–Liouville operator is self-adjoint with respect to the scalar product $(\varphi, \psi) = \int_{[a,b]} \varphi^*(x)\psi(x)\cdot r(x)\cdot dx$.*
b. *Its eigenvalues constitute a discrete set, with a smallest element and no point of accumulation. None of the eigenvalues is degenerate.*

The Sturm–Liouville eigenvalue problem can be solved analytically only in very special situations, e.g. if $p(x), q(x)$, and $r(x)$ are constant functions. In general, a numerical method is needed to solve Sturm–Liouville problems.

16.3.1   *Finite difference method*

The *finite difference method* (FDM) discretizes the continuous function $\psi(x)$ into values on a lattice. In one dimension, for instance, $\psi_n = \psi(x_n)$, defined on discrete points x_n along a segment. The derivatives of $\psi(x)$ are approximated, in the simplest way, by

$$\psi'(x_n) \approx \frac{\psi_{n+1} - \psi_n}{a}$$

$$\psi''(x_n) \approx \frac{\psi'(x_n) - \psi'(x_{n-1})}{a} \approx \frac{\psi_{n+1} - 2\psi_n + \psi_{n-1}}{a^2}$$

where a is the interval between the discrete points. In three dimensions, the discrete Laplace operator is written as

$$\begin{aligned}
\nabla^2 \psi(x_l, y_m, z_n) = \frac{1}{a^2} [&\psi(x_{l-1}, y_m, z_n) + \psi(x_{l+1}, y_m, z_n) \\
&+ \psi(x_l, y_{m-1}, z_n) + \psi(x_l, y_{m+1}, z_n) \\
&+ \psi(x_l, y_m, z_{n-1} + \psi(x_l, y_m, z_{n+1}) \\
&- 6\psi(x_l, y_m, z_n)]
\end{aligned}$$

The equations above have an error of order $\mathscr{O}(a)$, so one can reduce a to increase the accuracy of the solution at the cost of introducing more discrete points x_n and a larger linear space.

To illustrate how the finite difference method works, we apply it to the one-dimensional Schrödinger equation

$$-\frac{\hbar^2}{2m} \psi''(x) + V(x)\psi = E\psi$$

where $x \in [0, L]$. This equation is of the Sturm–Liouville type:

$$\left[-\frac{d^2}{dx^2} + \underbrace{\frac{2m}{\hbar^2} V(x)}_{v(x)} \right] \psi(x) = k^2 \psi(x)$$

($k^2 = \frac{2mE}{\hbar^2}$). We introduce into the segment L a lattice consisting of $n + 1$ equidistant points with coordinates

$$x_0 = 0, x_1 = 1 \cdot a, x_2 = 2 \cdot a, \ldots, x_n = n \cdot a \quad a = \frac{L}{n}$$

$\psi(x)$ is approximated by an n-dimensional vector, $\psi_0, \psi_1, \ldots, \psi_n$, expressing the amplitude $\psi(x)$ at the lattice points. We approximate the derivatives with differences to obtain the set of algebraic equations ($v_n = v(x_n)/a^2$, $\lambda = k^2/a^2$):

$$(2 + v_n)\psi_n - \psi_{n-1} - \psi_{n+1} = \lambda \psi_n$$

These equations involve only the active lattice points: The points at the boundary carry the boundary conditions and are not active ones. These

set of equations can be cast into matrix form. For instance, we impose the
Dirichlet boundary conditions $\psi(0) = \psi(L) = 0$ to obtain ($i = 1, 2, \ldots, $n-1)

$$
\begin{pmatrix}
2 + v_1 & \bar{1} & 0 & 0 & & & \cdots \\
\bar{1} & 2 + v_2 & \bar{1} & 0 & 0 & & \cdots \\
0 & \bar{1} & 2 + v_3 & \bar{1} & 0 & 0 & \cdots \\
\vdots & \vdots & \vdots & \vdots & \vdots & \vdots & \bar{1} \\
0 & 0 & & & & \bar{1} & 2 + v_{n-1}
\end{pmatrix}
\begin{pmatrix}
\psi_1 \\
\psi_2 \\
\\
\vdots \\
\\
\psi_{n-1}
\end{pmatrix}
= \lambda \cdot
\begin{pmatrix}
\psi_1 \\
\psi_2 \\
\\
\vdots \\
\\
\psi_{n-1}
\end{pmatrix}
$$

The FDM produces a "conventional" eigenvalue problem for a finite matrix.
The matrix is symmetric and tridiagonal. It has only a relatively small
number of non-vanishing matrix elements (so called "sparse matrices").
Its diagonalization is performed numerically. Depending on the symmetry
of $V(x)$, one can hope that the diagonalization process is simplified. When
periodic boundary conditions are used, the points at the edge of the interval
become active. When applied to x_0, the operator involves the lattice point
x_{-1}, which, because of the boundary conditions, behaves as x_{n-1}. When
applied to x_n, the lattice point x_{n+1} behaves as x_1. In all equations, x_0 and
x_n must appear with exactly the same coefficients. The discrete equations
must be constructed accordingly, and the matrix describing these equations
contains the corresponding matrix elements.

Let us consider a concrete example with $n = 4$. We have the following
five discrete points:

$$
x_0 = 0, \; x_1 = \frac{L}{4}, \; x_2 = \frac{L}{2}, \; x_3 = \frac{3L}{4}, \; x_4 = L
$$

x_0 and x_4 represents the boundaries. Under the Dirichlet boundary
conditions $\psi_0 = \psi_4 = 0$, we obtain the 3×3 matrix

$$
\begin{pmatrix}
2 + v_1 & -1 & 0 \\
-1 & 2 + v_2 & -1 \\
0 & -1 & 2 + v_3
\end{pmatrix}
\begin{pmatrix}
\psi_1 \\
\psi_2 \\
\psi_3
\end{pmatrix}
= \lambda
\begin{pmatrix}
\psi_1 \\
\psi_2 \\
\psi_3
\end{pmatrix}
$$

If we choose periodic boundary condition, $\psi_4 = \psi_0$. Of the five active points,
the equation for, say, ψ_4 can be omitted, as it is the same as that for ψ_0.
The problem becomes

$$
\begin{pmatrix}
2 + v_0 & -1 & 0 & -1 \\
-1 & 2 + v_1 & -1 & 0 \\
0 & -1 & 2 + v_2 & -1 \\
-1 & 0 & -1 & 2 + v_3
\end{pmatrix}
\begin{pmatrix}
\psi_0 \\
\psi_1 \\
\psi_2 \\
\psi_3
\end{pmatrix}
= \lambda
\begin{pmatrix}
\psi_0 \\
\psi_1 \\
\psi_2 \\
\psi_3
\end{pmatrix}
$$

There are other FDMs that do not involve the diagonalization of matrices. One of them is the Numerov algorithm (https://en.wikipedia.org/wiki/Numerov%27s_method). The Numerov algorithm produces a recursion formula for computing ψ_{n+1}, starting from two initial values (two values are required for solving second-order differential equations). For instance, the Schrödinger equation

$$\psi''(x) + k^2(x) \cdot \psi(x) = 0$$

$(k^2(x) = \frac{2m}{\hbar^2}(E - V(x)))$ becomes the recursion formula

$$\psi_{n+1} = \frac{2\left(1 - \frac{5a^2}{12}k_n^2\right)\psi_n - \left(1 + \frac{a^2}{12}k_{n-1}^2\right)\psi_{n-1}}{\left(1 + \frac{a^2}{12}k_{n+1}^2\right)} + O(a^6)$$

The eigenvalues of interest, E, are found by considering that the desired solution $\{\psi_n\}$ must obey the boundary conditions. At the start of a computation, a value of E is assumed, and all $k_n^2 = k^2(x_n)$ can be determined. The wave function is then computed by means of the recursion formula, successively computing ψ_{n+1} from ψ_n and ψ_{n-1}. Whether E is accepted as an eigenvalue depends on how precisely the boundary conditions are fulfilled.

Let us consider, as an example, the case of a harmonic potential:

$$V(x) = \begin{cases} c(x^2 - x) & 0 \leq x \leq 1 \\ 0 & \text{otherwise} \end{cases}$$

with c a constant set to be 400 in the current simulation (lengths are in units of L and energies are in units of $\frac{\hbar^2}{m \cdot L^2}$). We choose $a = 0.001$. E is somewhere between the bottom of the potential (-100) and 0. The potential used in this example is symmetric with respect to $x = 0.5$, so that the final wave function will be either symmetric or antisymmetric with respect to $x = 0.5$. We start from two points around $x = 0.5$ and solve all ψ_n values for $x \in [0.5, 1.0]$ through the iterative approach. For $x > 1$, $V(x) = 0$, and negative energies, the exact solution is

$$\psi(x > 1) = \psi(x = 1) \cdot e^{-\sqrt{2 \cdot E} \cdot x}$$

If the recursively computed wave function converges to this exponential decay above $x = 1$, it is accepted as an eigenfunction, and the value of E is accepted as an eigenvalue. For $x < 0.5$, the wave function is continued symmetrically or antisymmetrically. After finding the correct eigenvalues and eigenfunctions, we normalize the wave functions to 1. Some eigenfunctions are shown in the following figure.

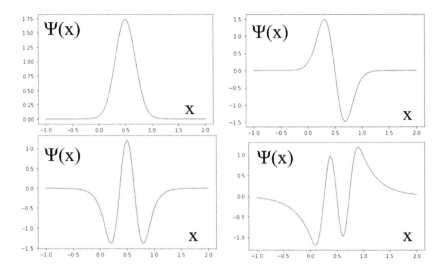

Eigenfunctions for the parabolic potential obtained using the Numerov recursion method.

16.3.2   *Finite element method*

The starting point of the finite element method is the rewriting of the Sturm–Liouville eigenvalue problem into an integral over test functions:

$$\int_{[a,b]} dx \left[\frac{d}{dx} \left[p(x) \frac{d\psi}{dx} \right] + q(x) \cdot \psi + \lambda \cdot r(x) \cdot \psi \right] \cdot \varphi(x) = 0$$

$\varphi(x)$ is any smooth "test-function" satisfying the boundary conditions. To be specific, we consider the Dirichlet boundary conditions again. For practical purposes, one relaxes the strong requirement on the desired solution that the second derivative exists and transforms the integral by partial integration to obtain the weak (variational) form of the eigenvalue problem:

$$\int_{[a,b]} dx \left[(-(p(x) \cdot \psi') \cdot \varphi'(x)) + (q(x) \cdot \psi + \lambda \cdot r(x) \cdot \psi) \cdot \varphi(x) \right] = 0$$

This relaxation allows for more practical test functions (see later). The Ritz–Rayleigh principle consists of:

A. Using a set of n trial functions $\{\Psi_i(x)\}$ within which the eigenvalue problem is solved. Accordingly, the *ansatz*

$$\psi(x) = \sum_j c_j \Psi_j(x)$$

is inserted into the integral, with the coefficients c_j being the desired quantities:

$$\int_{[a,b]} dx \left[\left(-(p(x) \cdot \sum_j c_j \cdot \Psi_j') \cdot \varphi'(x) \right) + \left(q(x) \cdot \sum_j c_j \Psi_j + \lambda \cdot r(x) \cdot \sum_j c_j \Psi_j \right) \right. $$
$$\left. \cdot\, \varphi(x) \right] = 0$$

B. The test functions are replaced by each one of the n trial functions, giving rise to a set of n linear equations for the n coefficients $\{c_j\}$:

$$\sum_j c_j \cdot \underbrace{\int_{[a,b]} dx \left[(p(x)\Psi_j(x)')\Psi_i'(x) - q(x) \cdot \Psi_j(x) \cdot \Psi_i(x) \right]}_{A_{ij}}$$

$$= \lambda \cdot \sum_i c_j \cdot \underbrace{\int_{[a,b]} dx \cdot r(x) \cdot \Psi_j(x)\Psi_i(x)}_{S_{ij}}$$

($i = 1, 2, \ldots, n$). These n equations are called the Ritz–Rayleigh–Galerkin (RRG) equations (see e.g. https://en.wikipedia.org/wiki/Galerkin_method). They can be summarized in a matrix form ($A = [A_{ij}]$, $S = S_{ij}$):

$$A\vec{c} = \lambda S \vec{c}$$

With the RRG equations, we have reduced the Sturm–Liouville eigenvalue problem to an n-dimensional matrix equation. If the basis functions are chosen to be orthonormal, $\mathbf{S} = \mathbb{1}$

C. In the finite element method, the RRG equations are formulated for a special set of trial functions, defined on a lattice. As in the FDM, a set of $n + 1$ points is defined in the continuous interval $[a, b]$. The trial functions consist of, for example, a Gaussian centered on each lattice point. A simpler *ansatz* is to use "hat" functions centered at each lattice point, see the following figure for an example. As a specific example, we now consider the differential equation

$$\frac{d^2\psi(x)}{dx^2} + \frac{2m}{\hbar^2} \cdot E \cdot \psi(x) = 0$$

which, in physics, is the Schrödinger equation for a particle in a box. This is a Sturm–Liouville eigenvalue problem with

$$p(x) \equiv 1$$
$$q(x) \equiv 0$$

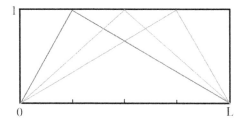

Possible hat functions used as trial functions in the interval $[0, L]$. The interval contains five points, two of which are at the edges to account for the boundaries.

$$r(x) \equiv \frac{2m}{\hbar^2}$$

The Dirichlet boundary conditions are written as

$$\psi(0) = \psi(L) = 0$$

The problem can be solved exactly, providing a good reference to check the accuracy of the numerical simulations:

$$E_n = \frac{\hbar^2 n^2 \pi^2}{2mL^2}$$
$$\psi_n(x) = \sin\left(\frac{n\pi}{L}x\right)$$

- To apply the finite element method, we establish in the interval, $N + 1$ lattice points, $x_0 = 0$, and $x_{N+1} = L$. We then choose $N - 1$ *hat functions* centered at the remaining $N - 1$ lattice points as trial functions:

$$\Psi_i(x) = \begin{cases} \frac{x}{x_i}, & \text{if } x \leq x_i \\ \frac{L-x}{L-x_i}, & \text{else} \end{cases} \quad \forall i \in \{1, \ldots, N-1\}, \ x_i = i\frac{L}{N}$$

- We compute the matrix elements A_{ij} and S_{ij}. This can be done analytically for the hat functions as the integrand is at most a polynomial of degree 2. Alternatively, one can use numerical methods provided by standard software. Having calculated all elements of the matrices $\mathbf{A}$ and

S, we can numerically solve the eigenvalue problem. One must take into account the fact that the hat functions are not orthogonal to each other. The eigenvalues E_n are then given by the eigenvalues of the matrix $\mathbf{AS}^{-1}$, and the corresponding eigenfunctions are

$$\psi_n(x) = \sum_{i=1}^{N-1} c_i^{(n)} \Psi_i(x)$$

where $c^{(n)}$ is the eigenvector corresponding to the eigenvalue E_n. The numerically calculated eigenvalues and eigenfunctions are depicted together with the exact values in the following figures. Note that the numerical solutions converge toward the exact solutions for $N \to \infty$, as expected. As the center of the interval is a point of inversion symmetry, the eigenfunctions can be classified according to the irreducible representations of the group C_i, i.e. they are either *gerade* (do not change sign upon inversion) or *ungerade* (change sign upon inversion) with respect to the center of inversion.

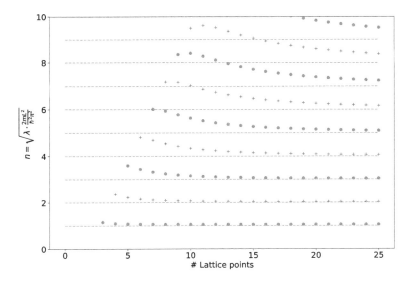

Numerically calculated eigenvalues for different numbers of lattice points: The exact eigenvalues are given by dashed lines. The red dots correspond to levels with wave function symmetric (*gerade*) with respect to the center of the interval. The green crosses correspond to levels with u (*ungerade*) symmetry. There are as many eigenvalues as lattice points in the interior of the lattice, as this corresponds to the number of basis functions. A smooth convergence to the correct eigenvalues can be observed with increasing number of lattice points.

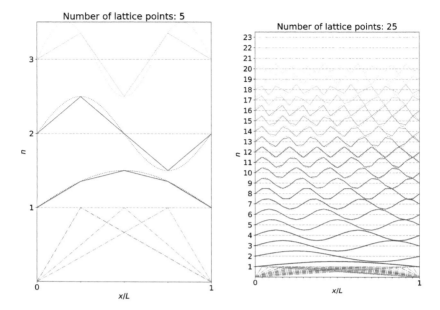

The exact eigenfunctions ψ_n (dotted) are shown together with the numeric approxima-
tions (solid) for 5 (left) and 25 (right) lattice points. At the bottom of each figure, the
hat functions used as trial basis functions are also shown (dash–dot). The eigenfunctions
are shifted in the vertical direction by their respective quantum numbers. Note that the
numerical functions are symmetric or antisymmetric with respect to $\frac{L}{2}$. For 25 lattice
points, the difference between numerical and analytical eigenfunctions of the lowest
energy levels is almost unrecognizable on the scale shown on the right-hand side of
the figure.

17

PROJECT FOR LECTURE 4: SYMMETRY ARGUMENTS IN CLASSICAL MECHANICS

The symmetry group of classical mechanics is the so-called Galilei group. It is a continuous group of elements that transform time and spatial coordinates. It is the scope of this project to show that this symmetry group has two important applications in classical mechanics: the principle of relativity of Galilei and a set of quantities that remain invariant during classical motion.

17.1 The Symmetry Groups of Classical Mechanics

Time reversal and time translation group: In Newtonian mechanics, time is described with a continuum real parameter t. Those transformations that preserve time distances are given by

$$\hat{t} = \sigma \cdot t + \tau \quad \sigma = \pm 1$$

where $\sigma = -1$ describes the element of time reversal and τ is a time translation. The multiplication law reads

$$(\sigma_2, \tau_2) \circ (\sigma_1, \tau_1) = (\sigma_2 \cdot \sigma_1, \sigma_2 \tau_1 + \tau_2)$$

Isometries (The Euclidean group): In an Euclidean space, one can introduce a linear mapping ("transformation") that transforms the coordinates $\mathbf{x} = (x_1, \ldots, x_n)$ to $\hat{\mathbf{x}} = (\hat{x}_1, \ldots, \hat{x}_n)$ by the relation

$$\hat{x}_i = \sum_j \Lambda_{ij} x_j + t_i$$

or, as in matrix form,

$$\hat{\mathbf{x}} = \Lambda \mathbf{x} + \mathbf{t}$$

where $\mathbf{t}$ are called translations. If

$$\Lambda^t \Lambda = \mathbb{1}$$

with $\det(\Lambda) = 1$ (the so-called "proper rotations") or $\det(\Lambda) = -1$) (the so-called "improper rotations"), the mapping $(\Lambda \mid \mathbf{a})$ has the important property that

$$d(\hat{P}, \hat{Q}) = d(P, Q)$$

i.e. it preserves the distance between points and are called isometries of the Euclidean space. The set of isometries

$$\{(\Lambda \mid \mathbf{t})\}$$

represented with a Seitz symbol constitutes the so-called Euclidean group (in n-dimensions: $\mathrm{Euc}(n)$).

In order to define the rule of multiplication in $\mathrm{Euc}(n)$, we compose two successive transformations:

$$\hat{\mathbf{x}} = \Lambda_1 \mathbf{x} + \mathbf{a}_1$$
$$\hat{\hat{\mathbf{x}}} = \Lambda_2 \hat{\mathbf{x}} + \mathbf{a}_2$$

to obtain

$$\hat{\hat{\mathbf{x}}} = \Lambda_2[\Lambda_1 \mathbf{x} + \mathbf{a}_1] + \mathbf{a}_2 = \Lambda_2 \Lambda_1 \mathbf{x} + \Lambda_2 \mathbf{a}_1 + \mathbf{a}_2$$

We use the Seitz symbols to express this law of composition between the elements $(\Lambda_2, \mathbf{a}_2)$ and $(\Lambda_1, \mathbf{a}_1)$:

$$(\Lambda_2, \mathbf{a}_2)\,(\Lambda_1, \mathbf{a}_1) = (\Lambda_2 \Lambda_1, \Lambda_2 \mathbf{a}_1 + \mathbf{a}_2)$$

The Galilei group: The Galilei transformation reads

$$\hat{\mathbf{x}} = \Lambda \mathbf{x} + \mathbf{v} \cdot t + \mathbf{a}$$
$$\hat{t} = \sigma \cdot t + \tau$$

where Λ is a proper or improper rotation and $\sigma = \pm 1$. The group $\mathrm{Gal}(n)$ is characterized by the elements $g = (\Lambda, \mathbf{v}, \mathbf{a}, \tau, \sigma)$. The element corresponding to the identity is $(\mathbb{1}, \mathbf{0}, \mathbf{0}, 0, 1)$. The multiplication rule reads

$$g_1 g_2 = (\Lambda_1 \Lambda_2, \Lambda_1 \mathbf{v}_2 + \mathbf{v}_1, \Lambda_1 \mathbf{a}_2 + \mathbf{v}_1 \tau_2 + \mathbf{a}_1, \tau_1 + \tau_2, \sigma_1 \cdot \sigma_2)$$

17.2 The Relation Between Symmetry, Galilei Invariance, and Conservation Laws

In the following section, we apply symmetry transformations to Newton's classical law of motion for a point of mass m moving with a potential energy $U(\mathbf{x}, t)$ for Euclidean, Cartesian coordinates:

$$m \frac{d^2 \mathbf{x}(t)}{dt^2} = -\nabla_x U(\mathbf{x}, t)$$

(∇_x represents the gradient with respect to the variable $\mathbf{x}$). Let the initial conditions be $\mathbf{x}(t_0) = \mathbf{x}_0$, $\dot{\mathbf{x}}(t_0) = \mathbf{v}_0$. The Newton equation represents a second-order differential ordinary equation for the desired function $\mathbf{x}(t)$ with some given initial conditions. We assume that we have found the desired trajectory $\mathbf{x}(t)$ given the initial conditions $\mathbf{x}(t_0) = \mathbf{x}_0$, $\dot{\mathbf{x}}(t_0) = \mathbf{v}_0$.

The invariance of the potential energy with respect to symmetry elements has, in the first place, an impact on the structure of the classical trajectories. These results fall under the label of "Galilei invariance". In addition, symmetry often produces conserved quantities of classical motion. These results fall under the label of "conservation laws". The following arguments are developed to show explicitly these two kinds of results.

Time translation: We transform the variable time with the law

$$\hat{t} = t + \tau$$

The trajectory transforms as[1]

$$\hat{\mathbf{x}}(\hat{t}) = \mathbf{x}(\hat{t} - \tau)$$

The initial conditions become

$$\hat{\mathbf{x}}(\hat{t}_0) = \mathbf{x}(\hat{t}_0 - \tau) \quad \dot{\hat{\mathbf{x}}}(\hat{t}_0) = \dot{\mathbf{x}}(\hat{t}_0 - \tau)$$

Proposition. We show now that, provided U does not explicitly depend on t, the function $\hat{\mathbf{x}}(\hat{t})$ is a solution to the formally identical Newton equation.

---

[1] When time is transformed, we need to learn how $\frac{d}{dt}$ and $\frac{d^2}{dt^2}$ transform. We compute

$$\frac{d}{d\hat{t}} \hat{\mathbf{x}}(\hat{t}) = \frac{d}{d\hat{t}} \mathbf{x}(t(\hat{t})) = \frac{d}{dt} \mathbf{x}(t) \cdot \frac{dt}{d\hat{t}}$$

and accordingly,

$$\frac{d^2}{d\hat{t}^2} \hat{\mathbf{x}}(\hat{t}) = \frac{d}{dt} \frac{d}{dt} \mathbf{x}(t) \cdot \frac{dt}{d\hat{t}} = \frac{d^2}{dt^2} \mathbf{x}(t) \cdot \left(\frac{dt}{d\hat{t}} \right)^2$$

Proof. We start with a time-dependent U for the sake of generality and write

$$m\frac{d^2\hat{\mathbf{x}}(\hat{t})}{d\hat{t}^2} = m\frac{d^2\mathbf{x}(t)}{dt^2} = -\nabla_{x(t)}U(\mathbf{x}(t),\hat{t}-\tau) = -\nabla_{\hat{x}}U(\hat{\mathbf{x}},\hat{t}-\tau)$$

For the Newton equation to be invariant with respect to the time translations τ, it is necessary that U does not depend explicitly on time.

Comment. This result proves that the trajectory $\mathbf{x}(t)$ and the time translated $\hat{\mathbf{x}}(t-\tau)$ (with the corresponding initial conditions) are both solutions to the same Newton equation and that they coincide when they both coexist.

Proposition. Provided U is invariant with respect to time translation, the total energy

$$\frac{1}{2}\cdot m\cdot\dot{\mathbf{x}}^2 + U(\mathbf{x})$$

is a constant of the motion.

Proof. Invariance with respect to time translations is equivalent to its derivative with respect to the variable t vanishing. Accordingly, by multiplying the Newton equation

$$m\frac{d^2\mathbf{x}(t)}{dt^2} = -\nabla_x U(\mathbf{x},t)$$

from both sides with $\dot{\mathbf{x}}$, we obtain

$$\underbrace{\underbrace{m\cdot\dot{\mathbf{x}}\cdot\ddot{\mathbf{x}}}_{\frac{d}{dt}(\frac{1}{2}\cdot m\cdot\dot{\mathbf{x}}^2)} + \underbrace{\dot{\mathbf{x}}\cdot\nabla_x U(\mathbf{x},t)}_{\frac{d}{dt}U(\mathbf{x},t)-\underbrace{\frac{\partial}{\partial t}U(\mathbf{x},t)}_{0}}}_{\frac{d}{dt}(\frac{1}{2}\cdot m\cdot\dot{\mathbf{x}}^2+U(\mathbf{x}))} = 0$$

Time reversal: We transform the time variable according to

$$\hat{t} = -t$$

Proposition. Provided U does not explicitly depend on t, the function

$$\hat{\boldsymbol{x}}(\hat{t}) \doteq \mathbf{x}(-\hat{t})$$

is a solution to the formally identical Newton equation with the initial conditions $\hat{\mathbf{x}}(\hat{t}_0) = \mathbf{x}(-\hat{t}_0)$, $\dot{\hat{\boldsymbol{x}}} = -\dot{\mathbf{x}}(-\hat{t}_0)$.

Proof.

$$m\frac{d^2\hat{\mathbf{x}}(\hat{t})}{dt^2} = m\frac{d^2\mathbf{x}(-t)}{dt^2} = -\nabla_{x(-t)}U(\mathbf{x}(-t)) = -\nabla_{\hat{x}}U(\hat{\mathbf{x}})$$

Comment. This result proves that the trajectory $\mathbf{x}(t)$ and $\hat{\mathbf{x}}(-t)$ are both solutions to the Newton equation with the corresponding initial conditions and that they coincide. They are run through, however, with the opposite chirality. The symmetry against time reversal does not introduce any constant of motion.

Spatial translations: We transform the spatial variable according to

$$\hat{\mathbf{x}} = \mathbf{x} + \mathbf{a}$$

Proposition. Provided U is invariant with respect to a translation $\mathbf{a}$ along a direction specified by $\vec{n}$, the trajectory

$$\hat{\mathbf{x}}(t) = \mathbf{x}(t) + \mathbf{a}$$

is a solution to the formally identical Newton equation with the initial conditions $\hat{\mathbf{x}}(t_0) = \mathbf{x}(t_0) + \mathbf{a}$, $\dot{\hat{x}}(t_0) = \dot{\mathbf{x}}(t_0)$.

Proof.

$$m\frac{d^2\hat{\mathbf{x}}(t)}{dt^2} = m\frac{d^2\mathbf{x}(t)}{dt^2} = -\nabla_x U(\mathbf{x}(t)) = -\nabla_x U(\mathbf{x}(t) + \mathbf{a})$$
$$= -\nabla_{\hat{x}}U(\hat{\mathbf{x}}) \cdot \underbrace{\nabla_x(\mathbf{x} + \mathbf{a})}_{=1}$$

Comment. $\hat{\mathbf{x}}(t)$ is the same trajectory $\mathbf{x}(t)$, translated along $\vec{n}$ by $|\vec{a}|$. The spatial translations produce a set of translated classical trajectories. If the translation vector is given by $\mathbf{v} \cdot t$, the resulting trajectory is affected by a time-dependent drift along the coordinate specified by $\mathbf{v}$.

Proposition. Provided U is invariant with respect to a translation $\mathbf{a}$ along a direction specified by $\vec{n}$, the quantity

$$m \cdot (\dot{\mathbf{x}} \cdot \mathbf{n})$$

is a constant of the motion. It is called the momentum of the mass along the direction **n**.

Proof.

$$\frac{d}{dt}\left(m\cdot\dot{\mathbf{x}}\cdot\mathbf{n}\right) = m\cdot\left(\frac{d}{dt}\dot{\mathbf{x}}\right)\cdot\mathbf{n} = m\cdot\left(\frac{d^2}{dt^2}\mathbf{x}\right)\cdot\mathbf{n} = -m\cdot\underbrace{\nabla_x U(\mathbf{x})\cdot\mathbf{n}}_{=\lim_{h\to 0}\frac{U(\mathbf{x}+h\cdot\mathbf{n})-U(\mathbf{x})}{h}}$$

Because of the required translational invariance along $\mathbf{n}$, $U(\mathbf{x}+h\cdot\mathbf{n})-U(\mathbf{x})$ vanishes exactly.

Comment. Using the same argument, one can prove momentum conservation along a certain direction when the translation vector $\mathbf{a}$ is given by $\mathbf{v}\cdot t$.

Inversion: We transform the spatial coordinate according to

$$\hat{\mathbf{x}} = -\mathbf{x}$$

Proposition. Provided U is invariant with respect to inversion, the function

$$\hat{\mathbf{x}}(t) = -\mathbf{x}(t)$$

is a solution to the formally identical Newton equation with the initial conditions $\hat{\mathbf{x}}(t_0) = -\mathbf{x}(t_0)$, $\dot{\hat{\mathbf{x}}}(t_0) = -\dot{\mathbf{x}}(t_0)$.

Proof.

$$m\frac{d^2\hat{\mathbf{x}}(t)}{dt^2} = -m\frac{d^2\mathbf{x}(t)}{dt^2} = \nabla_x U(\mathbf{x}(t)) = \nabla_x U(-\mathbf{x}(t)) = -\nabla_{\hat{x}} U(\hat{\mathbf{x}})$$

Comment. The trajectory $\hat{\mathbf{x}}(t)$ is the same as the one traced by $\mathbf{x}(t)$ and is traced with the same chirality. It starts at the point $-\mathbf{x}_0$. The inversion symmetry does not introduce any constant of motion.

Spatial rotations: We transform the spatial coordinate according to

$$\hat{\mathbf{x}} = \Lambda\mathbf{x}$$

Proposition. Provided U is invariant with respect to a rotation Λ, the function

$$\hat{\mathbf{x}}(t) = \Lambda\mathbf{x}(t)$$

is a solution to the formally identical Newton equation with the initial conditions $\hat{\mathbf{x}}(t_0) = \Lambda\mathbf{x}(t_0)$, $\dot{\hat{x}}(t_0) = \Lambda\dot{\mathbf{x}}(t_0)$.

Proof.

$$m\frac{d^2\hat{\mathbf{x}}_{\mathbf{i}}(t)}{dt^2} = m\sum_j \Lambda_{ij}\frac{d^2\mathbf{x}_j(t)}{dt^2} = -\sum_j \Lambda_{ij}\partial_j U(\mathbf{x})$$

$$= -\sum_j \Lambda_{ij}\partial_j U\left(\sum_l \Lambda_{1l}x_l, \sum_l \Lambda_{2l}x_l, \sum_l \Lambda_{3l}x_l\right)$$

$$= -\sum_j \Lambda_{ij}\sum_l \Lambda_{lj}\partial_{\hat{l}}U(\Lambda\mathbf{x}) = -\sum_l \underbrace{\sum_j \Lambda_{ij}\Lambda_{jl}^t}_{\delta_{il}}\partial_{\hat{l}}U(\Lambda\mathbf{x})$$

$$= -\sum_l \delta_{il}\partial_{\hat{l}}U(\hat{\mathbf{x}}) = -\partial_{\hat{i}}U(\hat{\mathbf{x}})$$

Comment. The rotational symmetry produces a set of trajectories that are similar but rotated with respect to each other around the origin of the coordinate system.

Proposition. Provided U is invariant with respect to rotations about an axis specified by the unit vector $\mathbf{n}$, the quantity

$$\mathbf{n}\cdot(\mathbf{x}\times m\cdot\dot{\mathbf{x}})$$

is a constant of the motion. It is called the angular momentum of the mass about the axis $\mathbf{n}$, $\mathbf{n}\cdot\mathbf{L}$.

Proof.

$$\frac{d}{dt}\mathbf{n}\cdot(\mathbf{x}\times m\cdot\dot{\mathbf{x}}) = \mathbf{n}\cdot\mathbf{x}\times m\cdot\frac{d^2\mathbf{x}}{dt^2}$$

$$= -\mathbf{n}\cdot\mathbf{x}\times\nabla U(\mathbf{x})$$

$$= -\nabla U(\mathbf{x})\cdot\mathbf{n}\times\mathbf{x}$$

$$= -\lim_{\varphi\to 0}\frac{U(\mathbf{x}+\mathbf{n}\times\mathbf{x}\cdot\varphi) - U(\mathbf{x})}{\varphi}$$

$$= 0$$

The vanishing of the directional derivative $\lim_{\varphi\to 0}\frac{U(\mathbf{x}+\mathbf{n}\times\mathbf{x}\cdot\varphi)-U(\mathbf{x})}{\varphi}$ is a consequence of the invariance of U with respect to rotations of $\mathbf{x}$ around $\mathbf{n}$: Recall that $\mathbf{x}+\mathbf{n}\times\mathbf{x}\cdot\varphi$ is, for small φ, the rotated vector.

17.3 Scaling Transformations

Scaling (dilation) transformations are described by

$$\hat{\mathbf{x}} = \lambda \cdot \mathbf{x}$$

$$\hat{t} = \lambda^b \cdot t$$

Definition. A function $U(\mathbf{x})$ is called a homogeneous function of degree k if it fulfills the relation

$$U(\lambda \mathbf{x}) = \lambda^k \cdot U(\mathbf{x})$$

Proposition. Provided $U(\mathbf{x})$ is a homogeneous function, the Newton equations are invariant with respect to scaling transformations.

Proof.

$$m \cdot \frac{d^2 \hat{\mathbf{x}}}{d\hat{t}^2} = m \cdot \frac{\lambda}{\lambda^{2b}} \frac{d^2 \mathbf{x}}{dt^2} = -\frac{\lambda}{\lambda^{2b}} \cdot \nabla U(\mathbf{x}) = -\frac{\lambda}{\lambda^{2b}} \cdot \lambda^{-k} \nabla U(\lambda \mathbf{x})$$

$$= \frac{\lambda^{2-k}}{\lambda^{2b}} \cdot (-\nabla_{\hat{x}} U(\hat{\mathbf{x}}))$$

Requiring

$$2 - k = 2b$$

produces the scaling invariance.

Comment. The scaling of spatial coordinates and time is not independent. The trajectory $\hat{\mathbf{x}}(\hat{t})$ is similar to the trajectory $\mathbf{x}(t)$: It only occurs over spatial coordinates that are magnified by λ and times that are magnified by $\lambda^{\frac{2-k}{2}}$. For instance, for $k = -1$ (the Kepler problem), one has that the times required to trace a trajectory with all dimensions changed by λ are magnified by $\lambda^{\frac{3}{2}}$. This is Kepler's third law.

Using the Euler theorem for homogeneous functions, one can construct an integral of the motion related to the scaling symmetry.

Proposition. Provided the function $U(\mathbf{x})$ is an homogeneous function of degree k, the quantity

$$Q \doteq m \cdot \dot{\mathbf{x}} \cdot \mathbf{x} - 2 \cdot t \cdot E + (2 + k) \int_0^t d\tau \cdot U(\mathbf{x}(\tau))$$

is a constant of the motion.

Proof. Take the time derivative

$$\frac{d}{dt}Q = m \cdot \mathbf{x} \cdot \ddot{\mathbf{x}} + m \cdot \dot{\mathbf{x}}^2 - 2 \cdot E + (2 + k)U(\mathbf{x})$$

$$= -m \cdot \mathbf{x} \cdot \nabla U(\mathbf{x}) + k \cdot U(\mathbf{x}) = 0$$

The existence of a constant of motion Q can be rewritten as the well-known virial theorem for closed trajectories (with T being the period):

$$\frac{1}{T}\int_0^T \frac{d}{dt}Q(t) \equiv 0 = m \cdot \underbrace{\frac{1}{T}\int_0^T \frac{d}{dt}\dot{\mathbf{x}} \cdot \mathbf{x}}_{=0 \quad \text{for closed paths}} - \frac{1}{T}\int_0^T dt \cdot m \cdot \dot{\mathbf{x}}^2$$

$$+ k \cdot \frac{1}{T}\int_0^T dt \cdot U(\mathbf{x}(t))$$

Accordingly, the time average of twice the kinetic energy equals k times the time average of the potential energy (the so called "Virial Theorem"):

$$\frac{1}{T}\int_0^T dt \cdot m \cdot \dot{\mathbf{x}}^2 = k \cdot \frac{1}{T}\int_0^T dt \cdot U(\mathbf{x}(t))$$

Lecture 18

PROJECT FOR LECTURE 4: CRYSTAL FIELD SPLITTING AND THE JAHN–TELLER EFFECT

18.1 Crystal Field Splitting

Crystal field splitting is an important mechanism in chemistry and solid-state physics.[1] It describes the partial removal of degeneracy of energy levels when an atom (which typically carries a potential with spherical symmetry) is embedded in an environment that breaks the spherical symmetry. It occurs in many compounds of theoretical and technological interest, but an important example is materials with a perovskite structure. These have been, for example, proposed for applications in solar cells, as data storage materials, variable capacitors, and many different sensors, and have been the focus of a strong effort in material science and physics research in the past few decades.[2] In the following, we illustrate the mechanism of crystal field splitting and the specific role of symmetry with a concrete example.

---

[1] S. V. Streltsov *et al.*, *Phys. Rev. B* **71**, 245114 (2005), https://journals.aps.org/prb/PhysRevB.71.245114;
M. W. Haverkort *et al.*, *Phys. Rev. B* **94**, 056401 (2005), https://journals.aps.org/prl/abstract/10.1103/PhysRevLett.94.056401.
[2] A. Wang *et al.*, *Adv. Funct. Mater.* **29**, 1808843 (2019), https://onlinelibrary.wiley.com/doi/abs/10.1002/adfm.201808843;
M. V. Raymond and D. M. Smyth, *J. Phys. and Chem. Sol.* **57**, 1507–1511 (1996), https://www.sciencedirect.com/science/article/abs/pii/0022369796000200;
A. S. Bhalla, "Ruyan Guo and Rustum Roy", *Mat. Res. Innov.* **4**, 3–26 (2000), https://www.tandfonline.com/doi/abs/10.1007/s100190000062.

18.1.1 *The role of symmetry in determining the potential*

We have in mind a d-electron in the spherically symmetric potential such that the Hamiltonian reads

$$H_0 = -\frac{\hbar^2}{2m}\triangle + \Phi_{MF}(r)$$

where $\Phi_{MF}(r)$ is an effective potential, computed, for example, in a self-consistent manner, taking into account the nucleus at the origin and the remaining electrons. $\hbar$ is the reduced Planck's constant, and m is the mass of the electron. An example would be the Ti^{3+} ion, where the atomic configuration of neutral Ti is $(3d)^2(4s)^2$. We take this ion to be the central one and embed it into an octahedral distribution of other atoms (which are called ligands) that surround it. The octahedral sites could be occupied by H_2O molecules that are slightly negatively charged, or, for example, negatively charged O ions (as is the case in perovskites[3]). These molecules or atoms produce a further potential energy term $\Phi(\mathbf{r})$ which breaks the full rotational symmetry of the Ti^{3+} ion, i.e. the full potential is no longer spherical symmetric but is only invariant with respect to the operations of the subgroup O_h. The potential $\Phi(\mathbf{r})$ produced by the octahedral distribution is the solution to the Laplace equation of electrostatics and can be represented, generally, using the superposition of spherical harmonics Y_l^m:

$$\Phi(\mathbf{r}) = \sum_l \sum_{m=-l}^{l} A_{lm} r^l Y_l^m(\theta, \phi)$$

The coefficients forthcoming in this sum are not arbitrary: They must be such that the potential is completely symmetric with respect to the operation of O_h, i.e. it transforms according to the representation Γ_1. Our aim is to find these coefficients on the basis of this symmetry requirement.

Selection rules: We would like to compute the change in the atomic one-electron d-levels in the presence of $\Phi(\mathbf{r})$ using first-order perturbation theory. Within a perturbative approach, we only need to consider matrix elements of the potential $\Phi(\mathbf{r})$ between the basis states building the five-dimensional degenerate d-levels. These matrix elements will be

$$\propto \left(Y_2^{m_f}, Y_l^m Y_2^{m_i}\right)$$

---

[3]Class of compounds with the crystal structure of $CaTiO_3$, see e.g. https://en.wikipedia.org/wiki/Perovskite.

where m_f, m_i vary from -2 to $+2$, and m ranges between $-l$ and l. Only the angular-dependent part of the matrix elements is displayed here, as this is the component of the matrix elements that can be determined by symmetry. In fact, because of

$$D_l(\Phi) \otimes D_{2,i} = D_{l+2} \oplus D_{l+1} \oplus D_l \oplus D_{l-1} \oplus D_{l-2}$$

and taking into account that, for the matrix elements to be non-vanishing, $D_{2,f}$ must be contained in the decomposition of $D_l(\Phi) \otimes D_{2,i}$ (Wigner–Eckart–Koster selection rule), the component l of the potential energy that can contribute a non-vanishing matrix element is restricted to $l = 0, 1, 2, 3, 4$. Inversion symmetry causes the matrix elements involving the spherical harmonics $l = 1, 3$ to vanish. Accordingly, the components of the octahedral potential that contribute to the matrix elements must be searched among the spherical harmonics transforming according to the D_0^+, D_2^+, and D_4^+ representations of O_3.

Symmetry-adapted spherical harmonics: We can constrain the l values even further in the following way. We observe that the irreducible representations $D^{l\pm}$ of O_3, when limited to O_h, have the following compatibility table (given for $l = 0, 1, 2, 3, 4, 5$) with O_h.

D^{0+}	D^{0-}	D^{1+}	D^{1-}	D^{2+}	D^{2-}
Γ_1	Γ_1'	Γ_{15}'	Γ_{15}	$\Gamma_{12} + \Gamma_{25}'$	$\Gamma_{12}' + \Gamma_{25}$
D^{3+}	D^{3-}	D^{4+}	D^{4-}	D^{5+}	D^{5-}
$\Gamma_2 + \Gamma_{25}'$ $+\Gamma_{15}'$	$\Gamma_2' + \Gamma_{25}$ $+\Gamma_{15}$	$\Gamma_1 + \Gamma_{12}$ $+\Gamma_{25}' + \Gamma_{15}'$	$\Gamma_1' + \Gamma_{12}'$ $+\Gamma_{25} + \Gamma_{15}$	$\Gamma_{12} + \Gamma_{25}'$ $+2\Gamma_{15}'$	$\Gamma_{12}' + \Gamma_{25}$ $+2\Gamma_{15}$

From the table, we deduce that D_0^+ and D_4^+ decompose into irreducible representations containing Γ_1, but not D_2^+. Accordingly, we need to search for Γ_1 symmetry-adapted linear combinations of spherical harmonics only within the space spanned by Y_0^0 and $Y_4^4, Y_4^3, Y_4^2, Y_4^1, Y_4^0, Y_4^{-1}, Y_4^{-2}, Y_4^{-3}, Y_4^{-4}$. The $l = 2$ sector cannot appear in the potential.

- $Y_0^0 = \sqrt{\frac{1}{4\pi}}$ is trivially symmetry adapted to Γ_1: Being a constant, it remains invariant under all operations of O_h.

- Γ_1 appears only once in the decomposition of D_4^+ restricted to O_h: We must find the linear combination of $\{Y_4^m\}$ that transforms according to Γ_1. For this purpose, we use the projector method:

$$P_{\Gamma_1} Y_l^m = \sum_g \cdot \underbrace{1}_{\text{character of } \Gamma_1} \cdot U(g) Y_l^m$$

applied to the spherical harmonics

$$Y_4^{-4} = \frac{3}{16}\sqrt{\frac{35}{2\pi}} \cdot \frac{(x-iy)^4}{r^4}, \quad Y_4^4 = \frac{3}{16}\sqrt{\frac{35}{2\pi}} \cdot \frac{(x+iy)^4}{r^4}$$

$$Y_4^{-3} = \frac{3}{8}\sqrt{\frac{35}{\pi}} \cdot \frac{(x-iy)^3 z}{r^4}, \quad Y_4^3 = \frac{-3}{8}\sqrt{\frac{35}{\pi}} \cdot \frac{(x+iy)^3 z}{r^4}$$

$$Y_4^{-2} = \frac{3}{8}\sqrt{\frac{5}{2\pi}} \cdot \frac{(x-iy)^2 \cdot (7z^2-r^2)}{r^4}, \quad Y_4^2 = \frac{3}{8}\sqrt{\frac{5}{2\pi}} \cdot \frac{(x+iy)^2 \cdot (7z^2-r^2)}{r^4}$$

$$Y_4^{-1} = \frac{3}{8}\sqrt{\frac{5}{\pi}} \cdot \frac{(x-iy)\cdot z\cdot(7z^2-3r^2)}{r^4}, \quad Y_4^1 = \frac{-3}{8}\sqrt{\frac{5}{\pi}} \cdot \frac{(x+iy)\cdot z\cdot(7z^2-3r^2)}{r^4}$$

$$Y_4^0 = \frac{3}{16}\sqrt{\frac{1}{\pi}} \cdot \frac{(35z^4-30z^2r^2+3r^4)}{r^4}$$

$U(g)$ is the operator that implements the rotation $g \in O_h$ onto the spherical harmonics.

Transformation of Cartesian coordinates (x, y, z) under the symmetry operations of the octahedral group O_h.

Type	Operation	Coordinate	Type	Operation	Coordinate
C_1	C_1	xyz	I	I	$\bar{x}\bar{y}\bar{z}$
C_4^2	C_{2z}	$\bar{x}\bar{y}z$	IC_4^2	IC_{2z}	$xy\bar{z}$
	C_{2x}	$x\bar{y}\bar{z}$		IC_{2x}	$\bar{x}yz$
	C_{2y}	$\bar{x}y\bar{z}$		IC_{2y}	$x\bar{y}z$
C_4	C_{4z}^{-1}	$\bar{y}xz$	IC_4	IC_{4z}^{-1}	$y\bar{x}\bar{z}$
	C_{4z}	$y\bar{x}z$		IC_{4z}	$\bar{y}x\bar{z}$
	C_{4x}^{-1}	$x\bar{z}y$		IC_{4x}^{-1}	$\bar{x}z\bar{y}$
	C_{4x}	$xz\bar{y}$		IC_{4x}	$\bar{x}\bar{z}y$
	C_{4y}^{-1}	$zy\bar{x}$		IC_{4y}^{-1}	$\bar{z}\bar{y}x$
	C_{4y}	$\bar{z}yx$		IC_{4y}	$z\bar{y}\bar{x}$
C_2	C_{2xy}	$yx\bar{z}$	IC_2	IC_{2xy}	$\bar{y}\bar{x}z$
	C_{2xz}	$z\bar{y}x$		IC_{2xz}	$\bar{z}y\bar{x}$
	C_{2yz}	$\bar{x}zy$		IC_{2yz}	$x\bar{z}\bar{y}$
	$C_{2x\bar{y}}$	$\bar{y}\bar{x}\bar{z}$		$IC_{2x\bar{y}}$	yxz
	$C_{2\bar{x}z}$	$\bar{z}\bar{y}\bar{x}$		$IC_{2\bar{x}z}$	zyx
	$C_{2y\bar{z}}$	$\bar{x}\bar{z}\bar{y}$		$IC_{2y\bar{z}}$	xzy
C_3	C_{3xyz}^{-1}	zxy	IC_3	IC_{3xyz}^{-1}	$\bar{z}\bar{x}\bar{y}$
	C_{3xyz}	yzx		IC_{3xyz}	$\bar{z}\bar{x}\bar{y}$
	$C_{3x\bar{y}z}^{-1}$	$z\bar{x}\bar{y}$		$IC_{3x\bar{y}z}^{-1}$	$\bar{z}xy$
	$C_{3x\bar{y}z}$	$\bar{y}\bar{z}x$		$IC_{3x\bar{y}z}$	$yz\bar{x}$
	$C_{3x\bar{y}\bar{z}}^{-1}$	$\bar{z}\bar{x}y$		$IC_{3x\bar{y}\bar{z}}^{-1}$	$zx\bar{y}$
	$C_{3x\bar{y}\bar{z}}$	$\bar{y}z\bar{x}$		$IC_{3x\bar{y}\bar{z}}$	$y\bar{z}x$
	$C_{3xy\bar{z}}^{-1}$	$\bar{z}x\bar{y}$		$IC_{3xy\bar{z}}^{-1}$	$z\bar{x}y$
	$C_{3xy\bar{z}}$	$yz\bar{x}$		$IC_{3xy\bar{z}}$	$\bar{y}zx$

Considering the transformation properties of the (x, y, z) coordinates, we are able to find that some of the spherical harmonics are orthogonal to the Γ_1-sector. In fact, we observe that $Y_4^{\pm 3}$ and $Y_4^{\pm 1}$ yield zero when the projector is applied to them, as they are all odd in z. Furthermore, $Y_4^{\pm 2}$ are also orthogonal: They change sign under the transformation $(x, y) \leftrightarrow (\bar{y}, x)$, which generates an extra $(-i)^2 = -1$. When projected out, this change in sign causes the sum to the vanish. The only spherical harmonics which are not obviously orthogonal are those with $m \in \{0, +4, -4\}$. In fact, we find

$$P_{\Gamma_1} Y_4^{-4} \sim P_{\Gamma_1} (x - iy)^4$$
$$= 16[(x^4 + y^4 + z^4) + 3(x^2 y^2 + y^2 z^2 + z^2 x^2)]$$
$$P_{\Gamma_1} Y_4^4 \sim P_{\Gamma_1} (x + iy)^4$$
$$= 16[(x^4 + y^4 + z^4) + 3(x^2 y^2 + y^2 z^2 + z^2 x^2)]$$
$$P_{\Gamma_1} Y_4^0 \sim P_{\Gamma_1} (35 z^4 - 30 z^2 r^2 + 3 r^4)$$
$$= 16[35(x^4 + y^4 + z^4) - 21 r^4]$$

Accordingly, the potential will be, in the $l = 4$ sector, a linear combination of the type

$$f \doteq Y_4^0 + (a Y_4^4 + b Y_4^{-4})$$

where the coefficients a, b must be determined. First, we note that the potential must be a real quantity, i.e. f must be real, so we must set $a = b$. Then, we define $f = Y_4^0 + (a Y_4^4 + a Y_4^{-4})$ and compute

$$f \sim 3x^4 + 3y^4 + 8z^4 + 24 x^2 y^2 + 6 y^2 z^2 - 24 z^2 x^2$$
$$+ \sqrt{70} a (x^4 + 6 x^2 y^2 + y^4)$$
$$P_{\Gamma_1} f \sim 14(x^4 + y^4 + z^4) - 42(x^2 y^2 + y^2 z^2 + z^2 x^2)$$
$$+ \sqrt{70} a [(x^4 + y^4 + z^4) + 3(x^2 y^2 + y^2 z^2 + z^2 x^2)]$$

where we neglected the constant term $\frac{3}{16}\sqrt{\frac{1}{\pi}}$, multiplying all three Y_l^m. As the space transforming according to Γ_1 is one-dimensional, we must require

$$P_{\Gamma_1} f \sim f$$

which produces

$$a = \sqrt{\frac{5}{14}}$$

This result allows the *ansatz*

$$\Phi(\mathbf{r}) = A + C \cdot r^4 \cdot \left[Y_4^0 + \sqrt{\frac{5}{14}} (Y_4^4 + Y_4^{-4}) \right]$$

for that component of the potential which produces non-vanishing matrix elements between *d*-states.

For the sake of completeness, we provide an estimate of the coefficients A and C, which cannot be determined from symmetry arguments, as they also depend on the radial component. Assuming point-like charge distributions, the potential reads, exactly

$$\Phi(\mathbf{r}) = \frac{e^2}{4\pi\epsilon_0} \cdot Z_{\text{ligand}} \cdot \sum_{i=1}^{6} \frac{1}{|\mathbf{r} - \mathbf{R}_i|}$$

This potential describes the interaction of the *d*-electron with ligands having an effective negative charge of eZ_{ligand}. We can expand the $\mathbf{r}$ dependence of this exponential in terms of the spherical harmonics in the following way:

$$\frac{1}{|\mathbf{r} - \mathbf{R}_i|} = 4\pi \cdot \sum_l \sum_m \frac{1}{2l+1} \frac{r^l}{t^{l+1}} \overline{Y_l^m}(\theta_i, \phi_i) Y_l^m(\theta, \phi)$$

which is valid when the distance of interest is smaller than the distance t between the central and ligand atoms. The constants A and C can be expressed in terms of the properties of the ligand atoms, namely Z_{ligand} and t. In fact, recalling the orthonormality relation of spherical harmonics,

$$\int d\Omega \overline{Y_l^m}(\theta, \phi) Y_{l'}^{m'}(\theta, \phi) = \delta_{ll'} \delta_{mm'}$$

we obtain

$$A = \int d\Omega \overline{Y_0^0}(\theta, \phi) \Phi(\mathbf{r}) Y_0^0 = \frac{e^2}{4\pi\epsilon_0} Z_{\text{ligand}} \sum_{i=1}^{6} \left(\frac{1}{2} \sqrt{\frac{1}{\pi}} \frac{4\pi}{t} \right)^2$$

$$= \frac{e^2}{4\pi\epsilon_0} Z_{\text{ligand}} \frac{6}{t}$$

$$C r^4 = \int d\Omega \overline{Y_4^0}(\theta, \phi) \Phi(\mathbf{r}) Y_4^0 = \frac{e^2}{4\pi\epsilon_0} Z_{\text{ligand}} \sum_{i=1}^{6} \frac{4\pi}{9} \frac{r^4}{t^5} Y_4^0(\theta_i, \phi_i)$$

$$C = \frac{e^2}{4\pi\epsilon_0} Z_{\text{ligand}} \frac{7\sqrt{\pi}}{3t^5}$$

This last step is achieved by plugging in explicitly the (angular) coordinates of the ligands in the octahedral configuration: $(\theta_i, \phi_i) \in \{(\frac{\pi}{2}, 0), (\frac{\pi}{2}, \pi), (\frac{\pi}{2}, \frac{\pi}{2}), (\frac{\pi}{2}, \frac{3\pi}{2}), (0, 0), (\pi, 0)\}$.

18.1.2 Symmetry-assisted solution of the eigenvalue problem

The Hamilton operator to be diagonalized within the d-space consisting of the basis functions

$$f_2(r)Y_2^m(\theta,\phi)$$

with $m = 2, \ldots, -2$ is

$$H_0 + A + \underbrace{C \cdot r^4 \cdot \left[Y_4^0 + \sqrt{\frac{5}{14}}(Y_4^4 + Y_4^{-4})\right]}_{\Phi(\mathbf{r})}$$

where H_0 is diagonal within this space and the energy level is the five times degenerate level E_d. The matrix entailing $\Phi(\mathbf{r})$ is not diagonal but is diagonalized by the symmetry-adapted wave functions transforming according to the irreducible representations of O_h appearing in the decomposition of $D_2^+ = \Gamma_{12} + \Gamma_{25}'$; these are known in the literature as $(\Gamma_{12} \equiv E_g)$

$$Y_2^0 \qquad \frac{1}{\sqrt{2}}(Y_2^2 + Y_2^{-2})$$

and $(\Gamma_{25}' \equiv T_{2g})$

$$-\frac{1}{\sqrt{2}}(Y_2^1 - Y_2^{-1}) \qquad \frac{i}{\sqrt{2}}(Y_2^1 + Y_2^{-1}) \qquad -\frac{i}{\sqrt{2}}(Y_2^2 - Y_2^{-2})$$

The level E_d will split into two sublevels with symmetry E_g (twofold degenerate) and symmetry T_{2g} (threefold degenerate):

$$E(E_g) = E_d + \left(f_2(r)Y_2^0, \Phi(\mathbf{r})f_2(r)Y_2^0\right)$$

$$E(T_{2g}) = E_d + \left(f_2(r)\left(-\frac{i}{\sqrt{2}}(Y_2^2 - Y_2^{-2})\right),\right.$$

$$\left.\Phi(\mathbf{r})f_2(r)\left(-\frac{i}{\sqrt{2}}(Y_2^2 - Y_2^{-2})\right)\right)$$

To evaluate the matrix elements, we use

$$\int_0^\pi d\theta \int_0^{2\pi} d\phi \sin\theta \overline{Y_{l_1}^{m_1}} \cdot Y_{l_2}^{m_2} \cdot Y_{l_3}^{m_3} = (-1)^{m_1} \cdot \sqrt{\frac{(2l_1 + 1) \cdot (2l_2 + 1) \cdot (2l_3 + 1)}{4\pi}} \cdot$$

$$\begin{pmatrix} l_1 & l_2 & l_3 \\ 0 & 0 & 0 \end{pmatrix} \cdot \begin{pmatrix} l_1 & l_2 & l_3 \\ -m_1 & m_2 & m_3 \end{pmatrix}$$

The Wigner 3j-symbols can be computed, for example, using the call "Wigner 3j" in WolframAlpha. Some useful values of the 3j-symbol are listed here:

$$\begin{pmatrix} 2 & 4 & 2 \\ 0 & 0 & 0 \end{pmatrix} = \sqrt{\frac{2}{35}} \qquad \begin{pmatrix} 2 & 4 & 2 \\ 0 & \pm 4 & 0 \end{pmatrix} = 0$$

$$\begin{pmatrix} 2 & 4 & 2 \\ \pm 2 & 0 & \mp 2 \end{pmatrix} = \frac{1}{3\sqrt{70}} \qquad \begin{pmatrix} 2 & 4 & 2 \\ \pm 2 & \mp 4 & \pm 2 \end{pmatrix} = \frac{1}{3}$$

Accordingly, the values of the desired sublevels are

$$E(E_g) = E_d + \frac{e^2}{4\pi\epsilon_0} \cdot Z_{\text{ligand}} \cdot \left[\frac{6}{t} + \frac{1}{t^5}(f_2(r), r^4 f_2(r)) \right]$$

$$E(T_{2g}) = E_d + \frac{e^2}{4\pi\epsilon_0} \cdot Z_{\text{ligand}} \cdot \left[\frac{6}{t} - \frac{2}{3t^5}(f_2(r), r^4 f_2(r)) \right]$$

The energy scheme resulting from the octahedral crystal field splitting is given in the following figure.

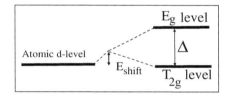

Figure displaying schematically the change in the energy levels of the atomic d-level in an octahedral symmetric crystal field. Δ is the so-called crystal field splitting energy $\Delta \doteq E(E_g) - E(T_{2g})$, and E_{shift} is a constant energy shift originating from the Y_0^0 component of the potential. The ordering of the energy levels is as expected for the d^1 and d^6 configurations. The symmetry-adapted polynomials to E_g are $2z^2 - x^2 - y^2; x^2 - y^2$. The symmetry-adapted polynomials to T_{2g} are xy, xz, yz.

Our symmetry analysis tells us how the degenerate energy levels split, but it does not reveal to us which of the new energy levels in the reduced symmetry is the lowest. For the specific case displayed in the figure, the T_{2g} energy level is lower than the E_g level. One expects such an ordering result for the d^1 and d^6 configurations. The reversed ordering of levels is expected for the d^4 and d^9 configurations. In the latter configurations, we can consider one electron to be "missing" from the half-filled case of the completely filled shell and describe it as a positively charged "hole". Experimental results on the observation of the crystal field splitting by pulse EPR are reported, for example, in S. Maurelli, S. Livraghi, M. Chiesa, E. Giamello, S. Van Doorslaer, C. Di Valentin and G. Pacchioni, *Inorg. Chem.* **50**, 2385–2394 (2011), https://pubs.acs.org/doi/10.1021/ic 1021802.

18.2   Jahn–Teller Effect

18.2.1   *Introduction*

The octahedral arrangement we have discussed until now might be unstable with respect to geometrical static distortions of the molecule or crystal that are likely to further reduce the symmetry of the system. The instability is driven by the possibility that, upon distortion, the total energy of the system is reduced with respect to the energy of the more symmetric environment. Such a "Jahn–Teller distortion" is indeed most often observed in transition metal ions subjected to an octahedral symmetry. Symmetry arguments allow us to determine whether a distortion is bound to reduce the value of the energy levels: This is the content of the Jahn–Teller "theorem".

In the formal framework for describing the Jahn–Teller effect, one begins with a system invariant under the operations of a symmetry group G_0. This system, which provides a potential landscape for the propagation of electrons, is described by a charge density, $\rho(\mathbf{R})$. $\rho(\mathbf{R})$ has the full symmetry G_0, i.e. it can be written as a sum of invariant polynomials transforming according to the irreducible representation "1" of G_0. We also explore charge density distribution which are distorted away from the G_0-invariant distribution, so that, in general, we deal with a distribution which is written as a Fourier series of the irreducible invariant basis functions transforming according to the irreducible representations of the symmetry group G_0 of the system:

$$\rho(X,Y,Z) = \sum_{s,i} \eta_{i,s} \varphi_{i,s}(X,Y,Z)$$

The index s runs over all irreducible representations Γ_s of G_0, whereas the index i labels the various partner functions transforming according to the d_s-dimensional representation Γ_s. One can think of the potential energy arising in the Hamilton operator as a functional of $\rho(\mathbf{R})$. Integrating this function over the coordinates (X,Y,Z) leaves behind an Hamilton operator containing the kinetic energy of the electron and a functional over the set of coefficients $\{\eta_{i,s}\}$ describing the potential energy term:

$$H(\mathbf{r}) = \underbrace{-\frac{\hbar^2}{2m}\triangle + V_{s=0}(\mathbf{r})}_{H_0} + \sum_{i,s\neq0} V_{i,s}(\mathbf{r}) \cdot \eta_{i,s} + \sum_{i,j,r,s} V_{i,j,r,s}(\mathbf{r}) \cdot \eta_{i,s}\eta_{jr} + \cdots$$

$V_{i,s}(\mathbf{r})$ transforms according to the ith column of the irreducible representation Γ_s. $V_{s=0}$ labels that component of the potential which transforms according to the "1" irreducible representation of G_0. H_0, accordingly, is the Hamilton operator of the "unperturbed" system, i.e. the one with G_0-symmetry. Assume that one coefficient $\eta_{i,s}$ with $s \neq'' 1''$ is non-vanishing. Let us call Γ_s the "active" representation. The presence of an active representation in the Hamiltonian is equivalent to the original charge distribution being distorted away from the one with the symmetry G_0: **Some subgroup** G **of** G_0 is realized as the symmetry group of the distorted system, as a consequence of the distortion. Possibly, the distortion produces a situation where the ground-state energy of the system is lowered with respect to the unperturbed one. If this happens, the system undergoes this distortion and assumes a state with the symmetry group G. Jahn and Teller investigated, using only symmetry arguments, the selection rules for the presence of an active representation to produce an energetically favorable state.

Let E_0 be the (possibly degenerate) energy level with

$$H_0 \psi_n = E_0 \psi_n$$

where ψ_n is some symmetry-adapted wave function transforming according to the irreducible representation Γ_0. The first-order perturbation theory correction to E_0 produced by the Γ_s-like distortion is given by the eigenvalues $\{E_1\}$ of the matrix with elements

$$\sum_{i,s} \eta_{i,s} \left(\psi_m, V_{i,s} \psi_n \right)$$

Note that if the set $\{E_1\}$ are the eigenvalues of this matrix for a given set of values $\{\eta_{i,s}\}$, $\{-E_1\}$ will be the eigenvalues of the matrix for the set of values $\{-\eta_{i,s}\}$. This means that, if any of these eigenvalues is non-vanishing, either the distortions $\{\eta_{i,s}\}$ or the distortions $\{-\eta_{i,s}\}$ will lower the energy value with respect to the undistorted value E_0: The high symmetric phase is unstable with respect to a lowering of the symmetry. A necessary condition for stability is therefore the vanishing of all matrix elements. Accordingly, the selection rule for stability of the high-symmetric geometry against the distortion with irreducible representation Γ_s is

$$\Gamma_0 \notin \Gamma_s \otimes \Gamma_0$$

This selection rule is, for example, realized when Γ_0 is once degenerate, as Γ_s is not the completely symmetric irreducible representation. Jahn and Teller demonstrated by analyzing all possible cases of point groups that, for any point group, any **degenerate** irreducible representation Γ_0 and any representation Γ_s describing distortions, the equation

$$\Gamma_0 \in \Gamma_s \otimes \Gamma_0$$

is satisfied. This means that, provided the matrix elements do not vanish accidentally, a **degenerate** energy level in a point group environment is always unstable against some symmetry lowering distortion (Jahn–Teller theorem, see H. A. Jahn, E. Teller, Stability of polyatomic molecules in degenerate electronic states, *Proc. R. Soc. Lond.* **A161**, 220–235 (1937), https://royalsocietypublishing.org/doi/10.1098/rspa.1937.0142).

Comment.

1. The irreducible representation Γ_s realized in the symmetry-broken potential is related to the subgroup G. The algorithm proceeds as follows:

   i. For a given subgroup, observe the characters of the representation Γ_s when restricted to the elements of G.
   ii. Find the compatibility

   $$\Gamma_s \mid_G = \oplus_i \Gamma_i(G)$$

   between Γ_s and the irreducible representations of G.
   iii. If "1" of G is in the compatibility series of Γ_s, then Γ_s is an active representation, leading to the lowering of symmetry.

2. A further aspect of the application of symmetry to the Jahn–Teller effect is provided by the fact that a given distortion realizes a subgroup of the original group G_0. The energy levels resulting from the Jahn–Teller distortion can be labeled according to the irreducible representations of the subgroup. This second aspect is helpful in classifying the energy levels resulting from the Jahn–Teller distortion.

18.2.2 *A simple example*

We apply the symmetry arguments to the distortion of the octahedral environment depicted in the following figure.

The distortion that produces an elongation or compression along the z-axis generates a molecule with the symmetry group D_{4h}. To apply the Jahn–Teller argument, we need, therefore, the character tables of both O_h and D_{4h}. To deal with this specific example, we use the labeling of irreducible representation and elements, which is very common in chemistry. We refer to the *Tables for group theory*, by P. W. Atkins, M. S. Child and C. S. Phillips, Oxford University Press, https://global.oup.com/uk/orc/chemistry/qchem2e/student/tables/.

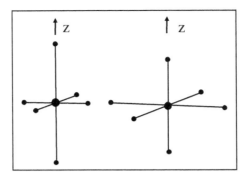

Figure displaying two possible Jahn–Teller distortions of the octahedral environment. The distortion either consists of elongation (left) or compression (right) of the octahedron along the z-axis.

Character table for O_h.

O_h	E	$8C_3$	$6C_2$	$6C_4$	$3C_2$	I	$6S_4$	$8S_6$	$3\sigma_h$	$6\sigma_d$	Basis
A_{1g}	1	1	1	1	1	1	1	1	1	1	$x^2 + y^2 + z^2$
A_{2g}	1	1	-1	-1	1	1	-1	1	1	-1	
E_g	2	-1	0	0	2	2	0	-1	2	0	$(x^2 - y^2)$ $2z^2 - x^2 - y^2$
T_{1g}	3	0	-1	1	-1	3	1	0	-1	-1	
T_{2g}	3	0	1	-1	-1	3	-1	0	-1	1	xy yz zx
A_{1u}	1	1	1	1	1	-1	-1	-1	-1	-1	
A_{2u}	1	1	-1	-1	1	-1	1	-1	-1	1	
E_u	2	-1	0	0	2	-2	0	1	-2	0	
T_{1u}	3	0	-1	1	-1	-3	-1	0	1	1	(x, y, z)
T_{2u}	3	0	1	-1	-1	-3	1	0	1	-1	

The first task of the Jahn–Teller algorithm is to find the active representation. The subgroup we are dealing with is the group D_{4h}. We need to find the compatibilities between O_h and D_{4h}. In Atkins *et al.*, these tables are called "descent tables". A look at the suitable descent table allows us to find that the only representation of O_h which (when restricted to D_{4h} and decomposed) contains the representation A_{1g} of D_{4h} is the irreducible representation E_g. E_g is therefore the active representation.

Character table of the tetragonal symmetry group D_{4h}.

	E	$2C_4$	C_2	$2C_2'$	$2C_2''$	I	$2S_4$	σ_h	$2\sigma_v$	$2\sigma_d$
A_{1g}	1	1	1	1	1	1	1	1	1	1
A_{2g}	1	1	1	-1	-1	1	1	1	-1	-1
B_{1g}	1	-1	1	1	-1	1	-1	1	1	-1
B_{2g}	1	-1	1	-1	1	1	-1	1	-1	1
E_g	2	0	-2	0	0	2	0	-2	0	0
A_{1u}	1	1	1	1	1	-1	-1	-1	-1	-1
A_{2u}	1	1	1	-1	-1	-1	-1	-1	1	1
B_{1u}	1	-1	1	1	-1	-1	1	-1	-1	1
B_{2u}	1	-1	1	-1	1	-1	1	-1	1	-1
E_u	2	0	-2	0	0	-2	0	2	0	0

The second task is to study the first-order perturbation behavior of the energy levels E_g and T_{2g}, computed in the section on crystal field splitting, upon introducing the distortion with the symmetry E_g. The levels E_g and T_{2g} are only stable against the perturbation if

$$E_g \notin E_g \otimes E_g \quad T_{2g} \notin E_g \otimes T_{2g}$$

We compute, using the character tables, that

$$E_g \otimes E_g = E_g \oplus A_{1g} \oplus A_{2g} \quad E_g \otimes T_{2g} = T_{1g} \oplus T_{2g}$$

According to these results, the conditions for the vanishing of the matrix elements are not fulfilled. In agreement with the Jahn–Teller theorem, the molecule is unstable against deformations along the vertical axis.

Having ascertained the instability, the third task is to figure out what will be the outcome of the representations E_g and T_{2g} of O_h when limited to the operations of D_{4h}. We find that they are reducible and split into irreducible representations of the group D_{4h}, according to the descent tables given by Atkins *et al.* (and according to our computations based on the characters):

$$E_g(O_h) = A_{1g}(D_{4h}) \oplus B_{1g}(D_{4h})$$

$$T_{2g}(O_h) = B_{2g}(D_{4h}) \oplus E_g(D_{4h})$$

We summarize the results of both the crystal field splitting in an octahedral crystal field and the (possible) subsequent Jahn–Teller distortion to the tetragonal symmetry in the following figure.

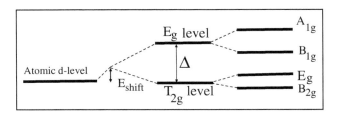

Figure displaying schematically the change in the energy levels of the atomic d-level under octahedral and tetragonal symmetries. Δ is the so-called crystal field splitting energy in the octahedral symmetry, and E_{shift} is a constant energy shift. For the octahedral symmetry, the ordering of the energy levels is demonstrated in the case of d^1 or d^6 filling, while for the tetragonal symmetry, it is shown in the case of compression along the z-axis.

As mentioned above, our symmetry analysis tells us how the energy levels change when the system adopts a new symmetry, but it does not provide their order (i.e. which one is the lowest energy level, which one follows, etc.). In the figure, we display an example of a Jahn–Teller distortion in the form of compression along the z-axis, lowering the octahedral symmetry to tetragonal. If we instead consider elongation along the z-axis, the levels A_{1g} and B_{1g} would switch, as would E_g and B_{2g}. The order would then be from low energy to high energy: E_g, B_{2g}, A_{1g}, and B_{1g}.

19

PROJECT FOR LECTURE 4: VIBRATIONAL MODES OF THE NH₃ MOLECULE

The following figure shows a schematic diagram of the NH_4 molecule, projected onto the basal plane. The N atom of ammonia (atom number 4, located at the center represented in green in the following figure) constitutes the tip of a pyramid with a triangular equilateral base. The H atoms (numbered 1, 2, and 3, and shown in blue in the figure) reside at the tips of the triangles. Each atom has three degrees of freedom. We have attached to each atom three coordinates to express these degrees of freedom. For the purpose of this project, the coordinates are along the bonds between the atoms, with the x and y coordinates along the in-plane bonds, and the z-coordinates along the out-of-plane bonds:

$$(x_1, y_1, z_1, x_2, y_2, z_2, x_3, y_3, z_3, z_4^1, z_4^2, z_4^3)$$

A Cartesian coordinate system is also indicated, with XY in the basal plane and Z perpendicular to it. The computation we will perform later for the vibrational modes assumes a simplified harmonic model, where the pairwise interactions, coupling the atoms and represented by "springs" with some "spring constants" f and g, are strictly along the bonds.

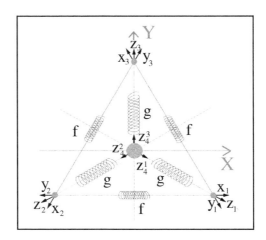

The computation of the vibrational frequencies of this molecule implies solving the determinantal equation of some force matrix. Before attempting this for a simplified harmonic model, we use symmetry arguments to simplify the problem.

19.1 Symmetry analysis of the vibrational problem of NH_3

The point group of NH_3 is C_{3v}.

A. The labeling of oscillatory modes with irreducible representations. We first construct the representation of C_{3v} within a 12-dimensional space spanned by the basis vectors indicated in the figure. We call this representation the "vectorial tower representation" or "vibrational representation" Λ_v. For a preliminary symmetry analysis, we need to consider some representatives of each class and find their characters. We must pay attention to the fact that when some symmetry operations are applied, the molecule is transformed into itself, whereas the basis vectors are not only rotated but might be transferred to a different atom.

- The identity element has a character of 12.
- A rotation by 120° does not leave any vector invariant, so it has a character of 0. For instance, x_1 is moved to x_2, y_1 is moved to y_2, z_1 is moved to z_2, and so on.
- Upon reflection at the plane containing the atoms 3 and 4, we observe that only z_3 and z_3^4 remain invariant, so that the character of the

Character table for the group C_{3v}. The characters of the vibration representations and its decomposition are shown in the bottom line.

C_{3v}	E	$2C_3$	$3\sigma_v$	
A_1	1	1	1	
A_2	1	1	-1	
E	2	-1	0	
Λ_v	12	0	2	$= 3A_1 \oplus A_2 \oplus 4E$

reflection elements in the representation Λ^v is 2. The table summarizes the characters of the irreducible representations and the characters of Λ_v.

The decomposition is performed using methods such as the decomposition theorem. The most notable result of this analysis is that the diagonalization of the 12×12 force matrix will produce three different blocks, a 3×3 block with A_1 symmetry, a one-dimensional block with A_2 symmetry, and an 8×8 block with E-symmetry. Accordingly, we expect vibrational eigenmodes with three different symmetry characters.

Before continuing with the search for the symmetry of vibrational frequencies, we observe that three of these eigenmodes will have a translational character and three will have a rotational character. Accordingly, their vibrational frequency vanishes. One of the translational modes is described by the displacement of each atom by exactly the same amount along the Z-direction. This mode has symmetry A_1. There will be two further translational modes describing coherent displacement along directions in the basal plane containing the basis of the pyramid. Let one translation be a common displacement of each atom along a direction specified by one of the sides of the triangle, and the second one along one of the other sides. In virtue of the in-plane triangular symmetry, the two basal translational modes must be degenerate and transform into themselves by the operations of C_{3v}, i.e. they transform as E. One rotational mode can be guessed based on our experience with the H₂O molecule. One can envisage it as a curl-like displacement configuration, where the displacement of the N atom is set to 0 and the in-plane displacement vectors are set to circulate in such a way that the displacement vector of one atom is transformed exactly into the displacement of the neighboring atom by a 120° degree rotation. This represents a rotational mode with a rotation axis along z. In virtue of the curl mode, the rotations transform the total displacement vector into itself, and the reflections transform it into "$-$" itself; hence, this rotational mode transforms according to A_2. The remaining two rotational modes must represent rotations with a rotation axis in the basal plane. They must be orthogonal to the A_2 rotational mode, i.e. they must belong either to A_1 or E. Because of the in-plane symmetry, the rotational modes with a rotation axis in the basal plane transform into themselves by the operations of C_{3v} and must belong to the irreducible representation E.

These results produce an "active" (with frequencies not necessarily 0) oscillatory representation $\Lambda_{\text{active}}^v$ of dimension 6, which must decompose into the irreducible representations left behind after subtracting the translations and the rotations, i.e.

$$\Lambda_{\text{active}}^v = 2A_1 \oplus 2E$$

As a result of this symmetry analysis, one expects two active vibrational modes with symmetry A_1 and two, with each one twice degenerate, active vibrational modes with symmetry E for a total of four separate vibrational frequencies. The comparison with quantitative results (see, for example, "Computational Chemistry Comparison and Benchmark DataBase Release 22 (May 2022), Standard Reference Database 101, National Institute of Standards and Technology", https://cccbdb.nist.gov/exp2x.asp?casno=76 64417) confirms that this analysis, based on symmetry, is a "robust" one, i.e. it produces results that are not dependent on the actual complexity of the electronic configuration involved in the vibrational motion.

B. Symmetry-adapted vectors: The next step in the symmetry analysis is the construction of the symmetry-adapted vectors. The preferred technique for this is the projector technique. However, one can also proceed by solving explicitly a very simple model that assumes axial linear restoring forces along the direction specified by the springs. This is a simplified version of the harmonic approximation that can be solved analytically because of its simplicity. Referring to the previous figure, the Newton equations of motion are written as (their order is chosen so that opposite degrees of freedom are described sequentially):

$$m_H \cdot \ddot{x}_1 = -f(x_1 + y_2) \quad m_H \cdot \ddot{y}_2 = -f(y_2 + x_1)$$
$$m_H \cdot \ddot{y}_1 = -f(y_1 + x_3) \quad m_H \cdot \ddot{x}_3 = -f(x_3 + y_1)$$
$$m_H \cdot \ddot{x}_2 = -f(x_2 + y_3) \quad m_H \cdot \ddot{y}_3 = -f(y_3 + x_2)$$
$$m_H \cdot \ddot{z}_1 = -g(z_1 + z_4^1) \quad m_N \cdot \ddot{z}_4^1 = -g(z_4^1 + z_1)$$

$$m_H \cdot \ddot{z}_2 = -g(z_2 + z_4^2) \quad m_N \cdot \ddot{z}_4^2 = -g(z_4^2 + z_2)$$

$$m_H \cdot \ddot{z}_3 = -g(z_3 + z_4^3) \quad m_N \cdot \ddot{z}_4^3 = -g(z_4^3 + z_3)$$

Inserting the eigenmode *ansatz* leads to the eigenvalue problem for the amplitudes:

$$\omega^2 \cdot \begin{pmatrix} x_1^0 \\ y_2^0 \\ y_1^0 \\ x_3^0 \\ x_2^0 \\ y_3^0 \\ z_{1,0} \\ z_4^0 \\ z_2^0 \\ z_{2,0} \\ z_3^0 \\ z_4^{3,0} \end{pmatrix} = \begin{pmatrix} \frac{f}{m_H} & \frac{f}{m_H} & 0 & 0 & 0 & 0 & 0 & 0 & 0 & 0 & 0 & 0 \\ \frac{f}{m_H} & \frac{f}{m_H} & 0 & 0 & 0 & 0 & 0 & 0 & 0 & 0 & 0 & 0 \\ 0 & 0 & \frac{f}{m_H} & \frac{f}{m_H} & 0 & 0 & 0 & 0 & 0 & 0 & 0 & 0 \\ 0 & 0 & \frac{f}{m_H} & \frac{f}{m_H} & 0 & 0 & 0 & 0 & 0 & 0 & 0 & 0 \\ 0 & 0 & 0 & 0 & \frac{f}{m_H} & \frac{f}{m_H} & 0 & 0 & 0 & 0 & 0 & 0 \\ 0 & 0 & 0 & 0 & \frac{f}{m_H} & \frac{f}{m_H} & 0 & 0 & 0 & 0 & 0 & 0 \\ 0 & 0 & 0 & 0 & 0 & 0 & \frac{g}{m_H} & \frac{g}{m_H} & 0 & 0 & 0 & 0 \\ 0 & 0 & 0 & 0 & 0 & 0 & \frac{g}{m_N} & \frac{g}{m_N} & 0 & 0 & 0 & 0 \\ 0 & 0 & 0 & 0 & 0 & 0 & 0 & 0 & \frac{g}{m_H} & \frac{g}{m_H} & 0 & 0 \\ 0 & 0 & 0 & 0 & 0 & 0 & 0 & 0 & \frac{g}{m_N} & \frac{g}{m_N} & 0 & 0 \\ 0 & 0 & 0 & 0 & 0 & 0 & 0 & 0 & 0 & 0 & \frac{g}{m_H} & \frac{g}{m_H} \\ 0 & 0 & 0 & 0 & 0 & 0 & 0 & 0 & 0 & 0 & \frac{g}{m_N} & \frac{g}{m_N} \end{pmatrix} \begin{pmatrix} x_1^0 \\ y_2^0 \\ y_1^0 \\ x_3^0 \\ x_2^0 \\ y_3^0 \\ z_{1,0} \\ z_4^0 \\ z_2^0 \\ z_{2,0} \\ z_3^0 \\ z_4^{3,0} \end{pmatrix}$$

The determinantal equation is written as

$$(\omega^2)^3 \cdot \left(\omega^2 - 2\frac{f}{m_H}\right)^3 \cdot (\omega^2)^3 \cdot \left(\omega^2 - \frac{g}{m_N} - \frac{g}{m_H}\right)^3 = 0$$

B1: Basal plane displacements.

• The solution $\omega^2 = 0$: Three of the $\omega = 0$ solutions carry in-plane displacements with basis vectors (only the basal plane coordinates are given explicitly):

$$(1,0,0,\bar{1},0,0) \quad (0,1,0,0,\bar{1},0) \quad (0,0,1,0,0,\bar{1})$$

Summing the second and third vectors, we obtain a displacement mode that represents a translation along $-Y$:

$$(0,1,1,0,\bar{1},\bar{1})$$

Subtracting the third from the first, we obtain a displacement along a direction rotated by 120° with respect to $-Y$:

$$(1,0,\bar{1},\bar{1},0,1)$$

These vectors transform as E. Subtracting the third and first vectors from the second gives

$$(\bar{1},1,\bar{1},1,\bar{1},1)$$

which is a vector representing a curling mode and transforms as A_2. These three vectors exhaust the $\omega = 0$-sector in the basal plane displacements.

- The thrice degenerate mode with finite frequency $\omega = \sqrt{2\frac{f}{m_H}}$. The basis vectors in this sector read

$$(1,0,0,1,0,0) \quad (0,1,0,0,1,0) \quad (0,0,1,0,0,1)$$

Addition of these basis vectors gives one eigenmode with A_1 symmetry:

$$(1,1,1,1,1,1)$$

where the basal plane hydrogen atoms are vibrating along the bisectrices (see the following figure, left) of the triangle sides (we refer to a similar mode in the H_2O molecule).

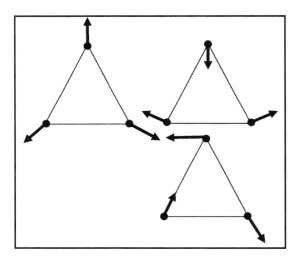

Left: In-plane A_1-mode. Right: The two degenerate E-modes.

Our previous analysis indicates that we need to search for two additional degenerate eigenmodes that transform like E as oscillatory modes. We obtain these modes by the Gram–Schmidt orthogonalization algorithm. We orthogonalize the vector

$$(1,0,0,1,0,0) \equiv w_2 \quad \text{to} \quad (1,1,1,1,1,1) \equiv w_1 \equiv v_1$$

to obtain

$$v_1 \perp v_2 = (1,0,0,1,0,0) - \frac{1}{3}(1,1,1,1,1,1) \propto (2,-1,-1,2,-1,-1)$$

The second mode is also found using the same algorithm and is written as

$$(0,1,-1,0,1,-1)$$

The E modes are drawn schematically on the right-hand side of the previous figure. In this simplified model, the eigenfrequency of the A_1 and E oscillatory modes is $\sqrt{2\frac{f}{m_H}}$. The degeneracy between the A_1 and E modes is accidental and can be removed by performing a more precise harmonic approximation. This involves considering the existence of a harmonic interaction between degrees of freedom which are not exactly along the spring axis. A coupling of the basal plane oscillations with the perpendicular direction is suppressed in this simplified model but appears when the harmonic approximation is refined. In virtue of this coupling, the N atom is involved in the in-plane oscillatory mode as well, while in our simplified model, the N atom is not displaced.

B2: z-displacements.

We now tackle the z-sector:
- The eigenvectors in the subspace to $\omega = 0$ are written as (only the z-coordinates are listed for simplicity)

$$(-1, 1, 0, 0, 0, 0) \quad (0, 0, -1, 1, 0, 0) \quad (0, 0, 0, 0, -1, 1)$$

One translational mode, describing a coherent translation along z, is obtained immediately by summing the three vectors

$$(-1, 1, -1, 1, -1, 1)$$

and has A_1-symmetry. The remaining two modes with vanishing frequency must be rotational modes and transform according to E. One can find them by searching (through trial and error) for a linear combination of the basis vectors such that one of the H atom is not displaced. If such a mode is constructed to be orthogonal to the z-translation, then it is a rotational one. We seek two such modes, which we obtain, for example, by subtracting pairs of basis vectors:

$$(-1, 1, 1, -1, 0, 0) \quad \text{and} \quad (-1, 1, 0, 0, 1, -1)$$

- The vibrational active modes in the vertical direction are

$$\omega = \sqrt{\frac{g}{m_H} + \frac{g}{m_N}}$$

The symmetry-adapted eigenmodes must be searched within the eigenspace spanned by the basis vectors

$$(1, 1, 0, 0, 0, 0) \quad (0, 0, 1, 1, 0, 0) \quad (0, 0, 0, 0, 1, 1)$$

The sum of these vectors provides, manifestly, an oscillatory mode with A_1-symmetry (left part of the following figure).

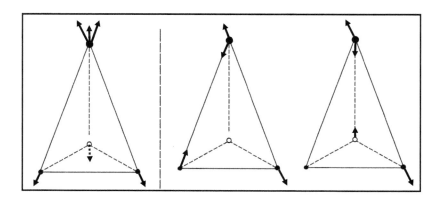

Left: Vertical oscillatory A_1-mode. Right: The two degenerate E-modes.

We find one of the E-modes by subtracting the second from the first vector:

$$(1, 1, -1, -1, 0, 0)$$

(middle part of the figure). This mode is perpendicular to the A_1-mode and can be taken as a possible basis vector in the E-space. The next basis vector within the E-space is constructed by analogy:

$$(1, 1, 0, 0, -1, -1)$$

(right part of the figure).

In this simplified model, the eigenfrequency of the A_1 and E oscillatory modes is degenerate. The degeneracy between the A_1 and E modes is accidental and can be removed by performing a more precise harmonic approximation.

By solving this model, we have produced a set of symmetry-adapted vectors that can be used to solve a more refined model of the ammonia vibrations problem, where a coupling exists between degrees of freedom which are not exactly along the "spring" axis. Such a model requires a detailed knowledge of the potential landscape $V(x_1, \ldots, z_4^3)$ so that its second derivative at the equilibrium sites can be computed (typically numerically). Once this 12×12 matrix is computed (with many of its matrix elements being identical by virtue of symmetry), one can use the 6 oscillatory, symmetry-adapted eigenmodes we have just computed to produce a 6×6 matrix that only contains the non-vanishing frequencies we are seeking: The translational and rotational modes are eliminated by symmetry arguments from the problem. In addition, the 6×6 matrix will also have a simplified structure. Besides being symmetric, it will contain two separate blocks — one carrying the two active symmetry-adapted vectors with A_1 symmetry, and the other carrying the four symmetry-adapted vectors with E symmetry. Accordingly, using symmetry arguments (and a simple analytical model), we have considerably reduced the size of matrices that have to be diagonalized (possibly numerically). The following website provides an animation of the eigenmodes of vibrations in ammonia: https: //www.chem.purdue.edu/jmol/vibs/nh3. One can recognize the A_1-mode

$$(-1, 1, -1, 1, -1, 1)$$

on the top left (called "N–H symmetric stretching") and the two E-modes

$$(1, 1, -1, -1, 0, 0) \quad (1, 1, 0, 0, -1, -1)$$

(called "N–H asymmetric stretching") on the top right. On the bottom right, we find the

$$(1, 1, 1, 1, 1, 1, 0, \ldots, 0)$$

mode with A_1 symmetry (called "N–H wagging"). The two E-modes on the bottom left are in the space spanned by

$$(2, -1, -1, 2, -1, -1)$$

(bottom left) and

$$(0, 1, -1, 0, 1, -1, 0, \ldots)$$

(bottom middle) They are called "H–N–H scissoring" modes. Note that the simple model used in this lecture does not incorporate the coupling of the planar and vertical modes. The modes displayed on the website quoted are based on a more refined numerical model that includes this coupling, so that vertical and in-plane oscillations are coupled (within the respective irreducible representation) and the more refined eigenmodes are a "mixing" of vertical and in-plane eigenmodes.

20

PROJECT FOR LECTURE 5: FRIEZE PATTERNS

Definition. A frieze pattern is a subset of the plane that repeats along one in-plane direction, specified by the primitive vector $\mathbf{t}$. The set of points $T_{\mathbf{t}} = \{\mathbf{t}_n = n \cdot \mathbf{t}\}$ with $n = 0, \pm 1, \ldots$ constitutes a lattice, spanned by the basis vector $\mathbf{t}$, called the primitive translation lattice of the pattern.

Comments. Frieze patterns are two-dimensional patterns with periodicity in one planar direction. $\mathbf{t}_n$ are the primitive translation vectors of the pattern. Patterns with translational symmetry along one direction might contain further symmetry elements that are compatible with the one-dimensional translational symmetry. The holohedral group of a frieze pattern is D_2 (i.e. C_{2v}). The subgroups of the holohedral group define the crystal classes which are sustained by the lattice. The crystal classes of frieze patterns are the point groups C_1, C_2, C_{1v}, C_{1h}, and C_{2v}. Any frieze group belongs to one of these frieze classes. Accordingly, the symmetry elements of a frieze pattern are of the type

$$(g \mid \mathbf{t}_n + \mathbf{t}_g)$$

g are the elements of the point group of the pattern. $\mathbf{t}_n$ are the lattice translations. $\mathbf{t}_g$ are the point group vectors, i.e. fractional translations, which are necessary to render g a symmetry element of the pattern. There is an infinite number of frieze patterns, but a finite set of frieze groups determined by the crystal class and the permissible point group vectors, which are unique modulo additions of an element of $T_{\mathbf{t}}$ and must satisfy the relation

$$\mathbf{t}_g + g(\mathbf{t}_h) - \mathbf{t}_{gh} \in T_{\mathbf{t}}$$

We now construct all possible space groups into which frieze patterns can be classified. For labeling the frieze groups, we use a notation standardized by the International Union of Crystallography, known as the IUC notation. These names all begin with "p" (for primitive lattice), followed by three characters. The first is "m" if there is a vertical reflection, or "1" if not. The second is "m" if there is a horizontal reflection, "g" if there is a glide reflection, or "1" otherwise. The third is "2" if there is a 180° rotation, or "1" if not. The patterns we use as examples have been redrawn freely from the summary of Korean art given in Hyunyong Shin, Shilla Sheen, Hyeyoun Kwon and Taeseon Mun, *Korean Traditional Patterns: Frieze and Wallpaper*, Springer International Publishing AG (Part of Springer Nature 2018), Editor: B. Sriraman, Handbook of the Mathematics of the Arts and Sciences, https://doi.org/10.1007/978-3-319-70658-0_17-1.

Frieze class C_1: The only element of the point group is $\mathbb{1}$. The point group vector $\mathbf{t}_\mathbb{1}$ must fulfill the relation

$$\underbrace{\mathbf{t}_\mathbb{1} + \mathbb{1}(\mathbf{t}_\mathbb{1}) - \mathbf{t}_\mathbb{1}}_{\mathbf{t}_\mathbb{1}} \in T_\mathbf{t}$$

and can be set to be vanishing. Accordingly, there is only one possible frieze group. Its symmetry elements

$$p111 = \{\mathbb{1}, n\mathbf{t}\}$$

$n \in \mathscr{Z}$ are the primitive translations. A possible pattern with this symmetry group is given in the following figure.

Frieze pattern with p111 space group. The unit cell is shown in red.

Frieze class C_2: The point group vector $\mathbf{t}_{C_2}$ must obey the relations

$$\underbrace{\mathbf{t}_{C_2} + C_2\mathbf{t}_{C_2}}_{0} - \mathbf{t}_\mathbb{1} = n\mathbf{t} \quad \mathbf{t}_{C_2} + 0 - \mathbf{t}_{C_2} = n\mathbf{t}$$

As these relations are fulfilled for any $\mathbf{t}_{C_2}$, this rotational element can be assigned any point group vector. But the equation $C_2\mathbf{x} + \mathbf{t}_{C_2} = \mathbf{x}$ always has a unique solution, a fixed point of a half turn, and we may as well take the fixed point of this half turn as the origin of the lattice, so that the point group vector can be chosen to be vanishing. The elements of the groups have the form

$$p112 = \{(\mathbb{1}, n\mathbf{t}), (C_2, m\mathbf{t})\}$$

The equation

$$C_2 \mathbf{x} + n\mathbf{t} = \mathbf{x}$$

has solutions

$$\mathbf{x} = \frac{n}{2} \cdot \mathbf{t}$$

i.e. there are twofold rotation axes going through each lattice site and halfway between them. A possible pattern with this symmetry group is given in the following figure.

Pattern with p112 space group. The unit cell is indicated by dashed red lines. The position of a twofold symmetry element is indicated.

Frieze class $C_{1v} = (\mathbb{1}, \sigma_v)$: The point group contains the elements $\{\mathbb{1}, \sigma_v\}$, with σ_v being the reflection at the vertical line. The point group vector $\mathbf{t}_{\sigma_v}$ must obey the relations

$$\mathbf{t}_{\sigma_v} + \sigma_v(\mathbf{t}_{\sigma_v}) = n\mathbf{t} \quad \mathbf{t}_{\sigma_v} + \sigma_v(0) - \mathbf{t}_{\sigma_v} = n\mathbf{t}$$

As these relations are fulfilled for any $\mathbf{t}_{\sigma_v}$, this element can be assigned any point group vector. But the equation $\sigma_v \mathbf{x} + \mathbf{t}_{\sigma_v} = \mathbf{x}$ has always a unique solution, a fixed point of a reflection in the vertical line, and we may as well take this fixed point as the origin of the lattice, so that the point group vector can be chosen to be vanishing. The elements of the groups have the form

$$pm11 = \{(\mathbb{1}, n\mathbf{t}), (\sigma_v, m\mathbf{t})\}$$

The equation

$$\sigma_v \mathbf{x} + n\mathbf{t} = \mathbf{x}$$

has solutions

$$\mathbf{x} = \frac{n}{2} \cdot \mathbf{t}$$

i.e. the symmetry reflection lines go through each lattice site and halfway between them. A possible pattern with this symmetry group is given in the following figure.

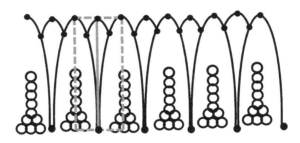

Pattern with pm11 space group. The unit cell is indicated by dashed red lines. The position of one vertical reflection line is indicated.

Crystal class $C_{1h} = (\mathbb{1}, \sigma_h)$: The point group contains the elements $\{\mathbb{1}, \sigma_h\}$, with σ_h being the reflection at the horizontal line. The point group vector $\mathbf{t}_{\sigma_h}$ must obey the relations

$$\mathbf{t}_{\sigma_h} + \sigma_h(\mathbf{t}_{\sigma_h}) = n\mathbf{t} \quad \mathbf{t}_{\sigma_h} + \sigma_h(0) - \mathbf{t}_{\sigma_h} = n\mathbf{t}$$

The first relation is non-trivial: It reads

$$2 \cdot \mathbf{t}_{\sigma_h} = n\mathbf{t}$$

This equation has at least two different solutions. The first one is

$$\mathbf{t}_{\sigma_h} = 0$$

This solution produces the frieze group

$$p1m1 = \{(\mathbb{1}, n\mathbf{t}), (\sigma_h, m\mathbf{t})\}$$

A possible pattern with this symmetry group is given in the following figure. The second solution is

$$\mathbf{t}_{\sigma_h} = \frac{1}{2}\mathbf{t}$$

There is a non-vanishing, fractional point group vector. The frieze group contains the elements

$$p1g1 = \left\{ (\mathbb{1}, n\mathbf{t}), \left(\sigma_h, \frac{1}{2}m\mathbf{t}\right) \right\}$$

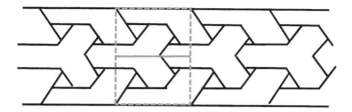

Pattern with p1m1 frieze group. The unit cell is indicated by dashed red lines. The position of a horizontal reflection line is indicated.

The horizontal reflection is accompanied by a horizontal glide amounting to half the lattice vector. A possible pattern with this symmetry group is given in the following figure.

Pattern with p112 space group. The unit cell is indicated by dashed red lines. The position of a horizontal glide line is given as dashed-dotted red lines.

Crystal class C_{2v}: The point group contains the elements $\{1, C_2, \sigma_h, \sigma_v\}$ The origin is chosen so that the vertical reflection line and the half-turn axis go through, so that these elements do not have finite point group vectors. The point group element σ_h has two possible point group vectors, and this produces two different frieze groups:

$$pmm2 = \{(1, n\mathbf{t}), (C_2, m\mathbf{t}), \sigma_v, p\mathbf{t}), \sigma_h, q\mathbf{t})\}$$

A possible pattern with this symmetry group is given in the following figure.

Pattern with pmm2 frieze group. The unit cell is indicated by dashed red lines. The positions of the horizontal and vertical reflection lines and of one twofold rotation center are indicated.

$$pmg2 = \left\{ (\mathbb{1}, n\mathbf{t}), (C_2, m\mathbf{t}), \sigma_v, p\mathbf{t}), \sigma_h, \frac{1}{2}\mathbf{t} + q\mathbf{t}) \right\}$$

A possible pattern with this symmetry group is given in the following figure.

Pattern with pmg2 frieze group. The unit cell is indicated by dashed red lines. The positions of the vertical reflection line (continuous red lines), horizontal glide line (dotted red lines), and one twofold axis are indicated.

Spin chain: As a final example, we consider an antiferromagnetically ordered chain of spins ("arrows"), and find the frieze group.

Let the spin-up arrows occupy the lattice points $2n \cdot a$ and the spin down arrows the lattice points $(2n+1) \cdot a$ along the x-direction, $n = 0, \pm1, \ldots$. Let the origin of the horizontal x-coodinate be at the beginning of one spin-up arrow.

Antiferromagnetic spin chain: One vertical reflection line in red, the horizontal glide line in dashed red, and a twofold rotation axis.

- The unit vector $\mathbf{t}$ is pointing in the positive x-direction and has the length a. The primitive translation vectors are $\mathbf{t}_n = (0, n \cdot a)$.
- Further, less-trivial elements of symmetry are reflections at vertical lines:

$$\left\{ \left(\begin{pmatrix} \bar{1} & 0 \\ 0 & 1 \end{pmatrix} \mid \mathbf{t}_n \right) \right\}$$

These operations are reflections at a vertical line going through the origin (the yz-plane), followed by primitive translations. These operations can also be described by reflection at vertical planes going through $n \cdot \frac{a}{2}$.

- Glide reflections:

$$\left\{ \left(\begin{pmatrix} 1 & 0 \\ 0 & \bar{1} \end{pmatrix} \mid (2n+1)\frac{a}{2} \cdot \mathbf{t} \right) \right\}$$

These elements consist of a reflection at a horizontal line comprising the x-line, followed by translations by odd integer multiples of half the lattice constant. These elements are the so-called glide reflections.

- Twofold rotation axes:

$$\left\{ \left(\begin{pmatrix} \bar{1} & 0 \\ 0 & \bar{1} \end{pmatrix} \mid (2n+1)\frac{a}{2} \cdot \mathbf{t} \right) \right\}$$

These elements consist of a half-turn followed by translations by odd integer multiples of half the lattice constant. Alternatively, one could describe these elements as half-turns about rotation axes parallel to z and intersecting the x-axis at $(2n+1)\frac{a}{4}$.

- The symmetry group of this frieze pattern is *pmg2*.

21

PROJECT FOR LECTURE 5: AN ALGEBRAIC PROOF OF THE HESSEL THEOREM

In the lecture, we stated the most important results concerning point groups and discussed their properties. The results of point groups are based on an exact mathematical theorem known as the Hessel theorem (Johann Friedrich Christian Hessel, 1796–1872). We propose an algebraic proof of this theorem based on the following paper: F. Seitz, "A matrix-algebraic development of the crystallographic groups. I", *Z. Krist. - Cryst. Mater.* **88**, 433–459 (1934), https://doi.org/10.1524/zkri.1934.88.1.433.

Theorem (Hessel). Any finite subgroup of $SO(3)$ is either one of the cyclic groups C_p, one of the dihedral groups D_p, or one of the point groups T, O, and Y.

Comment. C_p is the proper rotation group of a pyramid with a regular plane polygon as its base. D_p is the proper rotations group of a double pyramid on a regular planar polygon. T, O, and Y are the proper rotation groups of a regular tetrahedron, a regular octahedron, and a regular icosahedron, respectively (all three are Platonic solids).

Proof:

Part a. We first construct the groups with just one rotation axis. As the plane perpendicular to the rotation axis is transformed into itself, the finite subgroups of SO_3 with a single rotation axis coincide with the finite subgroups of SO_2, i.e. they are one of the cyclic groups C_p (Leonardo da Vinci).

Part b. We then allow for more than one rotation axis, say a q-fold rotation axis at an angle ω with respect to the z-axis and lying in the xz-plane. If ω is different from 0 or π, the existence of a second rotation axis implies the existence of a third one, which we call an n-fold axis. For the possible values of p, q, and n, there exists a sum rule, which we are going to prove. To formulate the sum rule, let us think of the destiny of a point P residing at the intersection of the p-fold axis with the unit sphere under the effect of the group elements. There are some elements that do not change the position of P, while some will move P on the surface of the unit sphere and drive it to a set of points $G_p(P)$. This set, including the original point, constitutes the orbit of $P(p)$.

Lemma. $G_p(P)$ consists of $\frac{g}{p}$ elements, with g being the order of the desired point group G.

Proof of the Lemma: The subgroup H of G, which leaves P fixed, has p elements. So, G is a union of cosets $g_1H, g_2H, \ldots, g_rH$, with $r = g/p$. All the elements in a coset g_iH will move P to the same point g_iP for $i = 1, 2, \ldots, r$. Hence, the orbit of $P(p)$ is the set $g_1P, g_2P, \ldots, g_rP$, as required. ◇

What now follows is some counting of the elements of the orbits. If we start from a point $P(p)$, the number of extra points generated on the unit sphere is $\frac{g}{p} - 1$. Considering that we are dealing with a group having three different rotation axes, we can start from $P(q)$ and produce $\frac{g}{q} - 1$ extra points, or we can start from $P(n)$ and produce $\frac{g}{n} - 1$ extra points. The sum rule states that the sum of these extra points must be the same as the number of extra points that we produce, starting from a general point of the unit sphere, which is $g - 1$:

$$\frac{g}{p} - 1 + \frac{g}{q} - 1 + \frac{g}{n} - 1 = g - 1$$

or

$$\boxed{\frac{g}{p} + \frac{g}{q} + \frac{g}{n} = g + 2}$$

We can use this sum rule to construct all possible combinations of p, q, n that can appear in a finite subgroup of SO_3 with three rotation axes:

There are no other possible combinations because any other triplet of numbers p, q, n does not fulfill the stringent inequality

$$\frac{1}{p} + \frac{1}{q} + \frac{1}{n} > 1$$

Possible values for p, q, n.

p	q	n	g
2	2	2	4
3	2	2	6
4	2	2	8
$\vdots$	$\vdots$	$\vdots$	$\vdots$
p	2	2	$2 \cdot p$
3	2	3	12
4	3	2	24
5	3	2	60

Part c. We now provide some useful formulas for the explicit construction of the groups. A useful formula due to Seitz allows us to compute ω for given p, q, n. The p-fold rotation around z is described by the matrix

$$\begin{pmatrix} \cos \frac{2\pi}{p} & -\sin \frac{2\pi}{p} & 0 \\ \sin \frac{2\pi}{p} & \cos \frac{2\pi}{p} & 0 \\ 0 & 0 & 1 \end{pmatrix}$$

A rotation by an angle $\frac{2\pi}{q}$ about the axis z' can be described as

$$R_{z'}\left(\frac{2\pi}{q}\right) = R(\omega) \cdot R_z \cdot R(\omega^{-1})$$

with

$$R(\omega) = \begin{pmatrix} \cos \omega & 0 & \sin \omega \\ 0 & 1 & 0 \\ -\sin \omega & 0 & \cos \omega \end{pmatrix}$$

For the matrix $R_{z'}\left(\frac{2\pi}{q}\right)$, we obtain

$$\begin{pmatrix} \cos \omega & 0 & \sin \omega \\ 0 & 1 & 0 \\ -\sin \omega & 0 & \cos \omega \end{pmatrix} \cdot \begin{pmatrix} \cos \frac{2\pi}{q} & -\sin \frac{2\pi}{q} & 0 \\ \sin \frac{2\pi}{q} & \cos \frac{2\pi}{q} & 0 \\ 0 & 0 & 1 \end{pmatrix} \cdot \begin{pmatrix} \cos \omega & 0 & -\sin \omega \\ 0 & 1 & 0 \\ \sin \omega & 0 & \cos \omega \end{pmatrix}$$

The combination of the rotation $R_z(\frac{2\pi}{p})$ around z and the rotation $R_{z'}(\frac{2\pi}{q})$ around z' is also a rotation, represented by the matrix $R_{z'}(\frac{2\pi}{q}) \cdot R_z(\frac{2\pi}{p})$.

The trace of the resulting matrix amounts to

$$\cos^2 \omega \left(4 \sin^2 \frac{\pi}{q} \cdot \sin^2 \frac{\pi}{p} \right) - \cos \omega \left(2 \sin \frac{2\pi}{q} \sin \frac{2\pi}{p} \right) + 4 \cos^2 \frac{\pi}{p} \cos^2 \frac{\pi}{q} - 1$$

We set this trace to be the trace of an n-fold rotation about a third axis:

$$1 + 2 \cos \frac{2\pi}{n} = \cos^2 \omega \left(4 \sin^2 \frac{\pi}{q} \cdot \sin^2 \frac{\pi}{p} \right) - \cos \omega \left(2 \sin \frac{2\pi}{q} \sin \frac{2\pi}{p} \right)$$

$$+ 4 \cos^2 \frac{\pi}{p} \cos^2 \frac{\pi}{q} - 1$$

Solving this equation for $\cos \omega$, we find

$$\cos \omega = \frac{\cos \frac{\pi}{q} \cos \frac{\pi}{p} \pm \cos \frac{\pi}{n}}{\sin \frac{\pi}{p} \sin \frac{\pi}{q}}$$

Part d. We now construct the finite subgroups explicitly.

• D_2. Take, for example, $p = q = n = 2$. Solving for ω, we find $\omega = \frac{\pi}{2}$, indicating that there is a twofold rotation axis along x. The two operations we have constructed so far are

$$C_{2z} = \begin{pmatrix} \bar{1} & 0 & 0 \\ 0 & \bar{1} & 0 \\ 0 & 0 & 1 \end{pmatrix} \quad C_{2x} = \begin{pmatrix} 1 & 0 & 0 \\ 0 & \bar{1} & 0 \\ 0 & 0 & \bar{1} \end{pmatrix}$$

We find all further elements by matrix multiplication, e.g.

$$C_{2x} C_{2z} = \begin{pmatrix} 1 & 0 & 0 \\ 0 & \bar{1} & 0 \\ 0 & 0 & \bar{1} \end{pmatrix} \cdot \begin{pmatrix} \bar{1} & 0 & 0 \\ 0 & \bar{1} & 0 \\ 0 & 0 & 1 \end{pmatrix} = \begin{pmatrix} \bar{1} & 0 & 0 \\ 0 & 1 & 0 \\ 0 & 0 & \bar{1} \end{pmatrix} \equiv C_{2y}$$

Together with the identity element, these three elements form a group called D_2 (or "Vierergruppe").

• D_p: The elements of D_p are constructed for $p, 2, 2$ using the same procedure. One finds again that the $q = 2$ axis constructs an angle $\frac{\pi}{2}$ with the p-fold axis. Furthermore, we have

$$C_{2x} C_{pz} = \begin{pmatrix} 1 & 0 & 0 \\ 0 & \bar{1} & 0 \\ 0 & 0 & \bar{1} \end{pmatrix} \cdot \begin{pmatrix} \cos \frac{2\pi}{p} & -\sin \frac{2\pi}{p} & 0 \\ \sin \frac{2\pi}{p} & \cos \frac{2\pi}{p} & 0 \\ 0 & 0 & 1 \end{pmatrix}$$

$$= \begin{pmatrix} \cos \frac{2\pi}{p} & -\sin \frac{2\pi}{p} & 0 \\ -\sin \frac{2\pi}{p} & -\cos \frac{2\pi}{p} & 0 \\ 0 & 0 & \bar{1} \end{pmatrix}$$

The trace of the resulting matrix is -1, i.e. it describes a rotation by π. As z is transformed to $-z$, the rotation axis is a line $y = m \cdot x$ in the

xy-plane. To find the rotation axis, we solve the equation

$$\begin{pmatrix} \cos\frac{2\pi}{p} & -\sin\frac{2\pi}{p} & 0 \\ -\sin\frac{2\pi}{p} & -\cos\frac{2\pi}{p} & 0 \\ 0 & 0 & \bar{1} \end{pmatrix} \begin{pmatrix} x \\ mx \end{pmatrix} = \begin{pmatrix} x \\ mx \end{pmatrix}$$

for the desired slope m and find $m = -\tan\frac{\pi}{p}$. This is a twofold rotation axis forming an angle $\frac{\pi}{p}$ with the twofold rotation axis along x. Further twofold rotation axes are found in the xy-plane by rotating the axes along x and $-\frac{\pi}{p}$ around the p-fold main rotation axis (alternatively, to find further elements, continue multiplying the resulting matrices until one has produced all $2p$ elements). The resulting group is called D_p and consists of one main p-fold rotation axis and p twofold rotation axes perpendicular to it.

- T: The next group is the $p = 2, q = 3, n = 3$-group. The angle between the twofold axis and one threefold axis is one solution of the equation

$$\cos\omega = \pm\frac{1}{\tan\frac{\pi}{3}} = \frac{\sqrt{3}}{3}$$

Let us take the twofold axis along $(0,0,1)$ and the threefold axis along $(\frac{\sqrt{3}}{3}, \frac{\sqrt{3}}{3}, \frac{\sqrt{3}}{3})$ so that the matrices describing the operations are

$$C_{2z} = \begin{pmatrix} \bar{1} & 0 & 0 \\ 0 & \bar{1} & 0 \\ 0 & 0 & 1 \end{pmatrix}$$

and

$$C_{3xyz} = \begin{pmatrix} 0 & 0 & 1 \\ 1 & 0 & 0 \\ 0 & 1 & 0 \end{pmatrix}$$

respectively. By taking all possible products, one obtains 12 matrices, which describe the proper symmetry elements of a tetrahedron. In terms of the xyz-coordinates, we have the elements

$$C_1 = (xyz) = \mathbb{1} \quad C_{2x} = (x\bar{y}\bar{z}) \quad C_{2y} = (\bar{x}y\bar{z}) \quad C_{2z} = (\bar{x}\bar{y}z)$$

$$C_{3xyz} = (yzx) \quad C_{xyz} = (yz\bar{x}) \quad C_{3x\bar{y}\bar{z}} = (\bar{y}z\bar{x}) \quad C_{3\bar{y}z} = (\bar{y}\bar{z}x)$$

$$C_{3xyz}^{-1} = (zxy) \quad C_{3xyz}^{-1} = (\bar{z}x\bar{y}) \quad C_{3x\bar{y}\bar{z}}^{-1} = (\bar{z}\bar{x}y) \quad C_{3x\bar{y}z}^{-1} = (z\bar{x}\bar{y})$$

- O: The next group of interest is $p = 2, q = 4, n = 3$, with $\cos\omega = \frac{\sqrt{2}}{2}$. We take the fourfold rotation axis along $(0,0,1)$ and the twofold rotation axis along $(1,1,0)$. The respective elements are represented by the matrices

$$C_{4z} = \begin{pmatrix} 0 & \bar{1} & 0 \\ 1 & 0 & 0 \\ 0 & 0 & 1 \end{pmatrix}$$

and

$$C_{2xy} = \begin{pmatrix} 0 & 1 & 0 \\ 1 & 0 & 0 \\ 0 & 0 & \bar{1} \end{pmatrix}$$

Performing all possible multiplications, we obtain 24 matrices describing the proper symmetry elements of an octahedron.

- Y: The final subgroup is the $p = 5, q = 2, n = 3$-group. This set of parameters produces the proper rotation group of an icosahedron.

22

PROJECT FOR LECTURE 6:
EMPTY-LATTICE BAND STRUCTURE

In Lecture 6, we defined the so-called empty-lattice band structure and discussed it for some simple, representative cases. In this project, we perform some more explicit computations. In Section 22.1, we explore the empty-lattice band structure at the Γ- and X-points of the square planar lattice and perform a symmetry analysis. In Section 22.2, we apply symmetry arguments to explicitly compute a model band structure for a primitive tetragonal lattice.

22.1 The Γ- and X-points of the Square Planar Lattice

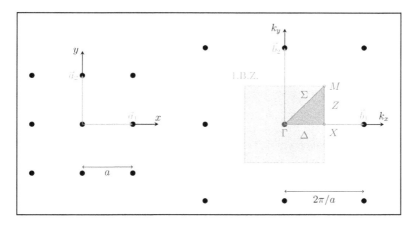

Sketch of the 2D square lattice (left) and the corresponding reciprocal lattice (right). The dark-shaded region is the reduced Brillouin zone.

The empty-lattice bands for a two-dimensional (2D) square lattice are given by

$$E_{mn}(\vec{k}) = (\hbar^2/2m)(\vec{k} + \vec{G})^2$$

with $\vec{k}$ in the first Brillouin zone (BZ) and $\vec{G} = (2\pi/a)(m\vec{b}_1 + n\vec{b}_2)$ $(m, n \in \mathbb{Z})$. The eigenfunctions are written as

$$\phi_{\vec{k}\vec{G}} = (1/\sqrt{Na^2}) \cdot e^{i(\vec{k}+\vec{G})\cdot\vec{r}}$$

Our tasks are the following:

1. Consider the empty-lattice energy level $E = 1.0$ (in units of $h^2/2ma^2$) at the Γ-point of the 2D square lattice and compute the plane waves that belong to this state.
2. Use these plane waves to build a representation of the group of Γ carried by the empty-lattice energy level.
3. Decompose the representation into irreducible representations of the group of Γ.
4. Find the corresponding symmetrized plane waves.
5. Compute the first two low-lying empty-lattice energy levels at the X-point of the 2D square lattice and compute the plane waves that belong to these states.
6. Use these plane waves to build representations of the group of X carried by the corresponding empty-lattice energy levels.
7. Decompose these representations into irreducible representations of the group of X.
8. Find the corresponding symmetrized plane waves.

Here, one can find the answers to the assigned tasks:

1. The empty-lattice energy level $E_\Gamma = 1$ at Γ is fourfold degenerate and contains the plane waves:

$$|\pm 1, 0\rangle = e^{\pm i(2\pi/a)x} \qquad |0, \pm 1\rangle = e^{\pm i(2\pi/a)y}$$

2. These plane waves carry the four-dimensional reducible representation $\Gamma_{E=1}$ of C_{4v} with matrices (only some point group elements are presented)

$$\Gamma_{E=1}(E) = \begin{pmatrix} 1 & 0 & 0 & 0 \\ 0 & 1 & 0 & 0 \\ 0 & 0 & 1 & 0 \\ 0 & 0 & 0 & 1 \end{pmatrix}, \quad \Gamma_{E=1}(C_4) = \begin{pmatrix} 0 & 0 & 0 & 1 \\ 0 & 0 & 1 & 0 \\ 1 & 0 & 0 & 0 \\ 0 & 1 & 0 & 0 \end{pmatrix},$$

$$\Gamma_{E=1}(C_2) = \begin{pmatrix} 0 & 1 & 0 & 0 \\ 1 & 0 & 0 & 0 \\ 0 & 0 & 0 & 1 \\ 0 & 0 & 1 & 0 \end{pmatrix}, \quad \Gamma_{E=1}(\sigma_x) = \begin{pmatrix} 1 & 0 & 0 & 0 \\ 0 & 1 & 0 & 0 \\ 0 & 0 & 0 & 1 \\ 0 & 0 & 1 & 0 \end{pmatrix},$$

$$\Gamma_{E=1}(\sigma_d) = \begin{pmatrix} 0 & 0 & 1 & 0 \\ 0 & 0 & 0 & 1 \\ 1 & 0 & 0 & 0 \\ 0 & 1 & 0 & 0 \end{pmatrix}.$$

3. We obtain

$G_\Gamma = C_{4v}$	E	$2C_4$	C_2	$2\sigma_v$	$2\sigma_d$
Γ_1	1	1	1	1	1
Γ_2	1	1	1	-1	-1
Γ_3	1	-1	1	1	-1
Γ_4	1	-1	1	-1	1
Γ_5	2	0	-2	0	0
$\Gamma_{E=1}$	4	0	0	2	0

One recognizes from the character table of G_Γ that

$$\Gamma_{E=1} = \Gamma_1 \oplus \Gamma_3 \oplus \Gamma_5.$$

4. We use the projector method:

$$P_{\Gamma_1} |1,0\rangle = \frac{1}{8} \cdot (2|1,0\rangle + 2|-1,0\rangle + 2|0,1\rangle + 2|0,-1\rangle)$$

$$= \left[\cos\left(\frac{2\pi}{a}x\right) + \cos\left(\frac{2\pi}{a}y\right)\right],$$

$$P_{\Gamma_3} |1,0\rangle = \frac{1}{8} \cdot (|1,0\rangle - |0,1\rangle - |0,-1\rangle + |-1,0\rangle$$

$$+ |-1,0\rangle + |1,0\rangle - |0,1\rangle - |-1,0\rangle)$$

$$= \left[\cos\left(\frac{2\pi}{a}x\right) - \cos\left(\frac{2\pi}{a}y\right)\right],$$

$$P_{\Gamma_5} |1,0\rangle \approx 2|1,0\rangle - 2|0,-1\rangle \propto \sin\left(\frac{2\pi}{a}x\right)$$

$$\Rightarrow \quad C_4 \sin\left(\frac{2\pi}{a}x\right) \propto \sin\left(\frac{2\pi}{a}y\right).$$

5. The lowest-lying level at X has energy $E_{00,\bar{1}0} = 0.25$ (in units of $h^2/2ma^2$), with $m = n = 0$ or $m = -1, n = 0$. It is twofold degenerate and carries the wave functions $|0,0\rangle = e^{i(\pi/a)x}$ and $|-1,0\rangle = e^{-i(\pi/a)x}$. The next higher energy level at X has the energy $E = 1.25$. It is fourfold degenerate and carries the wave functions $|0,\pm 1\rangle = e^{i(\pi/a)(x\pm 2y)}$ and $|-1,\pm 1\rangle = e^{i(\pi/a)(-x\pm 2y)}$.

6. The group of X is C_{2v}. Applying all the symmetry operations of C_{2v} upon these two states, we obtain a 2D representation $X_{E=0.25}$ with the matrices

$$X_{E=0.25}(E) = \begin{pmatrix} 1 & 0 \\ 0 & 1 \end{pmatrix}, \quad X_{E=0.25}(C_2) = \begin{pmatrix} 0 & 1 \\ 1 & 0 \end{pmatrix}$$

$$X_{E=0.25}(\sigma_x) = \begin{pmatrix} 1 & 0 \\ 0 & 1 \end{pmatrix}, \quad X_{E=0.25}(\sigma_y) = \begin{pmatrix} 0 & 1 \\ 1 & 0 \end{pmatrix}.$$

The next higher energy level at X carries a 4D representation of X with matrices

$$X_{E=1.25}(E) = \begin{pmatrix} 1 & 0 & 0 & 0 \\ 0 & 1 & 0 & 0 \\ 0 & 0 & 1 & 0 \\ 0 & 0 & 0 & 1 \end{pmatrix}, \quad X_{E=1.25}(C_2) = \begin{pmatrix} 0 & 0 & 0 & 1 \\ 0 & 0 & 1 & 0 \\ 0 & 1 & 0 & 0 \\ 1 & 0 & 0 & 0 \end{pmatrix}$$

$$X_{E=1.25}(\sigma_x) = \begin{pmatrix} 0 & 1 & 0 & 0 \\ 1 & 0 & 0 & 0 \\ 0 & 0 & 0 & 1 \\ 0 & 0 & 1 & 0 \end{pmatrix}, \quad X_{E=1.25}(\sigma_y) = \begin{pmatrix} 0 & 0 & 1 & 0 \\ 0 & 0 & 0 & 1 \\ 1 & 0 & 0 & 0 \\ 0 & 1 & 0 & 0 \end{pmatrix}.$$

7. Using the character table, we obtain

$G_X = C_{2v}$	E	C_2	σ_x	σ_y
X_1	1	1	1	1
X_2	1	1	-1	-1
X_3	1	-1	1	-1
X_4	1	-1	-1	1
$X_{E=0.25}$	2	0	2	0
$X_{E=1.25}$	4	0	0	0

By guessing or using the decomposition formula, we conclude that

$$X_{E=0.25} = X_1 \oplus X_3 \quad \text{and} \quad X_{E=1.25} = X_1 \oplus X_2 \oplus X_3 \oplus X_4$$

8. To find the linear combination of plane waves transforming according to the irreducible representations X_1, X_2, X_3, X_4 of C_{2v}, we apply the projector technique. For the 2D representation $X_{E=0.25}$, we use the trial function $e^{i(\pi/a)x}$ to obtain

$$P_{X_1} e^{i\pi/ax} = \frac{1}{2}(e^{i\pi/ax} + e^{-i\pi/ax}) = \cos\left(\frac{\pi}{a}x\right)$$

$$P_{X_3} e^{i\pi/ax} \approx \sin\left(\frac{\pi}{a}x\right)$$

For the 4D representation $X_{E=1.25}$, we obtain

$$P_{X_1} e^{i\pi/a(x\pm 2y)} = \cos\left(\frac{\pi}{a}x\right) \cdot \cos\left(\frac{2\pi}{a}y\right)$$

$$P_{X_2} e^{i\pi/a(x\pm 2y)} = \cos\left(\frac{\pi}{a}x\right) \cdot \sin\left(\frac{2\pi}{a}y\right)$$

$$P_{X_3} e^{i\pi/a(x\pm 2y)} \approx \sin\left(\frac{\pi}{a}x\right) \cdot \sin\left(\frac{2\pi}{a}y\right)$$

$$P_{X_4} e^{i\pi/a(x\pm 2y)} \approx \sin\left(\frac{\pi}{a}x\right) \cdot \cos\left(\frac{2\pi}{a}y\right)$$

One can now label the states with the proper symmetrized plane waves corresponding to the irreducible representations at the X-point:

$$|E_{X=0.25}, X_1\rangle = \frac{1}{2} \cdot (|0,0\rangle + |0,-1\rangle) = \cos\left(\frac{\pi}{a}x\right)$$

$$|E_{X=0.25}, X_3\rangle = \frac{1}{2i} \cdot (|0,0\rangle - |0,-1\rangle) = \sin\left(\frac{\pi}{a}x\right)$$

$$|E_{X=1.25}, X_1\rangle = \frac{1}{2} \cdot (|0,1\rangle + |-1,-1\rangle) = \cos\left(\frac{\pi}{a}x\right)\cos\cdot\left(\frac{2\pi}{a}y\right)$$

$$|E_{X=1.25}, X_2\rangle = \frac{1}{4i} \cdot (|0,1\rangle - |-1,-1\rangle + |-1,1\rangle - |0,-1\rangle)$$

$$= \cos\left(\frac{\pi}{a}x\right) \cdot \sin\left(\frac{2\pi}{a}y\right)$$

$$|E_{X=1.25}, X_3\rangle = \frac{1}{2} \cdot (|0,1\rangle - |-1,-1\rangle) = \sin\left(\frac{\pi}{a}x\right) \cdot \sin\left(\frac{2\pi}{a}y\right)$$

$$|E_{X=1.25}, X_4\rangle = \frac{1}{4i} \cdot (|0,1\rangle - |-1,-1\rangle - |-1,1\rangle + |0,-1\rangle)$$

$$= \sin\left(\frac{\pi}{a}x\right) \cdot \cos\left(\frac{2\pi}{a}y\right)$$

In the following figure, on the left, the empty-lattice band structure is shown. On the right, a realistic model band structure, which includes a finite potential, is displayed.

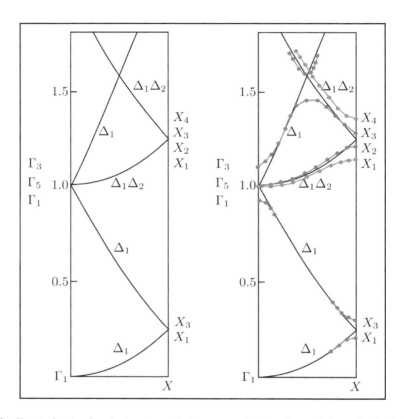

Left: Empty-lattice band structure of the square lattice along Γ, Δ, and X. Right: Realistic model band structure, after switching on the strength of the crystal potential. Δ_1 bands are in blue, and Δ_5 bands are in red.

The connection between the band structure along Δ with the endpoints of BZ can be directly obtained in general terms from the compatibility relations given in the following table and is made clear by the colors used in the previous figure.

A comprehensive, interactive summary of BZs and the corresponding empty-lattice band structures can be found in http://lampx.tugraz.at/ ss1/empty/~hadley/empty.php.

Compatibility table for the square lattice.

Γ_1	Γ_2	Γ_3	Γ_4	Γ_5
Δ_1	Δ_2	Δ_1	Δ_2	$\Delta_1\Delta_2$
X_1	X_2	X_3	X_4	

22.2 Model Band Structure for a Tetragonal Lattice

22.2.1 *Empty-lattice eigenvalues*

The primitive tetragonal lattice (see the following figure, left)

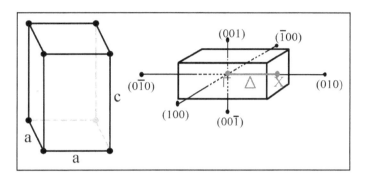

Left: The primitive tetragonal lattice with $\frac{a}{c} < 1$. Right: The Brillouin zone of the primitive tetragonal lattice. The reciprocal lattice points surrounding the Brillouin zone are also indicated. The high symmetry points $\Gamma = (0,0,0)$ and $X = (0,\frac{\pi}{a},0)$ and the line ($\Delta = \frac{\pi}{a} \cdot (0,\delta,0)$, $\delta \in [0,1]$) are also indicated.

sustains a number of space groups. We impose a D_{4h} point group symmetry, i.e. we consider the D_{4h} (Schönflies) crystal class ($4/m2/m2/m$, Hermann–Mauguin). A possible space group within this crystal class is $P4/mmm$ (D_{4h}^1, Schönflies). The BZ (right-hand part in the figure) is a rectangular parallelepiped with a volume of $\frac{2\pi}{a} \times \frac{2\pi}{a} \times \frac{2\pi}{c}$. The coordinates of the reciprocal lattice points immediately surrounding the first BZ are also given in the figure. The empty-lattice energy eigenvalues are calculated from the relation

$$E_{\mathbf{b}}(\vec{k}) = (\hbar^2/2m)(\mathbf{k} + \mathbf{b})^2$$

with $\mathbf{k}$ in the first BZ and

$$\mathbf{b} = \left(\frac{2\pi}{a}\right)(n_1\vec{b}_1 + n_2\vec{b}_2) + \frac{2\pi}{c}n_3 \cdot \mathbf{b}_3$$

$n_i \in \mathbb{Z}$ and

$$\mathbf{b_1} = \frac{2\pi}{a}(1,0,0) \quad \mathbf{b_2} = \frac{2\pi}{a}(0,1,0) \quad \mathbf{b_3} = \frac{2\pi}{c}(0,0,1)$$

For the specific case of the Δ-direction, we obtain some low-energy empty bands, as illustrated in the following figure.

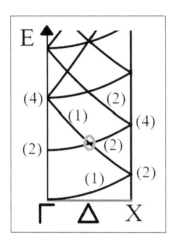

The low-energy empty-lattice band structure of a primitive tetragonal lattice, computed for $\frac{a}{c} = 0.7$. The energy is in units of $\frac{\hbar^2 \cdot a^2}{2m} \cdot (\frac{\pi}{a})^2$. The degeneracy of the energy levels is given in brackets. A particular band crossing is encircled and will be discussed later.

All energies in units of $\frac{\hbar^2 \cdot a^2}{2m} \cdot (\frac{\pi}{a})^2$ are:

- the lowest band with degeneracy (1):

$$E_{\mathbf{b}=0}(\delta) = (\delta)^2$$

  a higher band with degeneracy (1):

$$E_{\mathbf{b}=\frac{2\pi}{a}(0,\bar{1},0)}(\delta) = (\delta - 2)^2$$

- a lower, twofold degenerate band:

$$E_{\mathbf{b}=\frac{2\pi}{c}(0,0,\pm1)}(\delta) = \delta^2 + \left(\frac{2a}{c}\right)^2$$

- an energy level with degeneracy (2) at

$$E_{\mathbf{b}=\frac{2\pi}{c}(0,\bar{1},\pm1)}(\delta) = (\delta - 2)^2 + \left(\frac{2a}{c}\right)^2$$

These bands join the Γ-point at:

- an energy level with degeneracy (1) and energy

$$E_{\mathbf{b}=0}(\delta = 0) = 0$$

- an energy level with degeneracy (2) at

$$E_{\mathbf{b}=\frac{2\pi}{c}(0,0,\pm 1)}(\delta = 0) = \left(\frac{2a}{c}\right)^2$$

- an energy level with degeneracy (4) and energy

$$E_{\mathbf{b}=\frac{2\pi}{a}(\pm 1,0,0),\,\frac{2\pi}{a}(0,\pm 1,0)}(\delta = 0) = 4$$

At the X-point, we observe the following energy values in the low-energy range:

- an energy level with degeneracy (2) and energy

$$E_{\mathbf{b}=(0,0,0),\,\frac{2\pi}{a}(0,\bar{1},0)}(\delta = 1) = 1$$

- an energy level with degeneracy (4) at

$$E_{\mathbf{b}=\frac{2\pi}{c}(0,\bar{1},\pm 1)}(\delta = 1) = 1 + \left(\frac{2a}{c}\right)^2$$

22.2.2 *Symmetry analysis*

We proceed with the symmetry analysis of the bands. The point group of the crystal contains the following elements (characterized here by their action on the (x, y, z) coordinates):

D_{4h}	E (xyz)	$2C_2^{x,y}$ $(x\bar{y}\bar{z})$ $(\bar{x}y\bar{z})$	C_2^z $(\bar{x}\bar{y}z)$	$2C_4^z$ $(\bar{y}xz)$ $(y\bar{x}z)$	$2C_2^{a,b}$ $(yx\bar{z})$ $(\bar{y}\bar{x}\bar{z})$
	I $(\bar{x}\bar{y}\bar{z})$	$2IC_2^{x,y}$ $(\bar{x}yz)$ $(x\bar{y}z)$	IC_2^z $(xy\bar{z})$	$2IC_4^z$ $(y\bar{x}\bar{z})$ $(\bar{y}x\bar{z})$	$2IC_2^{a,b}$ $(\bar{y}\bar{x}z)$ (yxz)

Δ-line: We choose those elements that keep y invariant and obtain the point group of the Δ direction: C_{2v}. Its character table is given as follows (the elements are characterized by their action on the (x, y, z) coordinates). For the irreducible representations, we use the labeling scheme that is common in chemistry.

C_{2v}	E (xyz)	C_2^y $(\bar{x}y\bar{z})$	IC_2^z $(xy\bar{z})$	IC_2^x $(\bar{x}yz)$
A_1	1	1	1	1
A_2	1	1	$\bar{1}$	$\bar{1}$
B_1	1	$\bar{1}$	1	$\bar{1}$
B_2	1	$\bar{1}$	$\bar{1}$	1

- The plane wave carrying the lowest (1)-degenerate band is $e^{i\frac{\pi}{a}\delta \cdot y}$ and transforms according to the irreducible representation A_1.
- The plane wave carrying the next (1)-degenerate band is $e^{i\frac{\pi}{a}(\delta-2)\cdot y}$ and transforms again according to A_1.
- The lowest twofold degenerate band is carried by the plane waves

$$e^{i\frac{\pi}{a}\delta x+\frac{2\pi}{c}\cdot z} \quad ; \quad e^{i\frac{\pi}{a}\delta x-\frac{2\pi}{c}\cdot z}$$

This space carries a 2D representation of the group C_{2v} with the following characters.

C_{2v}	E (xyz)	C_2^y $(\bar{x}y\bar{z})$	IC_2^z $(xy\bar{z})$	IC_2^x $(\bar{x}yz)$
$\Delta_{low,2}$	2	0	0	2

$\Delta_{\text{low},2}$ is reducible, i.e. the band hosts an accidental degeneracy

$$\Delta_{\text{low},2} = A_1 \oplus B_2$$

which will be removed when the crystal potential is switched on.

- The higher twofold degenerate level (see the figure) hosts the plane waves

$$e^{i\frac{\pi}{a}(\delta-2)y+i\frac{2\pi}{c}z} \quad ; \quad e^{i\frac{\pi}{a}(\delta-2)y-i\frac{2\pi}{c}z}$$

which also split in the presence of a finite crystal potential into $A_1 \oplus B_2$.

Γ-point: The point group of the Γ-point is D_{4h}. The character table is given as follows (the elements are characterized by their action on the (x,y,z) coordinates). For the irreducible representations, we use the labeling scheme that is common in chemistry (see, for example, http://symmetry.jacobs-university.de/ for extensive tables on point groups).

D_{4h}	E (xyz)	$2C_2^{x,y}$ $(x\bar{y}\bar{z})$ $(\bar{x}\bar{y}z)$	C_2^z $(\bar{x}\bar{y}z)$	$2C_4^z$ $(\bar{y}xz)$ $(y\bar{x}z)$	$2C_2^{a,b}$ (yxz) $(\bar{y}\bar{x}z)$	I $(\bar{x}\bar{y}\bar{z})$	$2IC_2^{x,y}$ $(\bar{x}yz)$ $(x\bar{y}z)$	IC_2^z $(xy\bar{z})$	$2IC_4^z$ $(y\bar{x}\bar{z})$ $(\bar{y}x\bar{z})$	$2IC_2^{a,b}$ $(\bar{y}\bar{x}z)$ $(yx\bar{z})$
A_{1g}	1	1	1	1	1	1	1	1	1	1
B_{1g}	1	1	1	$\bar{1}$	$\bar{1}$	1	1	1	$\bar{1}$	$\bar{1}$
A_{2g}	1	$\bar{1}$	1	1	$\bar{1}$	1	$\bar{1}$	1	1	$\bar{1}$
B_{2g}	1	$\bar{1}$	1	$\bar{1}$	1	1	$\bar{1}$	1	$\bar{1}$	1
E_g	2	0	$\bar{2}$	0	0	2	0	$\bar{2}$	0	0
A_{1u}	1	1	1	1	1	$\bar{1}$	$\bar{1}$	$\bar{1}$	$\bar{1}$	$\bar{1}$
B_{1u}	1	1	1	$\bar{1}$	$\bar{1}$	$\bar{1}$	$\bar{1}$	$\bar{1}$	1	1
A_{2u}	1	$\bar{1}$	1	1	$\bar{1}$	$\bar{1}$	1	$\bar{1}$	$\bar{1}$	1
B_{2u}	1	$\bar{1}$	1	$\bar{1}$	1	$\bar{1}$	1	$\bar{1}$	1	$\bar{1}$
E_u	2	0	$\bar{2}$	0	0	$\bar{2}$	0	2	0	0

- The basis function of the lowest level at $E = 0$ is a constant and carries the irreducible representation A_{1g}.
- The next level is the twofold level at $E = \frac{2a}{c}^2$, carrying the plane waves

$$e^{i\frac{2\pi}{c}z} \quad ; \quad e^{-i\frac{2\pi}{c}z}$$

They carry a 2D representation of D_{4h} with the following characters.

D_{4h}	E	$2C_2^{x,y}$	C_2^z	$2C_4^z$	$2C_2^{a,b}$	I	$2IC_2^{x,y}$	IC_2^z	$2IC_4^z$	$2IC_2^{a,b}$
Γ_2	2	0	2	2	0	0	2	0	0	2

The representation is a reducible one:

$$\Gamma_2 = A_{1g} \oplus A_{2u}$$

and the empty-lattice level will split into two separate levels when the crystal potential is switched on.

- The next energy level is the fourfold level at $E = 4$. The plane waves building this subspace are

$$e^{i\frac{2\pi}{a}x} \quad ; \quad e^{-i\frac{2\pi}{a}x} \quad ; \quad e^{i\frac{2\pi}{a}y} \quad ; \quad e^{-i\frac{2\pi}{a}y}$$

These plane waves carry a 4D representation of D_{4h} with the following characters.

D_{4h}	E	$2C_2^{x,y}$	C_2^z	$2C_4^z$	$2C_2^{a,b}$	I	$2IC_2^{x,y}$	IC_2^z	$2IC_4^z$	$2IC_2^{a,b}$
Γ_4	4	2	0	0	0	0	2	4	0	0

In a finite crystal field, the level will split into the irreducible components

$$A_{1g} \oplus B_{1g} \oplus E_u$$

X-point: The point group of the X-point D_{2h}. Its character table is given as follows (the elements are characterized by their action on the (x, y, z) coordinates). For the irreducible representations, we again use the labeling scheme that is common in chemistry.

| D_{2h} | E | $2C_2^x$ | C_2^y | C_2^z | I | IC_2^x | IC_2^y | IC_2^z |
	(xyz)	$(x\bar{y}\bar{z})$	$(\bar{x}y\bar{z})$	$(\bar{x}\bar{y}z)$	$(\bar{x}\bar{y}\bar{z})$	$(\bar{x}yz)$	$(x\bar{y}z)$	$(xy\bar{z})$
A_g	1	1	1	1	1	1	1	1
B_{1g}	1	$\bar{1}$	$\bar{1}$	1	1	$\bar{1}$	$\bar{1}$	1
B_{2g}	1	$\bar{1}$	1	$\bar{1}$	1	$\bar{1}$	1	$\bar{1}$
B_{3g}	1	1	$\bar{1}$	$\bar{1}$	1	1	$\bar{1}$	$\bar{1}$
A_u	1	1	1	1	$\bar{1}$	$\bar{1}$	$\bar{1}$	$\bar{1}$
B_{1u}	1	$\bar{1}$	$\bar{1}$	1	$\bar{1}$	1	1	$\bar{1}$
B_{2u}	1	$\bar{1}$	1	$\bar{1}$	$\bar{1}$	1	$\bar{1}$	1
B_{3u}	1	1	$\bar{1}$	$\bar{1}$	$\bar{1}$	$\bar{1}$	1	1

• The lowest, twofold level contains the plane waves

$$e^{i\frac{\pi}{a}y} \quad ; \quad e^{-i\frac{\pi}{a}y}$$

The representation X_2 built using these plane waves as basis states has the following characters.

D_{2h}	E	$2C_2^x$	C_2^y	C_2^z	I	IC_2^x	IC_2^y	IC_2^z
X_2	2	0	2	0	0	2	0	2

It is reducible:

$$X_2 = A_g \oplus B_{2u}$$

and the accidental degeneracy of X_2 will be removed when a finite crystal potential is switched on.

The next level is the fourfold degenerate one with the energy $1 + (\frac{2a}{c})^2$. The basis states are

$$e^{i\frac{\pi}{a}y + i\frac{2\pi}{c}z} \quad ; \quad e^{i\frac{\pi}{a}y - i\frac{2\pi}{c}z} \quad ; \quad e^{i\frac{\pi}{a}(-y) + i\frac{2\pi}{c}z} \quad ; \quad e^{i\frac{\pi}{a}(-y) - i\frac{2\pi}{c}z}$$

They are the basis states for a fourfold-dimensional representation of D_{2h} with the following characters.

D_{2h}	E (xyz)	$2C_2^x$ $(x\bar{y}\bar{z})$	C_2^y $(\bar{x}y\bar{z})$	C_2^z $(\bar{x}\bar{y}z)$	I $(\bar{x}\bar{y}\bar{z})$	IC_2^x $(\bar{x}yz)$	IC_2^y $(x\bar{y}z)$	IC_2^z $(xy\bar{z})$
X_4	4	0	0	0	0	4	0	0

This representation is reducible:

$$X_4 = A_g \oplus B_{3g} \oplus B_{1u} \oplus B_{2u}$$

22.2.3 *Model calculation*

The plane wave states (building a CONS for the periodic wave functions) are suitable basis states for computing the band structure in the nearly free electron approximation. This approximation predicts that the desired wave function is expanded into plane waves:

$$\psi(\mathbf{k}, \mathbf{b}, \mathbf{r}) = \sqrt{\frac{1}{V}} \sum_{\mathbf{b}} c(\mathbf{k}, \mathbf{b}) e^{i(\mathbf{k}+\mathbf{b})\cdot\mathbf{r}}$$

containing the coefficients of interest:

$$c(\mathbf{k}, \mathbf{b})$$

and the potential is also expanded into a sum of plane waves:

$$V(\mathbf{r}) = \sqrt{\frac{1}{V}} \sum_{\mathbf{b}} V(\mathbf{b}) e^{i\mathbf{b}\cdot\mathbf{r}}$$

containing the known coefficients

$$V(\mathbf{b})$$

The reality of the potential requires

$$V(-\mathbf{b}) = \bar{V}(\mathbf{b})$$

When these expansions are inserted into the Schrödinger equation

$$-\frac{\hbar^2}{2m}\triangle\psi(\mathbf{k},\mathbf{b},\mathbf{r}) + V(\mathbf{r})\cdot\psi(\mathbf{k},\mathbf{b},\mathbf{r}) = E(\mathbf{k},\mathbf{b})\psi(\mathbf{k},\mathbf{b},\mathbf{r})$$

one obtains

$$+\frac{\hbar^2}{2m}\sum_{\mathbf{b}}(\mathbf{k}+\mathbf{b})^2 c(\mathbf{k},\mathbf{b})e^{i(\mathbf{b})\cdot\mathbf{r}} + \sum_{\mathbf{b},\mathbf{b}'}V(\mathbf{b}')c(\mathbf{k},\mathbf{b})e^{i(\mathbf{b}+\mathbf{b}')\cdot\mathbf{r}}$$

$$= E(\mathbf{k},\mathbf{b})\sum_{\mathbf{b}}V(\mathbf{b})e^{i\mathbf{b}\cdot\mathbf{r}}$$

Multiplying on both sides with

$$e^{-\mathbf{b}''\mathbf{r}}$$

integrating over the crystal, and taking into account that the plane waves build a CONS, one obtains

$$+\frac{\hbar^2}{2m}\sum_{\mathbf{b}}(\mathbf{k}+\mathbf{b})^2 c(\mathbf{k},\mathbf{b})\delta(\mathbf{b}''-\mathbf{b}) + \sum_{\mathbf{b},\mathbf{b}'}V(\mathbf{b}')c(\mathbf{k},\mathbf{b})\delta(\mathbf{b}+\mathbf{b}'-\mathbf{b}'')$$

$$= E(\mathbf{k},\mathbf{b})\sum_{\mathbf{b}}V(\mathbf{b})\delta(\mathbf{b}-\mathbf{b}'')$$

The convenient Kronecker deltas originate from the orthonormality property of the plane waves. Making use of the Kronecker deltas, the eigenvalue equation in terms of the desired coefficients simplifies to

$$+\frac{\hbar^2}{2m}(\mathbf{k}+\mathbf{b})^2 c(\mathbf{k},\mathbf{b}) + \sum_{\mathbf{b}'}V(\mathbf{b}-\mathbf{b}')c(\mathbf{k},\mathbf{b}') = E(\mathbf{k},\mathbf{b})c(\mathbf{k},\mathbf{b})$$

This is an exact Schrödinger eigenvalue equation for the energy bands we are seeking, involving the desired coefficients for the eigenvectors.

We use our symmetry arguments to solve this equation, at least approximately. The approximation consists of solving this equation within the plane wave eigenspaces provided by the empty-lattice solutions. For example, we consider the fourfold-degenerate level E_{Γ_4}. The reciprocal lattice vectors in this space are

$$\frac{2\pi}{a}(1,0,0) \quad \frac{2\pi}{a}(\bar{1},0,0) \quad \frac{2\pi}{a}(0,1,0) \quad \frac{2\pi}{a}(0,\bar{1},0)$$

The eigenvalue equation, restricted to this space, requires the vanishing of

$$
\begin{pmatrix}
E_{\Gamma_4} - E & V((1,0,0) - (\bar{1},0,0)) & V((1,0,0) - (0,1,0)) & V((1,0,0) - (0,\bar{1},0)) \\
V((\bar{1},0,0) - (1,0,0)) & E_{\Gamma_4} - E & V((\bar{1},0,0) - (0,1,0)) & V((\bar{1},0,0) - (0,\bar{1},0)) \\
V((0,1,0) - (1,0,0)) & V((0,1,0) - (\bar{1},0,0)) & E_{\Gamma_4} - E & V((0,1,0) - (0,\bar{1},0)) \\
V((0,\bar{1},0) - (1,0,0)) & V((0,\bar{1},0)) - (\bar{1},0,0)) & V((0,\bar{1},0) - (0,1,0)) & E_{\Gamma_4} - E
\end{pmatrix}
$$

$$
\times
\begin{pmatrix}
c(1,0,0) \\
c(\bar{1},0,0) \\
c(0,1,0) \\
c(0,\bar{1},0)
\end{pmatrix}
$$

Taking into account the restriction on the coefficients of the crystal potential, we have

$$
\begin{pmatrix}
E_{\Gamma_4} - E & V((2,0,0)) & V((1,\bar{1},0)) & V((1,1,0)) \\
\overline{V}((2,0,0)) & E_{\Gamma_4} - E & V((\bar{1},\bar{1},0)) & V((\bar{1},1,0)) \\
\overline{V}((1,\bar{1},0)) & \overline{V}((\bar{1},\bar{1},0)) & E_{\Gamma_4} - E & V((0,2,0)) \\
\overline{V}((1,1,0)) & \overline{V}((\bar{1},1,0)) & \overline{V}((0,2,0)) & E_{\Gamma_4} - E
\end{pmatrix}
\begin{pmatrix}
c(1,0,0) \\
c(\bar{1},0,0) \\
c(0,1,0) \\
c(0,\bar{1},0)
\end{pmatrix}
= 0
$$

For this specific lattice, the inversion is a symmetry element, i.e.

$$
V(-\mathbf{b}) = V(\mathbf{b})
$$

Combined with the requirement that $V(-\mathbf{b}) = \overline{V}(\mathbf{b})$, we have

$$
V(\mathbf{b}) = \overline{V}(\mathbf{b}) = V(-\mathbf{b})
$$

and the eigenvalue problem is written as

$$
\begin{pmatrix}
E_{\Gamma_4} - E & V((2,0,0)) & V((1,\bar{1},0)) & V((1,1,0)) \\
V((2,0,0)) & E_{\Gamma_4} - E & V((\bar{1},\bar{1},0)) & V((\bar{1},1,0)) \\
V((1,\bar{1},0)) & V((\bar{1},\bar{1},0)) & E_{\Gamma_4} - E & V((0,2,0)) \\
V((1,1,0)) & V((\bar{1},1,0)) & V((0,2,0)) & E_{\Gamma_4} - E
\end{pmatrix}
\begin{pmatrix}
c(1,0,0) \\
c(\bar{1},0,0) \\
c(0,1,0) \\
c(0,\bar{1},0)
\end{pmatrix}
= 0
$$

with all matrix elements being real numbers. Further symmetry elements of the space group provide the relations

$$
V((2,0,0)) = V((0,2,0)) \equiv U
$$

and

$$
V(\pm 1, \pm 1, 0) \equiv V
$$

so that the eigenvalue problem becomes

$$
\begin{pmatrix}
E_{\Gamma_4} - E & U & V & V \\
U & E_{\Gamma_4} - E & V & V \\
V & V & E_{\Gamma_4} - E & U \\
V & V & U & E_{\Gamma_4} - E
\end{pmatrix}
\begin{pmatrix}
c(1,0,0) \\
c(\bar{1},0,0) \\
c(0,1,0) \\
c(0,\bar{1},0)
\end{pmatrix}
= 0
$$

So far, we have used symmetry to simplify the operator itself. Symmetry can be further used to find the eigenvalues without solving the determinantal equation.

Γ-point: We find the symmetry-adapted linear combinations of the four plane waves: They will provide the eigenvectors of this eigenvalue problem. The method of choice is the projector method.

- We find

$$P_{A_{1g}} e^{i\frac{2\pi}{a}x} \propto \cos\frac{2\pi}{a}x + \sin\frac{2\pi}{a}x$$

  This result entails that

$$c(1,0,0) = c(\bar{1},0,0) = c(0,1,0) = c(0,\bar{1},0) = 1$$

  is one of the desired eigenvectors. Inserting this eigenvector into the eigenvalue problem produces an algebraic equation for the desired eigenvalue:

$$(E_{\Gamma_4} - E_{A_{1g}}) \cdot 1 + U \cdot 1 + V \cdot 1 + V \cdot 1 = 0$$

  i.e.

$$E_{A_{1g}} = E_{\Gamma_4} + U + 2V$$

- We find further that

$$P_{B_{1g}} e^{i\frac{2\pi}{a}x} \propto \cos\frac{2\pi}{a}x - \sin\frac{2\pi}{a}x$$

  i.e.

$$c(1,0,0) = c(\bar{1},0,0) = 1 \quad c(0,1,0) = c(0,\bar{1},0) = \bar{1}$$

  Inserting into the algebraic system of equations produces

$$(E_{\Gamma_4} - E_{B_{1g}}) \cdot 1 + U \cdot 1 - V \cdot 1 - V \cdot 1 = 0$$

  i.e.

$$E_{B_{1g}} = E_{\Gamma_4} + U - 2V$$

- We find finally one eigenvector in the 2D space E_u:

$$P_{E_u} e^{i\frac{2\pi}{a}x} \propto \sin\frac{2\pi}{a}x$$

  i.e.

$$c(1,0,0) = 1 \quad c(\bar{1},0,0) = \bar{1} \quad c(0,1,0) = c(0,\bar{1},0) = 0$$

and accordingly,

$$E_{E_u} = E_{\Gamma_4} - U$$

(twofold essential degeneracy).

• In the space of Γ_2, the eigenvalue equation is written as

$$\begin{pmatrix} E_{\Gamma_2} - E & \underbrace{V((0,0,1)) - (0,0,\bar{1}))}_{V((0,0,2))} \\ \underbrace{V((0,0,\bar{1})) - (0,0,1))}_{V((0,0,2))} & E_{\Gamma_2} - E \end{pmatrix} \begin{pmatrix} c(0,0,1) \\ c(0,0,\bar{1}) \end{pmatrix} = 0$$

With

$$P_{A_{1g}} e^{\frac{2\pi}{c} z} \propto \cos \frac{2\pi}{c} z$$

and

$$P_{A_{2u}} e^{\frac{2\pi}{c} z} \propto \sin \frac{2\pi}{c} z$$

we find that the empty-lattice level E_{Γ_2} splits into

$$E_{A_{1g}} = E_{\Gamma_2} + V((0,0,2)) \quad E_{A_{2u}} = E_{\Gamma_2} - V((0,0,2))$$

Δ-direction: Band crossing (encircled with light gray in the figure of the empty-lattice band structure): This is a particularly interesting point, as it reveals the presence of the von Neumann–Wigner phenomenon. Exactly at the crossing point, we observe a three-dimensional representation of C_{2v}, carried by the plane waves

$$e^{i\frac{\pi}{a}\delta^* y + i\frac{2\pi}{c}\cdot z} \quad e^{i\frac{\pi}{a}\delta^* y - i\frac{2\pi}{c} z\cdot z} \quad e^{i\frac{\pi}{a}(\delta^* - 2)\cdot y}$$

with characters

C_{2v}	E (xyz)	C_2^y $(\bar{x}y\bar{z})$	IC_2^z $(xy\bar{z})$	IC_2^x $(\bar{x}yz)$
Δ^*	3	1	1	3

and accordingly,

$$\Delta^* = 2A_1 \oplus B_2$$

We anticipate that, when a finite crystal potential is switched on, the accidental A_1-degeneracy will be lifted, producing a gap in the A_1-band structure (the so-called von Neumann–Wigner gap or hybridization gap). We study this phenomenon more quantitatively by solving the eigenvalue problem

$$\begin{pmatrix} E_{\Delta^*} - E & V((0,0,2)) & V((0,1,1)) \\ V((0,0,2)) & E_{\Delta^*} - E & V((0,1,1)) \\ V((0,1,1)) & V((0,1,1)) & E_{\Delta^*} - E \end{pmatrix} \begin{pmatrix} c(0,0,1) \\ c(0,0,\bar{1}) \\ c(0,\bar{1},0) \end{pmatrix} = 0$$

This matrix can be block-diagonalized into

- a one-dimensional sector for the eigenvalue E_{B_2} by inserting the suitable eigenvector:

$$P_{B_2} e^{i\frac{\pi}{a}\delta^* y + i\frac{2\pi}{c}\cdot z} \propto e^{i\frac{\pi}{a}\delta^* y} \sin\frac{2\pi}{c} \cdot z$$

i.e.

$$c((0,0,1) = 1 \quad c(0,0,\bar{1}) = \bar{1} \quad c(0,\bar{1},0) = 0$$

and

$$E_{B_2} = E^* - V((0,0,2))$$

- a 2D sector with A_1-symmetry:

$$P_{A_1} e^{i\frac{\pi}{a}\delta^* y + i\frac{2\pi}{c}\cdot z} \propto e^{i\frac{\pi}{a}\delta^* y} \cos\frac{2\pi}{c} \cdot z$$

$$P_{A_1} e^{i\frac{\pi}{a}(\delta^* - 2)\cdot y} \propto e^{i\frac{\pi}{a}(\delta^* - 2)\cdot y}$$

This means that the basis vectors within the A_1-sector can be chosen to be

$$I \equiv \frac{1}{\sqrt{2}}\begin{pmatrix} 1 \\ 1 \\ 0 \end{pmatrix} \quad II \equiv \begin{pmatrix} 0 \\ 0 \\ 1 \end{pmatrix}$$

and the eigenvalue problem built onto these basis vectors reads

$$\begin{pmatrix} E_{\Delta^*} - E + V((0,0,2)) & \sqrt{2}V((0,1,1)) \\ \sqrt{2}V((0,1,1)) & E_{\Delta^*} - E \end{pmatrix} \begin{pmatrix} c_1^I \\ c_2^{II} \end{pmatrix} = 0$$

This last eigenvalue problem cannot be solved by symmetry arguments: We must solve explicitly the determinantal equation. This is, however, a second-order algebraic equation (we started with a 3×3 matrix, which would have given a more nasty third-order determinantal equation), which is solved by a standard formula. The symmetry arguments provided a mathematical simplification of the eigenvalue problem and allowed us to predict a von Neumann–Wigner gap in the band structure.

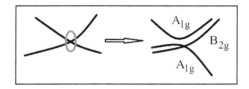

The von Neumann–Wigner mechanism along the Δ-direction.

23

PROJECT FOR LECTURE 6: SYMMETRY ANALYSIS OF GRAPHENE BAND STRUCTURE AND DIRAC CONES

23.1 Description of Structure and Symmetry Elements of the Graphene Lattice

Graphene (see e.g. https://www.nobelprize.org/uploads/2018/06/novose lov_lecture.pdf) is a two-dimensional (2D) material made of identical carbon atoms arranged in a honeycomb lattice. The picture of a honeycomb structure, as seen in nature by bees, is given in the following figure.

Picture of a honeycomb lattice. Courtesy of D. Conconi, (foto: apiservice).

The honeycomb lattice belongs to the 2D hexagonal crystal system. It has point group D_{6h} and space group $P6/mmm$. As shown in the following figure, the lattice can be created by translating the unit cell entailed by the basis vectors **a** and **b**. The unit cell contains two atoms, located at positions τ_A and τ_B. In terms of the xy-coordinate system the vectors used here are given by

$$\mathbf{a} = \frac{a}{2}(\sqrt{3}\mathbf{e}_x + \mathbf{e}_y) \quad \mathbf{b} = \frac{a}{2}(-\sqrt{3}\mathbf{e}_x + \mathbf{e}_y)$$

$$\tau_A = \frac{a}{2}\left(\frac{1}{\sqrt{3}}\mathbf{e}_x + \mathbf{e}_y\right) \quad \tau_B = \frac{a}{\sqrt{3}} \cdot \mathbf{e}_x$$

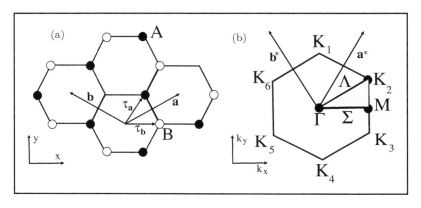

(a) The honeycomb lattice of graphene. (b) The reciprocal lattice. The reduced Brillouin zone is shown in dark gray.

The reciprocal lattice of graphene is also a hexagonal lattice with the reciprocal basis vectors

$$\mathbf{a}^* = \frac{2\pi}{a}\left(\frac{1}{\sqrt{3}}\mathbf{e}_x + \mathbf{e}_y\right) \quad \mathbf{b}^* = \frac{2\pi}{a}\left(-\frac{1}{\sqrt{3}}\mathbf{e}_x + \mathbf{e}_y\right)$$

The primitive translations of the graphene structure are given by the integer linear combinations of **a** and **b**. Further symmetry elements are the proper rotations described in the following figure.

C_2' and C_2'' represent twofold rotations perpendicular to the sixfold rotational axis. The axes of C_2'' are rotated by $\pi/6$ away from the C_2'-axes. We call $C_2^{(1)'}$ and $C_2^{(1)''}$ the twofold rotations about the axes indicated in the figure (we will use later these operations explicitly). The full point group is obtained by multiplying the proper rotations with the inversion I.

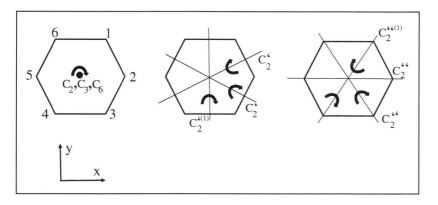

The proper rotations of D_{6h}.

We will discuss the band structure at the high-symmetric points $\Gamma = 0$ and $K = 2\pi/3a(\sqrt{3}\mathbf{e}_x + \mathbf{e}_y)$.

23.2   The π-bands of Graphene

The outer configuration of each carbon atom is $2s^2 2p^2$. In the honeycomb lattice, the $2s^2 2p^2$ configuration hybridizes to form three sp^2-hybridized states in the plane and a p_z orbital perpendicular to the plane. Each atom in the unit cell acquires three sp^2 orbitals and one p_z-orbital. The p_z energy level is twice degenerate, as it contains the two orbitals centered at each atom. The twice degenerate p_z atomic level develops into, possibly, two separate π-bands in graphene. We designate the p_z basis state centered at the atom A with p_{zA} and the p_z basis state centered at the atom B with p_{zB}.

A possible set of symmetry-adapted basis states that describes the p_z electrons in graphene and takes into account the translational symmetry of the lattice is provided by the Bloch sums

$$\Phi_A(\mathbf{k}, \mathbf{r}) = \sqrt{\frac{1}{N}} \cdot \sum_{\mathbf{t}} e^{i\mathbf{k}\cdot\mathbf{t}} \cdot p_z\left(\mathbf{r} - (\tau_A + \mathbf{t})\right)$$

$$\Phi_B(\mathbf{k}, \mathbf{r}) = \sqrt{\frac{1}{N}} \cdot \sum_{\mathbf{t}} e^{i\mathbf{k}\cdot\mathbf{t}} \cdot p_z\left(\mathbf{r} - (\tau_B + \mathbf{t})\right)$$

$\mathbf{t}$ are integer translation vectors. The functions $p_z\left(\mathbf{r}-(\tau_A+\mathbf{t})\right)$ and $p_z\left(\mathbf{r}-(\tau_B+\mathbf{t})\right)$ are p_z-like atomic orbitals centered at $(\tau_A+\mathbf{t})$ and $(\tau_B+\mathbf{t})$, respectively.

23.2.1 π-energy levels at Γ point

To analyze the behavior of the p_z level in the honeycomb lattice, we need to build the matrix representation of the group of Γ in the space spanned by the two Bloch functions. The Bloch functions are adapted to translational symmetry, so we only need to consider their transformation with respect to the elements of the small point group. The small point group of Γ is the entire point group $D_{6h} = D_6 \otimes I$. The character table of D_{6h} is given next.

	E	$2C_6$	$2C_3$	C_2	$3C_2'$	$3C_2''$	I	$2S_3$	$2S_6$	σ_h	$3\sigma_d$	$3\sigma_v$
A_{1g}	1	1	1	1	1	1	1	1	1	1	1	1
A_{2g}	1	1	1	1	-1	-1	1	1	1	1	-1	-1
B_{1g}	1	-1	1	-1	1	-1	1	-1	1	-1	1	-1
B_{2g}	1	-1	1	-1	-1	1	1	-1	1	-1	-1	1
E_{1g}	2	1	-1	-2	0	0	2	1	-1	-2	0	0
E_{2g}	2	-1	-1	2	0	0	2	-1	-1	2	0	0
A_{1u}	1	1	1	1	1	1	-1	-1	-1	-1	-1	-1
A_{2u}	1	1	1	1	-1	-1	-1	-1	-1	-1	1	1
B_{1u}	1	-1	1	-1	1	-1	-1	1	-1	1	-1	1
B_{2u}	1	-1	1	-1	-1	1	-1	1	-1	1	1	-1
E_{1u}	2	1	-1	-2	0	0	-2	-1	1	2	0	0
E_{2u}	2	-1	-1	2	0	0	-2	1	1	-2	0	0

The subscripts g, u refer to representations in which the inversion has a positive (negative) character.

Let us compute, as an example, the matrix representation of I. We need to find out how the basis states transform when I is applied to them. Let us compute, step by step:

$$I\Phi_A(\mathbf{\Gamma}, \mathbf{r}) = \Phi_A(\mathbf{\Gamma}, -\mathbf{r})$$

$$= \sqrt{\frac{1}{N}} \cdot \sum_{\mathbf{t}} e^{i\mathbf{\Gamma}\cdot\mathbf{t}} \cdot p_z\left(-\mathbf{r}-(\tau_A+\mathbf{t})\right)$$

Insert $\Gamma = \vec{0}$ and shift the sum to $\mathbf{t} = \mathbf{t}' - \mathbf{a}$. Since we sum over all lattice points, this leaves the total sum invariant. Next, we identify $\tau_A - \mathbf{a}$ as $-\tau_B$. Accordingly, we find

$$\sqrt{\frac{1}{N}} \cdot \sum_{\mathbf{t}'} p_z \left(-\mathbf{r} - (\tau_A + \mathbf{t}' - \mathbf{a})\right) = \sqrt{\frac{1}{N}} \cdot \sum_{\mathbf{t}'} p_z \left(-\mathbf{r} + \tau_B + \mathbf{t}'\right)$$

Using the symmetry of the p_z orbitals, $p_z(-\mathbf{r}) = -p_z(\mathbf{r})$, we find

$$I\Phi_A(\Gamma, \mathbf{r}) = \sqrt{\frac{1}{N}} \cdot \sum_{\mathbf{t}'} p_z \left(-\mathbf{r} + \tau_B + \mathbf{t}'\right))$$

$$= \sqrt{\frac{1}{N}} \cdot \sum_{\mathbf{t}'} p_z \left(\mathbf{r} - (\tau_B + \mathbf{t}'\right))$$

$$= -\Phi_B(\Gamma, \mathbf{r})$$

Similarly,

$$I\Phi_B(\Gamma, \mathbf{r}) = -\Phi_A(\Gamma, \mathbf{r})$$

The representation matrices of all remaining elements are found in a similar way (it is a tedious but straightforward job!). We list some of them, which will turn out to be sufficient for the symmetry analysis:

$$I : \begin{pmatrix} 0 & -1 \\ -1 & 0 \end{pmatrix}, \quad \sigma_h : \begin{pmatrix} -1 & 0 \\ 0 & -1 \end{pmatrix},$$

$$C_2'^{(1)} : \begin{pmatrix} 0 & -1 \\ -1 & 0 \end{pmatrix}, \quad C_2''^{(1)} : \begin{pmatrix} -1 & 0 \\ 0 & -1 \end{pmatrix}$$

Their characters are summarized in the following table.

Characters of some symmetry elements in the p_z representation at the Γ point.

	I	σ_h	C_2'	C_2"
p_z	0	−2	0	−2

Using the characters of the representation "p_z" and comparing them with the character table, we observe that the Bloch representation decomposes into

$$B_{1g} \oplus A_{2u}$$

Consequently, there will be two separate energy eigenvalues at the Γ point, originating from the originally twice degenerate p_z level.

As a next step, we determine the symmetry-adapted vectors at the Γ point. We bypass the building of the projector by noting that symmetry-adapted vectors must be eigenvectors of all matrix representations. We therefore diagonalize, for example, the matrix representation of the inversion

$$I = \begin{pmatrix} 0 & -1 \\ -1 & 0 \end{pmatrix}$$

to obtain the eigenvectors

$$\vec{v}_1 = \begin{pmatrix} 1 \\ 1 \end{pmatrix}, \quad \vec{v}_2 = \begin{pmatrix} 1 \\ -1 \end{pmatrix}.$$

In order to associate the vectors to an irreducible representation, we calculate the eigenvalues of the representation of I. These must correspond to the character of the irreducible representation for the inversion operation:

$$\lambda_1 = -1, \quad \lambda_2 = 1$$

We deduce that

$$\Phi_A(\mathbf{r}) + \Phi_B(\mathbf{r})$$

is the symmetry-adapted vector transforming according to the representation A_{2u}, and

$$\Phi_A(\mathbf{r}) - \Phi_B(\mathbf{r})$$

is the symmetry-adapted vector transforming according to the representation B_{1g}.

23.2.2 π-energy levels at K-point

The K-point is given by $K = 2\pi/3a(\sqrt{3}\mathbf{e}_x + \mathbf{e}_y)$. The character table of D_{3h} (corresponding to the point symmetry group at the K-point) is given next.

D_{3h}	E	$2C_3$	$3C_2'$	σ_h	$2S_3$	$3\sigma_v$
A_1'	1	1	1	1	1	1
A_2'	1	1	-1	1	1	-1
E'	2	-1	0	2	-1	0
A_1''	1	1	1	-1	-1	-1
A_2''	1	1	-1	-1	-1	1
E''	2	-1	0	-2	1	0

In order to find the decomposition of the p_z-Bloch representation into irreducible representations of D_{3h}, we need the characters of the p_z-Bloch representation for one element of each conjugacy class. We work out the example of the element C_3^{-1} applied to Φ_A:

$$C_3^{-1}\Phi_A(\mathbf{K},\mathbf{r}) = \sqrt{\frac{1}{N}} \cdot \sum_{\mathbf{t}} e^{i\mathbf{K}\cdot\mathbf{t}} \cdot p_z\left(C_3\mathbf{r} - (\tau_A + \mathbf{t})\right)$$

Next, we apply a counter-rotation to the arguments of the p_z orbitals (which has rotational symmetry in the XY-plane). Using

$$C_3^{-1}\tau_A = \tau_A - \mathbf{a} - \mathbf{b}$$

and

$$\mathbf{t}' = C_3^{-1}\mathbf{t} - \mathbf{a} - \mathbf{b}$$

we obtain

$$C_3^{-1}\Phi_A(\mathbf{K},\mathbf{r}) = \sqrt{\frac{1}{N}} \cdot \sum_{\mathbf{t}} e^{i\mathbf{K}\cdot\mathbf{t}} \cdot p_z\left(\mathbf{r} - (\tau_A - \mathbf{a} - \mathbf{b} + \mathbf{t})\right)$$

$$= \sqrt{\frac{1}{N}} \cdot \sum_{\mathbf{t}'} e^{i\mathbf{K}\cdot C_3(\mathbf{t}'+\mathbf{a}+\mathbf{b})} \cdot p_z\left(\mathbf{r} - (\tau_A + \mathbf{t}')\right)$$

As $C_3^{-1}\mathbf{K} = \mathbf{K}$, we obtain

$$\mathbf{K}\cdot C_3(\mathbf{t}' + \mathbf{a} + \mathbf{b}) = C_3^{-1}\mathbf{K}\cdot(\mathbf{t}' + \mathbf{a} + \mathbf{b})$$

$$= \mathbf{K}\cdot\mathbf{t}' + \mathbf{K}\cdot(\mathbf{a} + \mathbf{b})$$

$$= \mathbf{K}\cdot\mathbf{t}' + \frac{2\pi}{3}$$

so that

$$C_3^{-1}\Phi_A(\mathbf{K},\mathbf{r}) = e^{i\frac{2\pi}{3}} \cdot \Phi_A(\mathbf{K},\mathbf{r})$$

We use the same procedure for Φ_B and obtain

$$C_3^{-1} \rightarrow \begin{pmatrix} e^{i\frac{2\pi}{3}} & 0 \\ 0 & e^{-i\frac{2\pi}{3}} \end{pmatrix}$$

The remaining matrix representations are found in a similar way:

$$E \rightarrow \mathbb{1}, \quad C_3^{-1} \rightarrow \begin{pmatrix} e^{i\frac{2\pi}{3}} & 0 \\ 0 & e^{-i\frac{2\pi}{3}} \end{pmatrix}, \quad C_2'' \rightarrow \begin{pmatrix} 0 & -e^{i\frac{2\pi}{3}} \\ -e^{-i\frac{2\pi}{3}} & 0 \end{pmatrix},$$

$$\sigma_h \rightarrow -\mathbb{1}, \quad S_3 \rightarrow \begin{pmatrix} -e^{i\frac{2\pi}{3}} & 0 \\ 0 & -e^{-i\frac{2\pi}{3}} \end{pmatrix}, \quad \sigma_v \rightarrow \begin{pmatrix} 0 & e^{-i\frac{2\pi}{3}} \\ e^{i\frac{2\pi}{3}} & 0 \end{pmatrix},$$

The characters of the "p_z" representation at K are listed in the last line of the following table.

	E	$2C_3$	$3C_2'$	σ_h	$2S_3$	$3\sigma_v$
A_1'	1	1	1	1	1	1
A_2'	1	1	-1	1	1	-1
E'	2	-1	0	2	-1	0
A_1''	1	1	1	-1	-1	-1
A_2''	1	1	-1	-1	-1	1
E''	2	-1	0	-2	1	0
p_z	2	-1	0	-2	1	0

Accordingly, the representation constructed with the p_z orbitals at K is identical to the 2D irreducible representation E'' and the Bloch functions used are symmetry adapted. The following figure summarizes the symmetry analysis at Γ and K.

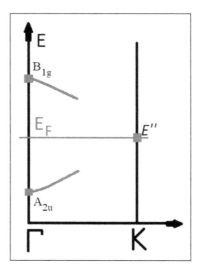

Summary of symmetry analysis at Γ and K. The ordering of the levels at Γ is according to the quantitative computation in F. Bassani and G. Pastori-Parravicini, Il Nuovo Cimento, **L**, B, 95 (1967).

The p_z-atomic level splits into bonding and antibonding levels at the Γ-point. The split bands merge into a twofold degenerate level when one moves from Γ toward K, e.g. along the Λ line.

The 2D degeneracy of the π-bands at K is of profound significance to the electronic properties of graphene: It can host four electrons (including the spin states), but only two are provided by the involved atoms (one originating from A, the other from B). This pins the E''-level **exactly** at the Fermi level and renders the graphene gapless. The pinning of the Fermi level is produced by symmetry and is referred to in modern research as a **topologically protected** feature of the electronic structure of graphene.

23.2.3   *The π-bands in the vicinity of K*

To estimate the band structure of the π-bands in the vicinity of K, we solve the determinantal equation of the perturbative Hamilton operator

$$\begin{pmatrix} E_K & A \\ A^* & E_K \end{pmatrix}$$

E_K is the energy value at the K-point: It is doubly degenerate and thus appears on both diagonal matrix elements. A is the non-diagonal matrix element that must appear in the vicinity of K, e.g. along Λ, as the double degeneracy is removed. The solution of the determinantal equation is

$$E_{\pm} = E_K \pm \sqrt{A \cdot A^*}$$

where $+$ denotes the antibonding linear combination and $-$ the bonding linear combination. A has been estimated, for example, using the density functional theory (DFT), but we can make an educated guess about its k-dependence along the Λ-line. A is a quantity that must vanish at K by symmetry and therefore can be written as some power of the component of the vector $\mathbf{k}_\Lambda - \mathbf{K}_\Lambda$ along Λ

$$A = \underbrace{A_0}_{0} + A_1 \cdot \underbrace{(\mathbf{k}_\Lambda - \mathbf{K}_\Lambda) \cdot \mathbf{n}_\Lambda}_{\text{component along the direction specified by } \mathbf{n}_\Lambda} + \cdots$$

From the point of view of symmetry, there is no reason that would forbid the presence of a linear term, which is certainly a scalar transforming according to the $\mathbb{1}$-irreducible representation on the group of Λ. Accordingly, the energy bands will scale linearly along Λ when they depart from the K-point.

A more quantitative computation provides proof of this remarkable linear dependence and also gives the complete band structure based on the p_z orbitals. This computation (called tight-binding model[1]) starts with the basis functions

$$\Phi_A(\mathbf{k}, \mathbf{r}) = \sqrt{\frac{1}{N}} \cdot \sum_{\mathbf{t}} e^{i\mathbf{k} \cdot \mathbf{t}} \cdot p_z\left(\mathbf{r} - (\tau_A + \mathbf{t})\right)$$

$$\Phi_B(\mathbf{k}, \mathbf{r}) = \sqrt{\frac{1}{N}} \cdot \sum_{\mathbf{t}} e^{i\mathbf{k} \cdot \mathbf{t}} \cdot p_z\left(\mathbf{r} - (\tau_B + \mathbf{t})\right),$$

Within the space provided by these basis functions, the eigenvalue problem of the Hamiltonian reads, formally,

$$\begin{pmatrix} H_{11} & H_{12} \\ H_{21} & H_{22} \end{pmatrix} \begin{pmatrix} c_1 \\ c_2 \end{pmatrix} = E \begin{pmatrix} S_{11} & S_{12} \\ S_{21} & S_{22} \end{pmatrix} \begin{pmatrix} c_1 \\ c_2 \end{pmatrix}.$$

with the matrix elements defined by the integrals

$$H_{mn} \equiv \frac{1}{N} \sum_{t,s} e^{i\mathbf{k}(\mathbf{t}-\mathbf{s})} \int d\mathbf{r} \; p_z(\mathbf{r} - \tau_n - \mathbf{t})^* \mathscr{H} p_z(\mathbf{r} - \tau_m - \mathbf{s})$$

$$S_{mn} \equiv \frac{1}{N} \sum_{t,s} e^{i\mathbf{k}(\mathbf{t}-\mathbf{s})} \int d\mathbf{r} \; p_z(\mathbf{r} - \tau_n - \mathbf{t})^* p_z(\mathbf{r} - \tau_m - \mathbf{s}).$$

The desired coefficients c_1, c_2 determine those linear combinations,

$$c_1(\mathbf{k})\Phi_A(\mathbf{k}, \mathbf{r}) + c_2(\mathbf{k})\Phi_B(\mathbf{k}, \mathbf{r}),$$

that build the eigenstates to the desired energy eigenvalue E. The S-matrix appears in the tight-binding model, which uses a non-orthogonal linear combination of atomic orbitals. The diagonal matrix elements read

$$H_{11} = \frac{1}{N} \sum_{s,t} e^{i\mathbf{k}(\mathbf{t}-\mathbf{s})} \langle p_z(\mathbf{r} - \tau_A - \mathbf{s})|\mathscr{H}|p_z(\mathbf{r} - \tau_A - \mathbf{t})\rangle$$

$$H_{22} = \frac{1}{N} \sum_{s,t} e^{i\mathbf{k}(\mathbf{t}-\mathbf{s})} \langle p_z(\mathbf{r} - \tau_B - \mathbf{s})|\mathscr{H}|p_z(\mathbf{r} - \tau_B - \mathbf{t})\rangle,$$

H_{11} must be equal to H_{22} because the atomic orbitals of the two lattice sites are identical (both are $2p_z$ orbitals of the carbon atoms). The expression of H_{11} can be further simplified by assuming that the largest contributions to

---

[1] See, for example, E. Kogan and V. U. Nazarov, "Symmetry classification of energy bands in graphene", *Phys. Rev. B* **85**, 115418 (2012), https://journals.aps.org/prb/abstract/10.1103/PhysRevB.85.115418 and the references therein.

the sum come from terms with $\mathbf{s} = \mathbf{t}$. This assumption is justified by the relatively weak orbital overlap between neighboring unit cells. In this case, the sum reduces to

$$H_{11} \approx \frac{1}{N} \sum_{\mathbf{t}} \langle p_z(\mathbf{r} - \tau_A - \mathbf{t})| \mathscr{H} |p_z(\mathbf{r} - \tau_A - \mathbf{t}) \rangle$$

$$= \langle p_z(\mathbf{r} - \tau_A)| \mathscr{H} |p_z(\mathbf{r} - \tau_A) \rangle.$$

The single-particle Hamiltonian $\mathscr{H}$ for an electron in the graphene lattice is given by

$$-\frac{\hbar^2}{2m} \nabla^2 + \sum_{\mathbf{u}} [V(\mathbf{r} - \tau_A - \mathbf{u}) + V(\mathbf{r} - \tau_B - \mathbf{u})]$$

$$= \left[-\frac{\hbar^2}{2m} \nabla^2 + V(\mathbf{r} - \tau_A) \right] + \sum_{\mathbf{u} \neq 0} V(\mathbf{r} - \tau_A - \mathbf{u}) + \sum_{\mathbf{u}} V(\mathbf{r} - \tau_B - \mathbf{u}).$$

Since the magnitude of the lattice potential $V(\mathbf{r})$ decreases with increasing distance (consider the screened Coulomb potential), we assume that the largest contribution to the matrix element H_{11} is given by

$$H_{11} = H_{22} \approx \langle p_z(\mathbf{r} - \tau_A)| -\frac{\hbar^2}{2m} \nabla^2 + V(\mathbf{r} - \tau_A) |p_z(\mathbf{r} - \tau_A) \rangle$$

$$= E_{2p_z}$$

with $E_{2p_z} = -0.66 \ Ry$ being the energy of the atomic orbital $2p_z$ of the carbon atom.

The non-diagonal matrix elements of the Hamilton operator represent hopping between the lattice sites A and B. In the following, we only consider hopping between the nearest neighbors. Since each atom A has three neighboring atoms B, the sum over N unit cells reduces to

$$H_{12} = \frac{1}{N} \sum_{\mathbf{s},\mathbf{t}} e^{i\mathbf{k}(\mathbf{t}-\mathbf{s})} \langle p_z(\mathbf{r} - \tau_B - \mathbf{s})| \mathscr{H} |p_z(\mathbf{r} - \tau_A - \mathbf{t}) \rangle$$

$$\approx \frac{1}{N} \sum_{\mathbf{t},\delta_j} e^{-i\mathbf{k}(\delta_\mathbf{j}+\tau_A-\tau_B)} \langle p_z(\mathbf{r} - (\mathbf{t} + \tau_\mathbf{A} + \delta_\mathbf{j}))| \mathscr{H} |p_z(\mathbf{r} - (\mathbf{t} + \tau_\mathbf{A})) \rangle$$

where δ_js are given by

$$\delta_1 \equiv a \left(-\frac{1}{\sqrt{3}}, 0 \right) \qquad \delta_2 \equiv \frac{a}{2} \left(\frac{1}{\sqrt{3}}, 1 \right) \qquad \delta_3 \equiv \frac{a}{2} \left(\frac{1}{\sqrt{3}}, -1 \right)$$

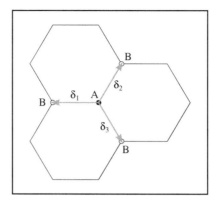

Graphics for determining δ_j.

Because of the translational symmetry, we obtain

$$H_{12} \approx \frac{1}{N} \sum_{\delta_j} e^{-i\mathbf{k}(\delta_j + \tau_\mathbf{A} - \tau_\mathbf{B})}$$

$$\times \sum_{\mathbf{t}} \langle p_z(\mathbf{r} - (\tau_\mathbf{A} + \delta_j)| \mathscr{T}^*(\mathbf{t}) \mathscr{H} \mathscr{T}(\mathbf{t}) | p_z(\mathbf{r} - \tau_\mathbf{A}) \rangle$$

$$= \sum_{\delta_j} e^{-i\mathbf{k}(\delta_j + \tau_\mathbf{A} - \tau_\mathbf{B})} \langle p_z(\mathbf{r} - (\tau_\mathbf{A} + \delta_j)| \mathscr{H} | p_z(\mathbf{r} - \tau_\mathbf{A}) \rangle.$$

Because of the rotational symmetry of $p_z(\mathbf{r})$ around the $\hat{\mathbf{z}}$ axis, we have

$$H_{12} \approx \left[\sum_{\delta_j} e^{-i\mathbf{k}(\delta_j + \tau_\mathbf{A} - \tau_\mathbf{B})} \right] \langle p_z(\mathbf{r} - (\tau_\mathbf{A} + \delta_1)| \mathscr{H} | p_z(\mathbf{r} - \tau_\mathbf{A}) \rangle$$

$$\equiv -\gamma f(\mathbf{k}),$$

where $f(\mathbf{k})$ and γ are defined as

$$f(\mathbf{k}) \equiv \sum_{\delta_j} e^{-i\mathbf{k}(\delta_j + \tau_\mathbf{A} - \tau_\mathbf{B})}$$

$$\gamma \equiv - \langle p_z(\mathbf{r} - (\tau_\mathbf{A} + \delta_1)| \mathscr{H} | p_z(\mathbf{r} - \tau_\mathbf{A}) \rangle.$$

γ is called the hopping parameter and can be determined numerically, whereas $f(\mathbf{k})$ is a kind of band structure factor that contains the $\mathbf{k}$-dependence. Note, finally, that $H_{21} = H_{12}^*$.

One important drawback of the tight-binding model is that the Bloch sums are not necessarily orthogonal and normalized. One must therefore include the overlap integrals S in the calculation of the eigenvalues of the matrix. For normalized orbital wave functions

$$\langle p_z(\mathbf{r} - \tau_A - \mathbf{t})|p_z(\mathbf{r} - \tau_A - \mathbf{t})\rangle = \langle p_z(\mathbf{r} - \tau_B - \mathbf{t})|p_z(\mathbf{r} - \tau_B - \mathbf{t})\rangle = 1$$

the diagonal terms of the overlap integral matrix are given by

$$S_{11} = S_{22} = \frac{1}{N} \sum_{\mathbf{t},\mathbf{s}} e^{i\mathbf{k}(\mathbf{t}-\mathbf{s})} \langle p_z(\mathbf{r} - \tau_A - \mathbf{s})|p_z(\mathbf{r} - \tau_A - \mathbf{t})\rangle$$

$$\approx \frac{1}{N} \sum_{\mathbf{t}} \langle p_z(\mathbf{r} - \tau_A - \mathbf{t})|p_z(\mathbf{r} - \tau_A - \mathbf{t})\rangle$$

$$= \frac{1}{N} \sum_{\mathbf{t}} 1$$

$$= 1$$

The off-diagonal terms of the overlap integral matrix are given by

$$S_{12} = \frac{1}{N} \sum_{\mathbf{t},\mathbf{s}} e^{i\mathbf{k}(\mathbf{t}-\mathbf{s})} \langle p_z(\mathbf{r} - \tau_B - \mathbf{s})|p_z(\mathbf{r} - \tau_A - \mathbf{t})\rangle$$

$$\approx \frac{1}{N} \sum_{\mathbf{t},\delta_j} e^{-i\mathbf{k}(\delta_j + \tau_A - \tau_B)} \langle p_z(\mathbf{r} - (\mathbf{t} + \tau_A + \delta_j)|p_z(\mathbf{r} - (\mathbf{t} + \tau_A))\rangle$$

$$= \left[\sum_{\delta_j} e^{-i\mathbf{k}(\delta_j + \tau_A - \tau_B)} \right] \langle p_z(\mathbf{r} - (\tau_A + \delta_1)|p_z(\mathbf{r} - \tau_A))\rangle$$

$$\equiv \sigma f(\mathbf{k}),$$

where we introduce a new quantity,

$$\sigma \equiv \langle p_z(\mathbf{r} - (\tau_A + \delta_1)|p_z(\mathbf{r} - \tau_A))\rangle$$

which describes the $2p_z$ orbital overlapping the nearest neighboring atoms. Again, the other off-diagonal term is obtained by $S_{21} = S_{12}^*$. The quantity σ is provided by explicit numerical computations, e.g. within the framework of DFT. For the purpose of our qualitative discussion, we set the overlap integral S_{12} to zero. Accordingly, the eigenvalue problem reduces to solving the determinantal equation of the matrix

$$\begin{pmatrix} E_{2p_z} - E & -\gamma \cdot f(\mathbf{k}) \\ -\gamma^* \cdot \overline{f(\mathbf{k})} & E_{2p_z} - E \end{pmatrix}$$

The eigenvalues for the π-bands read

$$E_\pm = E_{2p_z} \pm \mid \gamma \mid \cdot \mid f(\mathbf{k}) \mid.$$

with

$$f(\mathbf{k}) = \sum_{\delta_j} e^{-i\mathbf{k}(\delta_\mathbf{j}+\tau_\mathbf{A}-\tau_\mathbf{B})}$$

$$= 1 + e^{-\frac{ia}{2}\left(\sqrt{3}k_x+k_y\right)} + e^{-\frac{ia}{2}\left(\sqrt{3}k_x-k_y\right)}$$

$$= 1 + 2e^{-i\frac{\sqrt{3}ak_x}{2}} \cos\left(\frac{ak_y}{2}\right),$$

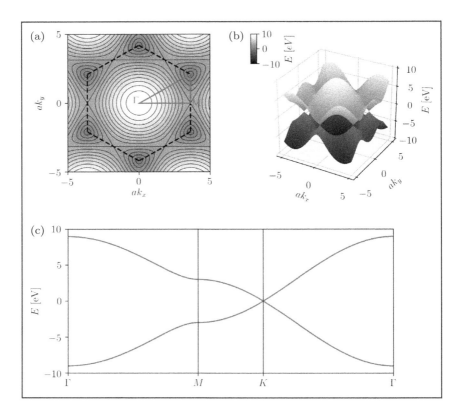

(a) Contour plot of the conduction band of graphene. The dashed line outlines the first Brillouin zone. (b) Plot of the π-energy band dispersion, obtained with the tight-binding calculation. (c) Linecut of the π-bands along the $\Gamma - M - K - \Gamma$ line.

which yields

$$|f(\mathbf{k})| = \sqrt{1 + 2e^{-iak_x\sqrt{3}/2}\cos\left(\frac{ak_y}{2}\right) + 2e^{iak_x\sqrt{3}/2}\cos\left(\frac{ak_y}{2}\right) + 4\cos^2\left(\frac{ak_y}{2}\right)}$$

$$= \sqrt{1 + 4\cos\left(\frac{ak_x\sqrt{3}}{2}\right)\cos\left(\frac{ak_y}{2}\right) + 4\cos^2\left(\frac{ak_y}{2}\right)}$$

The previous figure contains a visualization of this tight-binding result, computed with $|\gamma| = 3$ eV. The conduction and valence bands touch at the $\pm K$-points of the first Brillouin zone. The low-energy dispersion relation is linear, forming a conical dispersion in two dimensions — the so-called **Dirac cones**.

23.3   The sp^2-bands at the Γ-point within the Tight-Binding Model: Symmetry Analysis

23.3.1   *Part A: The labeling of energy levels with irreducible representations*

The atomic orbitals s, p_x, p_y hybridize to form directed bonds residing at the A and B sites of the graphene lattice. At A, we find the three directed sp_2 orbitals denoted by A_0, A_1, A_2. At B, the three sp_2 orbitals are denoted B_0, B_1, B_2. The relations between the "A" and "B" orbitals with the atomic s, p_x, and p_y orbitals are, respectively, given by the equations

$$|A_0\rangle = \frac{1}{\sqrt{3}}|s\rangle_A - \frac{1}{\sqrt{2}}|p_x\rangle_A + \frac{1}{\sqrt{6}}|p_y\rangle_A$$

$$|A_1\rangle = \frac{1}{\sqrt{3}}|s\rangle_A + \frac{1}{\sqrt{2}}|p_x\rangle_A + \frac{1}{\sqrt{6}}|p_y\rangle_A$$

$$|A_2\rangle = \frac{1}{\sqrt{3}}|s\rangle_A - \frac{2}{\sqrt{6}}|p_y\rangle_A$$

$$|B_0\rangle = \frac{1}{\sqrt{3}}|s\rangle_B + \frac{2}{\sqrt{6}}|p_y\rangle_B$$

$$|B_1\rangle = \frac{1}{\sqrt{3}}|s\rangle_B + \frac{1}{\sqrt{2}}|p_x\rangle_B - \frac{1}{\sqrt{6}}|p_y\rangle_B$$

$$|B_2\rangle = \frac{1}{\sqrt{3}}|s\rangle_B - \frac{1}{\sqrt{2}}|p_x\rangle_B - \frac{1}{\sqrt{6}}|p_y\rangle_B$$

and the resulting bond direction is schematically represented by arrows in the following figure.

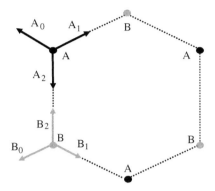

Bond direction of hybrid A and B orbitals.

In the absence of overlapping, these states are degenerate and belong to the same atomic energy level. When the graphene lattice is formed, these atomic-like $sp2$-orbitals overlap, and in the spirit of the tight-binding model, a band structure develops, with energy levels distributed over the Brillouin zone. Symmetry arguments can be used, in the first place, to label the bands that develop from the sixfold degenerate atomic level atomic level according to the irreducible representations of the suitable group of $\mathbf{k}$. We perform this symmetry analysis, as an example, for the Γ-point.

We include, for completeness, the p_z-orbitals that we have already separately analyzed in the previous section. Together with the p_{zA}- and p_{zB}-basis functions, the A-s and B-s orbitals constitute a eight-dimensional representation of the group of the Γ-point, i.e. D_{6h}. Let us call this representation Γ^8. This representation will split into irreducible components that will finally label the desired energy levels at the Γ point. To find the irreducible components, we need the matrices of the representatives of the conjugacy classes of D_{6h} in the representation Γ^8, so that we can obtain the characters of Γ^8.

As an example of the computation of these matrices, we construct explicitly the matrix $\Gamma^8(I)$. We order the basis states as given here:

$$(A_0 \quad A_1 \quad A_2 \quad B_0 \quad B_1 \quad B_2 \quad \Phi_A \quad \Phi_B)$$

Upon inversion, A_0 is transformed into a B_1 orbital residing at the B site opposite the A site. This places a "1" at the fifth matrix element along the first column of the 8×8 matrix and a "0" elsewhere along this column. A_1 is mapped onto a B_0 at the opposite B site. This produces a "1" at the fourth matrix element of the second column and a "0" elsewhere along the second column. A_2 is mapped onto a B_2 orbital at the B site: This places a "1 at the sixth matrix element of the third column and a "0" elsewhere along this column. The block of the matrix representing the two p_z orbitals has already been computed in the previous section about the π-bands of graphene. In general, we observe that the p_z orbitals are transformed into themselves, so that the 8×8 matrices consist of six-dimensional and two-dimensional, separated blocks. Finally, we obtain

$$\Gamma^8(I) = \begin{pmatrix} 0 & 0 & 0 & 0 & 1 & 0 & 0 & 0 \\ 0 & 0 & 0 & 1 & 0 & 0 & 0 & 0 \\ 0 & 0 & 0 & 0 & 0 & 1 & 0 & 0 \\ 0 & 1 & 0 & 0 & 0 & 0 & 0 & 0 \\ 1 & 0 & 0 & 0 & 0 & 0 & 0 & 0 \\ 0 & 0 & 1 & 0 & 0 & 0 & 0 & 0 \\ 0 & 0 & 0 & 0 & 0 & 0 & 0 & -1 \\ 0 & 0 & 0 & 0 & 0 & 0 & -1 & 0 \end{pmatrix}$$

The entire set of matrices was computed for this project and is available on request. We summarize in the following table their characters.

	E	C_6	C_3	C_2	C_2'	C_2''	I	S_3	S_6	σ_h	σ_d	σ_v
Γ^8	8	0	2	0	0	0	0	−2	0	4	4	0

Using the character table of D_{6h}, we find

$$\Gamma^8 = A_{1g} \oplus B_{1g} \oplus E_{2g} \oplus A_{2u} \oplus B_{2u} \oplus E_{1u}$$

The following table provides a check of the validity of the decomposition.

	E	C_6	C_3	C_2	C_2'	C_2''	I	S_3	S_6	σ_h	σ_d	σ_v
A_{1g}	1	1	1	1	1	1	1	1	1	1	1	1
B_{1g}	1	−1	1	−1	1	−1	1	−1	1	−1	1	−1
E_{2g}	2	−1	−1	2	0	0	2	−1	−1	2	0	0
A_{2u}	1	1	1	1	−1	−1	−1	−1	−1	−1	1	1
B_{2u}	1	−1	1	−1	−1	1	−1	1	−1	1	1	−1
E_{1u}	2	1	−1	−2	0	0	−2	−1	1	2	0	0
sum:	8	0	2	0	0	0	0	−2	0	4	4	0

This symmetry analysis tells us that the Γ-point sustains two single degenerate energy levels and two double degenerate energy levels, hosting the electrons originating from the sp^2 orbitals. A sketch of the symmetry results obtained for the sp^2 sector is shown in the following figure.

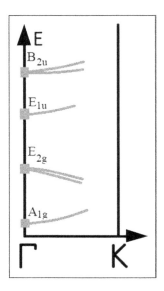

Sketch of the band structure of the in-plane sp^2 electrons in graphene. The ordering is according to the quantitative results by F. Bassani and G. Pastori-Parravicini, Il Nuovo Cimento, **L**, B, 95 (1967).

For a realistic band structure of graphene, see, for example, E. Kogan, and V. U. Nazarov, *Phys. Rev. B* **85**, 115418 (2012), https://journals.aps.org/prb/abstract/10.1103/PhysRevB.85.115418.

23.3.2 Part B: Symmetry-adapted tight-binding orbitals at the Γ-point

Using the matrix representations for the sp^2-block and knowing the characters of the irreducible representations, one can construct the projector matrices:

$$
P_{A_{1g}} = \frac{1}{6}
\begin{pmatrix}
1 & 1 & 1 & 1 & 1 & 1 \\
1 & 1 & 1 & 1 & 1 & 1 \\
1 & 1 & 1 & 1 & 1 & 1 \\
1 & 1 & 1 & 1 & 1 & 1 \\
1 & 1 & 1 & 1 & 1 & 1 \\
1 & 1 & 1 & 1 & 1 & 1
\end{pmatrix}
\qquad
P_{B_{2u}} = \frac{1}{6}
\begin{pmatrix}
1 & 1 & 1 & -1 & -1 & -1 \\
1 & 1 & 1 & -1 & -1 & -1 \\
1 & 1 & 1 & -1 & -1 & -1 \\
-1 & -1 & -1 & 1 & 1 & 1 \\
-1 & -1 & -1 & 1 & 1 & 1 \\
-1 & -1 & -1 & 1 & 1 & 1
\end{pmatrix}
$$

$$PE_{1u} = \frac{1}{12} \begin{pmatrix} 2 & -1 & -1 & 1 & -2 & 1 \\ -1 & 2 & -1 & -2 & 1 & 1 \\ -1 & -1 & 2 & 1 & 1 & -2 \\ 1 & -2 & 1 & 2 & -1 & -1 \\ -2 & 1 & 1 & -1 & 2 & -1 \\ 1 & 1 & -2 & -1 & -1 & 2 \end{pmatrix}$$

$$PE_{2g} = \frac{1}{12} \begin{pmatrix} 2 & -1 & -1 & -1 & 2 & -1 \\ -1 & 2 & -1 & 2 & -1 & -1 \\ -1 & -1 & 2 & -1 & -1 & 2 \\ -1 & 2 & -1 & 2 & -1 & -1 \\ 2 & -1 & -1 & -1 & 2 & -1 \\ -1 & -1 & 2 & -1 & -1 & 2 \end{pmatrix}$$

To determine the symmetry-adapted vectors (i.e. the symmetry-adapted linear combinations of the "A" and "B" orbitals), one can either apply the projector matrices to a trial vector or determine the eigenvectors of the projector matrices to eigenvalues other than 0, using a suitable software (here, we used Mathematica). We find

$$v_{A_{1g}} = (1,1,1,1,1,1) \qquad v_{B_{2u}} = (\bar{1}, \bar{1}, \bar{1}, 1, 1, 1)$$

$$v_{E_{1u},1} = (\bar{1}, 1, 0, \bar{1}, 1, 0) \qquad v_{E_{1u},2} = (0, 1, \bar{1}, \bar{1}, 0, 1)$$

$$v_{E_{2g},1} = (1, \bar{1}, 0, \bar{1}, 1, 0) \qquad v_{E_{2g},2} = (0, \bar{1}, 1, \bar{1}, 0, 1)$$

In terms of the atomic orbitals $\{|s\rangle_A, |p_x\rangle_A, |p_y\rangle_A, |s\rangle_B, |p_x\rangle_B, |p_y\rangle_B\}$, we obtain

$$v_{A_{1g}} = (1,0,0,1,0,0) \qquad v_{B_{2u}} = (\bar{1},0,0,1,0,0)$$

$$v_{E_{1u},1} = \left(0, 2, 0, 0, 1, \overline{\sqrt{3}}\right) \qquad v_{E_{1u},2} = \left(0, 1, \sqrt{3}, 0, \bar{1}, \overline{\sqrt{3}}\right)$$

$$v_{E_{2g},1} = \left(0, \bar{2}, 0, 0, 1, \overline{\sqrt{3}}\right) \qquad v_{E_{2g},2} = \left(0, \bar{1}, \overline{\sqrt{3}}, 0, \bar{1}, \overline{\sqrt{3}}\right)$$

Within a perturbational approach, the Bloch sums of either of these two sets are bound to diagonalize a realistic H matrix describing the in-plane sector of the electronic structure at the Γ-point.[2] The following figure plots the charge density of these states.

---

[2] The Bloch sums just translate the symmetry-adapted orbitals to the entire lattice.

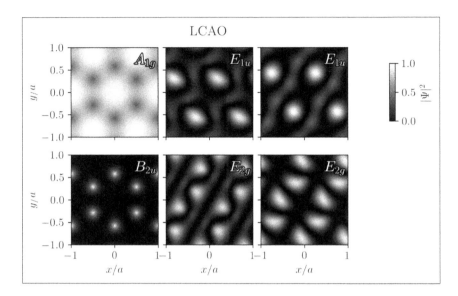

Contour plots of the charge density (the square of the absolute value of the wave function) of the symmetry-adapted wave functions generated using the tight-binding approach.

23.4   In-plane Band Structure with Symmetrized Plane Waves

23.4.1   *Part A: Symmetry analysis of the energy levels at the Γ-point*

In this section, we approach the computation of the energy levels using the empty-lattice model. The energy levels are generated by

$$E_{\mu\upsilon} = \frac{\hbar^2}{2m}(\mathbf{k} + \mathbf{G}_{\mu\upsilon})^2$$

$\mathbf{G}_{\mu\upsilon}$ are the reciprocal lattice vectors. Referring to the following figure, we choose as basis vectors for the hexagonal reciprocal lattice the vectors

$$\mathbf{b_1} = \frac{2\pi}{a}\left(\frac{\sqrt{3}}{3}\mathbf{e_x} - \frac{1}{3}\mathbf{e_y}\right) \quad \mathbf{b_2} = \frac{2\pi}{a}\left(\frac{2}{3}\mathbf{e_y}\right)$$

so that

$$\mathbf{G}_{\mu\upsilon} = n \cdot \mathbf{b_1} + m \cdot \mathbf{b_2}$$

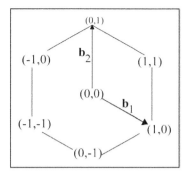

The basis vectors of the reciprocal lattice and the reciprocal lattice vectors used to generate the lowest-energy empty-lattice energy bands, labeled by their coordinates (n, m).

We focus on the lowest-energy levels at the Γ-point, i.e. we set $\mathbf{k} = 0$.

- The lowest empty-lattice energy level is

$$E_{0,0} = 0$$

  Its wave function is

$$|00\rangle = \frac{1}{\sqrt{N}a}$$

  and transforms according to A_{1g} of D_{6h}.

- The next level is sixfold degenerate, provided by the six reciprocal lattice vectors building the hexagon pictured in the previous figure:

$$E_{\bar{1}0} = E_{\bar{1}\bar{1}} = E_{10} = E_{11} = E_{0\bar{1}} = E_{01} = \frac{8\pi^2\hbar}{9ma^2}$$

The basis states hosted by this level are

$$|01\rangle = \frac{1}{\sqrt{N}a}e^{\frac{2\pi i}{a}\frac{2}{3}y}$$

$$|11\rangle = \frac{1}{\sqrt{N}a}e^{\frac{2\pi i}{a}\left(\frac{\sqrt{3}}{3}x+\frac{1}{3}y\right)}$$

$$|10\rangle = \frac{1}{\sqrt{N}a}e^{\frac{2\pi i}{a}\left(\frac{\sqrt{3}}{3}x-\frac{1}{3}y\right)}$$

$$|0\bar{1}\rangle = \frac{1}{\sqrt{N}a}e^{-\frac{2\pi i}{a}\frac{2}{3}y}$$

$$|\bar{1}\bar{1}\rangle = \frac{1}{\sqrt{N}a} e^{\frac{2\pi i}{a}\left(-\frac{\sqrt{3}}{3}x - \frac{1}{3}y\right)}$$

$$|\bar{1}0\rangle = \frac{1}{\sqrt{N}a} e^{\frac{2\pi i}{a}\left(-\frac{\sqrt{3}}{3}x + \frac{1}{3}y\right)}$$

These basis states carry a six-dimensional representation of the group D_{6h} — let us call it Γ^6. We compute their corresponding matrices with the aim of obtaining the characters (and later the projector matrices). We do not reproduce the 24 matrices explicitly here, for space considerations. As an example, we seek the matrix representation $\Gamma^6(I)$. We refer to the previous figure and assume that the basis functions are ordered as

$$\{|01\rangle, |11\rangle, |10\rangle, |0\bar{1}\rangle, |\bar{1}\bar{1}\rangle, |\bar{1}0\rangle\}$$

Upon inversion, the basis state $|01\rangle$ is mapped onto $|0\bar{1}\rangle$. This places a "1" in the fourth matrix element of the first column and a "0" elsewhere in the column. The ket $|11\rangle$ is mapped onto $|\bar{1}\bar{1}\rangle$. This places a "1" in the fifth matrix element of the second column and a "0" elsewhere in the column. This process is repeated. Finally, we have

$$\Gamma^6(I) = \begin{pmatrix} 0 & 0 & 0 & 1 & 0 & 0 \\ 0 & 0 & 0 & 0 & 1 & 0 \\ 0 & 0 & 0 & 0 & 0 & 1 \\ 1 & 0 & 0 & 0 & 0 & 0 \\ 0 & 1 & 0 & 0 & 0 & 0 \\ 0 & 0 & 1 & 0 & 0 & 0 \end{pmatrix}$$

The characters are summarized in the following table.

	E	C_6	C_3	C_2	C_2'	C_2''	I	S_3	S_6	σ_h	σ_d	σ_v
Γ^6	6	0	0	0	0	2	0	0	0	6	2	0

Using these characters, we find

$$\Gamma^6 = A_{1g} \oplus \oplus B_{2u} \oplus E_{1u} \oplus E_{2g}$$

The sixfold degenerate empty-lattice energy level at Γ splits, when a finite crystal potential is switched on, into four separate levels, corresponding to the decomposition of Γ^6. Their ordering is reported in the section on the tight-binding model. The symmetry-adapted vectors can be computed with the projector technique, using the representation matrices. For example,

$$
P_{A_{1g}} = \frac{1}{6}
\begin{pmatrix}
1 & 1 & 1 & 1 & 1 & 1 \\
1 & 1 & 1 & 1 & 1 & 1 \\
1 & 1 & 1 & 1 & 1 & 1 \\
1 & 1 & 1 & 1 & 1 & 1 \\
1 & 1 & 1 & 1 & 1 & 1 \\
1 & 1 & 1 & 1 & 1 & 1
\end{pmatrix}.
$$

and

$$
P_{A_{1g}} |01\rangle = \frac{1}{6} \left[|01\rangle + |11\rangle + |10\rangle + |0\bar{1}\rangle + |\bar{1}\bar{1}\rangle + |\bar{1}0\rangle \right]
$$

$$
\propto \frac{1}{6\sqrt{N}a} \left[e^{\frac{2\pi i}{a} \frac{2}{3} y} + e^{\frac{2\pi i}{a} (\frac{\sqrt{3}}{3} x + \frac{1}{3} y)} + e^{\frac{2\pi i}{a} (\frac{\sqrt{3}}{3} x - \frac{1}{3} y)} \right.
$$

$$
\left. + e^{-\frac{2\pi i}{a} \frac{2}{3} y} + e^{\frac{2\pi i}{a} (-\frac{\sqrt{3}}{3} x - \frac{1}{3} y)} + e^{\frac{2\pi i}{a} (-\frac{\sqrt{3}}{3} x + \frac{1}{3} y)} \right]
$$

$$
= \frac{1}{3\sqrt{N}a} \left[\cos\left(\frac{2\pi}{a} \left(\frac{\sqrt{3}}{3} x + \frac{1}{3} y \right) \right) + \cos\left(\frac{2\pi}{a} \left(\frac{\sqrt{3}}{3} x - \frac{1}{3} y \right) \right) \right.
$$

$$
\left. + \cos\left(\frac{2\pi}{a} \frac{2}{3} y \right) \right]
$$

In summary,

$$
P^{A_{1g}} |11\rangle \propto \cos\left(\frac{2\pi}{a} \left(\frac{\sqrt{3}}{3} x + \frac{1}{3} y \right) \right) + \cos\left(\frac{2\pi}{a} \left(\frac{\sqrt{3}}{3} x - \frac{1}{3} y \right) \right)
$$

$$
+ \cos\left(\frac{2\pi}{a} \frac{2}{3} y \right)
$$

$$
P^{B_{2u}} |11\rangle \propto \sin\left(\frac{2\pi}{a} \left(\frac{\sqrt{3}}{3} x + \frac{1}{3} y \right) \right) - \sin\left(\frac{2\pi}{a} \left(\frac{\sqrt{3}}{3} x - \frac{1}{3} y \right) \right)
$$

$$
- \sin\left(\frac{2\pi}{a} \frac{2}{3} y \right)
$$

$$
P^{E_{1u}} |11\rangle \propto 2\sin\left(\frac{2\pi}{a} \left(\frac{\sqrt{3}}{3} x + \frac{1}{3} y \right) \right) + \sin\left(\frac{2\pi}{a} \left(\frac{\sqrt{3}}{3} x - \frac{1}{3} y \right) \right)
$$

$$
+ \sin\left(\frac{2\pi}{a} \frac{2}{3} y \right)
$$

$$P^{E_{1u}}\,|10\rangle \propto \sin\left(\frac{2\pi}{a}\left(\frac{\sqrt{3}}{3}x+\frac{1}{3}y\right)\right)+2\sin\left(\frac{2\pi}{a}\left(\frac{\sqrt{3}}{3}x-\frac{1}{3}y\right)\right)$$

$$-\sin\left(\frac{2\pi}{a}\frac{2}{3}y\right)$$

$$P^{E_{2g}}\,|11\rangle \propto 2\cos\left(\frac{2\pi}{a}\left(\frac{\sqrt{3}}{3}x+\frac{1}{3}y\right)\right)-\cos\left(\frac{2\pi}{a}\left(\frac{\sqrt{3}}{3}x-\frac{1}{3}y\right)\right)$$

$$-\cos\left(\frac{2\pi}{a}\frac{2}{3}y\right)$$

$$P^{E_{2g}}\,|10\rangle \propto -\cos\left(\frac{2\pi}{a}\left(\frac{\sqrt{3}}{3}x+\frac{1}{3}y\right)\right)+2\cos\left(\frac{2\pi}{a}\left(\frac{\sqrt{3}}{3}x-\frac{1}{3}y\right)\right)$$

$$-\cos\left(\frac{2\pi}{a}\frac{2}{3}y\right)$$

The following figure plots the charge density of the symmetry-adapted linear combinations of plane waves.

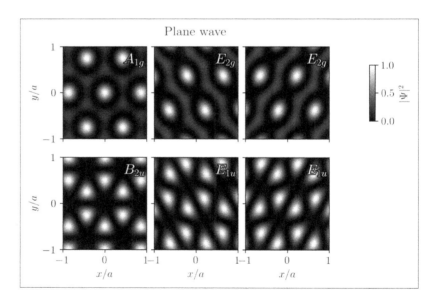

Plot of the charge densities generated by the symmetry-adapted plane waves.

24

PROJECT FOR LECTURE 6:
TOPOLOGICAL PROTECTION BY
NON-SYMMORPHIC DEGENERACY

24.1 Introduction

In this project, we use the Herring method to analyze the electronic band structure in crystals with non-symmorphic space groups and find the specific ("topologically protected") electronic properties that are produced by the existence of symmetry elements with fractional translations. For the sake of illustration, we start with a two-dimensional crystal consisting of identical atoms arranged in a rectangular lattice. The lattice vectors are $\mathbf{a}_1 = a \cdot (1,0)$ and $\mathbf{a}_2 = b \cdot (0,1)$. We set the origin of our coordinate system at the center of the rectangle. The translational symmetry elements are described by

$$\mathbf{t} = t_1 \mathbf{a}_1 + t_2 \mathbf{a}_2$$

(where t_1 and t_2 are integers). The basis vectors of the reciprocal space read

$$\mathbf{b}_1 = \frac{2\pi}{a}(1,0) \quad \mathbf{b}_2 = \frac{2\pi}{b}(0,1)$$

i.e. the Brillouin zone is also rectangular.

There are twofold rotation axes through the centers of the rectangles and further twofold rotation axes through the corners. There are reflections at lines intersecting the centers of the rectangles and parallel to their sides and further reflections lines parallel to the sides of the rectangle. Accordingly, the point group of the pattern is given by the four elements $\{\mathbb{1}, C_2, m_x, m_y\}$ and is the group C_{2v}. The space group is $p2mm$.

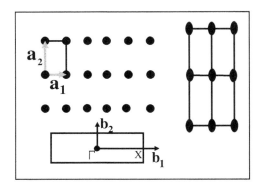

Top left: The rectangular lattice. Indicated are the basis vectors and the unit cell. Top right: International symbols for the group *p2mm*, summarizing all symmetry elements present in the unit cell. Bottom: The first Brillouin zone. The $\Gamma - X$-direction is indicated.

We focus our interest on the band structure along the $\Gamma - X \equiv \Delta = \frac{\pi}{a} \cdot (\delta, 0)$-direction, with $X = \frac{\pi}{a} \cdot (1, 0)$ being at the edge of the Brillouin zone. The general expression of the empty-lattice bands is

$$E = \frac{\hbar^2}{2m} \cdot (\mathbf{k} + n \cdot \mathbf{b_1} + m \cdot \mathbf{b_2})^2$$

For the specific line considered, the bands read

$$E_\Delta = \frac{\hbar^2}{2m} \cdot \left[\frac{\pi}{a} \cdot (\delta + 2n)^2 + \left(\frac{2\pi}{b} \cdot m \right)^2 \right]$$

This formula is applied to, for instance, the X-point itself ($\delta = 1$). The irreducible representations of the group of X are identical with the irreducible representations of the point group C_{2v}, and the energy levels are labeled accordingly. The lowest empty-lattice energy level at X ($n = 0$, $m = 0$ or $n = -1, m = 0$) is twice degenerate. Its degeneracy will be removed upon introducing a finite crystal potential and the level splits into two levels that each carry some irreducible representations of the group of X. The empty lattice band structure along the Delta-direction is given in the next figure.

24.2 The Space Group p2mg

One can obtain a crystal with the non-symmorphic space group p2mg by, for example, buckling down each second atom along horizontal rows, see the following figure.

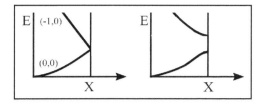

Left: Empty-lattice band structure with the two lowest energy bands along the $\Gamma - X$-direction. The band indices corresponding to the (n, m) values are indicated. Right: A finite crystal potential is introduced, and a gap opens at the X-point.

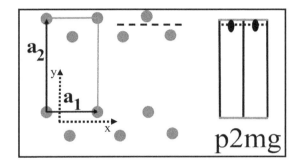

Left: Lattice with buckled-down atoms. The unit cell is indicated. Right: The symmetry elements in the unit cell, summarized using international symbols. Note the glide line (dashed).

The fundamental symmetry elements, described with respect to the coordinate system defined in the figure, read

$$e = \{\mathbb{1}|0\}$$
$$c = \{C_2|0\}$$
$$m = \left\{m_y \Big| \frac{a}{2} \cdot (1, 0)\right\}$$
$$g = \left\{m_x \Big| \frac{a}{2} \cdot (1, 0)\right\}$$

g is the glide reflection that renders the group non-symmorphic. The Brillouin zone is identical to the Brillouin zone of the p2mm lattice.

The non-symmorphicity has important consequences at the border of the Brillouin zone. We focus therefore on one of the points along the border, the X-point. C_2 and m_y transform the X into $X - \frac{2\pi}{a}(1, 0)$, i.e. into an equivalent X-point. The small point group of X is therefore

$$\{\mathbb{1}, C_2, m_x, m_y\}$$

but the group of X contains the fractional translation associated with the glide reflection. One could try to construct a representation of the elements of the group of X by taking the fractional translation into account through an exponential factor (which is certainly possible **inside** the Brillouin zone):

$$T^X(g) = T(m_x) \cdot e^{i\frac{\pi}{a}(1,0)\cdot\frac{a}{2}(1,0)} = T(m_x) \cdot e^{i\frac{\pi}{2}}$$

with $T(m_x)$ being the point group representation of m_x. However, this is not a viable choice. For instance,

$$\{C_2|\mathbf{0}\}\left\{m_x|\frac{a}{2}(1,0)\right\} = \left\{C_2 m_x|\frac{a}{2}\cdot(-1,0)]\right\}$$

is a possible operation between elements of the group of X. Assuming that we represent the elements with the exponential pre-factor, we have on the left-hand side

$$T(C_2) \cdot T(m_x) \cdot e^{i\frac{\pi}{2}}$$

On the right-hand side, we have

$$T(C_2) \cdot T(m_x) \cdot e^{-i\frac{\pi}{2}}$$

This result contradicts the rule of homomorphism and shows that the multiplication with the exponential pre-factor does not lead to a representation of the elements of the group of X. The Herring method provides a remedy for this issue. In this method, a suitable group, more manageable than the entire $G(X)$, is constructed. Some of the irreducible representations of this more manageable group can be used as irreducible representations of the group $G(X)$. For the specific case, the Herring method identifies the group of translations $T(X) \equiv \{\mathbb{1}|a\cdot(2\cdot n, 0)\}$ as an invariant subgroup that can be used to build the factor group $\frac{G(X)}{T(X)}$. The factor group consists of the following eight elements (equivalence classes):

$$e' = \{\mathbb{1}|2na\cdot(1,0)\}$$
$$e'' = \{\mathbb{1}|(a+2na)\cdot(1,0)\}$$
$$c' = \{C_2|(2na\cdot(1,0)\}$$

$$c'' = \{C_2 | (a + 2na) \cdot (1,0)\}$$

$$m' = \left\{ m_x \Big| \left(\frac{a}{2} + 2na\right) \cdot (1,0) \right\}$$

$$m'' = \left\{ m_x \Big| \left(\frac{3a}{2} + 2na\right) \cdot (1,0) \right\}$$

$$g' = \left\{ m_y \Big| \left(\frac{a}{2} + 2na\right) \cdot (1,0) \right\}$$

$$g'' = \left\{ m_y \Big| \left(\frac{3a}{2} + 2na\right) \cdot (1,0) \right\}$$

The question now is to identify this group as, possibly, a known one and to find its irreducible representations. The starting point is the following multiplication table, which summarizes the results of the product of elements.

Multiplication table for the factor group $\frac{G(X)}{T(X)}$.

	e'	e''	c'	c''	m'	m''	g'	g''
e'	e'	e''	c'	c''	m'	m''	g'	g''
e''	e''	e'	c''	c'	m''	m'	g''	g'
c'	c'	c''	e'	e''	g''	g'	m''	m'
c''	c''	c'	e''	e'	g'	g''	m'	m''
m'	m'	m''	g'	g''	e''	e'	c''	c'
m''	m''	m'	g''	g'	e'	e''	c'	c''
g'	g'	g''	m'	m''	c'	c''	e'	e''
g''	g''	g'	m''	m'	c''	c'	e''	e'

One recognizes that the factor group is isomorphic to C_{4v}, so that one can compile the following character table: We recognize five irreducible representations. According to the Herring method, only those that fulfills the Bloch condition are physical ones. This is the case only for Γ_5:

$$\chi_5(e'') = \chi_5(e') \cdot e^{-i\frac{\pi}{a}(1,0)\cdot(a+2na)\cdot(1,0)} = -\chi_5(e)$$

Character table for the factor group $\frac{G(X)}{T(X)}$.

	e'	e''	(c',c'')	(m',m'')	(g',g'')
Γ_1	1	1	1	1	1
Γ_2	1	1	1	$\bar{1}$	$\bar{1}$
Γ_3	1	1	$\bar{1}$	1	$\bar{1}$
Γ_4	1	1	$\bar{1}$	$\bar{1}$	1
Γ_5	2	$\bar{2}$	0	0	0

This is a two-dimensional representation. As a consequence, any energy level at the X-point must necessarily be twice degenerate on the grounds of symmetry. This is referred to as **non-symmorphic degeneracy**. For the specific case of the lowest level treated in the example of the p2mm-space group, we observe that the empty-lattice accidental degeneracy at X is replaced by a non-symmorphic, symmetry-induced degeneracy when a finite crystal potential is switched on: The original degeneracy is **topologically protected**.

Symmetry-adapted wave functions to Γ_5 can be constructed starting from, say, $e^{\pm i \frac{\pi}{a} \cdot x}$:

$$e' e^{\pm i \frac{\pi}{a} \cdot x} - \bar{e}' e^{\pm i \frac{\pi}{a} \cdot x} \propto e^{\pm i \frac{\pi}{a} \cdot x} - e^{\pm i \frac{\pi}{a} \cdot (x+a)} \propto e^{\pm i \frac{\pi}{a} \cdot x}$$

They build the basis states to the lowest energy once degenerate bands along the Δ direction and to the twice degenerate lowest-energy level at X. Without the fractional translation, the plane waves with indices $(0, 0)$ and $(-1, 0)$, originating from the separate bands along the $\Gamma - X$-direction, would entangle at X to form the usual bonding and antibonding states that produce the formation of a gap. The non-symmorphicity prevents the entanglement and, accordingly, the opening of the gap (one could call this a phenomenon of "disentanglement").

The gapless ending of the bands at the X-point has a profound impact on the band dispersion in the vicinity of X. When a gap opens, the separation of bands from the gapped energy levels follows a quadratic relationship with respect to the distance from the edge point. One can convince oneself that the non-symmorphic degeneracy entails (using the first-order perturbation theory) the appearance of a Dirac cone in the vicinity of X, i.e. a linear dependence of the band energy as a function of the distance (in k-space) from X. For this purpose, we compute a simple model based on the empty-lattice band structure. We operate within the space built by the basis functions $\{e^{i \cdot \frac{\delta}{a} \cdot x}, e^{i \cdot (\frac{\delta}{a} - \frac{2\pi}{a}) \cdot x}\}$. These basis functions describe the lowest empty-lattice bands in the vicinity of the X point. We take the Hamilton operator of the empty lattice and solve its determinantal equation along the $\Gamma - X$ under two conditions. In the first one, we let the basis functions entangle by imposing a non-vanishing matrix element A between them. In this case, the energy eigenvalues are the solutions of the determinantal equation of the matrix

$$\begin{pmatrix} \frac{\hbar^2}{2m} \left(\frac{\delta}{a} \right)^2 - E & A \\ A & \frac{\hbar^2}{2m} \left(\frac{\delta}{a} - \frac{2\pi}{a} \right)^2 - E \end{pmatrix}$$

A is the strength of the matrix element that produces entanglement. The following figure plots the energy as a function of δ in the interval $\delta = [2.5, \pi]$ (the value π identifying the X-point, and we set $\frac{\hbar^2}{2m}$ and a to 1).

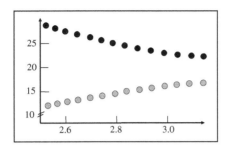

Solution of the determinantal equation for finite A.

The quadratic behavior of the energy bands when the bands approach the value of π is observed numerically.

The second condition involves solving the determinantal equation for $A = 0$. The following figure plots the bands in the vicinity of π for $A = 0$. The linearity of the dispersion when approaching the point π is numerically evident.

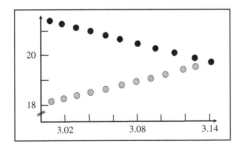

Solution of the determinantal equation for $A = 0$.

24.3   The Space Group D_{6h}^4 ($P6_3/mmc$)

Graphene monolayers evolve into a three-dimensional (3D) crystal through stacking. The simplest way of performing the stacking is to stack a further graphene honeycomb lattice on top of an existing one in such a way that the atoms in the two consecutive layers are perfectly aligned (called as AA stacking). The AA-stacking produces the symmorphic space group $P6/mmm$. A second way is called AB stacking, where the next layer is obtained by a rotation with respect to the underlying one by $60°$ and a translation by $\frac{c}{2}$, with c being the perpendicular lattice constant. The AB-stacking produces the non-symmorphic crystal with the space group D_{6h}^4 ($P6_3/mmc$); for a comprehensive treatment of this space group, see M. Schlüter, "Symmetry considerations", in *Electrons and Phonons in Layered Crystal Structures*, Editors: T. J. Wieting and M. Schlüter (1979). *Physics and Chemistry of Materials with Layered Structures*, Vol. 3, Springer, Dordrecht, https://doi.org/10.1007/978-94-009-9370-9_1. The band structure of the AB-stacked crystal is computed, for example, in E. Doni and G. Pastori Parravicini, *Il Nuovo Cimento*, **LXIV B**, 117–144 (1969), https://link.springer.com/article/10.1007/BF02710286. The bands of both AA and AB stacked bylayers are computed in N. V. Petrova and I. N. Yakovkin, *Surf. Rev. Lett.* **24**, 1750020 (2017), https://www.worldscientific.com/doi/10.1142/S0218625X17500202. Both papers identify a non-symmorphic degeneracy in the energy bands describing the p_z-orbitals perpendicular to the AB-stacking. This degeneracy occurs at the K-point of the 3D Brillouin zone. The degeneracy is lifted in the AA-stacked band structure.

25

PROJECT FOR LECTURE 7: TOPOLOGICAL ASPECTS OF CONTINUOUS GROUPS AND THE UNIVERSAL COVERING GROUPS

The relationship between SO_3 and SO_3^D was found on the basis of the necessity of taking into account half-integer Cartan weights. In mathematics, however, there is a more general framework that establishes SO_3^D as an example of what is known as a "universal covering group". This framework uses some tools of algebraic topology.

25.1 Connectivity of Groups

Continuous groups are characterized by their elements being a function of some parameters. These parameters span a so-called parameter space. For instance, the parameter space of the group SO_2 can be represented as points lying on a unit circle. These points identify the rotation angle φ. The topological aspects of the parameter space, i.e. its behavior under continuous deformations, can be translated to the continuous group, allowing us to define the topological aspects of the continuous group. For example, one of these properties that we quoted, referring to the group average, is the property of compactness. A group is said to be **compact** if the parameter space is bounded and closed (a set is closed if every Cauchy sequence of elements of the set has a limit element which also belongs to the set). For example, the axial group C_∞ is a compact group, as the parameter space is a compact set (the unit circle in the plane). The translation group, instead, is not compact, as the parameter space is unbounded.

A further topological feature of groups which plays an important role is their connectivity. In order to define this concept, we need to study the parameter space using the methods of algebraic topology.

25.1.1   *Some concepts of algebraic topology*

Algebraic topology studies the behavior of a set under continuous mappings. The set we have in mind is the parameter space S_r. One important concept of algebraic topology is that of a **loop**. Consider one point x_0 in the parameter space, and let $\alpha(t)$ be a continuous map of $t \in [0, 1]$ onto S_r with

$$\alpha(0) = x_0 \quad ; \quad \alpha(1) = x_0$$

Definition ("loop"). The continuous function $\alpha(t)$ is called a loop with the base point x_0.

Example: The map $\alpha(\varphi) = e^{i\varphi}$ with $\varphi \in [0, 2\pi]$ associates to every point in the interval $[0, 2\pi]$ a point lying on a circle in the complex plane. The base point is the point $1 + i \cdot 0$.

Within a parameter space, one can think of many loops which start from one single point. However, we would like to define an operation that allows us to relate them, so that we can choose one single loop as a representative. This operation might be considered a deformation of the various loops into each other, so that the base point remains fixed. This operation of "deformation" will be made more precise by the following definition.

Definition ("homotopy"). Two loops α_0, α_1 with a base point x_0 are said to be **homotopic**

$$\alpha_0 \sim \alpha_1$$

if there is a continuous map

$$F : [0, 1] \times [0, 1] \mapsto S_r$$

such that

$$F(t, 0) = \alpha_0(t), \quad t \in [0, 1]$$
$$F(t, 1) = \alpha_1(t), \quad t \in [0, 1]$$
$$F(0, s) = F(1, s) = x_0, \quad s \in [0, 1]$$

We can say that $\alpha_0(t)$ and $\alpha_1(t)$ are related via the homotopy $F(t, s)$.

Examples:

1. The map

$$F(\varphi, r) = r \cdot e^{i \cdot \varphi} + (1 - r)$$

shrinks the unit circle ($r = 1$) to the point $1 + i \cdot 0$ ($r = 0$). Actually, any loop in the complex plane with the base point $1 + i \cdot 0$ is homotopic to $1 + i \cdot 0$.

2. In general, the family of loops given by

$$F(s, t) = \alpha_s(t) = (1 - s)\alpha_0(t) + s\alpha_1(t)$$

deforms continuously from $\alpha_0(t)$ to $\alpha_1(t)$.

The homotopy relation is an equivalence relation because it is reflexive (α_0 is homotopic to itself), symmetric (if α is homotopic to β via the homotopy F_s, then β is homotopic to α via the homotopy F_{1-s}). It also has the property of transitivity: Assume that α_0 is homotopic to α_1 via the homotopy F_s and α_1 is homotopic to α_2 via the homotopy G_s. Then, α_0 is homotopic to α_2 via a homotopy H_s defined by F_{2s} for $s \in [0, \frac{1}{2}]$ and G_{2s-1} for $s \in [\frac{1}{2}, 1]$. Accordingly, the homotopic loops constitute a family of equivalent loops (equivalent with respect to the homotopy operation). The (equivalence) class to which the representative α belongs is indicated by $[\alpha]$. Among loops, one can define a composition operation. Consider two loops with common the base point x_0, called α and β.

Definition ("composition or concatenation of loops"). The operation

$$(\alpha \circ \beta)(t) = \begin{cases} \beta(2t) & \text{if } t \in [0, \frac{1}{2}] \\ \alpha(2t - 1) & \text{if } t \in [\frac{1}{2}, 1] \end{cases}$$

defines another loop $\alpha \circ \beta$, which consists in first following β and then α. This composition law means that the two loops are traversed sequentially twice as quickly so that both loops are traversed once in the unit interval. The operation of composition can only be defined if the two loops have the same base point.

Definition ("inverse loop"). The inverse loop α^{-1} is defined as

$$\alpha^{-1}(t) = \alpha(1 - t)$$

Definition ("constant loop"). A constant loop with base point x_0 can also be defined by

$$\alpha(t) = x_0$$

These three loop operations respect homotopy classes: If $\alpha_0 \sim \alpha_1$ via F_s and $\beta_0 \sim \beta_1$ via G_s, then $\alpha_0 \circ \beta_0 \sim \alpha_1 \circ \beta_1$ via $F_s \circ G_s$. We can use this operation to define a law of composition among homotopy classes as follows:

$$[\alpha] \circ [\beta] = [\alpha \circ \beta]$$

i.e. one can multiply classes via multiplying one of their representatives. Furthermore,

$$[\alpha]^{-1} = [\alpha^{-1}]$$

and the homotopy class of the constant loop define an identity element in the set of homotopy classes

$$[x_0] \doteq e$$

By virtue of these operations, the set of homotopy classes becomes a group.

Definition. The set of homotopy classes in S_r constitutes the so-called fundamental group of S_r at x_0 (or the first homotopy group):

$$\pi_1(S_r, x_0)$$

Comment.
We have related the fundamental group to a base point x_0, and one could ask whether the elements of this group depend on the choice of the base point. Consider the two base points x_0 and x_1, and suppose to construct $\pi_1(S_r, x_0)$ and $\pi_1(S_r, x_1)$. Assume that S_r is path connected, i.e. there exists a continuous map $f(t)$ such that $f(0) = x_0$ and $f(1) = x_1$. Then, one can show that $\pi_1(S_r, x_0)$ is isomorphic to $\pi_1(S_r, x_1)$, so that for a path-connected set S_r, one can use the symbol $\pi_1(S_r)$ to label the fundamental group.

Examples:

1. The fundamental group of the complex plane at $x_0 = 1 + i0$ is $[1 + i0]$, i.e. it is a trivial group consisting of only the identity element. One can also state that any loop in the plane is null-homotopic.

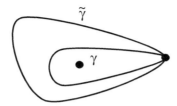

Two equivalent loops in the punctured plane.

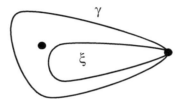

Non-equivalent loops in the punctured plane.

2. The fundamental group of the punctured plane $\mathbb{R}^2 \setminus \{(0,0)\}$. Consider the two loops γ and $\tilde{\gamma}$ in the punctured plane shown in the top figure.

The loops represent different maps of the same interval, but the difference in the loops is not topological: It is possible to continuously deform the two loops into each other. The two loops belong to the same equivalence class, consisting of those loops that go around the hole in the middle once.

Consider now the two loops γ and ξ in the punctured plane $\mathbb{R}^2 \setminus \{(0,0)\}$, as seen in the second figure from the top.

One loop, γ, encircles the hole in the plane, while ξ does not. It is not possible to continuously deform γ such that it does not encircle the hole anymore without passing through the hole. Thus, it is not possible to continuously deform γ into ξ. Imagine γ and ξ as two ropes lying on the floor and the hole as a stick extending perpendicular to the floor. The rope γ is wrapped around the stick; no matter which way you deform the rope, you cannot unwrap it from the stick without cutting the rope. Cutting the rope, however, is not continuous deformation, and thus, the rope γ is topologically different from the rope ξ. This topological difference is what lies at the heart of locking mechanisms for bicycles.

We continue with this example to find $\pi_1(\mathbb{R}^2 \setminus \{(0,0)\})$. The loop ξ can be rewrapped to a point without going through the hole: It is therefore equivalent to the null-loop. The loop γ, on the other hand, is equivalent to all loops wrapped around the hole once. Thus, we call the equivalence classes of ξ, and γ as $[0]$ and $[1]$, respectively. The numbers "0" and "1" are known as winding numbers (or chirality). We define a winding number as positive if the loop is run across anticlockwise, or else it is negative. The equivalence class of γ^{-1}, for instance, we call $[-1]$, meaning the loop γ is run through with the opposite winding number. We have, therefore,

$$[1] \circ [-1] = [0]$$

By concatenating γ with itself, we can expand on this notion to find $[\gamma] \circ [\gamma] =: [2]$ to be the equivalence class of all loops wrapped around the hole twice in the counterclockwise direction number. Following this scheme yields all different equivalence classes $[n]$, $n \in \mathbb{Z}$, and allows us to build the relation

$$[m] \circ [k] = [m + k]$$

This establishes that the fundamental group of the punctured plane is

$$\pi_1(\mathbb{R}^2 \setminus \{(0,0)\}) = \mathscr{Z}$$

3. The fundamental group of a circle: A circle can be viewed as a set of points surrounding a "large hole". Accordingly, the fundamental group is also

$$\pi_1(S_1) = \mathscr{Z}$$

To make this result rigorous, we refer to a general, exact theorem of complex analysis: The integral of the complex function $\frac{1}{z}$ over a path circumventing the pole n times is (by the residual theorem) exactly $2\pi \cdot n$. As path integrals are homotopy invariant, this result means that a path with the winding number n is not homotopic to a path with the winding number $n' \neq n$.

4. The fundamental group of the two-torus $T^2 = S^1 \times S^1$ is given by

$$\pi_1(T^2) = \pi_1(S^1) \times \pi_1(S^1) = \mathscr{Z} \times \mathscr{Z}$$

The fundamental group can be used to define the concept of connectivity of sets.

Definition. A set S_r (and the corresponding group) are **simply con-nected** if they are path connected, and the fundamental group is the trivial one. For "manifold connected sets", the fundamental group has elements other than the identity.

Examples:

1. The group of general linear transformations in one dimension defined by

$$x' = a_1 \cdot x + a_2$$

   with a_1, a_2 being both real numbers, is not simply connected because it is not path connected: The line $a_1 = 0$ is excluded from the a_1, a_2-plane; therefore, one cannot travel from the top half-plane to the bottom half-plane while being within the allowed parameter space.
2. The translation group is simply connected.
3. The punctured plane and the circle are manifold connected.

25.2   The Universal Covering Set and the Universal Covering Group

We show now that the unit circle S^1 in the plane provides a simple example of how to "connect" non-homotopic paths. Suppose that when one moves around the circle in order to follow a loop, one simultaneously lifts the path along the walls of a cylinder with a unit diameter, with the cylinder extending along a direction perpendicular to the plane where the unit circle resides. By doing this, the paths with different winding numbers end up as sections of a helix that climbs along the cylinder walls. This lifting effectively defines a new set — a helix in $\mathscr{R}^3$, parametrized by $s \mapsto (\cos 2\pi s, \sin 2\pi s, s)$, which is simply connected. This simply connected set is called the **universal covering set** of the circle.

   What are the consequences of having found the universal covering set of a parameter space for a group G? In other words, what group is associated with the universal covering set? Again, we look at the example of C_∞, whose elements are rotations $R(\varphi)$ by an angle $\varphi \in [0, 2\pi]$. The fact that the fundamental group of the circle is $\mathscr{Z}$ tells us that C_∞ is an infinitely manifold connected group. One can construct a new group C'_∞ which is simply connected and is parametrized by the universal covering set of the circle by defining a set of new operations as separate elements:

$R(+2\pi), R(-2\pi), R(+4\pi),\ldots$ The new group

$$C'_\infty = C_\infty \cup R(+2\pi) \cdot C_\infty \cup R(-2\pi) \cdot C_\infty \cdots$$

is parametrized by the helix and is, accordingly, called the **universal covering group of** C_∞.

25.2.1 *Representations of universal covering groups*

An important aspect of the representations of manifold connected continuous groups can be illustrated with the example of SO_2. We have shown that Γ^m, with $m = 0, \pm 1, \ldots$ and $\chi_m(\varphi) = e^{-im\varphi}$ are irreducible representations of C_∞ and that they represent all irreducible representations of C_∞. However, if we allow **multivalued** representations, one observes that

$$\Gamma^{\frac{m}{k}} = e^{-i\frac{m}{k}\varphi}$$

(with k being any integer) obeys also the homomorphism condition. Of course, $\Gamma^{\frac{m}{k}}$ is a k-valued homomorphism of C_∞: The rotations differing by $0, 2\pi, 4\pi, \ldots, (k-1) \cdot 2\pi$ are represented by different complex numbers, although they correspond to one and the same element of the group — if we impose the quite "natural" 2π-periodicity onto C_∞. We conclude that it is the property of k-fold connected continuous groups to have single-valued, double valued,..., and k-valued representations.

The question that arises naturally is the following: When multivalued representations exist, are they realized in physical systems? What is the meaning of considering them in physical problems?

There is a further, less mathematical and more physical problem hidden behind multivalued representations: Those representations which are constructed from $k \neq 1$ cannot have 2π-periodical wave functions as symmetry-adapted wave functions. The periodicity of the symmetry-adapted functions is $k \cdot 2\pi$. At first glance, such a periodicity is unphysical, e.g. for the wave functions of quantum mechanical particles in the standard Euclidean space, which must assume the same value when the angle φ is changed by 2π.

Bethe suggested a mathematical way of dealing with such multivalued representations, which we illustrate with the double-valued representations of C_∞. In the representation, $\Gamma^{\frac{m}{2}}$, E $(R(\varphi = 0))$ is represented by "1" and $R(\varphi = 2\pi)$ is represented by " -1 ". Bethe's strategy for eliminating the double valuedness of $\Gamma^{\frac{m}{2}}$ is to define a new (covering) group

$$SO_2^D \doteq SO_2 \cup R_{2\pi}SO_2$$

with $R_{2\pi}$ being a rotation by 2π, which is set to be a separate, new element. SO_2^D has twice as many elements as SO_2 and is called the "double group" of SO_2. The identity element of the new group SO_2^D is $E \equiv R_{4\pi}$, showing that a rotation by 4π is identical to a rotation by 0. The parameter space of SO_2^D is $[0, 4\pi]$, and the double-valued representation of SO_2 becomes a single-valued representation of SO_2^D. In addition, we have the relations $(k, k' = 1, 2)$

$$\frac{1}{4\pi} \int_0^{4\pi} \chi^{\frac{m}{k}*}(\varphi) \cdot \chi^{\frac{m'}{k'}}(\varphi) d\varphi = \frac{1}{4\pi} \int_0^{4\pi} e^{i\left(\frac{m}{k} - \frac{m'}{k'}\right)}(\varphi) d\varphi$$

This last integral is 1 for $m = m'$ and $k = k'$, or else it is vanishing, showing that the representations Γ^m and $\Gamma^{\frac{m}{k}}$ are irreducible representations of the new double group SO_2^D.

SO_2^D	E	R_φ	$R_{2\pi}$	$R_{2\pi}R_\varphi$
Γ^m	1	$e^{-im\varphi}$	1	$e^{-im\varphi}$
$\Gamma^{\frac{m}{2}}$	1	$e^{-i\frac{m}{2}\varphi}$	-1	$-e^{-i\frac{m}{2}\varphi}$

The representations Γ^m, $m = 0, \pm 1, \pm 2, \ldots$ are called the **single-group representations** of SO_2^D. The representations $\Gamma^{\frac{m}{2}}$, $m = 0, \pm 1, \pm 2, \ldots$ are called the **double-group (or extra) representations** of SO_2^D. The single-group representations of SO_2^D have symmetry-adapted functions with 2π-symmetry, the double-group representations have symmetry-adapted functions with a periodicity of 4π. Note that SO_2 is, in precise terms, an infinitely manifold connected group and its universal covering group consists of an infinite number of summands:

$$SO_2' \doteq SO_2 \cup R_{2\pi}SO_2 \cup R_{4\pi}SO_2 \cup R_{6\pi}SO_2 \cdots$$

The parameter space of the universal covering group of SO_2 spans therefore the entire positive real axis obtained by expanding the interval $[0, 2\pi]$ to $n \cdot [0, 2\pi]$, $n = \pm 1, \pm 2 \ldots$. Its irreducible representations are all representations with $m = 0, \pm 1, \ldots$ and $k = 1, 2, 3, \ldots$: The multivalued representations of

SO_2 become single-valued representations of the universal covering group SO_2'.

The question regarding multivalued representations can now be reformulated as follows: **Given a k-fold (k can even be infinite) connected group, which group is physically relevant: the group itself, some "double group" that we have introduced as a way of illustration, or its universal covering group?**

This question is a very crucial one. If one remembers that the index m of an irreducible, single-valued representation of SO_2 is related to the z-component of the quantum mechanical angular momentum by the relation $L_z = \hbar \cdot m$, should the covering group be the relevant one, then we expect further values of L_z to appear in a two-dimensional (2D) system with SO_2-symmetry, given by

$$L_z = \hbar \frac{m}{k}$$

with k being some integer. In particular, we expect fractional quantum numbers for the z-component of the angular momentum. This has implications for the particle statistics: In 2D, one could think of the appearance of particles which are neither Fermions nor Bosons and could be called **anyons**. The multivalued representations of SO_2 are called, accordingly, **anyon representations**. Quasiparticles of anyonic character have been observed in 2D systems known as quantum Hall systems (see S. Rao, "An anyon primer", https://arxiv.org/abs/hep-th/9209066).

25.3   The Universal Covering Group of SO_3

We now have all the tools of algebraic topology we need to construct the universal covering group of SO_3. In the axis–angle representation, defined in terms of a unit vector specifying the rotation axis and an angle φ specifying the rotation angle, the vector $\varphi \cdot \mathbf{n}$ defines the rotation uniquely: Every rotation is determined by the coordinates

$$n_1 \varphi, n_2 \varphi, n_3 \varphi$$

$n_1^2 + n_2^2 + n_3^2 = 1$. Accordingly, the parameter space of SO_3 can be represented by a sphere with a radius of π.

We consider first a loop with base point A residing within the sphere and going through points that remain within the sphere without ever touching the sphere's surface. Evidently, this loop is homotopic to the point A. We call this loop a "type a" loop. In a second instance ("type b" loops), we

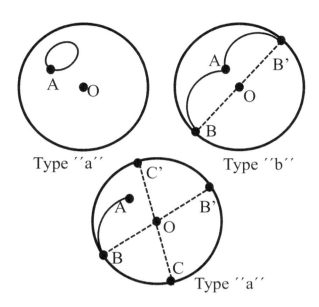

The two different homotopy classes in SO_3.

start again with our loop at A but move along a path that leads us to the point B residing on the surface of the sphere. B corresponds to a rotation by π around the axis OB. We construct the next piece of loop by following the rule that, when the surface is touched, the path jumps to the point B' diametrically opposite to B. Only after the jump can the path be detached from the surface and continued to A along points residing within the sphere. This second loop cannot be shrunk to A by continuous distortions because as we move the point B on the surface, B' remains diametrically opposite to B. We conclude that SO_3 has at least two homotopy classes, i.e. it is at least doubly connected.

We explore now a further path involving two jumps across the sphere at B and C. Our rule is that B and C cannot be detached from the surface of the sphere, but B' and C' can. In an attempt to find the homotopy class, we move B toward C', so that B' also moves toward C. Ultimately, B can be made to coincide with C' and B' with C. At this point, we follow our rule to detach B' and C' from the surface to shrink the loop onto A. Accordingly, a loop that makes an even number of jumps belongs to the same class of type-a loops, and a loop with an uneven number of jumps is of type b: SO_3 is exactly doubly connected. An educated guess for the universal covering group of SO_3 is, based on our knowledge of the subgroup SO_2 (Bethe),

$$SO_3^D = SO_3 \cup R(2\pi) \cdot SO_3$$

where SO_3^D indicates the double group of SO_3. SO_3^D has both single- and double-group representations. The double-group representations (such as the $D^{\frac{1}{2}}$ representation) carry states which have 4π-periodicity and therefore cannot be displayed in an Euclidean space.

It appears therefore that the answer to the question, "given a k-fold connected group, which group is relevant in physics: the group itself or its universal covering?" is (when referring to SO_3 and the experimental facts about the spin of the electron) "its universal covering".

A general introduction to the topic of using topological tools in science is given in N. D. Mermin, "The topological theory of defects in ordered media", *Rev. Mod. Phys.* **51**, 591–648 (1979), https://link.aps.org/doi/10.1103/RevModPhys.51.591

26

PROJECT FOR LECTURE 7: OPTICAL SPIN ORIENTATION IN ATOMS AND SOLIDS

26.1 Introduction

The interaction of light with atoms and solids can excite spin-polarized electrons, a phenomenon known as "optical (spin) orientation". In Lecture 7, we provided some examples of this phenomenon, which is mostly determined by the symmetry of the electron wave functions involved in the optical excitation. It is the scope of this project to go through the entire technical steps that result in the emergence of spin polarization, paying attention to the detailed computations, so that students can learn how to analyze similar situations. A detailed account of the results presented here can be found, for example, in F. Meier, D. Pescia, "Spin-polarized photoemission", *Optical Orientation*, Chapter 7 in *Modern Problems in Condensed Matter Sciences*, Vol. 8, Editors: F. Meier and B. P. Zakharchenya, Elsevier, 1964, pp. 295–351, https://www.sciencedirect.com/science/article/pii/B978044486740145 00123. In this project we refer to the Tables of Clebsch-Gordan coefficients for point groups, published in Koster, Dimmock, Wheeler and Statz, *Properties of the 32 Point Groups*, MIT Press, 1963, https://babel.hathit rust.org/cgi/pt?id=mdp.39015022422318&view=1up&seq=30. We refer to this book as KDWS.

26.2 Optical Orientation at the Γ-point of GaAs

Step 1: The relativistic band structure: The starting point is the relativistic band structure of GaAs. By "relativistic band structure", we mean that the spin–orbit interaction has been properly considered and the bands are labeled by the double group representations of the group of **k**.

At the Γ-point, the group of $\mathbf{k}$ is the point group T_d. We summarize here some properties of this group:

- The elements of T_d in terms of the coordinates (x, y, z) (see the following figure)

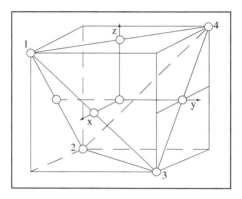

Coordinate system for the symmetry elements of T_d.

are given in the following table.

Type	Operation	Coordinate
C_1	C_1	xyz
C_4^2	C_{2z}	$\bar{x}\bar{y}z$
	C_{2x}	$x\bar{y}\bar{z}$
	C_{2y}	$\bar{x}y\bar{z}$
IC_4	IC_{4z}^{-1}	$y\bar{x}\bar{z}$
	IC_{4z}	$\bar{y}x\bar{z}$
	IC_{4x}^{-1}	$\bar{x}z\bar{y}$
	IC_{4x}	$\bar{x}\bar{z}y$
	IC_{4y}^{-1}	$\bar{z}\bar{y}x$
	IC_{4y}	$z\bar{y}\bar{x}$
IC_2	IC_{2xy}	$\bar{y}\bar{x}z$
	IC_{2xz}	$\bar{z}y\bar{x}$
	IC_{2yz}	$x\bar{z}\bar{y}$
	$IC_{2x\bar{y}}$	yxz
	$IC_{2\bar{x}z}$	zyx
	$IC_{2y\bar{z}}$	xzy

Continued

Type	Operation	Coordinate
C_3	C_{3xyz}^{-1}	zxy
	C_{3xyz}	yzx
	$C_{3x\bar{y}z}^{-1}$	$z\bar{x}\bar{y}$
	$C_{3x\bar{y}z}$	$\bar{y}\bar{z}x$
	$C_{3x\bar{y}\bar{z}}^{-1}$	$\bar{z}\bar{x}y$
	$C_{3x\bar{y}\bar{z}}$	$\bar{y}z\bar{x}$
	$C_{3xy\bar{z}}^{-1}$	$\bar{z}x\bar{y}$
	$C_{3xy\bar{z}}$	$y\bar{z}\bar{x}$

- The character table of T_d^D, taken from KDWS, is reported as follows.

Character table of T_d^D.

T_d^D	E	$\bar{E}$	$8C_3$	$8\bar{C}_3$	$3C_2, 3\bar{C}_2$	$6S_4(IC_4)$	$6\bar{S}_4$	$6\sigma_d(IC_2), 6\bar{\sigma}_d$	Basis
Γ_1	1	1	1	1	1	1	1	1	
Γ_2	1	1	1	1	1	$\bar{1}$	$\bar{1}$	$\bar{1}$	
$\Gamma_3(=\Gamma_{12})$	2	2	$\bar{1}$	$\bar{1}$	2	0	0	0	
$\Gamma_4(=\Gamma_{25})$	3	3	0	0	$\bar{1}$	1	1	$\bar{1}$	
$\Gamma_5(=\Gamma_{15})$	3	3	0	0	$\bar{1}$	$\bar{1}$	$\bar{1}$	1	(x,y,z)
Γ_6	2	$\bar{2}$	1	$\bar{1}$	0	$\sqrt{2}$	$\sqrt{2}$	0	
Γ_7	2	$\bar{2}$	1	$\bar{1}$	0	$\sqrt{2}$	$\sqrt{2}$	0	
Γ_8	4	$\bar{4}$	$\bar{1}$	1	0	0	0	0	

- The atomic configuration of the valence electrons of Ga is $4s^24p^1$. The configuration of As is $4s^24p^3$. The binding of the two atoms produces a sp^3-orbital, each occupied by a pair of electrons with opposite spins. The hybridized orbital can be used to produce a 4×4 representation of T_d which splits into $\Gamma_{15} \oplus \Gamma_1$. The Γ_1 level is at the bottom of the valence band and has s-character. The Γ_{15} level is at the top of the valence band and has p-character. It is followed by a gap amounting to about 1.4 eV. The bottom of the conduction band has s-character and transforms as Γ_1.

- We now proceed to the relativistic band structure, which involves assigning the bands to the double-group representations. We note that

$$D^{\frac{1}{2}}\big|_{T_d^D} \equiv \Gamma_6$$

One then computes

$$\Gamma_1 \otimes \Gamma_6$$

and

$$\Gamma_{15} \otimes \Gamma_6$$

either by the decomposition theorem or by using the suitable table provided by KDWS. Specifically, we refer to Table 82 in KDWS on p. 89. Inspection of the table produces

Direct products table for T_d^D.

Γ_1	Γ_2	Γ_3	Γ_4	Γ_5	Γ_6	Γ_7	Γ_8	
					Γ_6			Γ_1
$\vdots$	$\vdots$	$\vdots$	$\vdots$	$\vdots$	$\vdots$	$\vdots$	$\vdots$	$\vdots$
$\vdots$	$\vdots$	$\vdots$	$\vdots$	$\vdots$	$\vdots$	$\vdots$	$\vdots$	$\vdots$
$\vdots$	$\vdots$	$\vdots$	$\vdots$	$\vdots$	$\Gamma_7 + \Gamma_8$	$\vdots$	$\vdots$	Γ_5
$\vdots$	$\vdots$	$\vdots$	$\vdots$	$\vdots$	$\vdots$	$\vdots$	$\vdots$	$\vdots$

$$\Gamma_5 \otimes \Gamma_6 = \Gamma_7^5 \oplus \Gamma_8^5$$
$$\Gamma_1 \otimes \Gamma_6 = \Gamma_6^1.$$

Using this result, we can find the outcome of the single-group energy levels when spin–orbit splitting is switched on. The relativistic band structure in the vicinity of the gap is shown, schematically, in the following figure. The levels are labeled by two indices: The superscript corresponds to the single-group representation from which the level originates, while the subscript is the double-group representation. The ordering of the levels is taken from A. M. Gray, "Evaluation of electronic energy band structures of GaAs and GaP", *Phys. Stat. Sol.* **37**, 11–28 (1970). The fourfold top Γ_8^{15} level is reminiscent of the $P_{\frac{3}{2}}$ spin–orbit split level in atoms. The second spin–orbit split level has the symmetry of a $P_{\frac{1}{2}}$ atomic level.

The spin–orbit split valence has Γ_7^{15}-symmetry behaviour. The bottom of the conduction band is a Γ_6^1-band. The single-group assignement is of great help when making the transition from the non-relativistic band structure to the relativistic one within the first-order perturbation theory but is not an exact one: Using a higher-order perturbation theory, further single-group components might appear in the level with a given double-group representation (von Neumann–Wigner hybridization).

Step 2: The optical matrix elements: We use circularly polarized light propagating along $-z$. The interaction operator O between light and

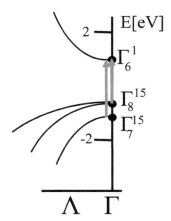

Relativistic band structure in the gap region of GaAs (at the Γ-point.) The initial band structure along the Λ-direction is also drawn, with the aim of linking it with the example treated in the lecture. The two optical transitions considered are indicated by red arrows. The bands are redrawn from A. M. Gray, "Evaluation of electronic energy band structures of GaAs and GaP", *Phys. Stat. Sol.* **37**, 11–28 (1970).

electrons is proportional to

$$x - i \cdot y.$$

i.e. transforms, in the notation of KWDS,[1] as $v_{yz}^5 - i \cdot v_{xz}^5$. We compute the matrix elements for the optical transition $\Gamma_8 \to \Gamma_6$ and $\Gamma_7 \to \Gamma_6$. The matrix elements are written as

$$\left\langle \psi_{\pm 1/2}^6 \mid O \mid \psi_{\pm ...}^{8,7} \right\rangle = \left\langle \psi_{\pm 1/2}^6 \mid v_{yz}^5 \mid \psi_{\pm ...}^{8,7} \right\rangle - i \cdot \left\langle \psi_{\pm 1/2}^6 \mid v_{xz}^5 \mid \psi_{\pm ...}^{8,7} \right\rangle$$

Let us compute, for example, the matrix element

$$\left\langle \psi_{-1/2}^6 \mid O \mid \psi_{1/2}^7 \right\rangle = \left\langle \psi_{-1/2}^6 \mid v_{yz}^5 \mid \psi_{1/2}^7 \right\rangle - i \cdot \left\langle \psi_{-1/2}^6 \mid v_{xz}^5 \mid \psi_{1/2}^7 \right\rangle$$

explicitly. For this purpose, we turn to Table 83, p. 94 (partly reproduced in the following) to find the expansion of the product functions on the

---

[1] KDWS uses "u" for the left-hand side partner of a direct product state, and "v" for the right-hand side partner. The subscript then labels the basis states transforming according to the irreducible representation given by the superscript. If the irreducible representation is one-dimensional, there is only a subscript.

right-hand side (top entry of the table) into symmetry-adapted functions (left entry of the table).

We use

Clebsch–Gordan coefficients (or "coupling coefficients") for $\Gamma_5 \otimes \Gamma_7$ in T_d^D.

	$u_{yz}^5 v_{-\frac{1}{2}}^7$	$u_{yz}^5 v_{\frac{1}{2}}^7$	$u_{xz}^5 v_{-\frac{1}{2}}^7$	$u_{xz}^5 v_{\frac{1}{2}}^7$	$u_{xy}^5 v_{-\frac{1}{2}}^7$	$u_{xy}^5 v_{\frac{1}{2}}^7$
$\psi_{-\frac{1}{2}}^6$	0	$-\frac{i}{\sqrt{3}}$	0	$-\frac{1}{\sqrt{3}}$	$\frac{i}{\sqrt{3}}$	0
$\psi_{+\frac{1}{2}}^6$	$-\frac{i}{\sqrt{3}}$	0	$\frac{1}{\sqrt{3}}$	0	0	$-\frac{i}{\sqrt{3}}$
$\vdots$	$\vdots$	$\vdots$	$\vdots$	$\vdots$	$\vdots$	$\vdots$

$$u_{yz}^5 v_{\frac{1}{2}}^7 \sim -\overline{\frac{i}{\sqrt{3}}} \psi_{-\frac{1}{2}}^6$$

($\sim$ meaning "transforming as") and

$$u_{xz}^5 v_{\frac{1}{2}}^7 \sim -\frac{1}{\sqrt{3}} \psi_{-\frac{1}{2}}^6$$

to obtain

$$\left\langle \psi_{-1/2}^6 \mid v_{yz}^5 \mid \psi_{1/2}^7 \right\rangle = c_{5,7,6} \cdot i \frac{1}{\sqrt{3}}$$

$$\left\langle \psi_{-1/2}^6 \mid v_{xz}^5 \mid \psi_{1/2}^7 \right\rangle = c_{5,7,6} \cdot \left(-\frac{1}{\sqrt{3}} \right)$$

and thus,

$$\left\langle \psi_{-1/2}^6 \mid O \mid \psi_{1/2}^7 \right\rangle = c_{5,7,6} \cdot \left(i \frac{1}{\sqrt{3}} + i \frac{1}{\sqrt{3}} \right) = c_{5,7,6} \cdot 2 \cdot i \frac{1}{\sqrt{3}}$$

$c_{5,7,6}$ is the Wigner–Eckart constant that depends on the details of the wave functions and operator and must be computed numerically. It is only dependent on the irreducible representations involved (the superscrips, not the subscripts). The remaining matrix elements from the ψ^7-states are all vanishing for this specific operator. In the same way, all the other matrix elements from the Γ_8-state are found using Table 83 on p. 95:

$$\left\langle \psi^6_{-1/2} \mid O \mid \psi^8_{-3/2} \right\rangle = c_{5,8,6} \cdot \left(-i\frac{1}{\sqrt{3}} \right)$$

$$\left\langle \psi^6_{1/2} \mid O \mid \psi^8_{-1/2} \right\rangle = c_{5,8,6} \cdot i$$

All remaining matrix elements vanish.

Step 3: Spin polarization: The electrons in the final state have an important physical property: They are spin polarized. We now compute the spin polarization and show that it is determined on the basis of symmetry properties alone. The spin polarization of an electron ensemble is defined as the expectation value of the spin polarization vector operator

$$(\sigma_x, \sigma_y, \sigma_z)$$

(σ being the Pauli matrices). The optical excitation $\Gamma_8 \to \Gamma_6$ can occur via two different channels: One with the matrix element $c_{5,8,6} \cdot (-i\frac{1}{\sqrt{3}})$ produces electrons with spin polarization

$$(0, 0, <\sigma_z>) = \left(0, 0, \left\langle \psi^6_{-1/2} \mid \psi^6_{-1/2} \right\rangle \right) = (0, 0, -100\%)$$

The second channel produces $+100\%$ spin-polarized electrons, but it has a different matrix element: $c_{5,8,6} \cdot i$. When there are an ensemble of atoms, one transition will occur in one atom, the other one in the other atom, so that the production of spin-polarized electrons is an incoherent process, and the spin polarization produced in the two channels must be summed incoherently by weighting the respective spin polarization with the intensity (i.e. the absolute square of the matrix element) of the optical transition:

$$<\sigma_z> = \frac{\mid c_{5,8,6} \cdot i \mid^2 - \mid c_{5,8,6} \cdot \left(-i\frac{1}{\sqrt{3}} \mid^2 \right)}{\mid c_{5,8,6} \cdot i \mid^2 \cdot + \mid c_{5,8,6} \cdot \left(-i\frac{1}{\sqrt{3}} \mid^2 \right)} = 50\%$$

The canceling of the Wigner–Eckart constant is the most remarkable result of this computation and leaves a physically measurable quantity as an algebraic result of the Clebsch–Gordan coefficients. The second most remarkable property is the fact that the spin polarization is the direct result of spin and orbital wave functions being **entangled**, see the initial state of the transition.

Comments.

There is only one transition from the Γ_7-split level, and this transition produces -100% spin-polarized electrons. An interesting situation arises when both transitions occur and the lowest conduction band hosts electrons originating from the excitation of spin–orbit split bands. In this case, the spin polarization is expressed as

$$
<\sigma_z> = \frac{\mid c_{5,8,6} \cdot i \mid^2 - \mid c_{5,8,6} \cdot \left(-i\frac{1}{\sqrt{3}}\mid^2\right) - \mid c_{5,7,6} \cdot 2 \cdot i\frac{1}{\sqrt{3}} \mid^2}{\mid c_{5,8,6} \cdot i \mid^2 + \mid c_{5,8,6} \cdot \left(-i\frac{1}{\sqrt{3}}\mid^2\right) + \mid c_{5,7,6} \cdot 2 \cdot i\frac{1}{\sqrt{3}} \mid^2}
$$

It contains two Wigner–Eckart constants which do not cancel out immediately. There is a method to eliminate one Wigner–Eckart constant and estimate the spin polarization resulting from the excitation with both channels: We explicitly compute, within the first-order perturbation theory, the symmetry-adapted functions transforming according to Γ_8 and Γ_7 from the basis states transforming according to Γ^{15}. We also compute the final state Γ_6-symmetry-adapted functions from the single-group states transforming according to Γ_1. We search for these functions in the basis state products arising in the spaces, supporting the representations

$$
\Gamma^1 \otimes \Gamma_6
$$

$$
\Gamma^{15} \otimes \Gamma_6
$$

Γ_6^1 are (not explicitly given in KWDS)

$$
\psi^6_{-1/2} = u_1 v^6_{-1/2},
$$

$$
\psi^6_{1/2} = u_1 v^6_{1/2}.
$$

To find the symmetry-adapted states with Γ_7^5, Γ_8^5 symmetries, we consider the the middle table in p. 94 of KDWS. The entries along the top line of the table are the six product functions spanning the $\Gamma_5 \otimes \Gamma_6$-space. The entries along the left-hand column are the labels of the symmetry-adapted basis states transforming as Γ_7^5, Γ_8^5. The bulk of the table contains the Clebsch–Gordan coefficient. The symmetry-adapted linear combinations obtained from this table are written as

$$
\psi^7_{-1/2} = -\frac{i}{\sqrt{3}} u^5_{yz} v^6_{1/2} - \frac{1}{\sqrt{3}} u^5_{xz} v_{1/2} + \frac{i}{\sqrt{3}} u^5_{xy} v^6_{-1/2},
$$

$$
\psi^7_{1/2} = -\frac{i}{\sqrt{3}} u^5_{yz} v^6_{-1/2} + \frac{1}{\sqrt{3}} u^5_{xz} v_{-1/2} - \frac{i}{\sqrt{3}} u^5_{xy} v^6_{1/2},
$$

$$\psi^8_{-3/2} = -\frac{i}{\sqrt{6}}u^5_{yz}v^6_{-1/2} + \frac{1}{\sqrt{6}}u^5_{xz}v_{-1/2} + \frac{i\sqrt{2}}{\sqrt{3}}u^5_{xy}v^6_{1/2},$$

$$\psi^8_{-1/2} = \frac{i}{\sqrt{2}}u^5_{yz}v^6_{1/2} - \frac{1}{\sqrt{2}}u^5_{xz}v_{1/2},$$

$$\psi^8_{1/2} = -\frac{i}{\sqrt{2}}u^5_{yz}v^6_{1/2} - \frac{1}{\sqrt{2}}u^5_{xz}v_{-1/2},$$

$$\psi^8_{3/2} = \frac{i}{\sqrt{6}}u^5_{yz}v^6_{1/2} + \frac{1}{\sqrt{6}}u^5_{xz}v_{1/2} + \frac{i\sqrt{2}}{\sqrt{3}}u^5_{xy}v^6_{-1/2}.$$

Here, $u^5_{yz}, u^5_{xz}, u^5_{xy}$ and $v^6_{1/2}, v^6_{-1/2}$ are basis functions belonging to the irreducible representation Γ_5 and Γ_6, respectively. The subscripts in the $u's$-basis states resemble monomials of the Cartesian coordinates that transform according to the representation Γ_5, but there are further monomials of the coordinates that have exactly the same transfomation properties, such as the monomials x, y, z themselves. One can verify this statement by working with the operations listed in the table on the monomials. For the outcome of this computation, only the transformation properties are relevant, not the precise composition of the monomials.

Once these symmetry-adapted wave functions have been found, we can proceed with the computation of the matrix elements between the basis states. The algorithm is the same as illustrated above. For instance,

$$\langle\psi^6_{-1/2}\mid O\mid\psi^7_{1/2}\rangle = \Big\langle u_1 v^6_{-1/2}|v^5_{yz} - iv^5_{xz}| - \frac{i}{\sqrt{3}}u^5_{yz}v^6_{-1/2}$$

$$+ \frac{1}{\sqrt{3}}u^5_{xz}v^6_{-1/2} - \frac{i}{\sqrt{3}}u^5_{xy}v^6_{1/2}\Big\rangle$$

$$= \Big\langle u_1|v^5_{yz} - iv^5_{xz}| + \frac{1}{\sqrt{3}}(-iu^5_{yz} + u^5_{xz})\Big\rangle$$

$$= \Big\langle u_1|v^5_{yz}| - i\cdot\frac{1}{\sqrt{3}}u^5_{yz}\Big\rangle$$

$$+ \Big\langle u_1|v^5_{yz}|\frac{1}{\sqrt{3}}u^5_{xz}\Big\rangle$$

$$+ \Big\langle u_1|-iv^5_{xz}| - i\cdot\frac{1}{\sqrt{3}}u^5_{yz}\Big\rangle$$

$$+ \Big\langle u_1|-iv^5_{xz}|\frac{1}{\sqrt{3}}u^5_{xz}\Big\rangle$$

To compute these matrix elements, we refer now to Table 83 at the top of p. 94 and find, finally,

$$\left\langle \psi^6_{-1/2} \mid O \mid \psi^7_{1/2} \right\rangle = c_{5,5,1} \cdot \left(-i \cdot \frac{2}{3} \right)$$

$$\left\langle \psi^6_{-1/2} \mid O \mid \psi^8_{-3/2} \right\rangle = c_{5,5,1} \cdot (-i\sqrt{2}/3)$$

$$\left\langle \psi^6_{1/2} \mid O \mid \psi^8_{1/2} \right\rangle = c_{5,5,1} \cdot (i\sqrt{2}/\sqrt{3})$$

with all remaining matrix elements vanishing. Using these matrix elements, we compute the spin polarization of the total ensemble produced by optical excitation from **both** the spin–orbit split levels:

$$< \sigma_z >= \frac{\mid c_{5,5,1} \cdot (i\sqrt{2}/\sqrt{3}) \mid^2 - \mid c_{5,5,1} \cdot (-i\sqrt{2}/3) \mid^2) - \mid c_{5,5,1} \cdot \left(-i\frac{2}{3} \right) \mid^2}{\mid c_{5,5,1} \cdot (i\sqrt{2}/\sqrt{3}) \mid^2 + \mid c_{5,5,1} \cdot (-i\sqrt{2}/3) \mid^2) + \mid c_{5,5,1} \cdot \left(-i\frac{2}{3} \right) \mid^2} = 0$$

This result shows that spin–orbit coupling is an essential element for optical spin orientation, together with the use of circularly polarized light. This is because the interaction Hamiltonian between light and charges does not contain (in the dipolar approximation) spin-dependent terms.

27

PROJECT FOR LECTURE 9: QUANTUM CHEMISTRY

Quantum chemistry aims at computing the electronic structure of molecules and the properties related to them. Suitable software are available and continuously developed, which can be used for practical computations. One of them is the software package "Gaussian" (https://gaussian.com/), which is used for the practical computations in this project. The background of "Gaussian" involves well-established algorithms of quantum chemistry, rooted in quantum mechanics. We explain this background briefly, as it further emphasizes the role of eigenvalue problems in science. There is one further motive of importance that justifies the introduction of quantum chemistry in this set of lectures. So far, we have solved one-particle problems, i.e. the wave functions and the operators we have used in most problems were functions of only one variable $q \equiv (\mathbf{r}, \sigma)$, with $\mathbf{r}$ being the position of an electron and σ the value of the z component of its spin. Quantum chemistry deals with many-electron systems: The wave functions and the operators appearing in them are functions of many variables q_i, where i counts the number of particle. This is a new aspect that we have to take into account when learning how to use symmetry arguments.

27.1 Introduction: The H_2-Molecule

The computation of the energy levels of the H_2 molecule ($i = 1, 2$) contains some essential notions that one can generalize within quantum chemistry. The first computation of the energy levels was performed by Heitler and London (W. Heitler and F. London, "Wechselwirkung neutraler Atome und homopolare Bindung nach der Quantenmechanik", *Z. Phys.* **44**, 455–472, http://dx.doi.org/10.1007/BF01397394). The Heitler–London computation was based on the Born–Oppenheimer

approximation. The Born–Oppenheimer operator describing the two electrons in the molecule subject to an electric field of two nuclei at positions
A, B at a distance R is given by

$$H_2 = \sum_{i=1,2} \left[\frac{-\hbar^2}{2m}(\triangle_i) - \frac{e^2}{4\pi\epsilon_0} \cdot \left(\frac{1}{r_{Ai}} + \frac{1}{r_{Bi}} \right) \right]$$

$$+ \frac{e^2}{4\pi\epsilon_0} \frac{1}{r_{12}}$$

$$+ \frac{e^2}{4\pi\epsilon_0} \cdot \frac{1}{R}$$

(r_{xy} stands for $\frac{1}{|\mathbf{x}-\mathbf{y}|}$). The first line is the Hamilton operator of the two
electrons moving in the field of the nuclei. The second line is the Hamilton
operator describing the Coulomb interaction between the electrons. The
third line is the Coulomb interaction between the nuclei. In the Born–
Oppenheimer approximation, their distance R enters the computation as
a parameter that must be optimized to minimize the total energy of the
system. This minimization will provide the equilibrium distance between
the molecules. As trial basis states for the Ritz space, the Heitler–London
(H–L) model uses the three spin triplet states

$$[2(1 - S^2)]^{-1/2} \cdot Y^+(1)Y^+(2) \otimes [\psi_A(1)\psi_B(2) - \psi_B(1)\psi_A(2)]$$

$$[4(1 + S^2)]^{-1/2} \cdot [Y^+(1)Y^-(2) + Y^-(1)Y^+(2)] \otimes [\psi_A(1)\psi_B(2) - \psi_B(1)\psi_A(2)]$$

$$[2(1 + S^2)]^{-1/2} \cdot Y^-(1)Y^-(2) \otimes [\psi_A(1)\psi_B(2) - \psi_B(1)\psi_A(2)]$$

and the spin singlet state

$$[4(1 + S^2)]^{-1/2} \cdot [Y^+(1)Y^-(2) - Y^-(1)Y^+(2)] \otimes [\psi_A(1)\psi_B(2) + \psi_B(1)\psi_A(2)]$$

The orbital states are the $1s$ orbitals centered at the sites A and B:

$$\psi_A(1) = (\pi a^3)^{-1/2} e^{\left(\frac{-r_{A1}}{a}\right)}$$

$$\psi_A(2) = (\pi a^3)^{-1/2} e^{\left(\frac{-r_{A2}}{a}\right)}$$

$$\psi_B(1) = (\pi a^3)^{-1/2} e^{\left(\frac{-r_{B1}}{a}\right)}$$

$$\psi_B(2) = (\pi a^3)^{-1/2} e^{\left(\frac{-r_{B2}}{a}\right)}$$

These orbitals have a cusp at the origin. Such cusp-like orbitals are called
in the chemistry literature Slater-type orbitals. Note that the use of atomic

orbitals introduces an overlap integral

$$S = \int dV_1 \psi_A(1)\psi_B(1) = \frac{1}{\pi a^3}\int dV_1 e^{-\frac{r_{A1}+r_{B1}}{a}}$$

between the orbitals centered at different sites, which appears, for instance, in the normalization of the basis states. If one considers the orthogonality properties of the spin states, the Ritz matrix within this trial vector space is diagonal. The diagonal matrix elements for the threefold-degenerate triplet spin states are the matrix elements

$$E_t \doteq (\psi_A(1)\psi_B(2) + \psi_B(1)\psi_A(2)], H[\psi_A(1)\psi_B(2) - \psi_B(1)\psi_A(2)])]$$

The diagonal matrix elements for the singlet spin state are the matrix elements

$$E_s \doteq (\psi_A(1)\psi_B(2) + \psi_B(1)\psi_A(2)], H[\psi_A(1)\psi_B(2) + \psi_B(1)\psi_A(2)])]$$

After some algebra and taking into account that

$$\left[\frac{-\hbar^2}{2m}\Delta_1 - \frac{e^2}{r_{1A}}\right]\psi_A(1) = E_{1s}\psi_A(1)$$

(E_{1s} is the atomic energy of the 1s orbital) one finds

$$E_t = 2E_{1s} + \frac{Q-J}{1-S^2}$$

$$E_s = 2E_{1s} + \frac{Q+J}{1+S^2}$$

The integral

$$Q = \int dV_1 dV_2 \psi_A^2(1)\psi_B^2(2)\left[\frac{-e^2}{r_{B1}} + \frac{-e^2}{r_{A2}} + \frac{e^2}{r_{12}} + \frac{e^2}{R}\right]$$

$$= -\int dV_1\psi_A^2(1)\frac{e^2}{r_{B1}} - \int dV_2\psi_B^2(2)\frac{e^2}{r_{A2}} + \int dV_1 dV_2 \psi_A^2(1)\psi_B^2(2)\frac{e^2}{r_{12}} + \frac{e^2}{R}$$

contains various Coulomb energies, such as the Coulomb energy of the first electron with the nucleus B, the Coulomb energy of the second electron with nucleus A, the Coulomb interaction of the two electrons, and the Coulomb repulsion of the two nuclei. Q is called the **Coulomb integral**.

The integral

$$J = \frac{e^2 \cdot S^2}{R} + \int dV_1 dV_2 \psi_A(1)\psi_B(2)\frac{e^2}{r_{12}}\psi_A(2)\psi_B(1)$$

$$-S\int dV_1\psi_A(1)\frac{e^2}{r_{B1}}\psi_B(1) - S\int dV_2 d\psi_B(2)\frac{e^2}{r_{A2}}\psi_A(2)$$

represents the **exchange energy**, arising from the requirement of antisymmetrizing the wave functions. Both Q and J depend on the distance R. Both can be computed exactly within the subspace chosen in the H–L model. $Q(R)$ is a small positive quantity, whereas $J(R)$ is negative. Accordingly, the singlet state has the lowest energy. Note that the negative sign of J is the essential ingredient for a stable chemical bond, which then occurs in the singlet state.

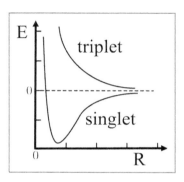

Schematic plot of E_t and E_s as a function of R. The singlet state has a minimum at a well-defined distance R_0, at which the chemical bond of the two H atoms is realized. R_0 is the equilibrium distance of the nuclei in the H_2 molecule. The experimental value is $0.7395 A°$. The experimental value for the binding energy of the singlet state is 4.74 eV.

However, a negative J leads to a preferred antiparallel coupling of the two spins and is a simple but very clear and robust demonstration that antiparallel spin alignment is the key coupling mechanism when atoms are assembled to form molecules and solids. We thus have a very complicated situation where the Pauli principle, essential to chemical bonds, works against parallel spin alignment — and ultimately against, for example, a ferromagnetic state.

In the presence of only one electron with the wave function $\psi(\mathbf{r})$, the charge density is given by $|\psi(\mathbf{r})|^2$. When the system has two or more particles, one must find a physical way of defining a charge density. This was done by Kohn with his "Density functional approach": Given the normalized wave function $\psi(\mathbf{r}_1, \mathbf{r}_2, \ldots, \mathbf{r}_N)$, the charge density $\rho(\mathbf{r})$ is given by

$$\rho(\mathbf{r}) = N \cdot \int dV_2 \cdot dV_3 \cdots dV_N \cdot |\psi(\mathbf{r}, \mathbf{r}_2, \ldots, \mathbf{r}_N)|^2$$

The charge distribution computed with the two-particle orbital singlet wave function is sketched in the following figure.

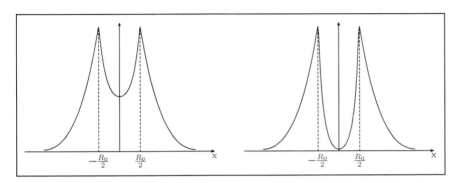

Left: The bonding charge density along the axis hosting the two protons. The bonding charge density has a strong charge accumulation halfway between the protons. Right: The antibonding charge density has a knot exactly at that location.

The singlet bonding charge density has an accumulation of charge halfway between the two protons: It is the accumulation of charge between the two nuclei that actually drives the chemical bond.

27.2 The Eigenvalue Problem of Quantum Chemistry

The H–L solution provides the fundamental understanding of the chemical bond that, ultimately, paved the way to quantum chemistry. It is a two-electron computation, involving a two-particle wave function. The definition of the charge density, as introduced by Kohn, allows us to predict the existence of a molecular orbital with a one-particle orbital coordinate that hosts two electrons, paired to form a singlet. Quantum chemistry actually developed around the idea of molecular orbitals.

Modern computational quantum chemistry was initiated by the seminal works of C. C. J. Roothaan (C. C. J. Roothaan, "New developments in molecular orbital theory", *Rev. Mod. Phys.* **23**, 69–89 (1951), https://doi.org/10.1103/RevModPhys.23.69) and G. G. Hall (G. G. Hall, "The molecular orbital theory of chemical valency. VIII. A method of calculating ionization potentials", *Proc. R. Soc. A.* **205** (1951), S. 541–552 https://doi.org/10.1098/rspa.1951.0048. The approach pursued in these works represents the starting point to modern software for quantum chemistry, such as "Gaussian".

The quantum chemistry approach by Roothaan and Hall (RH) aims at finding the n molecular orbitals $\varphi_1(\mathbf{r}), \ldots, \varphi_n(\mathbf{r})$ each hosting two electrons in the spin singlet state.

• To this end, RH starts with one Slater determinant as a trial wave function with $2n$ variables to describe the chemical bonds containing $2n$ electrons. The Slater matrix is written as

$$\begin{pmatrix} \varphi_1(\mathbf{r}_1)Y^+(1) & \varphi_1(\mathbf{r}_2)Y^+(2) & \varphi_1(\mathbf{r}_3)Y^+(3) & \varphi_1(\mathbf{r}_4)Y^+(4) & \ldots & \varphi_1(\mathbf{r}_{2n-1})Y^+(2n-1) & \varphi_1(\mathbf{r}_{2n})Y^+(2n) \\ \varphi_1(\mathbf{r}_1)Y^-(1) & \varphi_1(\mathbf{r}_2)Y^-(2) & \varphi_1(\mathbf{r}_3)Y^-(3) & \varphi_1(\mathbf{r}_4)Y^-(4) & \ldots & \varphi_1(\mathbf{r}_{2n-1})Y^-(2n-1) & \varphi_1(\mathbf{r}_{2n})Y^-(2n) \\ \varphi_2(\mathbf{r}_1)Y^+(1) & \varphi_2(\mathbf{r}_2)Y^+(2) & \varphi_2(\mathbf{r}_3)Y^+(3) & \varphi_2(\mathbf{r}_4)Y^+(4) & \ldots & \varphi_2(\mathbf{r}_{2n-1})Y^+(2n-1) & \varphi_2(\mathbf{r}_{2n})Y^+(2n) \\ \varphi_2(\mathbf{r}_1)Y^-(1) & \varphi_2(\mathbf{r}_2)Y^-(2) & \varphi_2(\mathbf{r}_3)Y^-(3) & \varphi_2(\mathbf{r}_4)Y^-(4) & \ldots & \varphi_2(\mathbf{r}_{2n-1})Y^-(2n-1) & \varphi_2(\mathbf{r}_{2n})Y^-(2n) \\ & & & \vdots & & & \\ \varphi_n(\mathbf{r}_1)Y^-(1) & \varphi_n(\mathbf{r}_2)Y^-(2) & \varphi_n(\mathbf{r}_3)Y^-(3) & \varphi_n(\mathbf{r}_4)Y^-(4) & \ldots & \varphi_n(\mathbf{r}_{2n-1})Y^-(2n-1) & \varphi_n(\mathbf{r}_{2n})Y^-(2n) \end{pmatrix}$$

For each desired molecular orbital, there are two lines, so that a quadratic $2n \times 2n$ Slater matrix can be constructed, as required by the formation of a Slater determinant. For a two-electron bond, for instance, we have one sought-after molecular orbital φ_1, and the Slater determinant

$$\det \begin{pmatrix} \varphi_1(\mathbf{r}_1)Y^+(1) & \varphi_1(\mathbf{r}_2)Y^+(2) \\ \varphi_1(\mathbf{r}_1)Y^-(1) & \varphi_1(\mathbf{r}_2)Y^-(2) \end{pmatrix}$$
$$= \varphi_1(\mathbf{r}_1)\varphi_1(\mathbf{r}_2)(Y^+(1)Y^-(2) - Y^-(1)Y^+(2))$$

describes precisely the ground-state spin singlet we are aiming to compute. This example shows that the Slater determinant *ansatz* used here encompasses the case where two electrons forming a molecular orbital are paired into a singlet spin state, in accordance with Heitler and London's original result on the state with the lowest energy.

• The next step in the Roothaan method is aimed at computing the expectation value of the Hamilton operator

$$
H_{2n} = \sum_{i=1,2,\ldots,2n} \left[\frac{-\hbar^2}{2m} (\triangle_i) - \sum_{\alpha=A,B,\ldots} \frac{e^2}{4\pi\epsilon_0} \cdot \frac{1}{r_{i\alpha}} \right]
$$
$$
+ \frac{1}{2} \sum_{i \neq j} \frac{e^2}{4\pi\epsilon_0} \frac{1}{r_{ij}}
$$
$$
+ \frac{1}{2} \frac{e^2}{4\pi\epsilon_0} \cdot \sum_{\alpha \neq \alpha'} \frac{1}{R_{\alpha\alpha'}}
$$

In H_{2n}, The indices i, j are used to specify the position vector of the electrons, and the indices α, α' specify the position vectors of the nuclei.

• The final step of the RH algorithm is minimizing this expectation value with respect to the n molecular orbitals entering the Slater determinant.

27.2.1   *Two-electron RH algorithm*

The computation of the expectation value for the n orbitals over the Slater determinant leads to quite cumbersome expressions that one can find in the original literature (Eq. (20) in the paper by Roothaan). We spell out this computation and the associated minimization algorithm for the case of $n = 1$ and $\alpha = A, B$ (i.e. two spin-singlet electrons in the desired molecular orbital $\varphi_1(\mathbf{r})$).

• The Slater determinant produces the trial wave function

$$
\varphi_1(\mathbf{r}_1)\varphi_1(\mathbf{r}_2)\frac{1}{\sqrt{2}}[Y^+(1)Y^-(2) - Y^-(1)Y^+(2)]
$$

$\varphi_1(\mathbf{r})$ is a single-particle trial wave function, subject to the condition of being normalized to 1. The Hamiltonian (neglecting spin–orbit coupling) is written as

$$H_2 = \sum_{i=1,2} \underbrace{\left[\frac{-\hbar^2}{2m}(\triangle_i) - \frac{e^2}{4\pi\epsilon_0} \cdot \left(\frac{1}{r_{Ai}} + \frac{1}{r_{Bi}} \right) \right]}_{H_{A,B}(\mathbf{r}_i)}$$

$$+ \underbrace{\frac{e^2}{4\pi\epsilon_0} \frac{1}{r_{12}}}_{H_{12}(\mathbf{r}_1,\mathbf{r}_2)}$$

$$+ \underbrace{\frac{e^2}{4\pi\epsilon_0} \cdot \frac{1}{R}}_{H_{A,B}(R)}$$

• The expectation value of H_2 over the Slater determinant (the spin function cancels out in virtue of being normalized to 1) is

$$(\varphi_1(\mathbf{r}_1)\varphi_1(\mathbf{r}_2), H_2\varphi_1(\mathbf{r}_1)\varphi_1(\mathbf{r}_2)) = \underbrace{2 \cdot (\varphi_1(\mathbf{r}), H_{A,B}(\mathbf{r})\varphi_1(\mathbf{r}))}_{I}$$

$$+ \underbrace{(\varphi_1(\mathbf{r}_1)\varphi_1(\mathbf{r}_2), H_{12}(\mathbf{r}_1,\mathbf{r}_2)\varphi_1(\mathbf{r}_1)\varphi_1(\mathbf{r}_2))}_{II}$$

$$+ \underbrace{\frac{e^2}{4\pi\epsilon_0} \cdot \frac{1}{R}}_{III}$$

The symbol $(.,.)$ implies two separate integrals when two spatial variables are involved and one integral when one spatial variable is involved.

• We seek to find the molecular orbital $\varphi_1(\mathbf{r})$ that minimizes $I + II + III$. For this purpose, we vary the sought-after orbital by a small amount

$$\delta\varphi_1 \doteq \epsilon(\mathbf{r})$$

We now compute δI to first order in ϵ and $\bar{\epsilon}$.

$$\delta I = 2\big(\varphi_1(\mathbf{r}), H_{A,B}(\mathbf{r})\epsilon(\mathbf{r})\big)$$

$$+ 2\big(\epsilon(\mathbf{r}), H_{A,B}(\mathbf{r})\varphi_1(\mathbf{r})\big)$$

$$= 2 \cdot \int d\mathbf{r} \cdot \epsilon(\mathbf{r}) \cdot H_{A,B}(\mathbf{r})\overline{\varphi_1(\mathbf{r})}$$

$$+ 2 \cdot \int d\mathbf{r} \cdot \overline{\epsilon(\mathbf{r})} \cdot H_{A,B}(\mathbf{r})\varphi_1(\mathbf{r})$$

We compute δII:

$$\delta II = 2\big(\varphi_1(\mathbf{r}') \cdot \varphi_1(\mathbf{r}), H_{12}(\mathbf{r}, \mathbf{r}')\varphi_1(\mathbf{r}') \cdot \epsilon(\mathbf{r})\big)$$

$$+ 2\big(\varphi_1(\mathbf{r}') \cdot \epsilon(\mathbf{r}), H_{12}(\mathbf{r}, \mathbf{r}')\varphi_1(\mathbf{r}') \cdot \varphi_1(\mathbf{r})\big)$$

$$= 2 \int d\mathbf{r}' \epsilon(\mathbf{r}) \cdot \overline{\varphi_1(\mathbf{r}')} \cdot \overline{\varphi_1(\mathbf{r})} H_{12}(\mathbf{r}, \mathbf{r}')\varphi_1(\mathbf{r}')$$

$$+ 2 \int d\mathbf{r}' \overline{\epsilon(\mathbf{r})\varphi_1(\mathbf{r}')} \cdot H_{12}(\mathbf{r}, \mathbf{r}')\varphi_1(\mathbf{r})\varphi_1(\mathbf{r}')$$

We compute δIII:

$$\delta III = 0$$

The variational principle of RH consists in requiring the variational equation

$$\delta(I + II + III) = 0$$

to first order in ϵ and $\bar{\epsilon}$. The variational equation is written as

$$\int d\mathbf{r} \cdot \epsilon(\mathbf{r}) \cdot \left[H_{A,B}(\mathbf{r})\overline{\varphi_1(\mathbf{r})} + 2 \int d\mathbf{r}' \overline{\varphi_1(\mathbf{r}')} \cdot \overline{\varphi_1(\mathbf{r})} H_{12}(\mathbf{r}, \mathbf{r}')\varphi_1(\mathbf{r}') \right] = 0$$

and

$$\int d\mathbf{r} \cdot \overline{\epsilon(\mathbf{r})} \cdot \left[H_{A,B}(\mathbf{r})\varphi_1(\mathbf{r}) + 2 \int d\mathbf{r}' \overline{\varphi_1(\mathbf{r}')} \cdot H_{12}(\mathbf{r}, \mathbf{r}')\varphi_1(\mathbf{r})\varphi_1(\mathbf{r}') \right] = 0$$

These two equations are the complex conjugates of each other, so, if φ is a solution to one equation, $\bar{\varphi}$ is necessarily the solution to the second. We therefore only need to consider one of them.

Boundary condition: The variational principle is subject to a boundary condition:

$$\delta(\varphi_1, \varphi_1)$$

must vanish. This implies

$$(\varphi_1 + \epsilon, \varphi_1 + \epsilon) - (\varphi_1, \varphi_1) = (\epsilon, \varphi_1) + (\varphi_1, \epsilon) \overset{!}{=} 0$$

i.e. $\epsilon(\mathbf{r})$ is to be chosen among those functions which are perpendicular to $\varphi_1(\mathbf{r})$. Owing to this boundary condition, a choice of φ_1 such that

$$\underbrace{\left[H_{A,B}(\mathbf{r}) + \int d\mathbf{r}' \overline{\varphi_1(\mathbf{r}')} \cdot H_{12}(\mathbf{r}, \mathbf{r}')\varphi_1(\mathbf{r}') \right]}_{H_{HF} \doteq \text{Hartree–Fock}} \varphi_1(\mathbf{r}) = E \cdot \varphi_1(\mathbf{r})$$

is bound to fulfill both the variational equations. This last equation is an eigenvalue problem for the so-called **Hartree–Fock** Hamiltonian.

Comments. The Hartree–Fock effective Hamiltonian has important features:

- It is Hermitic and real, i.e. both solutions φ and $\overline{\varphi}$ belong to the same eigenspace. If the eigenspace is once degenerate, then

$$\overline{\varphi} = e^{i\phi}\varphi$$

This means that in a one-dimensional eigenspace, one can always define a new function

$$\propto (\varphi + e^{i\phi}\varphi)$$

which is real. If there are two solutions, φ_1 and φ_2, in the eigenspace to E, then one can construct the two real functions

$$\varphi_1 + \overline{\varphi}_1; \quad i \cdot (\varphi_1 - i \cdot \overline{\varphi}_1)$$

- The Hamiltonian contains a non-local term that, in turn, contains the sought-after function $\varphi_1(\mathbf{r})$. On the other hand, the application of the variational principle has transformed the original two-particle eigenvalue problem into an effectively one-particle eigenvalue problem for an effectively one-particle molecular orbital.
- As the desired function is contained in the operator itself, the eigenvalue problem is a nonlinear one. Typically, it is solved self-consistently. One uses as *ansatz* some function for computing the non-local term and solving the "regular", linear eigenvalue problem. If the solution is "sufficiently close" to the *ansatz*, the function is taken as the "best one" in the sense of the variational principle.
- The physical meaning of the eigenvalue E is computed in the review paper by Roothaan: $-E$ is the energy required to remove one electron from the molecule by keeping the wave function of all other electrons the same, i.e. $-E$ is, for instance, the ionization energy observed in a photoemission process, where the electrons left behind after excitation do not have the time to relax in virtue of the photoemission process being a very quick one. This justifies the name of "single-particle energy levels" for those values of E that fulfill the Hartree–Fock equation. The lowest one, among them, is the ground-state energy level E_0, and the corresponding molecular orbital φ_1^0 describes the ground state.

- The eigenvalues E resulting from the solution of the Hartree–Fock equation depend explicitly on the parameter R, which appears in $\langle H_{AB} \rangle$ and, because of the non-local term, in the $\langle H_{12} \rangle$ term as well. Accordingly, the total energy of the molecule is expressed as

$$E_{\text{Slater}}(R) = \underbrace{2 \cdot (\varphi_1^0(\mathbf{r}), H_{A,B}(\mathbf{r})\varphi_1^0(\mathbf{r})) + 2(\varphi_1^0(\mathbf{r}_1)\varphi_1^0(\mathbf{r}_2), H_{12}(\mathbf{r}_1, \mathbf{r}_2)\varphi_1^0(\mathbf{r}_1)\varphi_1^0(\mathbf{r}_2))}_{2E_0(R)}$$

$$- (\varphi_1^0(\mathbf{r}_1)\varphi_1^0(\mathbf{r}_2), H_{12}(\mathbf{r}_1, \mathbf{r}_2)\varphi_1^0(\mathbf{r}_1)\varphi_1^0(\mathbf{r}_2))(R)$$

$$+ \frac{e^2}{4\pi\epsilon_0} \cdot \frac{1}{R}$$

The equation

$$\frac{dE_{\text{Slater}}(R)}{dR} = 0$$

determines the equilibrium molecular radius and completes the energy minimization process, at the core of which is the eigenvalue problem of the Hartree–Fock operator. The generalization of this last step to more complex molecules is that the total energy of the ground state, computed over the Slater determinant, contains the geometry of the molecule as a parametric input. The "best" geometry is the geometry that corresponds to a minimum of the total ground state energy. In a practical computation, an initial guess of the geometry must be provided as an input to start the computation. Subsequently, the geometry is optimized so that the total energy of the system is at a minimum. For the remaining part of this project, these general aspects are no longer considered, as they relate to specific aspects of quantum chemistry. The interested reader is referred to the paper by Roothaan. Our scope here is more limited: We would like to further analyze the eigenvalue problem of the Hartree–Fock operator itself and solve it for a model system.

- In the paper by Roothaan, one can find the generalization of the HF equation to n molecular orbitals $\varphi_1, \varphi_2, \ldots, \varphi_n$. The eigenvalue problem is a set of n coupled equations. For the ith orbital φ_i, the equation is written as

$$\underbrace{\left[H_\alpha(\mathbf{r}) + \sum_{j=1}^{n} (2J_j(\mathbf{r}) - K_j(\mathbf{r})) \right]}_{\text{Hartree–Fock operator}} \varphi_i(\mathbf{r}) = E_i \varphi_i(\mathbf{r})$$

$$H_\alpha(\mathbf{r}) \doteq \left[-\frac{\hbar^2}{2m}\nabla^2 - \sum_{\alpha=A,B,\ldots} \frac{Ze^2}{4\pi\epsilon_0} \cdot \frac{1}{|\mathbf{r} - \mathbf{R}_\alpha|} \right]$$

is a single-particle operator that incorporates the kinetic energy of the electrons and their interaction with the nuclei. J_j is called the Coulomb operator:

$$J_j(\mathbf{r})\varphi_i(\mathbf{r}) = \int dV' \varphi_j(\mathbf{r}')^2 \frac{1}{|\mathbf{r} - \mathbf{r}'|}\varphi_i(\mathbf{r})$$

The expectation value of J_j,

$$J_{ij} = (\varphi_i, J_j\varphi_i) = \int dV \int dV' \varphi_j(\mathbf{r}') \cdot \varphi_j(\mathbf{r}') \frac{1}{|\mathbf{r} - \mathbf{r}'|}\varphi_i(\mathbf{r}) \cdot \varphi_i(\mathbf{r})$$

is the Coulomb energy between the two charge densities $\varphi_i^2(\mathbf{r})$ and $\varphi_j^2(\mathbf{r}')$. K_j is called the exchange operator:

$$K_j(\mathbf{r})\varphi_i(\mathbf{r}) = \int dV' \varphi_j(\mathbf{r}')\varphi_i(\mathbf{r}') \frac{1}{|\mathbf{r} - \mathbf{r}'|}\varphi_j(\mathbf{r})$$

The expectation value K_{ij} of K_j is

$$K_{ij} = \int dV \int dV' \varphi_j(\mathbf{r}') \cdot \varphi_i(\mathbf{r}') \frac{1}{|\mathbf{r} - \mathbf{r}'|}\varphi_j(\mathbf{r}) \cdot \varphi_i(\mathbf{r})$$

This double integral has no classical analog. It arises by exchanging the roles of the orbitals i and j in two of the integrands of J_{ij} (and thus the name "exchange integral").

27.2.2 The LCAO computation of the molecular orbitals

The eigenvalue equation of the Hartree–Fock operator is typically solved within a set of limited basis functions into which the sought-after molecular orbital is expanded. The coefficients of this expansion become the sought-after quantities. A set of common basis functions are atomic-like orbitals centered at the nuclear sites. For the specific case of the H_2 molecule, one could use the two orbitals

$$\psi_A = \psi_{1s}(\mathbf{r} - \mathbf{R}_A) \quad \psi_B = \psi_{1s}(\mathbf{r} - \mathbf{R}_B)$$

with

$$(\psi_A, \psi_A) = (\psi_B, \psi_B) = 1 \quad (\psi_A, \psi_B) = S \neq 0$$

Each of these orbitals describes an electron in the $1s$ configuration of the hydrogen atom. This is the configuration with the lowest atomic energy level in the atom, so that one can safely assume that such orbitals might be good trial functions for the ground state of the molecule. One could, alternatively, start with, say, two $2s$ wave functions. These would be good trial functions for picking up, for example, the excited states. Note that the overlap integral S is generally non-vanishing for atomic orbitals centered at different sites.

Roothaan and Hall have shown how to deal mathematically with the LCAO ansatz for the molecular orbitals. In their work, the LCAO expansion is inserted into the expression for the expectation value of the Hamilton operator containing the molecular orbitals. This produces a functional defined onto the set of expansion coefficients and a set of integrals over known atomic orbitals. The functional over the expansion coefficients is then minimized using the method of variations, under the constraint that the molecular orbitals are orthonormal (this constraint is also expressed in terms of the expansion coefficients). The variational principle produces a set of algebraic equations for the coefficients.

One can obtain this set of equations using a more straightforward method, which consists of inserting the expansion ansatz into the Hartree–Fock equation and performing scalar multiplication on both sides of the Hartree–Fock equation from the left with each of the LCAO basis functions. As an example, for the two-LCAO problem of the H_2 molecule, the set of algebraic equations is written as the eigenvalue problem of a matrix:

$$\begin{pmatrix} (\psi_A, H_{HF}\psi_A) & (\psi_A, H_{HF}\psi_B) \\ (\psi_B, H_{HF}\psi_A) & (\psi_B, H_{HF}\psi_B) \end{pmatrix} \begin{pmatrix} c_A \\ c_B \end{pmatrix} = E \begin{pmatrix} 1 & S \\ S & 1 \end{pmatrix} \begin{pmatrix} c_A \\ c_B \end{pmatrix}$$

The matrix on the left is usually referred to as the Hartree–Fock matrix, and equations of this type involving the expansion coefficients are referred to as Roothaan–Hall equations. One important feature of this eigenvalue problem is, however, that the non-local term in H_{HF} contains the sought-after eigenfunctions, so that the matrix elements contain the sought-after coefficients. Precisely,

$$H_{HF}^{i,j} = F_{i,j}(c_a^2, c_A \cdot c_b, c_b^2)$$

This produces a set of nonlinear (precisely, cubic) equations for the desired coefficients. These equations are typically solved self-consistently: One assumes, to start with the computation, a set of coefficients and constructs the molecular orbitals that are normalized to "1". These tentative molecular orbitals are inserted in the matrix elements. This step renders the remaining algebraic equations linear. One subsequently solves the linear eigenvalue problem. The resulting coefficients are then used to recompute the matrix elements. The loop is repeated until the coefficients solving the linear equations do not differ essentially from one loop to the next.

One final remark: Atomic orbitals (of the Slater type) have a cusp at the origin, which renders the computation of the numerous two-, three-, and four-site integrals over the Coulomb kernel entailed by the Roothaan–Hall equations very time-consuming. Modern software for quantum chemistry computations uses Gaussian-type orbitals: They are less precise at the origin; therefore, one needs more of them to simulate the cusp-like behavior. However, the computation time of the many-site integrals with Gaussian-type orbitals is reduced by one or more orders of magnitude. The idea of using Gaussians for speeding up practical computations originates from the work of J. A. Pople (Nobel Lecture in https://www.nobelprize.org/uploads/2018/06/pople-lecture.pdf).

27.3 Symmetry Arguments in Quantum Chemistry Computations

The original Hamiltonian is invariant with respect to some group of isometries that transform the molecule into itself. The same invariance should be imposed on the Hartree–Fock operator. While the local component is indeed manifestly invariant with respect to those transformations that transform the molecule into itself, the non-local one requires a closer inspection.

We first study the transformation properties of the Coulomb operator

$$J_j(\mathbf{r}) = \int dV' \varphi_j(\mathbf{r}')^2 \frac{1}{|\mathbf{r} - \mathbf{r}'|}$$

We compute its transformation behavior under the isometry $\mathbf{y} = R\mathbf{r}$:

$$J_j(R^{-1}\mathbf{y}) = \int dV' \varphi_j(\mathbf{r}')^2 \frac{1}{|R^{-1}\mathbf{y} - \mathbf{r}'|}$$

Then, we perform the change of variable:

$$\mathbf{y}' = R\mathbf{r}'$$

and substitute it in the integral, which becomes

$$\int dV_{y'} \varphi_j^2(R^{-1}\mathbf{y}') \frac{1}{|R^{-1}\mathbf{y} - R^{-1}\mathbf{y}'|} = \int dV_y \varphi_j^2(R^{-1}\mathbf{y}) \frac{1}{|\mathbf{y} - \mathbf{y}'|}$$

because of the invariance of the kernel with respect to any isometry. The transformation property of the Coulmob integral depends on the transformation properties of the tensor product function

$$\varphi_j^2(\mathbf{r})$$

We have two alternatives:

Alternative A: The desired solution $\varphi_j(\mathbf{r})$ transforms according to a one-dimensional representation — let us call it Γ_j — of the symmetry group. Its characters can be assumed to be real, as the symmetry-adapted functions are expected to be real, i.e. its characters are either "1" or "$\bar{1}$". The character of the tensor product of the representation Γ_j with itself is therefore "1" for any element, i.e. the tensor product function $\varphi_j^2(\mathbf{r})$ and with it the Coulomb operator $J_j(\mathbf{r})$ are invariant with respect to all elements of the symmetry group.

Alternative B: Γ_j has a larger dimensionality and carries a set of basis functions $\varphi_{j,1}, \ldots, \varphi_{j,d}$. As $\Gamma_j \otimes \Gamma_j$ contains the one-dimensional "1" representation once, we ought to search for the symmetry-adapted linear combination of the product functions $\varphi_{j,n} \cdot \varphi_{j,m}$ that transform as the representation "1". The sought-after symmetry-adapted function is

$$\sum_{n=1}^{d} \varphi_{j,n} \cdot \varphi_{j,n}$$

The use of this symmetry-adapted function renders the corresponding Coulomb operator invariant with respect to the symmetry group of the Hamilton operator.

We now study the transformation properties of the exchange operator

$$K_j(\mathbf{r})\varphi(\mathbf{r}) = \int dV' \varphi_j(\mathbf{r}')\varphi(\mathbf{r}') \frac{1}{|\mathbf{r}-\mathbf{r}'|}\varphi_j(\mathbf{r})$$

We compute

$$K_j(R^{-1}\mathbf{y})\varphi(R^{-1}\mathbf{y}) = \int dV' \varphi_j(\mathbf{r}')\varphi(\mathbf{r}') \frac{1}{|R^{-1}\mathbf{y}-\mathbf{r}'|}\varphi_j(R^{-1}\mathbf{y})$$

Then, we perform the change of variable:

$$\mathbf{y}' = R\mathbf{r}'$$

and substitute it on the right-hand side to obtain

$$\int dV_{\mathbf{y}'} \varphi_j(R^{-1}\mathbf{y}')\varphi(R^{-1}\mathbf{y}) \frac{1}{|R^{-1}\mathbf{y}-R^{-1}\mathbf{y}'|}\varphi_j(R^{-1}\mathbf{y})$$

$$= \int dV_{\mathbf{y}'} \varphi_j(R^{-1}\mathbf{y}')\varphi(R^{-1}\mathbf{y}) \frac{1}{|\mathbf{y}-\mathbf{y}'|}\varphi_j(R^{-1}\mathbf{y})$$

Looking back at the alternatives a and b, we conclude that

$$\varphi_j(\mathbf{y})\varphi_j(\mathbf{y}')$$

can be rendered invariant using the procedure given in alternative a or alternative b. Accordingly,

$$K_j(\mathbf{r})\varphi(\mathbf{r})$$

transforms as $\varphi(\mathbf{r})$, i.e. the exchange operator is invariant with respect to the symmetry group of the Hamilton operator.

In summary, provided suitable symmetry-adapted functions are used in it, the Hartree–Fock Hamiltonian also becomes invariant with respect to the symmetry operations that transform the molecule into itself.

The use of symmetry in the molecular orbitals LCAO method boils down to finding symmetry-adapted linear combinations of atomic orbitals, given a suitable starting set. Symmetry-adapted linear combinations then enter quantum chemistry computations at two levels. The first level is the setting up of the matrix elements of the Hartree–Fock matrix. The use of symmetry-adapted linear combinations fixes the weight of the atomic orbitals and thus greatly reduces the uncertainty in setting up the matrix elements.

• For example, let us consider the case of the H_2 molecule. Without considering its symmetry, the Roothan–Hall equation is written as

$$\begin{pmatrix} F_{1,1}(c_a^2, c_A \cdot c_b, c_b^2) & F_{1,2}(c_a^2, c_A \cdot c_b, c_b^2) \\ F_{1,2}(c_a^2, c_A \cdot c_b, c_b^2) & F_{2,2}(c_a^2, c_A \cdot c_b, c_b^2) \end{pmatrix} \begin{pmatrix} c_A \\ c_B \end{pmatrix} = E \begin{pmatrix} 1 & S \\ S & 1 \end{pmatrix} \begin{pmatrix} c_A \\ c_B \end{pmatrix}$$

to produce a set of coupled algebraic equations containing the third powers of the sought-after coefficients. We now consider that the molecule has the $C_{v\infty}$ symmetry, so that, with the two orbitals ψ_A and ψ_B, one can construct two symmetry-adapted molecular orbitals belonging to two different, one-dimensional representations:

$$\psi_+ \doteq \frac{1}{\sqrt{2 \pm 2S}}(\psi_A + \psi_B)$$

and

$$\psi_- \doteq \frac{1}{\sqrt{2 \pm 2S}}(\psi_A - \psi_B)$$

This fixes the values of c_A and c_B and transforms the Roothan–Hall equations:

$$E_+ = (\psi_+, H_{HF}\psi_+) \quad E_- = (\psi_-, H_{HF}\psi_-)$$

These matrix elements involve only the computation of integrals over known atomic orbitals, with the sought-after coefficients appearing in the integrals as fixed numbers.

The second level where symmetry enters the problem is the more conventional one: Once the matrix elements have been set up with suitable symmetry-adapted trial functions, a conventional linear problem emerges, which is accessible to standard symmetry arguments.

27.4   The Borane Molecule BH$_3$

We now work out the case of a molecule that is sufficiently complex to expose the pitfalls of quantum chemistry but sufficiently simple so that one can understand the full role played by symmetry: the tetratomic molecule trihydridoboron (BH$_3$), also commonly known as borane or borine. The molecule is made of a central boron atom, surrounded by three hydrogen atoms. At equilibrium, i.e. neglecting all vibrational modes, all atoms are found in a single plane with the hydrogen atoms equally spaced around the central boron. The following figure is a scheme of the borane molecule and indicates its symmetry elements.

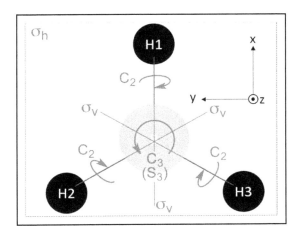

The symmetry elements are indicated in red.

The main C_3 axis is along the z-direction. This rotational element maps all three hydrogen atoms onto each other, leaving the boron at the origin unchanged. There are three σ_v mirror planes and three C_2 axes. The planar nature of the molecule produces a σ_h mirror plane (the xy-plane). Since the normal of the σ_h mirror plane is parallel to the C_3 axis, the C_3 main axis is also an S_3 axis. Taken all together, these symmetry elements, along with the identity, form the D_{3h} point group whose character table is shown as follows.

Character table of D_{3h}.

D_{3h}	E	$2C_3(z)$	$3C_2$	$\sigma_h(xy)$	$2S_3$	$3\sigma_v$
A_1'	1	1	1	1	1	1
A_2'	1	1	$\bar{1}$	1	1	$\bar{1}$
E'	2	$\bar{1}$	0	2	$\bar{1}$	0
A_1''	1	1	1	$\bar{1}$	$\bar{1}$	$\bar{1}$
A_2''	1	1	$\bar{1}$	$\bar{1}$	$\bar{1}$	1
E''	2	$\bar{1}$	0	$\bar{2}$	1	0

The lowest-energy atomic orbitals that are considered as trial basis states for the lowest-energy molecular orbitals are sketched in the following figure.

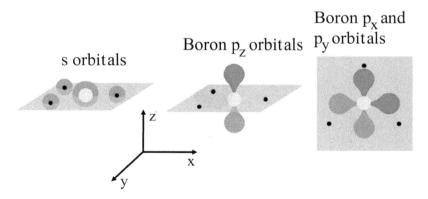

The figure shows the atomic orbitals used for the computation. Red and blue colors denote those sections of the orbitals with opposite signs.

This set includes the three $1s$ H orbitals constituting the $1s$ level at $-13.6\,\mathrm{eV}$ binding energy, the $2s$ B orbital constituting the B-$2s$ level at $-14.0\,\mathrm{eV}$ binding energy, the p_z B orbital aligned along the z-axis, and the p_x, p_y in-plane B orbitals constituting the B-$2p$ level at $-8.3\,\mathrm{eV}$. The lowest state of the B atom (the $1s$ level) is too low in energy to contribute to the chemical bond and must be considered as a "core" state and not taken into account for the computation.

The goal of symmetry arguments is to find the symmetrized linear combinations of the atomic orbitals that can be used to build the matrix elements of the HF matrix (and, afterward, to solve the linear eigenvalue problem). The first step toward realizing this goal is to establish what is technically called the "action table". This table gives the result of the symmetry elements acting on each individual basis state (in the current case, the seven atomic orbitals). The action table is shown as follows.

Table of the action of the operations of D_{3h} on the atomic orbitals of borane. The dihedral rotations and reflections are labeled according to the hydrogen atom that remains unchanged.

	E	$C_3(z)$	$C_3^2(z)$
H$_1$ 1s	H$_1$ 1s	H$_3$ 1s	H$_2$ 1s
H$_2$ 1s	H$_2$ 1s	H$_1$ 1s	H$_3$ 1s
H$_3$ 1s	H$_3$ 1s	H$_2$ 1s	H$_1$ 1s
B 2s	B 2s	B 2s	B 2s
B 2p_x	B 2p_x	$-\frac{1}{2}$(B 2p_x)$+\frac{\sqrt{3}}{2}$(B 2p_y)	$-\frac{1}{2}$(B 2p_x)$-\frac{\sqrt{3}}{2}$(B 2p_y)
B 2p_y	B 2p_y	$-\frac{\sqrt{3}}{2}$(B 2p_x)$-\frac{1}{2}$(B 2p_y)	$\frac{\sqrt{3}}{2}$(B 2p_x)$-\frac{1}{2}$(B 2p_y)
B 2p_z	B 2p_z	B2p_z	B2p_z

...	$\sigma_h(xy)$	$S_3(z)$	$S_3^2(z)$
H$_1$ 1s	H$_1$ 1s	H$_3$ 1s	H$_2$ 1s
H$_2$ 1s	H$_2$ 1s	H$_1$ 1s	H$_3$ 1s
H$_3$ 1s	H$_3$ 1s	H$_2$ 1s	H$_1$ 1s
B 2s	B 2s	B 2s	B 2s
B 2p_x	B 2p_x	$-\frac{1}{2}$(B 2p_x)$+\frac{\sqrt{3}}{2}$(B 2p_y)	$-\frac{1}{2}$(B 2p_x)$-\frac{\sqrt{3}}{2}$(B 2p_y)
B 2p_y	B 2p_y	$-\frac{\sqrt{3}}{2}$(B 2p_x)$-\frac{1}{2}$(B 2p_y)	$\frac{\sqrt{3}}{2}$(B 2p_x)$-\frac{1}{2}$(B 2p_y)
B 2p_z	-1(B 2p_z)	-1(B 2p_z)	-1(B 2p_z)

...	$C_2'(\text{H}_1)$	$C_2'(\text{H}_2)$	$C_2'(\text{H}_3)$
H$_1$ 1s	H$_1$ 1s	H$_3$ 1s	H$_2$ 1s
H$_2$ 1s	H$_3$ 1s	H$_2$ 1s	H$_1$ 1s
H$_3$ 1s	H$_2$ 1s	H$_3$ 1s	H$_1$ 1s
B 2s	B 2s	B 2s	B 2s
B 2p_x	B 2p_x	$-\frac{1}{2}$(B 2p_x)$-\frac{\sqrt{3}}{2}$(B 2p_y)	$-\frac{1}{2}$(B 2p_x)$+\frac{\sqrt{3}}{2}$(B 2p_y)
B 2p_y	-1(B 2p_y)	$-\frac{\sqrt{3}}{2}$(B 2p_x)$+\frac{1}{2}$(B 2p_y)	$\frac{\sqrt{3}}{2}$(B 2p_x)$+\frac{1}{2}$(B 2p_y)
B 2p_z	-1(B 2p_z)	-1(B 2p_z)	-1(B 2p_z)

From the action table, we observe that the seven-dimensional space determined by the selected atomic orbitals is closed with respect to the operations of D_{3h} and thus can be used to build a seven-dimensional permutation representation (designated here with Γ) of D_{3h}. This representation commutes with the HF operator for the borane molecule: Its reduction into irreducible representations of D_{3h} provides the labeling of the molecular orbitals. The corresponding symmetrized linear combinations are the most precise basis states for building the HF matrix and solving the eigenvalue problem. To find the decomposition of Γ into irreducible representations, we observe from the table that there are groups of orbitals that transform into themselves upon acting on them with the symmetry elements. This means that Γ is the direct sum of smaller matrices. Finding the characters of each of these smaller representations is a more manageable task than finding the characters of the entire Γ at once:

- All hydrogen $1s$ orbitals transform into each other. Accordingly, we can establish a $3\times$ block Γ_{3H}.
- The two p_x, p_y orbitals also transform into each other, generating the block Γ_{2pxy} of Γ.
- The $2p_z$ orbital generates the one-dimensional representation Γ_{2pz}
- The $2s$ orbital of boron generates the one-dimensional representation Γ_{2s}

In summary,

$$\Gamma = \Gamma_{3H} \oplus \Gamma_{2pxy} \oplus \Gamma_{2pz} \oplus \Gamma_{2s}$$

From the action table, we can also estimate the characters of the individual representations. For example, $\sigma_h(xy)$ leaves each individual hydrogen orbital invariant and its matrix representation carries three "1" along the diagonal, i.e. its character is "3". Spotting those orbitals that remain precisely unchanged allows us to assign the characters for Γ_{3H} given in the following table. $2p_z$ is left unchanged by some elements and transformed into -1 times itself by others. The characters of Γ_{2pz} are also given in the table. The boron $2s$ orbital is left unchanged by the actions of all elements. The 2×2 matrix representation of Γ_{2pxy} can also be determined from the action table and the characters assigned by summing the diagonal elements.

From the following table, the irreducible representations entering the blocks of Γ are apparent. We expect from solving the eigenvalue problem of the HF operator of the borane molecule within this seven-dimensional space:

- two twice-degenerate levels, each carrying the symmetry E'.
- two single-degenerate levels, each with symmetry A_1'.
- one single-degenerate level with symmetry A_2'.

(each level carrying a spin singlet).

Character table of D_{3h} and decomposition of the individual representations into irreducible components.

D_{3h}	E	$2C_3(z)$	$3C_2$	$\sigma_h(xy)$	$2S_3$	$3\sigma_v$	
A_1'	1	1	1	1	1	1	
A_2'	1	1	$\bar{1}$	1	1	$\bar{1}$	
E'	2	$\bar{1}$	0	2	$\bar{1}$	0	
A_1''	1	1	1	$\bar{1}$	$\bar{1}$	$\bar{1}$	
A_2''	1	1	$\bar{1}$	$\bar{1}$	$\bar{1}$	1	
E''	2	$\bar{1}$	0	$\bar{2}$	1	0	
Γ_{3H}	3	0	1	3	0	1	$A_1' \oplus E'$
Γ_{2pz}	1	1	$\bar{1}$	$\bar{1}$	$\bar{1}$	1	A_2''
Γ_{2s}	1	1	1	1	1	1	A_1'
Γ_{2pxy}	2	$\bar{1}$	0	2	$\bar{1}$	0	E'

To construct symmetry-adapted atomic orbitals from the three $1s$ H orbitals, we use the projector technique. With the help of the action table, we find

$$P_{A_1'}(H_1 1s) \propto H_1 1s + H_2 1s + H_3 1s$$

$$P_{E'}(H_1 1s) \propto \frac{1}{3}(H_1 1s) - \frac{1}{6}(H_2 1s) - \frac{1}{6}(H_3 1s)$$

$$P_{E'}(H_2 1s) \propto \frac{1}{3}(H_2 1s) - \frac{1}{6}(H_3 1s) - \frac{1}{6}(H_1 1s)$$

$$P_{E'}(H_3 1s) \propto \frac{1}{3}(H_3 1s) - \frac{1}{6}(H_1 1s) - \frac{1}{6}(H_2 1s)$$

Adding the two last functions with E' symmetry, we obtain two orthogonal basis functions with E' symmetry out of the three H atoms. The assignments of the further symmetry-adapted atomic orbitals are straightforward as the individual tower representations correspond to irreducible representations. In summary:

- Orbitals with E' symmetry are

$$\{Bp_x, Bp_y\}; \quad \left\{ \frac{1}{3}(H_1 1s) - \frac{1}{6}(H_2 1s) - \frac{1}{6}(H_3 1s), \frac{1}{2}(H_2 1s) - \frac{1}{2}(H_3 1s) \right\}$$

- Orbitals with A_1' symmetry are

$$B2s; \quad H_1 1s + H_2 1s + H_3 1s$$

- One orbital with A_2'' symmetry is

$$Bp_z$$

(all functions need to be normalized to "1" in order to use them as the non-local component of the HF matrix equations). The following figure is a graphical representation of the symmetry-adapted atomic functions obtained from the three H $1s$ orbitals.

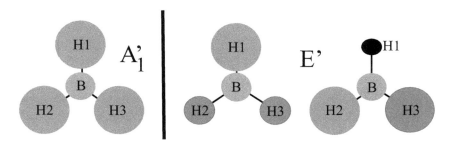

A graphical representation of the symmetry-adapted atomic orbitals obtained from the H $1s$ orbitals. The red and blue colors indicate sections of the wave functions with opposite signs.

The use of these symmetrized orbitals allows us to construct a block-diagonalized HF matrix, see the following figure. The block-diagonalization reduces the degrees of freedom of the problem by constraining many matrix elements to be vanishing. In addition, the number of sought-after parameters to be optimized is reduced from six to two: the relative weight of the two A_1' orbitals and the relative weight of the two pairs of E'-functions. There is still a large amount of one-, two-, three-, and four-site integrals over atomic orbitals to be computed, of course. Fortunately, most of them are available in libraries.

The values of the energy levels, their ordering, and the exact shape of the wave functions are the result of diagonalizing the self-consistent HF block matrix through quantum chemistry computations. The following figure shows the electronic structure of borane as computed with a state-of-the-art Gaussian-type software (see Thom H. Dunning Jr., "Gaussian basis sets for use in correlated molecular calculations. I. The atoms boron through neon and hydrogen", *J. Chem. Phys.* **90**, 1007–1023 (1989), https://doi.org/10.1063/1.456153). The basis set used in this computation is Gaussian. The trial space was extended to include further functions with higher l content to improve the precision of the computation.

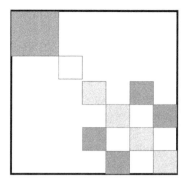

The block-diagonalized HF matrix: The matrix elements in white are vanishing by symmetry. The two A_1' symmetrized orbitals host a two-dimensional block (top, in green). Their relative weight must be included in the matrix as the sought-after coefficients to be optimized. The A_2'' orbital constitutes a one-dimensional block (yellow). The two pairs of E' orbitals constitute a four-dimensional block (bottom right). The relative weight of the pairs must be included in the matrix as the sought-after coefficients to be optimized.

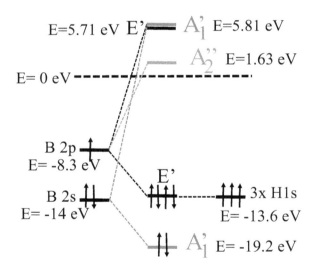

The molecular orbital diagram of borane showing the calculated energies of the molecular orbitals as well as the literature values for the atomic orbital energies. Atomic levels are in black. E' molecular levels are also in black. Further molecular symmetries are given in color. The vertical energy dimension of the diagram is to scale. Atomic energies have been taken from Harry B. Gray, Electrons and Chemical Bonding, W. A. Benjamin, New York, 1965, p. 218, https://archive.org/details/ost-chemistry-electrons_chemical_bonding.

The following two figures show the shape of the computed eigenstates. The top figure shows the bonding and antibonding A'_1 states. Note their resemblance with the states that we could intuitively anticipate by the $\pm$-hybridization of the s state of B with the symmetry-adapted A'_1 function built by the three s-states of H. We observe a triangular spheroid in the bonding case ($+$) and a B atom out of phase with respect to the H atoms in the anti-bonding case ($-$). On the left hand side of the bottom figure are the bonding E' states. Again, they have a resemblance with the E' states that we could intuitively anticipate by hybridizing the two symmetry-adapted three s-states of H with and the Bp_x, Bp_y states. On the right is the A''_2 orbital, which very much resembles the p_z orbital used in this work.

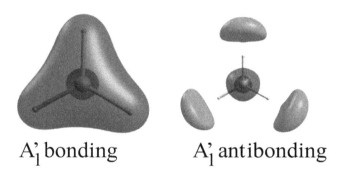

$$A'_1 \text{ bonding} \qquad A'_1 \text{ antibonding}$$

A visual representation of the computed eigenfunctions. An isosurface is plotted for all points in space for which the absolute value of the wave function has the same value.

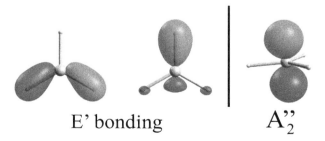

$$E' \text{ bonding} \qquad A''_2$$

A visual representation of the computed eigenfunctions: An isosurface is plotted for all points in space for which the absolute value of the wave function has the same value.

The final remark concerns the computational costs of quantum chemistry calculations. The following figure summarizes the time required for computing the electronic structure of some known molecules. Two HF-based computations were carried out sequentially on the same machine. In one computation, it was specified in the input file that no symmetry should be used (Symmetry = none). Switching off the symmetry-based input extended the computation time by a factor of two to four, depending on the size of the molecules.

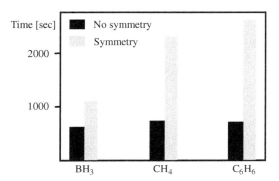

Computational cost of using symmetry vs. not using symmetry.

27.4.1 *Jahn–Teller-type distortion*

The highest occupied molecular orbital (HOMO) of the borane molecule has E' symmetry, see the next figure.

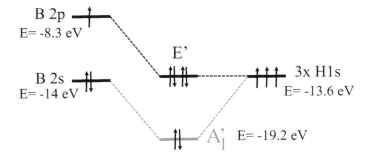

The occupied molecular orbitals in borane.

It has double degeneracy and is, accordingly, occupied by four electrons (counting the spin state). Upon removal of one electron from this state, the total energy (which depends on the geometry of the molecule by the electronic contribution and the potential energy of the nuclei) must be computed again in the presence of an asymmetric electronic potential distribution. It is possible that a phenomenon *à la* the Jahn–Teller effect — the distortion of the molecular geometry driven by the splitting of the doubly degenerate E' level (see the following figure) — produces a distortion of the molecule away from the planar, triangular arrangement.

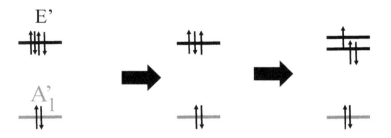

Illustration of the Jahn–Teller splitting in the electronic structure of borane.

Looking at the possible subgroups of D_{3h}, one finds, for example, C_{3v}, which corresponds to a lifting along the z-direction of the central B atom. In this subgroup, however, the E' irreducible representation remains doubly degenerate. A further subgroup is C_{2v}, which is constructed by selecting the elements E, σ_{xy}, one C_2 element, and one σ_v element. The molecular arrangement realizing this subgroup is shown in the following figure.

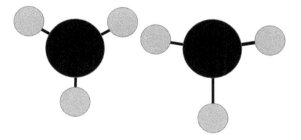

Distortion (right) of borane (left) realizing the C_{2v} subgroup.

When limited to the elements of this subgroup, the E' irreducible representation of D_{3h} becomes reducible and splits into the two irreducible representations $A_1 \oplus B_1$ of C_{2v} (see the following table),

Character table of C_{2v}, including the representation E'.

C_{2v}	E	$C_2(z)$	$\sigma_h(xy)$	$\sigma_v(xz)$	
A_1	1	1	1	1	
A_2	1	1	$\bar{1}$	$\bar{1}$	
B_1	1	$\bar{1}$	1	$\bar{1}$	
B_2	1	$\bar{1}$	$\bar{1}$	1	
E'	2	0	2	0	

producing a level splitting that might lower the total energy of the system. The total energy quantum chemistry computations of the BH_3^+ molecule find indeed that the distorted molecule represents an absolute minimum of the total energy.

Experimental observation

Borane is a very unstable molecule and not suitable for experimental spectroscopy. Experimentally, one uses the more stable CF_4 molecule with the T_d point group. It has an unoccupied (LUMO) orbital which is threefold degenerate and has symmetry T_2. A soft X-ray photon with suitable energy (about $300\,\mathrm{eV}$) can excite a core $1s$ electron into this LUMO state, producing a peak in the absorption spectrum. When an electron is removed from the molecule, such as by an intense, very short pulse of infrared light,

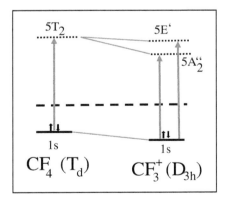

Left: The core-level electrons ($1s$) are excited into an unoccupied T_2 level in the CH_4 molecule. The symmetry is T_d. An intense short pulse removes one electron and the molecule, within about 100 fs, becomes a CF_3^+-ion with D_{3h} point group symmetry (right). The unoccupied level splits accordingly. A dashed horizontal thick line separates the deep-lying core levels from the unoccupied levels.

the molecule lowers its symmetry by extending one of its C–F bonds, resulting in the formation of a trigonal planar CF_3^+ ion with D_{3h} point group symmetry. In this environment, the T_2 irreducible representation decomposes into $E' \oplus A_2''$.

Accordingly, a splitting of LUMO level should occur, and the absorption peak should split into two separate peaks. Exactly, this splitting is observed in a pump (the infrared pulse) — probe (an ultrafast X-ray pulse following immediately the infrared pulse) spectroscopy experiment (Y. Pertot *et al.*, "Time-resolved x-ray absorption spectroscopy with a water window high-harmonic source", *Science*, **355**, 264–267 (2017), https://www.science.org/doi/10.1126/science.aah6114).

INDEX

1-representation, 63
3×3 unitary matrices, 305
Δ-line, 511
Γ point, xxiv, 512
Γ-point of GaAs, xxiv
Λ-direction, 132
Λ^0 baryon, 312
π-bands, xxiv, 525
π-energy band dispersion, 536
π-energy levels, xxiv
π–p scattering, 309
π-mesons, 307
$\pi_1(S_r)$, 558
$\pi_1(S_r, x_0)$, 558
σ_d, 9
σ_h, 9
σ_v, 9
Γ notation, 390
14 Bravais lattices, 178

AA stacking, 554
AB stacking, 554
Abel, Niels Henrik, 25
Abelian subgroup, 87
absorption peak, 604
absorption spectrum, 603
accidental degeneracy, 118, 216
accidental/essential degeneracies, 213
action table, 595
action–reaction, 140
active representation, 354, 474
active vibrational mode, 482

adjoint matrix, 43
adjoint operator, 443
Adjoint representation, 396
algebraic properties of groups, 23, xvii
algebraic structure, xvii
algebraic topology, xxiv, 555
analytic properties of continuous
    groups, 28
analytical and numerical methods, ix
angular momentum, 461
angular momentum operator, xiv
angular variable, ix
antibonding charge density, 581
anticlockwise rotations, 14
anticrossing of bands, 219
antiferromagnetically ordered chain of
    spins, 494
antihomomorphism, 48
antiparallel spin alignment, 580
antiprismatic point groups, 162
antisymmetric products, 295
antisymmetry with respect to
    coordinate exchange, 255
anyon, 564
anyon representations, 564
anyonic character, 564
associativity, 7
atomic chain with basis, 229
atomic levels, xx
atomic orbitals, 126, 590
atomic probability distribution, 348
atomic-like orbitals, 588

Au, Cu, and Ag, 195
averaging over the group, 31
avoided crossing, xx, 213, 219
axial group, 33
axial rotation group, 81

Baker, Henry Frederick, 92
band crossing, 510
band number, 204
band structure, 503
band structure at the Γ-point, 132
band structure factor, 534
band structure of graphene, 540
band structure of solids, xx, 131, 213
Bardeen, W., 318
baryon number, 312
baryons, 314
Basal plane displacements, 483
basal translational modes, 481
base point, 556
base states of the world, 433
basis functions, 80
basis set, 5
basis states, 40
basis vector, 6, 200, 431
benzene, 161
Bessel inequality, 60
Bethe Hypothesis, xx
Bethe, Hans, 126, 233, 238, 390
Bieberbach, Ludwig Georg, 168
bijective, 27
Bloch condition, 229
Bloch functions, 526
Bloch sums, 534
Bloch theorem, 203
Bloch, Felix, 203, 228
block matrices, 53
block-diagonalized HF matrix, 598
block-diagonalized matrices, 54
Boerner, Hermann, 296, 325, 336
Bohr circular orbits, 234
Bohr, Niels, 234
Bohr–Sommerfeld quantized model, 234
bonding charge density, 581

borane, 593
Borane Molecule, 593
Borane Molecule BH_3, xxv
borine, 593
Born, Max, 135, 257, 577
Born–Oppenheimer approximation, 578
Born–Oppenheimer operator, 578
Born–von Karman boundary conditions, 198
Boson, 564
Bouckaert, L. P., 69, 390
bound states, 444
boundary condition, x, 430
bounded and closed, 33
Bravais lattices, xix
Bravais, Auguste, xii
Brillouin zone (BZ), 132, 199
Brillouin, Léon Nicolas, 69, 503
buckled-down atoms, 549
buckling down, 548
building blocks, 49, 57
Burnside theorem, 68
Burnside, William, 64, 325

C_2, 8
C_3, 9
C_n, 8, 153
C_{2v}, 70
C_{3v}, 70
$C_{\infty h}$, 29
C_∞, 33
C_{nv}, 15
C, Si, and Ge, 195
Campbell, John Edward, 92
Cartan diagram, 252
Cartan multiplicators, 319
Cartan weights, xviii, 234
Cartan, Elie, 252
Cartan, Elie Joseph, 82
Cartesian coordinate system, 139
Casimir, Hendrik, 90
Cauchy sequence, 33
Cayley, Arthur, 96, 303, 379
Center of inversion, 9

center of the BZ, 208
centered rectangular lattice, 175, 185
centrosymmetric, 166
Character of a representation, 298
character table, 68, 83
characteristic equation, 96
charge accumulation, 581
charge density distribution, 473
chemical bond, 581
chemical notation, 390
Chevalley, C., 106
chirality, 560
circle group, 81, 83
circulant, 384
class function, 63
classes of mutually conjugate
  elements, 27
Classical mechanics, x, xxiii
classical physics, 328
Clebsch, Alfred, 75, 259
Clebsch–Gordan calculator, 262
Clebsch–Gordan Coefficients, xx
Clebsch–Gordan coefficients for
  double point groups, 271
Clebsch–Gordan series, xx, 251
closed loop, 411
closure, xi
color, 318
column vector, 19
commutation relations, 104
commutative, 25
commutator, 92, 102, 116
Commuting self-adjoint matrices,
  xxiii
compact group, 33
compatibility table, 127, 129
complete orthonormal set, 444
complete set, 41, 84
complete vector space, 432
completely reducible, 53
Completeness, 433
complex conjugate transpose, 43
complex numbers, 371
complex, non-singular $n \times n$ matrices,
  285

composition laws to first order, 98
Composition laws to second order,
  101
composition of functions, xi
composition of symmetry operations,
  7
composition table, 372
compositions of mappings, 18
compression, 476
Computation with YD, 295
computational costs, 601
concatenation of loops, 557
concentric spherical surfaces, 36
Condon and Shortley, 91
Condon, E. U., 91
configuration Hamiltonian, 254, 256,
  337
configuration space, 39
conjugacy classes, 27
conjugate, 22, 27
conjugation, 22
conjugation operation, 27
connectivity, xxiv, 556
CONS, 118, 433
Conservation Laws, xxiii, 457
conservation of charge, xv
constant loop, 558
constant of the motion, 458
constant of the quantum mechanical
  motion, 126
constants of the classical motion, 125
continuous curved line, 155
continuous deformation, 559
continuous group, 16, 233
continuous phase transition, 356
continuous variable, 81
Contour plot, 368
contra-gradient, 287
contra-variant vector, 287
contragredient, 114
conventional $n \times n$ matrices, 290
convolution, 56
coordinate transformation, 112
core-level electrons, 603
correspondence principle, 85

Cosets, xx, 223, 498
cosmic radiation, 312
Coulomb integral, 579
Coulomb interaction, 128, 255
Coulomb kernel, 590
Coulomb operator, 588
Coulomb repulsion, 128, 256
coupled algebraic system of
    equations, 366
coupling coefficients, 273, 572
coupling of degrees of freedoms, xx,
    249
coupling operator, xx
covariant vector, 287
crystal class, xix, 171
Crystal Field Splitting, xxiii
crystal potential, 519
crystal structure, 362
crystal systems, xix, 172
crystallization of a liquid, 347
crystallographic pyramidal groups,
    164
crystallographic restriction, 156, 172
crystallography, xii, xix, 151, 156
crystals, 3, 151
cubic system, 177
Curie temperature, 347
Curie, Pierre, 356, 384
curl-like displacement, 481
cusp-like behavior, 590
cutting the rope, 559
cycle, 322
cycled from left to right, 322
cyclic group, 27, 156, 371, 497
cyclic group of order n, 390
cyclic notation, 322

$D^{\frac{1}{2}}$, xx, 236
D_{6h}, 161
da Vinci, Leonardo, 153, 497
data storage material, 465
decomposition formula, 506
Decomposition theorem, 63
Dedekind, Julius Wilhelm Richard,
    384

deformation, 556
degeneracy, 127
degeneracy of the eigenvalue, 437
degenerate irreducible representation,
    474
degree of freedom, 240
delta function, 75
density function, 351
Density functional approach, 580
density functional theory (DFT), 531,
    535
density of symmetry elements, 32
descent tables, 476
determinant, 14
determinantal equation, xiii, 40
diagonal form, 437
diagonal matrix, 82
diagonal matrix elements, 437
diamond crystal structure, 195
diamond structure, 208
diatomic molecule, 407
differential equation, x, 84
diffraction pattern, 348
dihedral group D_n, 14, 153, 497
dilation, 21
Dimensionality of a YD, 297
dipolar approximation, 576
dipole approximation, 277
dipole selection rules, 277
Dirac cone in the vicinity of X, 552
Dirac cones, 197, 537
Dirac distribution, 75
Dirac, P. A. M., xxiv, 436
direct product of groups, xx, 233, 240
Direct sum of vectors and matrices,
    49
direct sum space, 50
director cosines, 266
Dirichlet boundary conditions, 450
Dirichlet, Johann Peter Gustav
    Lejeune, 135, 432
discontinuous (first-order) phase
    transition, 355
discrete mathematics, 297
disentanglement, 552

displacement, 139
distorted system, 372
distribution, xvi, 74, 436
dodecahedron, 154
domain of definition, 124
double n-pyramid, 157
double (or extra) group, 239
double group, 238, 563
double group representation, 264
double point groups, xxi, 263
double-group (or extra)
    representations, 563
double-group representation, 570
double-valued representation, 303
dual, 154
dual body, 71
dual pattern, 11
dynamical representations of the
    vibrational modes, 150

Eckart, C., 119
Eckart, Carl Henry, xxi
edge of the Brillouin zone, 548
effective potential, 466
eigenfrequencies, x, 429
eigenmode ansatz, 140, 406
eigenmodes, x
eigenstates, 40, 429
eigenvalue equation, x
eigenvalue problem, x, 430
eigenvalues, x
eightfold way, xxi
Einstein, Albert, 239
electron–electron interaction, 337
electronic structure of a crystal, 197
embedded atom, 126
Embedding in O_h, 372
empty-lattice approximation, 197
empty-lattice band structure, xx, 197,
    213, 503
empty-lattice eigenvalues, xxiii
empty-lattice energy levels, 504
enantiomorphic, 166
energy bands, 204
energy eigenvalues, 429

energy levels, x, 217
entangled, 243, 573
entangled states, 258
entanglement, 243, 259
entanglement between orbital and
    spin state, 280
entanglement of spin and orbital
    states, 233
entropy, 349
equatorial plane, 155
equilateral triangle, 361
equilibrium position, 139
equivalence relation, 557
equivalent representations, 53
Erhard, Schmidt, 123
Escher, Maurits Cornelis, 182
essential degeneracy, xx, 117, 219
Euclide, 239
Euclidean group, 20, 455
Euclidean space, xvii, 4, 6, 116
Euler, Leonhard, 97, 429, 462
even permutation, 323
exchange energy, 580
exchange integral, 588
exchange interaction, 328
exchange operator, 588
expectation value of the spin
    polarization vector operator, 573
experimental spectroscopy, 603
exponential map, 91
exponential mapping, 86
extra degree of freedom, 180
extremal values, 366

face-centered cubic lattice, 201
factor group, 226, 550
factor group of k, xx
family of equivalent loops, 557
Fermi level, 531
Fermion, 564
ferromagnetic state, 580
Feynman, Richard, 433
final state, 277
fine structure, xx, 235, 254
fine structure levels, 256

finite difference method (FDM), xxiii, 135, 430, 446
finite element, 405
finite element method, xxiii, 430, 450
finite irreducible representations, 83
finite number of constituents, 152
finite subgroups, 152, 500
finite subspace, 40
finite- and infinite-dimensional representations, xviii
finite-dimensional Hilbert space, xxii, 436
first BZ, 199
first homotopy group, 558
first-order perturbation theory, 466
fivefold axis, 166
Fjodorow, J. S., xii
flavors, 317
Fm3m, 194
Fock, Vladimir Aleksandrovich, 586
force constants, 140
force matrix, 403
Fourier integral, 435
Fourier series, 351
Fourier theorem, 84
Fourier transform of the convolution, 56
Fourier transform of the crystal potential, 219
Fourier, Jean Baptiste Joseph, 348, 352
Fourier, Joseph, 434
Fourier-transform, 46
fourth-order terms, 356
fractional quantum number, 564
fractional translation, 172, 209, 547, 550
free energy, 349
frieze groups, 490
frieze pattern, xxiii, 151, 489
Fritzsch, Harald, 318
Frobenius, Ferdinand Georg, 384
full proper rotational symmetry, 87
full rotational symmetry, xiv
function space, 45

functional, 31
functional analysis, 430
functional equation, 83
fundamental group, 558
fundamental group of a circle, 560
fundamental theorem of algebra, 58, 437

G-invariant, 49
$GL(n, \mathcal{C})$, 24, 285
$GL(n, \mathcal{R})$, 24
Galerkin, Boris, 451
Galilei group, 455, 456
Galilei Invariance, xxiii, 457
Galilei, Galileo, 455
gap, 217, 549
gapless, 531
garden furniture, 425
Gaussian-type orbitals, 590
gaussian.com, 577
Gel'fand, Israïl Moyseyovich, 436
Gell-Mann, Murray, 307, 312
general linear group, 24
Generalized eigenvalue problem, 440
generalized function, 436
Generalized periodic boundary conditions, 443
generator of the group, 371
geometrical static distortions, 473
Gerlach, Walther, 233
glide reflection, 20, 171
Google search algorithm, 39
Gordan, Paul, 75, 259
GOT, 59
Goudsmit, Samuel, 234
gradient of a function, 287
Gram, Jørgen Pedersen, 123, 432
Gram–Schmidt algorithm, 145
Grammatik der Ornamente, 416
Graphene, xxiv, 523
Graphene Band Structure, 197
Graphene Lattice, xxiv
Graphene monolayers, 553
great orthogonality theorem, 59
Greenberg, O., 318

ground-state energy, 121
group, 3, 7
group SU_3, 305
group average, 31
group of an icosahedron, 502
group of integers, 23
group of k, 206
group of the octahedron (O_h), 15
group theory, 3

H_2-Molecule, 577
H_2O molecule, xix
H_2O, 8
H–N–H scissoring, 487
Haar, Alfred, 31
hadron particle multiplets, xxi, 312
hadrons, 308
half-integer Cartan weights, 555
half-integer spherical harmonics, 239
half-integer spin quantum numbers, 236
half-integer spins, 236
Hall, George G., 582, 589
Hamilton function, 328
Hamilton operator, 39, 444, 473
Hamilton, William Rowan, 96
Hamiltonian, 249
Han, Moo-Young, 318
handedness of the electric field vector, 279
Harisch-Chandra character, 74
Harish-Chandra Mehrotra, 74
harmonic approximation, 139
harmonic coupling, 141
harmonic interaction, 485
harmonic potential, 449
harmonic vibrations, 40
Hartree, Douglas, 586
Hartree–Fock equation, 586
Hartree–Fock matrix, 589
Hausdorff, Felix, 92
Heisenberg classical ferromagnet, 359
Heisenberg exchange ferromagnet, 359
Heisenberg group, 48

Heisenberg, W., 48
Heisenberg, Werner, 307, 351
Heitler, Walter, 577
helix, 561
Helmholtz equation, x, 135
Hermann–Mauguin, 8
Hermite, Charles, 433, 436
Hermitic (Hilbert) scalar product, 431
Hermitic metric, 50
Hermitic operator, 85
Hermitic scalar product, 432
Herring, C., xx
Herring, William Conyers, 198, 203, 547
Hesse, Otto, 367
Hessel, J. C., 154
Hessel, Johann, xxiii
Hessel, Johann F. C., 151
Hessel, Johann Friedrich Christian, 497
hexagonal Bravais lattice, 175
hexagonal crystal system, 175
hexagonal system, 177
high symmetric phase, 367, 474
high-harmonic source, 604
high-symmetry locations, 205
high-temperature phase, 349
highest occupied molecular orbital (HOMO), 601
Hilbert space, xxii, 432
Hilbert, David, 168
Hilbert–von Neuman spectral theorem, xxii
holohedry group, 173
homogeneous function, 462
homogeneous linear differential equation, 446
homogeneous polynomials, 384
homomorphism, 26
homotopy, 556
homotopy invariant, 560
homotopy operation, 557
honeycomb lattice, 363, 523
hopping parameter, 534
horizontal and vertical boxes, 294

horizontal and vertical boxing, 295
horizontal glide reflection, 171
horizontal translational mode, 146
Hund's first and second rules, 256
Hund's rules, 337
Hund's third rule, 256
Hund, Friedrich, 255, 342
hybridization gap, 520
hybridization of wave functions, 223
hypercharge, 313
hypercharge quantum number, 313
hypothesis by Goudsmit and
  Uhlenbeck, 235

icosahedral point groups, 166
icosahedral quasi-crystals, 363
icosahedron, 154, 363
identical particles, 328
identity, 7
identity element, 63
improper crystallographic point
  groups, 159
improper point groups, xix, 159
Improper point groups of type a, 159
Improper point groups of type b, 163
Improper rotation axes, 10
improper rotation point groups, 161
incoherent process, 573
induction, 437
infinite-dimensional, 41
infinite-dimensional Hilbert spaces,
  438
infinitesimal generator, 85, 99
infinitesimal transformation, xviii, 85
infinitesimal transformation matrices,
  99
infinitesimal transformations for
  representations, 102
infrared pulse, 604
initial state, 277
injective, 27
integer numbers, 88
integer-order spots, 349
integral of the motion, 462
integral operator, 39

integrals over known atomic orbitals,
  589
International Union of
  Crystallography, 490
interpenetrating fcc sublattices, 209
interpenetrating lattices, 180
interval rule, 257
invariance group of the Hamilton
  operator, 115
invariant (normal) subgroup, 225
invariant polynomial, xxii, 352, 356,
  473
invariant subspaces, 53, 83
inverse loop, 557
inverse operation, 7
inverse phase transition, 367
inversion, 159
ionization energy, 586
irreducible representation, xviii, 49,
  54, 83
Ising magnet, 358
Ising model, 360
isometries, xvii, 17
isometry of the Euclidean space, 17
isomorphism, 27
isospin hypothesis, 308
isospin multiplet, 308
isosurface, 600
IUC, 490

Jacobi determinant, 34
Jacobi identity, 248, 401
Jacobi matrix, 32, 112
Jacobi, Carl Gustav Jacob, 288
Jahn, Hermann, 473
Jahn, Hermann Arthur, xxiii
Jahn–Teller distortion, 473
Jahn–Teller effect, 602
Jahn–Teller theorem, 475
Jahn–Teller-type distortion, 601
Jones, Owen, 416

k-fold connected continuous groups,
  562
K-point, xxiv, 528

k-valued representations, 562
K^0-meson, 312
k points on the surface of the BZ, 208
Kepler's third law, 462
Kepler, Johannes, 462
kernel, 27
kinetic energy, 39
Klein, Felix, 157, 379
Kohn, Walter, 580
Koster, G. F., 69, 119
Koster, George F., xxi
Kronecker (or tensor) product of
  vectors and matrices, 242
Kronecker product, 233, 244, 292
Kronecker product of two angular
  momentum vectors, 249
Kronecker product of vectors, 257
KWDS, 273

labeling of oscillatory modes, 480
labeling of the energy states, 127
ladder operator, 337
Lagrange, Joseph-Louis, 429
Lagrangian function, 429
Landau functional, 351
Landau, L. D., xxii, 239
Landau, Lev D., 348, 350, 355
Landé, Alfred, 257
Laplace operator, 135
Laplace, Pierre-Simon, Marquis de,
  447
largest eigenvalue, 39
largest weight, 88
lattice, 172, 415
lattice gas, 349
lattice points, 411
lattice potential, 533
LCAO, xxv, 588
LCAO ansatz, 589
LCAO basis functions, 589
LCAO expansion, 589
Lebesgue, Henri, 432
left and right invariances, 31
left coset, 224
left regular representation, 72

left translation, 35, 372
Lie algebra, 102, 396
Lie group, xviii, 30
Lie, Sophous, 89
Lifshitz, Evgeny, xxii, 239, 350, 355
ligand, 466
ligand atoms, 470
light electric field, 277
Line Δ, 208
Line Λ, 208
linear algebra, 40
linear combination, 5, 146
linear combination of plane waves,
  507
linear differential or integral operator,
  429
linear groups, 167
linear molecules, 167
linear operator, 6, 430
linear transformations, 29
Linearity, 431
linearized rotation operator, 90
lines and points of high symmetry,
  211
Liouville, Joseph, 446
liquid state, 351
liquid–solid phase transition, 360
little group of k, 206
London, Fritz, 577
loop, 556
low-energy electron diffraction, 347
lower symmetry, 126
LUMO, 603

m$\bar{3}$m, 195
magnetic phase transition, xxii, 358
main quantum number, 94
main quantum number n, 203
manifold connected sets, 561
many-body electron computations,
  321
many-body Hamiltonian, 337
Many-Body Problems, xxi
many-electron systems, 577
Maschke, Heinrich, 55

material science, 465
matrix, 6
matrix element, xviii, 7, 277
matrix multiplication, 42
matrix representation of a group, 42
mean field approximation, 255
mean field configuration Hamiltonian, 255
mean field Hamiltonian, 129
mean field potential energy, 337
measurable quantities, x
members of the star of k, 211
membrane of a drum, ix
Mendelejev, Dmitri Iwanowitsch, 254
Mermin, N. David, 566
midpoint of a hexagonal face, 208
midpoint of the square face, 208
Mirror planes, 9
missing row model, 348
Mittelwert, 31
mixed rank 2 tensor, 290
mixed tensors, 290
Mixed YDs, 294
mixing of product states, 259
mixing of states, 259
Model calculation, xxiii
modulation representation, 46
molecular orbitals, xxv, 589
molecular vibrations, xix, 139
molecule or a crystal, 126
momentum, 39
momentum conservation, 460
momentum vector, 125
monochromatic, linearly polarized light, 278
monoclinic system, 176
monomials, 132, 293
Moretti, Valter, 433
Mulliken, Robert S., 390
multi-particle problem, 128
multiplet, 255, 307
multiplet Hamiltonian, 256, 337
multiplets in atomic physics, 337
multiplication table, 26

multiplication theorem for characters, 66
multivalued representation, 562

n-pyramid, 157
n-sided pyramid, 154
(n, L) shell, 256
N–H asymmetric stretching, 487
N–H symmetric stretching, 487
N–H wagging, 487
Na flame, 235
Nambu, Y., 318
natural representation, 44, 285, 392
Ne'eman, Yuval, 315
neutron, xv, 307
Newton equation, 148, 405, 458
NH_3 Molecule, xxiii, 150
Nishijima, Kazuhiko, 312
non-centrosymmetric, 166
non-countable, 71
non-crystallographic point groups, xix, 166
non-enantiomorphic, 166
non-integer numbers, 88
non-local term, 586
non-orthogonal linear combination of atomic orbitals, 532
non-primitive unit cell, 170
non-reconstructed layers, 349
non-relativistic band structure, 266
non-relativistic electrons, 87
non-singular, 353
non-singular operator, 53
non-strange, 312
non-symmorphic, 181
non-symmorphic degeneracy, xxiv, 197, 231, 547, 552
Non-symmorphic Space Groups, xx
non-symmorphicity, 549
norm, 431
norm-wise convergence, 432
normal divisor, 225
northern hemisphere, 156
notation, xvii
nucleus, 128

null-homotopic, 558
Numerov, Boris Vasilyevich, 449

O_h, 71
O_h^5, 194
O. Stern and W. Gerlach, 234
oblique, 415
oblique Bravais lattice, 174
oblique crystal system, 174
octahedral crystal field splitting, 472
octahedral distribution, 466
octahedral symmetric crystal field,
    472
octahedron, 154
off-diagonal terms, 535
One-dimensional space groups, xix
one-dimensional systems, 4
one-parameter group, 396
Onofri, E., 444
ONS, 431
OP, 354
operator, xii
operator of infinitesimal rotations, 85
operator transformation, 116
Oppenheimer, J. Robert, 577
optical (spin) orientation, 567
optical matrix elements, 570
optical orientation along Λ, 283
Optical Spin Orientation, xxiv, 233
optical transition, 277
orbital angular momentum, xv
orbital overlap, 533
orbital quantum number, 203
order of a group, 23
order of the continuous group, 28
order parameter (OP), 347, 351
ordered partition, 295
orthogonal group, 24, 152
orthogonal matrices, 14
orthogonal sum, 73
orthogonality of the characters, 383
orthogonality theorem for characters,
    63
orthonormal basis functions, 41
orthonormal set (ONS), 431

orthonormality relation, 436
orthonormalized on the unit sphere,
    95
orthorhombic system, 176
overlap integral, 121, 579
overlap integral matrix, 535

p-character, 569
p-like electrons, 267
$p(1 \times 1)$-unit cell, 348
$p(2 \times 1)$-surface reconstruction, 348
$p \to s$ optical transitions, 279
p_z orbital, 525
P labels the primitive lattice, 193
P lattice, 174
P-crystal system, 174
pairing, 49
pairwise orthogonal, 76
Pais, Abraham, 312
parallel spin alignment, 580
parameter space, 28, 33, 556
parent energy bands, 270
Parseval identity, 435
Parseval, Marc-Antoine, 432
particle interchange, 328
particle physics, xv, xxi, 306
partition, 323
partner functions, 351
partner vectors, 79
path connected, 561
path integral, 560
pattern, 7, 151
Pauli matrices, 237
Pauli, Wolfgang, 235, 254, 340
periodic function, 84
periodic pattern in two dimensions,
    168
permutation group, xxi, 321
permutation operator, 336
permutations, 12, 321
perovskites, 466
perturbative approach, 124, 466
perturbed operator, 126
Peter, Fritz, 71
Peter–Weyl theorem, 83

phase transitions, xii, 347
photoemission experiment, 284
photoemission process, 586
physical continuity, 284
physics of crystals, xii
pinning, 531
pion–proton scattering, 309
planar polygon with n sides, 154
Plancherel, Michel, 436
plane waves, 504
Plato, 154, 497
plausibility arguments, xvi
Point Γ, 208
point group vector, 180
point groups, xix, 3, 151, 152
points within the BZ, 206
polar planar coordinates, 18
polarization vector of the radiation, 277
polyhedral geometry, 202
polynomials, 132
Pople, Sir John Anthony, 590
position operator, 47, 440
position vector, 39
Positivity, 431
positron, 307
potential energy, 39
potential energy landscape, 40
powers of the matrix, 96
practical computations, 121, 321
pre-selected basis functions, 121
primitive, 181
primitive lattice, 174
primitive monoclinic Bravais lattice, 176
primitive one-by-one, 347
primitive orthorhombic lattice, 177
primitive rectangular lattice, 184
primitive tetragonal lattice, 503
primitive unit cell, 170
prismatic point groups, 161
product functions, 259
product of cosets, 225
product of the Fourier transforms, 56
product of transpositions, 323

projector, 78
projector matrix, 78
projector technique, 406
projector technology, 127
propagation of electrons, 473
proper and improper rotations, 152
proper point groups, xix
proper rotation groups, 153
proper rotations, 14, 152
proton, xv, 307
proton–neutron scattering, xvi
pseudo-scalar mesons, 313
pseudovector, 248, 249
pulse EPR, 472
punctured plane, 559
pure displacement, 141
pure reflection, 20
pure rotation, 20, 141
pure translations, 20
pyramidal group, 164, 371

quantized, xv, 71
Quantum Chemistry, xxv, 321
quantum entangled states, xvi
quantum Hall systems, 564
quantum mechanical angular momentum, 85
Quantum Mechanical Conservation Laws, xix
quantum mechanical momentum, 47
quantum mechanical system, x, 429
quantum number k, 203
Quantum numbers, xviii
quantum numbers theorem, 117
quantum technologies, xvi, 436
quark, xxi
quark hypothesis, 307
quark quantum numbers, 320
Quasiparticles, 564

Racah, Giulio, 90
radial coordinate, 118
radial functions, 94
radial variable, ix
rank 0 representation, 293

rank 0 tensor, 288
rank 1 tensor, 288
rank 2 tensor, 288
rank 2 tensor representation, 292
rank 3, 290
rank of the tensor, 295
realistic model band structure, 508
rearrangement theorem, 32, 386
reciprocal lattice, 202, 524
reciprocal lattice numbers, 215
reciprocal lattice vectors, 200
reciprocal space, 199
reconstructed surface, 348
rectangular centered, 174
rectangular crystal system, 174
rectangular lattice, 169, 174, 175
rectangular matrix, 58
rectangular parallelepiped, 509
rectangular unit cell, 169
recursion formula, 449
reduced BZ, 211
reduced dimensionality, 4
reduced first BZ, 211
Reducible and irreducible
    representations, xviii
reducible representation, 53
reflection groups, 160
reflection lines, 171
reflexive, 557
regular n-gon, 13
regular icosahedron, 497
regular octahedron, 71, 497
regular representation, 71, 83
regular tetrahedron, 70, 497
relativistic band structure, xxi, 263,
    567
relativistic Dirac equation, 255
relativistic electronic structure, xx,
    233
relaxation, 353
representation Δ_5, 69
representation of the Lie algebra, 104
representation of the operator, 7
representation space, 41

representation theory of groups, xiv,
    xviii
Representations of Continuous
    Groups, xviii
representative of the class, 28, 63
resonant behavior, 311
RH algorithm, 583
rhobohedral unit cell, 177
rhombic, 174
rhombic lattice, 175
Riemann integral, xvi
Riemann, Georg Friedrich Bernhard,
    432
right coset, 224
right regular representation, 72
right-hand corkscrew rule, 19
ring-like molecule, 404
Ritz algorithm, 123
Ritz and Rayleigh, 121
Ritz matrix, 122
Ritz space, 578
Ritz, Walther, 121, 450, 451
rod-like structure, 363
Rodrigues, Benjamin Olinde, 35
Rodrigues, Olinde, 95
Roothaan, Clemens C. J., 581, 589
rotation axis, 8
rotation group, 85
rotations around a fixed axis, 81
rotatory inversion, 10
rotatory reflection, 10
RRG equations, 451

s-character, 569
S matrix, 308
S-matrix operator, 309
$SL(n, \mathcal{C})$, 24
$SL(n, \mathcal{R})$, 24
SO_2, 29
sp^2-bands, xxiv
sp^2-hybridized states, 525
sp^3-orbital, 569
$SU(2)$, 308
$SU(n)$, 24
S_n, 24

scalar, 5, 286
scalar operators, 120
scalar product, 5, 41, 431
scalar quantity, 431
Scalars, xxi
scaling, 21
scaling map, 22
scaling transformation, xxiii, 399, 462
scattering states, 444
Schönflies, A. M., xii, 8
Schmidt, Erhard, 432
Schrödinger equation, 214
Schrödinger representation, 46
Schrödinger, Erwin, 234
Schur, Issai, xxi, 58, 332
Schwartz, Laurent, 436
screened Coulomb potential, 533
screw reflection, 21
Screw rotation, 21
second derivative at the equilibrium
    sites, 486
second Hund's rule, 342
second-order perturbation theory, 223
second-order phase transition, 356
Seitz, Frederick, 17, 456, 497, 499
Selection rule, 466, 474
selection rule theorem, 276
self-adjoint matrices, 445
self-adjoint operator, 121, 442
self-consistent, 466
self-consistently, 586
sensor, 465
seven crystal systems, 178
shell, 128
shell model, 254
shell model of atomic physics, 337
short pulse of infrared light, 603
Shortley, G., 91
similar objects, 21
similarity, 21
similarity transformation, 53, 388
simply connected, 561
simultaneous excitation, 284
simultaneously diagonalized, 87
single group, 239

single group representation, 264
single-electron energy levels, 337
single-electron Hamiltonian, 337
single-group representation, 563, 570
single-particle energy levels, 254, 586
single-valued, 562
singlet state, 578
Slater determinant, 338, 342, 582
Slater matrix, 582
Slater, John C., xxii, 48, 254, 578
small perturbation, 117
small point group of k, 204
Smoluchowski, R., 69
Smoluchowsky, R., 390
soft X-ray photon, 603
solar cells, 465
solid phase, 347
solid-state physics, xix, 70
solidification process, 362
Sommerfeld, Arnold, 234
south pole of the sphere, 155
Space Group D_{6h}^4 ($P6_3/mmc$), xxiv
Space Group p2mg, xxiv
space groups, xix, 3, 151, 168
sparse matrices, 448
Spatial rotations, 460
Spatial translations, 459
spatial variables, ix
spatially dependent density, 46
spatially dependent temperature
    distribution, 46
special linear group, 24
special orthogonal group, 24, 152
special unitary group, 24
spectral lines, 235
Spectral theorem, xxii, 436
spectral theorem by Hilbert and von
    Neumann, 442
spectrometer, 235
spherical angles, 301
spherical harmonics, 76, 91
Spin, xx
Spin chain, 494
spin of the electron, 233
spin polarization, 279, 573

spin polarized, 573
spin rasse, 337
spin singlet, 341
spin space, 239
spin variable, 39
spin–orbit coupling, 255, 264
spin–orbit coupling operator, 255
spin-0 mesons, 313
spin-1 mesons, 313
spin-polarized electrons, 284, 567
spinors, 239
splitting of the original level, 127
spring constants, 141, 479
square crystal system, 175
square integrable, 40
square matrix, 53
square molecule, 4, 11
Square Planar Lattice, xxiii
square-based pyramid, 372, 377
square-integrable functions, 434
stacking, 553
star of k, 211
star operation, xx, 210
stereogram, 155
stereographic projection, 154
Stern, Otto, 233
STM, 347
strange particles, 312
strangeness hypothesis, 312
strangeness number, 312
strong interaction, 308
strong interaction is $SU(2)$-invariant, 308
structure constant theorem for representations, 103
structure constants, 101
structure constants theorem, 101
Strutt, John William, 3rd Baron Rayleigh, 450, 451
Strutt, John William, Baron Rayleigh, 121
Sturm, Jacques Charles François, 446
subgroup, 23
sum rule, 498
superposition principle, 430

surjective, 27
symmetric, 557
symmetric (permutation) group, 24
symmetric operators, 438
symmetrized orbitals, 598
symmetrized plane waves, xxiv, 504
Symmetry and Scientific Problems, xviii
symmetry elements, xvii, 3, 7
symmetry group of a cone, 16
symmetry group of the coupling operator, 249
symmetry group of the operator, 41, 111, 116
symmetry group of the Schrödinger equation, 115
symmetry group of the square, 12
symmetry groups, xvii, 7
symmetry in Newtonian mechanics, 125
symmetry in quantum mechanics, xiii
symmetry of composite systems, xx, 240
symmetry of the problem, xi
symmetry operation, 7
symmetry transformation, xi, xvii, 3, 7
symmetry-adapted, 329
symmetry-adapted linear combinations of atomic orbitals, 592
symmetry-adapted polynomial basis states, 270
symmetry-adapted polynomials, 132
symmetry-adapted vectors, xviii, 75
Symmetry-assisted computations, xviii
symmetry-assisted solution of eigenvalue problems, 121
symmetry-breaking environment, 129
symmetry-breaking field, 309
symmetry-breaking interactions, xix, 126
symmetry-breaking perturbation, 126

symmetry-conserving perturbation, 118
symmorphic, 181
symmorphic space groups, 182

T_d, 70
Taylor expansion, 86
Taylor, Brook, 352
technical, biological, and social sciences, 249
Teller, Edward, xxiii, 473
tensor, xxi, 288
tensor (or Kronecker) product of vector spaces, 240
tensor (or Kronecker) product of vectors and matrices, xx
tensor or Kronecker product space, 242
tensor product, 288
tensor product of states, 257
tensor product of the coordinates, 285
tensors with higher ranks, 290
terms, 256
test functions, 440
test integral, 440
test of irreducibility, 63
Tetragonal Lattice, xxiii
tetragonal system, 177
tetrahedron, 154
tetrahedron is self-dual, 154
tetratomic molecule, 593
the band index, 215
The character of a representation, 62
The classical group $GL(n, \mathcal{C})$, 285
the early 1960s, 312
The eightfold way, 315
the energy gap, 217
the factor group of k, 227
The Geometrical Aspects of Symmetry, 3
the group of k, xx, 204
The H₂-Molecule, xxv
the matrix elements theorem, 118
the quark hypothesis, 317
The space group 0_h^7, 208

The Symmetry Group of the Operator, xviii
The symmetry of the Ritz matrix, 123
The transformation operator, 115
The two-dimensional dihedral group, 13
theorem by Chevalley, 107
thermodynamic potential, 351
third-order terms, 355
three cubic Bravais lattices, 177
three pions, 308
three sub-bands, 269
three-dimensional dihedral groups D_n, 157
three-dimensional space groups, xix, 192
three-electron systems, 343
tight-binding model, xxiv, 532
tight-binding orbitals, xxiv
Time reversal, 455
time reversal symmetry, 265
time translation, 455
Time-resolved x-ray absorption spectroscopy, 604
Topological Aspects, xxiv, 233
topological group, 30
topological protected states, xvi
Topological Protection, xxiv, 197
topological theory of defects, 566
topological tools in science, 566
topologically protected, 231, 531, 547, 552
total angular momentum operator, 252
total angular momentum quantum number, 256
total cross-section, 311
total energy, 458, 603
total energy of the system, 473
total spin quantum number, 334
totally antisymmetric, 338
totally antisymmetric linear combinations, 338
tower (or permutation) representation, 79

tower representation, 79, 390, 412
trace of a matrix, 62
traffic jams, 249
transformation of the gradient, 113
transformation theory, 112
transforms according to the ith column of the representation, 117
transforms according to the mth column, 77
transition line, 350
transitivity, 557
translation, 17
Translation Group, xix
translational invariance, 125
translational mode, x
transposition, 322
trial functions, 121
trial vector, 78
trial wave function, 122
triangle in a cubic coordinate system, 132
triangular lattice, 363
trick, 256
triclinic system, 176
tridiagonal, 448
trigonal planar, 604
trigonal system, 177
trihydridoboron, 593
triplet spin states, 341
triplet states, 578
trivial representation, 44
two-, three-, and four-site integrals, 590
two-dimensional objects, 4
two-dimensional space groups, xix, 151, 415
Two-element group, 383
two-line notation, 322
two-torus, 560
twofold axis, 372
twofold rotation, 372
type-b improper rotation group, 165

$U(1)$, 308
$U(n)$, 24

Uhlenbeck, George Eugene, 234
uniform background, 349
unit cell, 169, 415
unit circle, 33
unitary equivalence, 47
unitary group, 24
unitary representation, 42
Universal covering group, xxiv, 233, 555, 561
Universal Covering Set, xxiv, 561
unperturbed operator, 126

Vandermonde polynomial, 323
variable capacitor, 465
variational principle, 121
vector, xxi, 5, 286
vector k, 204
vector space, 42, 430
vectorial tower (permutation) representation, 142
vectorial tower representation, 403, 480
vibrating string, 429
Vibrational Modes, 150
vibrational motion, ix
vibrational representation, 480
vibrationally active modes, 142
Vierergruppe, 500
Virial Theorem, 463
von Karman, Theodore, 135
von Neumann, John, 430, 446, 519
von Neumann, John Stone, Marshall, 47
Von Neumann–Wigner avoided crossing, 219, 223
von Neumann–Wigner gap, 520
von-Neumann–Wigner hybridization, 284

Waller, M. D., 138
wallpaper patterns, 167, 415
water window, 604
wave function, 40
weak interaction, 312
weight of a YT, 298

WEK, 119, 277
Weyl sequence, 440
Weyl, Hermann, xxi, 31, 41, 332, 444
Wigner 3j-symbols, 471
Wigner, E., 119, 390
Wigner, Eugene, xxi, 85, 519
Wigner, Eugene Paul, xiii
Wigner–Eckart constant, 311, 573
Wigner–Eckart–Koster, 233
Wigner–Eckart–Koster selection rule,
    467
Wigner–Eckart–Koster theorem, 274
Wigner–von Neumann anticrossing,
    222
winding number, 560
winding the linear lattice, 411
Wolfram Research, 471, 541

X-point, 231
$x \oplus y$, 49
X-ray pulse, 604
X-rays, 348

YD, 285
Young diagrams, 285
Young tableaux (YT), 285
Young, Alfred, xxi, 285, 334

z-component of the orbital angular
    momentum, 85
z-displacements, 485
Zeroth-order term, 353

Printed in the USA
CPSIA information can be obtained
at www.ICGtesting.com
CBHW070832180124
3287CB00045B/17

9 789811 280115